Experimental Design

with Applications in Management, Engineering, and the Sciences

Duxbury Titles of Related Interest

Albright, Winston & Zappe, *Data Analysis & Decision Making with Microsoft Excel*
Albright, Winston & Zappe, *Managerial Statistics*
Berk & Carey, *Data Analysis with Microsoft Excel*
Brightman, *Data Analysis in Plain English*
Canavos & Miller, *Introduction to Modern Business Statistics, 2nd Ed.*
Clemen, *Making Hard Decisions: An Introduction to Decision Analysis, 2nd Ed.*
Clemen & Reilly, *Making Hard Decisions with DecisionTools*
Davis, *Business Research for Decision Making, 5th Ed.*
Derr, *Statistical Consulting: A Guide to Effective Communication*
Devore, *Probability & Statistics for Engineering and the Sciences, 5th Ed.*
Dielman, *Applied Regression Analysis, 3rd Ed.*
Farnum, *Modern Statistical Quality Control and Improvement*
Hildebrand & Ott, *Statistical Thinking for Managers, 4th Ed.*
Hoerl & Snee, *Statistical Thinking: Improving Business Performance*
Johnson, *Applied Multivariate Methods for Data Analysts*
Keller & Warrack, *Statistics for Management & Economics, 5th Ed.*
Kenett & Zacks, *Modern Industrial Statistics: Design of Quality and Reliability*
Kirkwood, *Strategic Decision Making*
Lawson & Erjavec, *Modern Statistics for Engineering and Quality Improvement*
Lohr, *Sampling: Design and Analysis*
Lunneborg, *Data Analysis by Resampling*
McClelland, *Seeing Statistics®*
Middleton, *Data Analysis Using Microsoft Excel*
Minh, *Applied Probability Models*
Ramsey, *The Elements of Statistics with Applications to Economics and the Social Sciences*
Ramsey/Schafer, *The Statistical Sleuth, 2nd Ed.*
SAS Institute Inc., *JMP-IN: Statistical Discovery Software*
Schaeffer, Mendenhall & Ott, *Elementary Survey Sampling, 5th Ed.*
Schleifer & Bell, *Data Analysis, Regression, and Forecasting*
Schrage, *Optimization Modeling Using Lindo, 5th Ed.*
Shapiro, *Modeling the Supply Chain*
Vardeman & Jobe, *Basic Engineering Data Collection and Analysis*
Weiers, *Introduction to Business Statistics, 4th Ed.*
Winston, *Simulation Modeling Using @Risk*
Winston & Albright, *Practical Management Science, 2nd Ed.*

To order copies contact your local bookstore or call 1-800-354-9706.
For more information on these or over 100 additional titles contact
Duxbury Press at Forest Lodge Road, Pacific Grove, CA or go to: www.duxbury.com

Experimental Design
with Applications in Management, Engineering, and the Sciences

Paul D. Berger
Boston University

Robert E. Maurer
Boston University

DUXBURY
THOMSON LEARNING™

Australia • Canada • Mexico • Singapore • Spain
United Kingdom • United States

DUXBURY
THOMSON LEARNING

Sponsoring Editor: Curt Hinrichs
Assistant Editor: Ann Day
Editorial Assistant: Nathan Day
Marketing Manager: Tom Ziolkowski
Print/Media Buyer: Barbara Britton
Permissions Editor: Joohee Lee
Production Service: Penmarin Books
Text Designer: Hal Lockwood
Copy Editor: Anita Wagner
Illustrator: C. H. Wooley
Cover Designer: Lisa Langhoff
Cover Image: Zigy Kaluzny/Stone
Compositor: Atlis Graphics & Design
Printer: R. R. Donnelley & Sons/Crawfordsville

Printed in the United States of America
1 2 3 4 5 6 7 05 04 03 02 01

ISBN: 0-534-35822-5

Wadsworth/Thomson Learning
10 Davis Drive
Belmont, CA 94002-3098
USA

For more information about our products, contact us:
Thomson Learning Academic Resource Center
1-800-423-0563
http://www.duxbury.com

International Headquarters
Thomson Learning
International Division
290 Harbor Drive, 2nd Floor
Stamford, CT 06902-7477
USA

UK/Europe/Middle East/South Africa
Thomson Learning
Berkshire House
168-173 High Holborn
London WC1V 7AA
United Kingdom

Asia
Thomson Learning
60 Albert Street, #15-01
Albert Complex
Singapore 189969

Canada
Nelson Thomson Learning
1120 Birchmount Road
Toronto, Ontario M1K 5G4
Canada

Contents

CHAPTER 3

Some Further Issues in One-Factor Designs and ANOVA 59

CHAPTER 4

Multiple-Comparison Testing 87

CHAPTER 5

Orthogonality, Orthogonal Decomposition, and Their Role in Modern Experimental Design 129

CHAPTER 6

Two-Factor Cross-Classification Designs 155

CHAPTER 12

Designs with Factors at Three Levels 358

CHAPTER 13

Introduction to Taguchi Methods 378

PART THREE

Response-Surface Methods, Other Topics, and the Literature of Experimental Design 411

CHAPTER 14

Introduction to Response-Surface Methodology 413

Preface

The key objective of this book is to introduce and provide instruction on the design and analysis of experiments. We have tried to make this book special in two major ways. First, we have tried to provide a text that minimizes the amount of mathematical detail, while still doing full justice to the mathematical rigor of the presentation and the precision of our statements. Second, we have tried to focus on providing an intuitive understanding of the principles at all times. In doing so, we have filled the book with helpful hints, often labeled as ways to "practice safe statistics." Our perspective has been formed by decades of teaching, consulting, and industrial experience in the field of design and analysis of experiments.

Approach

Our approach seeks to teach both the fundamental concepts and their applications. Specifically, we include simple examples for understanding as well as larger, more challenging examples to illustrate their real-world nature and applications. Many of our numerical examples use simple numbers. This is a choice the authors consciously make, and it embraces a statement by C. C. Li, professor of biometry at the University of Pittsburgh, that the authors took to heart over thirty years ago and have incorporated into their teaching and writing: "How does one first learn to solve quadratic equations? By working with terms like $242.5189X^2 - 683.1620X + 19428.5149 = 0$, or with terms like $X^2 - 5X + 6 = 0$?" Our belief is that using simpler numerical calculations that students can more easily follow and verify aids them in the intuitive understanding of the material to a degree that more than offsets any disadvantage from using numbers that don't look like those in real cases. This does not mean that we focus solely on hand calculations (to us, this term includes the use of a calculator); we do not. We also have examples, as well as follow-up exercises at the end of chapters, that encourage, demonstrate, and, indeed, require the use of statistical software. (We illustrate the use of statistical software packages with Excel, SPSS, and JMP.) Nevertheless, we believe in the virtue of students' doing it at least once by hand or, at a minimum, seeing it done at least once by hand.

Background and Prerequisites

Most of our readers have some prior knowledge of statistics. However, as experienced teachers, we are aware that students often do not retain all the statistical knowledge they acquired previously. Since hypothesis testing is so fundamental to the entire text, we review it heavily, essentially repeating the depth of coverage the topic is accorded in an introductory course in statistics. Other useful topics from a typical introductory statistics course are reviewed on an ad hoc basis; an example of this is the topic of *confidence intervals.* With respect to topics such as probability and the student t distribution, we occasionally remind the student of certain principles that we are using (for example, reminding the student about the multiplication rule for independent events).

We have taught courses in the area of experimental design in which the audience varied considerably with respect to their application areas (for example, chemical engineering, marketing research, biology); we preface these courses by a statement we fervently believe to be true:

> The principles and techniques of experimental design transcend the area of their application; the only difference from one application area to another is that different situations arise with different frequency, and correspondingly, the use of various designs and design principles occurs with different frequency.

Still, it is always helpful for people to actually see applications in their area of endeavor. After all, many people beginning their study of experimental design don't know what they don't know; this includes envisioning the ways in which the material can be usefully applied.

We assume a working knowledge of high school algebra. On occasion, we believe it is necessary to go a small distance beyond routine high school algebra; we strive to minimize the frequency of these occasions, and when it is necessary we explain why it is in the most intuitive way that occurs to us. These circumstances exemplify where we aim to walk the fine line of minimal mathematical requirements without compromising the rigor of the material or the precision of our statements.

The second way in which we have tried to make this book special is to emphasize the application of the experimental design material in areas of management. We use the word *management* in a very general and broad sense, including traditional management disciplines such as marketing, finance, operations, management information systems, and organizational behavior, and including management in both the traditional business setting and nonprofit areas such as education, health care, and government. In addition, however, we interpret *management* to include some applications that could be placed in other categories as well—say, engineering and science.

For example, if management needs to test whether different brands of D-cell batteries differ with respect to average lifetime (with the same pattern of usage), in order to convince a television network to accept a promotion that claims one brand's superiority over other brands, we call this a management application. Even if the manager doesn't know in intimate detail how a battery works, he or she must have the ability to evaluate the validity of the experiment, and be able to understand the analysis, results, and implications.

Others may call this an engineering application, or a battery application, or label the application something else, but we view it as a management, perhaps a marketing, application. To a degree, the book disproportionately uses examples in the marketing field; this reflects the preponderance of the consulting experience (though far from exclusive) of one of the authors.

Organization and Coverage

We had some tough choices to make for which topics to cover. Our goal was to write a book that covered the most important and commonly used methods in the field of experimental design. We cover extensively the topics of two-level complete factorial designs, two-level fractional-factorial designs, and three-level complete factorial designs. In the interest of space, we prepare readers to study three-level fractional-factorial designs elsewhere, and we provide our favorite references on the topic. Again in the interest of space, we have included only limited discussion of the topics of outliers and missing values, and have omitted treatment of repeated-measures designs and mixture designs. However, for each of these topics, we do provide discussion of the issues involved and indicate our favorite references.

We have extensive experience that suggests that the book, in its entirety, provides an appropriate amount of material for a one-semester course. Indeed, the topics in this text compose the exact subject matter for an elective in the master of business administration (MBA) program at Boston University. Both authors have also successfully used parts of most of the chapters in this text in an undergraduate course in marketing research. Naturally, the 15 chapters in this book comprise *our* choice of topics; however, most of Chapters 7, 13, and 14 can be replaced by other material preferred by the instructor without compromising the integrity of the remaining chapters.

Part One, Chapters 2–8, covers the study of one-factor designs, two-factor designs—both cross-classification designs (including introduction to the concepts of blocking and interaction) and nested designs. It also includes designs having three or more factors—notably, Latin squares and Graeco-Latin squares. In Chapter 3, entitled Some Further Issues in One-Factor Designs and ANOVA, we introduce several topics that are with us throughout the text, such as underlying assumptions of the *F* test, hypothesis testing (encompassing the concept and calculation of power), and nonparametric tests (in this chapter, the Kruskal-Wallis test). Following this chapter, we cover the topics of multiple-comparison testing and orthogonal breakdown of sums of squares, topics that take the macro result of the *F* test and inquire more deeply into the message the data have for us. For the most part, the design and analysis concepts in Chapters 2–8 do not vary substantially as a function of the number of *levels* of the factor(s), but focus on the number of factors under study.

Part Two, Chapters 9–13, includes the introduction to two-level experimentation, confounding designs with factors at two levels, fractional-factorial designs with factors at two levels, and designs with factors at three levels. It also includes an introduction to Taguchi methods, in which we describe many aspects of Taguchi's philosophy, while focusing primarily on his design methodology (orthogonal arrays) for two-level designs.

Part Three wraps up with an introductory chapter on response-surface methods and a concluding chapter discussing the literature and resources in the field of experimental design, our choices of texts and other sources as references for specific topics, and the discussion of various topics not covered in the text. Although several of our references are quite recent, many references are from the 1980s and earlier. In our view, the best references for many of the fundamental topics are relatively older texts or journal articles, and we have included these excellent sources.

Exercises

The quality of a text in the area of design and analysis of experiments is, to an important extent, influenced by the end-of-chapter exercises. We present not only exercises that illustrate the basics of the chapter but also some more challenging exercises that go beyond the text examples. Many of the more challenging problems have appeared on take-home exams in courses we have taught. Although a few other texts also offer such challenging exercises, they are, sadly, still in the small minority.

Supplementary Material

The data sets for the text are available in several formats at www.duxbury.com. The data sets include data from many of the examples in the chapters, in addition to data from most end-of-chapter exercises.

Acknowledgments

Many people deserve our thanks for their contributions toward making this book what it is. First, we are grateful to the several individuals who gave us the opportunity to be exposed to a large variety of applications of experimental design to real-world situations. Most notable among these is Dr. Kevin Clancy, Chairman and CEO of Copernicus Marketing Consulting and Research. For the last fifteen years or so, working with Kevin throughout various incarnations of his company and with many excellent coworkers such as Dr. Steven Tipps,

Robert Shulman, Peter Krieg, and Luisa Flaim, among others, author PDB has observed more experimental-design application areas, discussed more experiments, and *designed* more experiments than would seem possible. Many of the examples in the book have their basis in this experience. Another person to be thanked is Douglas Haley, former Managing Partner of Yankelovich Partners, currently Chief Knowledge Officer of Harris Interactive, who also afforded PDB the chance to be exposed to a large variety of experimental-design application areas. Many other individuals—too numerous to list—have also provided PDB with consulting experience in the field of experimental design, which has contributed significantly to the set of examples in the book.

Author REM acknowledges the influence of his many colleagues at Bell Telephone Laboratories. Bell Labs was for many years the country's premier R&D organization, where the commitment to fundamental understanding was endemic. Many of the principles and techniques that constitute the essence of experimental design were developed at Bell Labs and its sister organization, Western Electric. Author REM also gratefully acknowledges the influence of his first teacher, his father, Edward, who showed by example the importance of a commitment to quality in all endeavors, and his mother, Eleanor, who was the inspiration for both father and son.

A very special thanks is due posthumously to Professor Harold A. Freeman of the economics department at MIT. Professor Freeman was one of "the great ones," both as a statistician and teacher of experimental design as well as, more importantly, a person. Professor Freeman, who died at age 88 in March 1998, was PDB's experimental-design teacher and mentor, instilling in him a love for the subject and offering him his first opportunity to teach experimental design in 1966, while a graduate student at MIT. Professor Freeman's teaching, as well as his *way* of teaching, have had a continuing and profound effect on PDB's teaching and writing in the field of experimental design. If the book is dedicated to any one individual, this individual is, indeed, Harold A. Freeman.

Finally, thanks are due to our families for affording us the ability to focus on writing this book. Susan Berger patiently waited for her husband to "tear himself away" from the computer to (finally) join her for dinner. She often wondered if he knew she was in the house. Mary Lou Maurer was never too busy to help her digitally impaired husband with the typing, along with providing copious amounts of encouragement and coffee.

Thanks to all of you.

Paul D. Berger and Robert E. Maurer

The thoughtful comments and suggestions of the following reviewers are gratefully acknowledged: David E. Booth, Kent State University; Thomas J. DiCiccio, Cornell University; Marie Gaudard, University of New Hampshire; Timothy C. Krehbiel, Miami University; Thomas E. Love, Case Western Reserve University; Jerrold H. May, University of Pittsburgh; Robert B. Miller, University of Wisconsin, Madison; Lisa Reagan, Allied Signal, Inc.; G. Bruce Schaalje, Brigham Young University; and Marietta Tretter, Texas A&M University.

About the Authors

DR. PAUL D. BERGER, Professor of Marketing and Quantitative Methods at the School of Management, Boston University, has been teaching experimental design for more than 30 years. He currently teaches it in the MBA and doctoral programs at Boston University. He has also taught at Massachusetts Institute of Technology (MIT) every summer since the mid-1960s, working until 1985 with Professor Harold Freeman of the Economics Department at MIT, one of the all-time greats in the field of statistics and experimental design, who, unfortunately, passed away at age 88 in early 1998. He has also conducted in-house courses at a large number of companies. He has consulted extensively in the field of experimental design at marketing research firms, manufacturing companies, and services providers. He has also published a large number of journal articles on statistics, experimental design, and mathematical modeling. He coauthored a textbook in the field of direct marketing (*Direct Marketing Management,* with Mary Lou Roberts, first edition 1989, second edition 1999) and a case book illustrating statistical applications (*Cases in Business Statistics,* with Ronald Klimberg and Peter Arnold, 1994). Dr. Berger also won the university-wide Metcalf prize for teaching excellence at Boston University. He has his bachelor of science, master of science, and doctoral degrees from the Sloan School of Management, MIT.

DR. ROBERT E. MAURER has more than 35 years of industrial experience at Bell Telephone Laboratories. In his last assignment, he was responsible for process and product design and manufacture of a several-hundred-million-dollar product line of hybrid integrated circuits. Through his initiative and guidance, the disciplines of statistical process control (SPC) and experimental design were deployed throughout his organization, leading to improved quality and reduced cost. Dr. Maurer has more than 20 years of experience teaching in the areas of statistical communication theory at the graduate school of engineering at Northeastern University and statistical analysis at the school of management at Boston University. He has published numerous papers and holds patents in the communications area. Dr. Maurer earned his bachelor and master of science and doctoral degrees in electrical engineering from Northeastern University, and an MBA from Boston University.

CHAPTER 1

Introduction to Experimental Design

1.1 What Is Experimentation?

Experimentation is part of everyday life. Will leaving 30 minutes earlier than usual in the morning make it easier to find a legal parking space at work? How about 20 minutes earlier? Or only 10 minutes earlier? Can I increase my gas mileage by using synthetic oil? Will my problem employees make more of an effort to be on time if I make it a practice to stop by their office to chat at the start of the day? We're frequently interested to learn if and how some measure of performance is influenced by our manipulation of the factors that might affect that measure. Usually we undertake these activities in an informal manner, typically not even thinking of them as experimentation, and the stakes are such that an informal, unstructured approach is quite appropriate. Not surprisingly, as the consequences grow, if the performance improvement means a substantial increase in profitability, or the running of the experiment involves a significant expenditure of time and resources, the adoption of a more structured experimental approach becomes more important.

An **experiment** is an inquiry in which an investigator chooses the levels (values) of one or more input, or **independent,** variables, and observes the values of the output, or **dependent,** variable(s). The purpose of experimental activity is to lead to an understanding of the relationship between input and output variables, often to further optimize the underlying process. An **experimental design** is, then, the aggregation of independent variables, the set of levels of each independent variable, and the combinations of these levels that are chosen for experimental purposes. That is, the core of an experimental design is to answer the three-part question, Which factors should we study, how should the levels of these factors vary, and in what way should these levels be combined? Of course, other, auxiliary issues need stipulation.

Frequently we have the latitude to select the levels of the factors under study in advance of data collection; in these instances, we realize advantages that increase the efficiency of the experimental effort. Sometimes, however, we cannot specify the levels of the independent variables—we have to take what we're given. For example, if we want to study the impact of corporate dividends on stock price, we generally cannot manipulate the companies' dividends; they would be what they would be. Such situations can also arise simply because the data are already collected, and an analysis can be done only ex post facto (that is, after the fact, or after the data are collected). However, in both situations, it might be possible to sort the data to find a subset of companies with the exact levels of dividends one would wish to choose.

1.2 The Growth in Experimental Design

Experimental design is a growing area of interest in an increasing number of applications. Initially, experimental design found application in agriculture, biology, and other areas of hard science. It has since spread through the engineering arenas to the social sciences of economics and behavioral analysis. Experimental design appears to have been used in traditional business and management applications only since the mid-1960s; more recently, experimental-design methodology has been applied in the nonprofit and government sectors. There are many reasons for this progression, but the principal one is the increased training in statistics and quantitative methods among professionals in management and the latter areas, along with the resultant widespread use of quantitative decision-making techniques.

This trend was further encouraged by the Total Quality Management (TQM) movement originating in the mid-1980s and continuing today. Indicative of the widespread acceptance of the virtues of experimental design was its being heralded by the esteemed "establishment" outlet *Forbes* magazine in a March 11, 1996, article entitled "The New Mantra: MVT." MVT stands for multivariable testing, a term the article uses along with experimental design, using the term MVT to distinguish factorial designs from the vilified (both in the article and this text) one-factor-at-a-time experiments.[1]

That the trend continues is attested to by a more recent article, "The Numbers Game: Multivariate Testing Scores Big," in the Executive Edge section of the April 2000 edition of *Continental,* the in-flight magazine of Continental Airlines. The article cites many companies that rely more and more on experimental design (including DuPont, American Express, Boise Cascade, Circuit City, and SBC Communications), and reports in detail on its use by the Deluxe Corporation.

Behind all the praise of experimental-design methods is the simple fact that it works. The two articles note many successes, and *Forbes* briefly described the history of the discipline. Experimental design is not, strictly speaking, a new field. It originated at the beginning of the 20th century with Sir Ronald

Fisher's work; indeed, Fisher is often referred to as "the father of experimental design."

The field is, in many ways, a simple one. The terminology and notation may appear strange to the uninitiated, and the mathematical connections may seem formidable, but as for any worthwhile new skill, one can overcome the barriers to entry with study and practice. With mastery one can admire the inherent beauty of the subject and appreciate how its success is enhanced by combining the discipline of statistics with the knowledge of process experts.

1.3 The Six Steps of Experimental Design

One can frame the experimental-design process as a six-step process, as seen in Figure 1.1.

Plan the Experiment

The planning process is vital to the success of the experiment. It is at this stage that close collaboration between the people knowledgeable about the process under study (the process experts) and those experienced in the design-of-experiments methodology (the designers) is required. Diligence at this stage can greatly increase the chance that appropriate, meaningful assumptions are made; giving short shrift to this stage of the process can make everything that follows a waste of time and resources. The planning stage itself consists of five steps:

1. Identify the dependent, or output, variable(s).
2. Translate output variables to measurable quantities.
3. Determine the factors (input, or independent, variables) that potentially affect the output variables that are to be studied.
4. Determine the number of levels or values for each factor and what those levels are.
5. Identify potential synergy between different factors.

The dependent, or output, variable to be studied should be chosen carefully. Sometimes the choice is obvious, but other times it is not. The general guideline is to ensure that the right performance measure is chosen—typically, the quantity that is really the important one; perhaps the one that tells how good a product is, or how well it does its job, or how it's received in the marketplace. For example, in an experiment to help develop a better cover for a specialty catalog, it's unlikely that a wise choice of dependent variable is the brightness of the cover; a wiser choice is the sales volume that results, or even a measure of the potential customer's attitude toward the cover.

One way to choose a dependent variable is to make sure that all vested interests are represented—those of the user, producer, marketing department, and relevant others—perhaps in a formal brainstorming session. If the goal is

1. *Plan* the experiment.
2. *Design* the experiment.
3. *Perform* the experiment.
4. *Analyze* the data from the experiment.
5. *Confirm* the results of the experiment.
6. *Evaluate* the conclusions of the experiment.

Figure 1.1 The process of experimental design

to produce popcorn that sells more, is it obvious to the consumer what the key qualities should be? Fluffiness? Color? Taste? Texture? Percent popped?

Once the dependent variable is selected, it must usually be transformed into a quantitative measure to make it useful. Many variables are subjective. How do you measure taste or appearance? What about the units of measurement? The choice of units is almost never important if those under consideration are linearly related to one another (such as inches versus feet, or dollars versus yen). However, how about the choice between circumference, area, and volume to measure the size of a spherical product? A known value for one measure determines the values of the other two measures, but if the value of one measure changes, the values of the other measures do not change proportionally. So although one measure may vary linearly with the level of a factor, another measure may vary nonlinearly. As we note later, varying linearly versus nonlinearly may be an important issue.

Choosing the factors to study is sometimes quite straightforward; however, other times it is not as easy as it might seem. In our consulting experience, the process experts sometimes propose an unworkably large number of possible factors for study. An effective way to raise for consideration all candidate input variables is to begin with a formal brainstorming session, perhaps using Ishikawa (often called Fishbone or cause-and-effect) diagrams, a technique developed in the context of quality control to help identify factors affecting product quality, but adaptable to any situation where one desires to identify factors that affect a dependent variable. This approach generally yields a nearly exhaustive list of choices in categories such as people factors, equipment factors, environmental factors, methods factors, and materials factors. This initial list may be pared down to its essentials through a Pareto analysis, which in this context classically states the now well-known concept that 80% of the identifiable impact will be associated with 20% of the factors, though the exact values of 20 and 80 are unlikely to be precisely realized. Of course, one should reduce the list in any obvious ways first. We recall one case in which (in simplistic terms) the factor "temperature greater than 212°F or not" was identified along with another factor, "presence or absence of steam" (with the experiment run at sea level in an unpressurized container).

One is usually motivated to minimize, to the degree possible, the number of factors under study. In general, everything else being equal, a higher number of factors under study is associated with an increased size of an experiment; this in turn increases the cost and time of the experiment. The connection between the number of factors in an experiment, and the size and efficiency of the experiment, is an issue that is discussed from numerous perspectives in the text.

At how many different levels (that is, settings) should a factor appear in an experiment? This is not always easily answered. A quick-and-dirty answer is that if the response of the dependent variable to the factor is linear (that is, the change in the dependent variable per unit change in the level of the factor is constant), two levels of the factor will suffice, but if the response is nonlinear (not a straight line), one needs more than two. In many nonlinear cases, if the factor is measured on a numerical scale (representing some unit of measurement, not simply categories), three levels will suffice. However, this answer has the major flaw of circular reasoning: the answer is clear if we know the response function (the true relationship between the dependent variable and the independent variable), but the reason for running the experiment generally is to find the response function. Naturally, if factors have a higher number of levels, the total number of combinations of levels of factors increases. Eight factors, each at two levels, have a total of 256 (2^8) different combinations of levels of factors, whereas eight factors, each at three levels, have 6561 (3^8) different combinations of levels of factors—a big difference! The issue of number of levels is addressed for various settings throughout the text.

The last of the planning steps noted earlier was that of identification of synergy (or of antisynergy) between factors. The more formal word used for synergy is *interaction.* An **interaction,** or **synergy,** is the combined effect of factor levels that is above and beyond the sum of the individual effects of the factors considered separately. That is, the total is greater than the sum of the parts. For example, if adding a certain amount of shelf space for a product adds 10 units of sales, and adding a certain amount of money toward promoting the product adds 8 units of sales, what happens if we add both the shelf space and the promotional dollars? Do we gain 18 (the sum of 10 and 8) units of sales? If so, the two factors, shelf space and promotion, are said to not interact. If the gain is more than 18 units of sales, we say that we have **positive interaction** or positive synergy; if the gain is less than 18 units, we say we have **negative interaction** or negative synergy. The text covers this concept in great depth; indeed, the number of interactions that might be present among the factors in the study has major implications for the experimental design that should be chosen.

Design the Experiment

Having completed the planning of the experiment, we undertake the design stage. Experimental design is the primary subject of the entire text. First, we make the choice of design type. A fundamental tenet of this text is that **factorial designs,** in which the experiment comprises varying combinations of levels of factors, are in general vastly superior to **one-at-a-time designs,** in which the level of each factor is varied, but only one factor at a time. This is a major theme of the *Forbes* article mentioned earlier; indeed, it leads directly to why the article is entitled "Multivariable Testing." We discuss this issue at length.

Having chosen the design type, we need to make the specific choice of design. A critical decision is to determine how much "fractionating" is appropriate. When there is a large number of combinations of levels of factors, inevitably only a fraction of them are actually run. Determining which fraction is appropriate, and the best subset of combinations of levels of factors to make up this fraction, is a large part of the skill in designing experiments. It is possible for the degree of fractionating to be dramatic. For example, if we study 13 different factors, each with three levels, we would have 1,594,323 different combinations of levels of factors. However, if we could assume that none of the factors interacted with others, a carefully selected, but not unique, subset of only 27 of these combinations, perhaps with modest replication, would be necessary to get a valid estimate of the impact of each of the 13 factors. The issue of the accuracy of the estimates would be determined in part by the degree of **replication,** the number of data values that are obtained under the same experimental conditions.

Another critical element of designing an experiment is the consideration of **blocking,** which is controlling factors that may not be of primary interest but that if uncontrolled will add to the variability in the data and perhaps obscure the true effects of the factors of real interest. For example, suppose that we wish to study the effect of hair color on the ability of a particular brand of shampoo to reduce the amount of a person's hair that has split ends. Further suppose that, independent of hair color, male and female hair react differently to the shampoo. Then we likely would want to introduce sex as a second factor (although some texts wouldn't label it a factor but simply a block, to distinguish it from the primary factor, hair color). By being controlled, or accounted for, the variability associated with the factor (or block) of sex could be calculated and extracted so that it does not count against and obscure differences due to hair color. Blocking is briefly discussed in Chapter 6, and is explored in greater detail in Chapter 10. Blocking is illustrated in some descriptions of experimental applications in Section 1.4.

The discussion in this section is necessarily somewhat superficial. Designing an experiment doesn't always follow such an easily describable set of separate substeps. Other considerations must be taken into account, at least tangentially if not more directly. For example, perhaps the experiment under consideration is to be one of a series of experiments; in this case, we might wish to design a somewhat different experiment—perhaps one that does not include every factor that might be needed for a stand-alone experiment. Issues of combining data from one experiment with data from another might also arise in choosing a design. These additional considerations and many others are discussed in various sections of the text.

Perform the Experiment

It goes without saying that once the experiment has been designed, it must be performed ("run") to provide the data that are to be analyzed. Although we do not spend a lot of time discussing the running of the experiment, we do not mean to imply that it is a trivial step. It is vital that the experiment that was designed is the experiment that is run. In addition, the order of running the combinations of levels of factors should be random (more about this later in the text). Indeed, some statistical software programs that use information decided during the planning stage to provide designs for the user also provide a worksheet in which the order of the combinations has been randomly generated.

Analyze the Data from the Experiment

Sometimes the conclusions from an experiment seem to stand out. However, that can be deceptive. Often the results are not clear-cut, even when they appear that way. It is important that we be able to tell whether an observed difference is indicating a real difference, or is simply caused by fluctuating levels of background noise. To make this distinction, we should go through a statistical analysis process called hypothesis testing.

Statistical analysis cannot prove that a factor makes a difference, but it can provide a measure of the consistency of the results with a premise that the factor has no effect on the dependent variable. If the results are relatively inconsistent with this premise (a term that gets quantified), we typically conclude that the factor does have an effect on the dependent variable, and consider the nature of the effect. The statistical analysis also provides a measure of how likely it is that a given conclusion would be in error.

Somewhat formidable mathematics are required to derive the methods of hypothesis testing that are appropriate for all but the simplest of experiments. However, the good news is that, for the most part, it is the application of the methodology, not its derivation, that is a core requirement for the proper analysis of the experiment. Most texts provide illustrations of the application. In addition, numerous statistical software packages do virtually all calculations for the user. Indeed, far more important than the mechanics of the calculations is the ability to understand and interpret the results. However, as we noted in the preface, we do believe that the ability to understand and interpret the results is enhanced by the competence to "crunch the numbers" by hand (and as noted earlier, to us this phrase includes the use of a calculator).

The principal statistical method used for the analysis of the data in experimental designs is called analysis of variance (ANOVA), a method developed by Sir Ronald Fisher. The primary question ANOVA addresses is whether the level of a factor (or interaction of factors) influences the value of the output variable. Other statistical analyses augment ANOVA to provide more detailed inquiries into the data's message.

Confirm the Results of the Experiment

Once we have reached the pragmatic conclusions from our analysis, it is often a good idea to try to verify these conclusions. If we are attempting to determine which factors affect the dependent variable, and to what degree, our analysis could include a determination as to which combination of levels of factors provides the optimal values of the dependent variable. "Practicing safe statistics" would suggest that we confirm that at this combination of levels of factors, the result is indeed what it is predicted to be.

Why? Well, it is very likely that while running only a fraction of the total number of combinations of levels of factors, we never ran the one that now seems to be the best; or if we did run it, we ran only one or a few replicates of it. The wisdom of the design we chose was likely based in part on assumptions about the existence and nonexistence of certain interaction effects, and the results derived from the analysis surely assumed that no results were misrecorded during the performance of the experiment, and that no unusual conditions occurred during the experiment that would harm the generalizability of the results. Thus, why not perform a confirmation test that (we hope) verifies our conclusions? If there is a discrepancy, better to identify it now rather than later!

Evaluate the Conclusions of the Experiment

Clearly, after any experiment, one evaluates whether the results make sense. If they do not make sense, further thinking is necessary to figure out what to do next. The particular kind of evaluation we have in mind as the sixth step of the experimental-design process is the economic evaluation of the results.

Not all situations lend themselves to an explicit utilization of this step. However, in our experience, a significant proportion of experiments applied in the management areas do lend themselves to, indeed mandate, a cost/benefit analysis to determine whether the solutions suggested by the results of the experiment are economically viable. Today, it is clear that companies generally cannot, and should not, embrace quality for quality's sake alone. Quality improvements need to be economically justified.

A fruitful application of designed experiments is in the area of product configuration, which involves running experiments to examine which combination of levels of factors yields the highest purchase intent, or sales. However, the combination of levels of factors that has the highest purchase intent (let's put aside the issue of if and how intent translates to sales) may be a big money loser. A three-scoop sugar-cone ice cream for a nickel would surely yield very high revenue, but not for long—the company would soon go out of business. As another example, suppose that a combination of certain levels of ingredients that has a variable cost higher than the current combination would result in the same average quality indicator value (such as battery life), but with a lower amount of variability from product to product. Everyone would agree that the lower variability is desirable, but does, say, a 20% drop in this variability warrant a 30% increase in variable cost? The answer lies with an economic evaluation, or cost/benefit analysis. It may or may not be easy to do such an evaluation; however, it is difficult to reach a conclusion without one.

1.4 Experimental-Design Applications in Management

In this section we present details of six case studies that reflect actual examples of the design and analysis of experiments on which the authors have consulted. The goal is to provide the reader with a variety of real-life illustrations of the use of the material covered in this text. Each subsequent chapter is introduced by one of these or a similar example on which at least one of the authors worked, to illustrate an application of the concepts in that chapter to an actual experimental-design problem in a management area. In most cases, the company name cannot be revealed; however, as noted earlier, each situation is real and the description of the specifics, although sometimes changed in minor ways, has not been altered in any way that would affect the design or analysis of the experiment.

Corporate Environmental Behavior

In some industries, the name or specific brand of a company is not a major selling point. This is true for most utility companies, except perhaps for some telecommunication companies. In this day and age of increased deregulation and actual competition, however, many utility companies are seeking to distinguish themselves from the pack. One particular energy company, Clean Air Electric Company (a fictitious name), decided to inquire whether it could achieve an advantage by highly publicizing promises of environmentally sound corporate behavior.

The company decided to study whether demand for its product in two newly deregulated states (Pennsylvania and California) would be influenced by a set of factors, notably including different levels of corporate environmental behavior; other factors included price, level of detail of information provided to customers about their pattern of use of the product, level of flexibility of billing options, and several others. The factor "corporate environmental behavior" had five levels for what should be publicized (and adhered to), as shown in Figure 1.2.

The experiment had several dependent-variable measures. Perhaps the most critical one for the company was an 11-point scale for likelihood of purchase. It is well known that one cannot rely directly on self-reported values for likelihood of purchase; they are nearly always exaggerated—if not for every single respondent, surely on average. Virtually every marketing research or other firm that designs and analyzes experiments involving self-reported likelihood of purchase has a transformation algorithm, often proprietary and dif-

1. Clean Air Electric Company will ensure that its practices are environmentally sound. It will always recycle materials well above the level required by law. It will partner with local environmental groups to sponsor activities good for the environment.
2. Clean Air Electric Company will ensure that its practices are environmentally sound. It will always recycle materials well above the level required by law. It will partner with local environmental groups to sponsor activities good for the environment. **Also, Clean Air Electric Company will donate 3% of its profits to environmental organizations.***
3. Clean Air Electric Company will ensure that its practices are environmentally sound. It will always recycle materials well above the level required by law. It will partner with local environmental groups to sponsor activities good for the environment. Also, Clean Air Electric Company will donate **6% of its profits** to environmental organizations.
4. Clean Air Electric Company will ensure that its practices are environmentally sound. It will always recycle materials well above the level required by law. It will partner with local environmental groups to sponsor activities good for the environment, **and will engage an unbiased third party to provide environmental audits of its operations. Also, Clean Air Electric Company will provide college scholarships to leading environmental colleges.**
5. Clean Air Electric Company will ensure that its practices are environmentally sound. It will always recycle materials well above the level required by law. It will partner with local environmental groups to sponsor activities good for the environment. **Also, Clean Air Electric Company will donate 3% of its profits to environmental organizations, and will actively lobby the government to pass environmentally friendly legislation.**

*Boldface typing is solely for ease of reader identification of the differences among the levels.

Figure 1.2 **Levels of factor: corporate environmental behavior**

ferent for different industries, to nonlinearly downscale the self-reported results. Certain firms have built up a base of experience with what these transformations should be, and that is a major asset of theirs. One would expect this to be important in the evaluation stage. Other measures included attitude toward the company, and the degree to which these company policies were environmentally friendly. The experiment was run for six mutually exclusive segments of potential customers, and the data were separately analyzed; the segments were formed by whether the potential customer was commercial or residential, and by the potential customer's past product usage. It might be noted that having the six segments of potential customers identified and analyzed separately is, in essence, using the segment of customer as a block, in the spirit of the earlier discussion of this topic.

Supermarket Decision Variables

The majority of supermarkets have a relatively similar physical layout. There are two primary reasons for this. One is that certain layouts are necessary to the functioning of the supermarket; for example, the location of the meat products is usually at the rear of the store, to allow the unloading (unseen by the public) of large quantities of heavy, bulky meat products, which are then cut up and packaged appropriately by the meat cutters working at the supermarket, eventually resulting in the wrapped packages that are seen by its patrons. The other reason for the similarity of layout is that the people that run supermarkets have a vast store of knowledge about superior product layout to enhance sales; for example, certain items are placed in locations known to encourage impulse purchases. As another example, products in the bread aisle will likely get more traffic exposure than products in the baby-food aisle. Placing necessities in remote locations ensures that customers have plenty of opportunity to select the more optional items on their way to the milk, eggs, and so on.

A large supermarket association wished to more scientifically determine some of its strategies concerning the allocation of shelf space to products, product pricing, product promotion, and location of products within the supermarket. In this regard, the association decided to sponsor an experiment that examined these "managerial decision variables" (as they put it). There was also concern that the strategies that might work in the eastern part of the United States might not be best in the stereotypically more laid-back atmosphere of the western part of the country. In addition, it was not clear that the impact of the level of these decision variables would be the same for each product; for example, promoting a seasonal product might have a different impact than the same degree of promotion for nonseasonal products, such as milk.

An experiment was set up to study the impact of eight factors, listed in Figure 1.3. The experiment involved 64 different supermarkets and varied the levels of the factors in Figure 1.3 for various products. The dependent variable was sales of a product in the test period of six weeks, in relation (ratio) to sales during a base period of six weeks immediately preceding the test period. This form of dependent variable was necessary, since different supermarkets have different sizes of customer bases and serve different mixes of ethnic groups. After all, a supermarket in an Asian community sells more of certain vegetables than supermarkets in some other neighborhoods, simply due to the neighborhood makeup—not due primarily to the level of promotion or to other factors in Figure 1.3.

1. Geography (eastern vs. western part of the U.S.)
2. Volume category of the product
3. Price category of the product
4. Degree of seasonality of the product
5. Amount of shelf space allocated to the product
6. Price of the product
7. Amount of promotion of the product
8. Location quality of the product

Figure 1.3 Factors for supermarket study

Financial Services Menu

A leading global financial institution, GlobalFinServe (fictitious name), wished to expand its services and both acquire new clients and sell more services to current clients. In these days of increased deregulation, financial service companies are allowed to market an expanding set of products. Their idea was to consider promoting a "special relationship with GlobalFinServe," allowing clients who join the "select client group" to take advantage of GlobalFinServe's experience and technological innovations.

In joining GlobalFinServe's select client group, a client would receive a set of benefits, including having a personal relationship with a manager who would provide certain services, depending on the results of the experiment, such as giving investment advice, insurance advice, and other types of more detailed financial planning than typically available from such a financial institution; these additional services include stock brokerage and foreign currency trading. Also, members of the select client group would receive several other "convenience" privileges: access to accounts 24 hours a day, seven days a week, by ATM or computer, from anywhere in the world; consolidated monthly statements; preferential treatment at any branch office (similar to a separate line for first-class passengers at an airline counter); and other possibilities, again depending on the results of the experiment.

GlobalFinServe hoped to resolve many questions using the results of the experiment. The primary issue was to determine which benefits and services would drive demand to join the special client group. Using the company's information on the different costs it would incur for the different levels of each factor, a cost/benefit analysis could then be done. Some of the factors and levels to be explored in the experiment are listed in Figure 1.4.

The experiment was conducted in many different countries around the world, and the data from each country were analyzed separately; in part, this was done in recognition that in some cases, different countries have different banking laws, as well as different attitudes toward saving and investing. (This

1. Dedicated financial relationship manager?
   - Yes; one specific person will be familiar with your profile, serve your needs, and proactively make recommendations to you
   - No; there is a pool of people, and you are served by whoever answers the phone; no proactive recommendations
2. Availability of separate dedicated centers at which clients and financial relationship managers can meet?
   - Yes
   - No
3. Financial services availability
   - About ten different levels of services available; for example, investment services and financial planning services available, but borrowing services not available
4. Cost to the special client
   - $20 per year
   - $200 per year
   - 0.5% of assets per year
5. Minimum account balance (total of investments and deposits)
   - $25,000
   - $50,000
   - $100,000

Figure 1.4 Financial services: factors and levels

is, in essence, introducing different countries as blocks.) The experiment for each country was carried out by assembling a large panel of different segments of clients and potential clients in that country. Segments included those who were already GlobalFinServe clients and heavy users; those who were already GlobalFinServe clients but nonheavy users; nonclients of GlobalFinServe with high income and/or net worth; and nonclients of GlobalFinServe with moderate income and net worth. Respondents were shown a profile containing a combination of levels of factors and then indicated their likelihood of joining the GlobalFinServe select client group. In addition, for each respondent, various demographic information was collected, and open-ended commentary was solicited concerning other services that respondents might like to have available in such a setting.

The Qualities of a Superior Motel

A relatively low-priced motel chain was interested in inquiring about certain factors that might play a large role in consumers' choice of motel. It knew that certain factors, such as location, couldn't be altered (at least in the short run). Price plays a significant role, but for the experiment it was considered to be already set by market forces for the specific location. The chain was interested in exploring the impact on customer satisfaction and choice of motel of a set of factors dealing with the availability and quality of food and beverages, entertainment, and business amenities. Some of the factors and their levels are listed in Figure 1.5.

The sample of respondents was separated into four segments (in essence, blocks). Two of the segments were frequent users of the motel chain (based on the proportion of stays in hotels/motels that are stays at the sponsoring

1. Breakfast (at no extra charge)
   - None available
   - Continental breakfast buffet—fruit juices, coffee, milk, fresh fruit, bagels, doughnuts
   - Enhanced breakfast buffet—add some hot items, such as waffles and pancakes, *that the patron makes himself*
   - Enhanced breakfast buffet—add some hot items, such as waffles and pancakes, *with a "chef" who makes them for the patron*
   - Enhanced breakfast buffet—add some hot items, such as waffles and pancakes, *that the patron makes himself,* **and also pastry (dough from a company like Sara Lee) freshly baked on premises**
   - Enhanced breakfast buffet—add some hot items, such as waffles and pancakes, *with a "chef" who makes them for the patron,* **and also pastry (dough from a company like Sara Lee) freshly baked on premises**
2. Business amenities available
   - Limited fax, printing, and copy services available at front desk, for a nominal fee
   - Expanded fax, printing, and copy services available 24 hours per day accessed by credit card
   - Expanded fax, printing, copy services, **and computers with Internet and email capability,** available 24 hours a day accessed by credit card
3. Entertainment
   - Three local channels and **five** of the more popular cable stations, plus pay-per-view movies
   - Three local channels and **fifteen** of the more popular cable stations, plus pay-per-view movies
   - Three local channels and fifteen of the more popular cable stations, plus pay-per-view movies **and Nintendo games**
   - Three local channels and fifteen of the more popular cable stations, plus pay-per-view movies **and VCR**
   - Three local channels and fifteen of the more popular cable stations, plus pay-per-view **movies and both Nintendo games and VCR**

Figure 1.5 Factors and levels for motel study

chain), split into business users and leisure users. The other two segments were infrequent or nonusers of the motel chain, split the same way. The key dependent variable (among other, more "intermediary" variables, such as attitude toward the chain) was an estimate of the number of nights in which the respondent would stay at the motel chain during the next 12 months.

Time and Ease of Seatbelt Use: A Public Sector Example

A government agency was interested in exploring why more people do not use seatbelts while driving although it appears that the increased safety so afforded is beyond dispute. One step the agency took was to sponsor an experiment to study the factors that might influence the time required to don (put on) and doff (take off) a seatbelt (*don* and *doff* were the words used by the agency). The concomitant ease of use of seatbelts was a second measure in the experiment.

The agency decided that two prime groups of factors could be relevant. One group had to do with the physical characteristics of the person using the seatbelt. The other group had to do with the characteristics of the automobile and with seatbelt type. For those not familiar with the terminology, a "windowshade" is just that: an inside shade that some windows have, sometimes factory installed, sometimes added, for the purpose of privacy or keeping out sunlight. The presence of a windowshade could affect the time and ease associated with donning and doffing the seatbelt. A "locking latchplate" arrangement is the name given to the type of latch common for seatbelts in virtually all automobiles today; that is, the seatbelt locks into a fixed piece of hardware, usually anchored on the floor or console. The nonlocking latchplate was used fairly often in the 1970s, when the study was conducted, but today it is rare except in race cars. With a nonlocking latchplate, the seatbelt may go through a latchplate, but it doesn't lock into any fixed hardware: it simply connects back onto itself.

The factors under study each had two levels, as noted in Figure 1.6. In essence, there were sixteen different automobiles in the experiment—2^4 combinations of levels of factors 4–7. There were eight people types. The male/female definition was clear; for the weight and height factors, a median split was used to define the levels.

Emergency Assistance Service for Travelers

An insurance company was developing a new emergency assistance service for travelers. Basically, the concept was to offer a worldwide umbrella of assistance and insurance protection in the event of most types of medical, legal, or financial emergency. The service would go well beyond what traditional travelers' insurance provided.

There would be a 24-hour, toll-free hot line staffed locally in every country in the free world (a list of countries with services available would be listed). A highly trained coordinator would answer your call, assess the situation, and refer you to needed services. The coordinator would call ahead to ready the

1. Sex
   - Male
   - Female
2. Weight
   - Overweight
   - Not overweight
3. Height
   - Tall
   - Short
4. Number of doors of automobile
   - Two
   - Four
5. Driver's side window has windowshade
   - Yes
   - No
6. Seatbelt's latchplate
   - Locking
   - Nonlocking
7. Front seat type
   - Bucket seats
   - Bench seats

Figure 1.6 **Seatbelt study: factors and levels**

services, and follow up appropriately. All medical expenses would be covered, and other service would be provided. The traveler would not be required to do any paperwork. The company wanted to explore the importance to "enrollment" of various factors and their levels with respect to these other services. Also, the importance of price needed to be explored.

The company wanted to determine one "best" plan at one fixed price, although it left open (for itself—not to be included in the experiment) the possibility that it might allow the level of several factors to be options selected by the traveler. All plans were to include the basic medical service and hot-line assistance. Some of the other factors whose levels were explored are listed in Figure 1.7.

The dependent variable was respondents' assessment of the number of days during the next two years that they would use the service. In addition to this measure, respondents were asked about their travel habits, including questions about with whom they typically traveled, the degree to which a traveler's destinations are unusual or "offbeat," and the countries often traveled to. In addition, respondents indicated their view of the anticipated severity of medical problems, legal problems, financial problems (such as a lost wallet), and travel problems (lost tickets and so on), using a five-point scale (ranging from 5 = extremely big problem to 1 = not a problem at all).

1. Personal liability coverage (property damage, bail, legal fees, etc.)
   - None
   - Up to $5000
   - Up to $10,000
2. Transportation home for children provided if parent becomes ill
   - Yes
   - No
3. Baggage insurance
   - Not provided
   - Up to $600
   - Up to $1200
4. Insurance for trip interruption or hotel or tour cancellation
   - A large variety of levels
5. Price
   - $2 per day for individual, family plans available
   - $6 per day for individual, family plans available
   - $10 per day for individual, family plans available
   - $14 per day for individual, family plans available

Figure 1.7 Travelers' emergency assistance: factor and levels

1.5 Perspective

As we noted in the beginning of the previous section, we use these applications and others throughout the text to illustrate the use of the techniques and thought processes discussed in the text. Often, further details about the applications are discussed to ensure that the reasoning for and logic of the techniques or processes used, and not solely the how-to and number crunching aspects of the application, are conveyed.

We view these applications to be management applications, but it has become clear that this is not a well-defined term. If we design an experiment to determine whether different brands of batteries have equal average lifetimes, is this a management application? Because the purpose of the actual experiment was to see if there was support for one brand's claim of superiority in terms of higher average battery life, without which a major network would not allow the claim of superiority ad to run, our view is that the application is indeed one of management (in particular, marketing). However, suppose that the factor under study was type of electrolyte (part of what makes a battery what it is); is it now a management application? Perhaps it depends on one's perspective. In some sense it's an engineering or science application. In another sense, however, a manager will supervise a group of engineers, and managers will likely be the decision makers who combine financial and other considerations to decide which electrolytes should no longer be tested, which look promising as research avenues, and so forth. Our view of the phrase "management application" is liberal, and goes well beyond the traditional departments of a school of management—marketing, accounting, finance, operations, economics, and strategy. Indeed, it includes most of the domain that many would label industrial experimentation. We believe that our way of categorizing applications is consistent with the majority of current thinking by managers about the expanding role of experimental design in industry, government, and other not-for-profit environments. Our objective is to encourage managers to better appreciate the utility of these techniques for many managerial problems.

Note

1. A factorial design consists of varying combinations of levels of factors. A full factorial design consists of all combinations of levels of factors, whereas a fractional factorial design consists of a carefully chosen subset of combinations. In a one-factor-at-a-time experiment the levels of factors are varied, but only one factor at a time. In Chapter 9 we describe and compare these designs in detail.

PART ONE

Primary Focus on Factors Under Study

CHAPTER 2

One-Factor Designs and the Analysis of Variance

We begin this and subsequent chapters by presenting a real-world problem in the design and analysis of experiments on which at least one of the authors consulted. At the end of the chapter we revisit the example and present analysis and results. As you read the chapter, think about how the principles discussed here can be applied to this problem.

EXAMPLE 2.1 Corporate Environmental Behavior at Clean Air Electric Co.

A number of states have deregulated the electric-power industry, and other states are considering doing so. As noted in Chapter 1, Clean Air Electric Company wondered whether it could achieve a competitive advantage by promising to provide electricity while conserving the environment. (Considering California utility companies' practical and political problems in early 2001, this was especially smart.)

So the company decided to study whether demand for its fuel would be influenced by a set of factors, notably including different levels of publicized corporate environmental behavior; other factors included price, level of detail of information provided to customers about their pattern of use of electricity, level of flexibility of billing options, and several more. The factor "corporate environmental behavior" had five levels, as shown in Figure 2.1. Because these five levels of the environmental factor have very different revenue and cost (and hence profit) implications, the company wanted to know whether and how the demand for its electricity would vary by the level implemented. We return to this example at the end of the chapter.

1. Clean Air Electric Company will ensure that its practices are environmentally sound. It will always recycle materials well above the level required by law. It will partner with local environmental groups to sponsor activities good for the environment.
2. Clean Air Electric Company will ensure that its practices are environmentally sound. It will always recycle materials well above the level required by law. It will partner with local environmental groups to sponsor activities good for the environment. **Also, Clean Air Electric Company will donate 3% of its profits to environmental organizations.***
3. Clean Air Electric Company will ensure that its practices are environmentally sound. It will always recycle materials well above the level required by law. It will partner with local environmental groups to sponsor activities good for the environment. Also, Clean Air Electric Company will donate **6% of its profits** to environmental organizations.
4. Clean Air Electric Company will ensure that its practices are environmentally sound. It will always recycle materials well above the level required by law. It will partner with local environmental groups to sponsor activities good for the environment, **and will engage an unbiased third party to provide environmental audits of its operations. Also, Clean Air Electric Company will provide college scholarships to leading environmental colleges.**
5. Clean Air Electric Company will ensure that its practices are environmentally sound. It will always recycle materials well above the level required by law. It will partner with local environmental groups to sponsor activities good for the environment. **Also, Clean Air Electric Company will donate 3% of its profits to environmental organizations, and will actively lobby the government to pass environmentally friendly legislation.**

*Boldface typing is solely for ease of reader identification of the differences among the levels.

Figure 2.1 Levels of factor: corporate environmental behavior

2.1 One-Factor Designs

In this chapter we consider studies involving the impact of a single factor on some performance measure. Toward this end, we need to define some notation and terminology. Careful thought has been put into how to do this, for the nature of the field of experimental design is such that, if we are to be consistent, the notation introduced here has implications for the notation throughout the remainder of the text.

We designate by Y the **dependent variable**—the quantity that is potentially influenced by some other factors (the independent variables). Other terms commonly used for the dependent variable are *yield, response variable,* and *performance measure.* We will, from time to time, use these as synonyms. In different fields, different terms are more common; it is no surprise, for example, that in agricultural experiments the term *yield* is prevalent.

Our one **independent variable**—the possibly influential factor under study—is designated as X. In this chapter, we consider the situation in which only one independent variable is to be investigated, but in subsequent chapters we extend our designs to include several independent variables.

In order to make things a bit more tangible, let us suppose that a national retailer is considering a mail-order promotional campaign to provide a reward (sometimes called a loyalty incentive) to members of their master file of active customers holding a company credit card; the retailer defines "active" as those company credit-card holders who have made a purchase during the last two years. The retailer wishes to determine the influence, if any, of the residence of the cardholder on the dollar volume of purchases during the past year. Company policy is to divide the United States into six different "regions of residence," and to have several other regions of residence outside the United States, representing parts of Europe, Asia, and other locations; there are over a dozen mutually exclusive regions of residence in total, covering the entire planet. Our variables are as follows:

$Y =$ dollar volume of purchases during the past year in \$100s (sales volume)
$X =$ region of residence of the cardholder

We might indicate our conjecture that a cardholder's sales volume is affected by the cardholder's region of residence by the following statement of a functional relationship:

$$Y = f(X, \varepsilon)$$

where ε is a random error component, representing all factors other than X having an influence on a cardholder's sales volume.

The equation says, "Y (sales volume) is a function of (depends on) X (region of residence) and ε (everything else)." This is a tautology; obviously sales volume depends on region of residence and everything else! How could it be otherwise? Nevertheless, this functional notation is useful and is our starting point. We seek to investigate further and with more specificity the relationship between sales volume and region of residence. We now develop the statistical model with which we will carry out our investigation.

Consider the following array of data:

	1	2	3	. . .	j	. . .	C
1	Y_{11}	Y_{12}	Y_{13}				Y_{1C}
2	Y_{21}	Y_{22}	Y_{23}				Y_{2C}
3	Y_{31}	Y_{32}	Y_{33}				Y_{3C}
⋮							
i	Y_{i1}	Y_{i2}	Y_{i3}		Y_{ij}		Y_{iC}
⋮							
R	Y_{R1}	Y_{R2}	Y_{R3}				Y_{RC}

Imagine that every element, Y_{ij}, in the array corresponds to a sales volume for person i (i indexes rows) whose region of residence is j (j indexes columns, and a column represents a specific region of residence). In general, Y_{ij} is an individual data value from an experiment at one specific "treatment" or **level** (value, category, or amount)[1] of the factor under study. The array is a depiction of the value of the dependent variable, or yield or response, for specific levels of the independent variable (factor under study). Specifically, the columns represent different levels of X; that is, all sales volumes in a specific column were obtained for active customers having the same (level of) region of residence. Column number is simply a surrogate for "level of region of residence"; there are C columns because C different regions of residence are being investigated in this experiment. (Naturally, there are many ways in which one could categorize parts of the United States and the world into specific regions of residence; for simplicity, let's assume that in this case it is merely based on company precedent.)

There are R rows in the array, indicating that R customers in each region of residence were examined with respect to their sales volume. In general, the array represents the fact that R values of the dependent variable are determined at each of the C levels of the factor under study. Thus, this is said to be a **replicated experiment,** meaning it has more than one data value at each level of the factor under study. (Some would also use the word *replicated* for a situation in which one or more, but not necessarily all, levels of the factor have more than one data value.) Here, R, the number of rows, is also the number of replicates, but this is usually true only when studying just one factor. The total number of experimental outcomes (data points) is equal to RC, the product of the number of rows (replicates) and the number of columns (factor levels). The depicted array assumes the same number of replicates (here, customers) for each level (here, region of residence) of the factor under study. This may not be the case, but for a fixed number of data points in total, having the same number of replicates per column, if possible, is the most efficient choice (that is, will yield *maximum reliability* of the results—a term defined more precisely later).

The *RC* positions in the array above are indexed; that is, each location corresponds to a specific row and a specific column. The subscripts i and j are used to designate the row and column position; that is, Y_{ij} is the data value that is in the ith row and jth column.[2]

The Statistical Model

It is useful to represent each data point in the following form, called a statistical model:

$$Y_{ij} = \mu + \tau_j + \varepsilon_{ij}$$

where

$i = 1, \ldots, R$
$j = 1, \ldots, C$
μ = overall average (mean)
τ_j = differential effect (response) associated with the jth level of X; this assumes that overall the values of τ_j add to zero (that is, $\Sigma\, \tau_j = 0$, summed over j from 1 to C)
ε_{ij} = noise or error associated with the particular ijth data value

That is, we envision (postulate or hypothesize, that is) an additive model that says every data point can be represented by summing three quantities: the true mean, averaged over all of the factor levels being investigated, plus an incremental component associated with the particular column (factor level), plus a final component associated with everything else affecting that specific data value. As we mentioned earlier, it is tautological to state that a quantity depends on the level of a particular factor plus everything else. However, in the model given, the relative values of the components are most important. To what degree is sales volume affected by region of residence, relative to everything else? As we shall see, that is a key question, perhaps the most important one.

Suppose we were randomly selecting adults and were interested in their weight as a function of their sex. Of course, in reality, we expect that a person's weight depends on a myriad of factors—age, physical activity, genetic background, height, current views of attractiveness, use of drugs, marital status, level of economic income, education, and others, in addition to sex. Suppose that the difference between the average weight of men and the average weight of women is 25 pounds, that the mean weight of all adults is 160 pounds, and that the number of men and women in the population is equal. Our experiment would consist of selecting R men and R women. Our data array would have two columns ($C = 2$), one for men and one for women, and R rows, and would contain $2R$ weights (data points). In the relationship $Y_{ij} = \mu + \tau_j + \varepsilon_{ij}$, $\mu = 160$, $\tau_1 = 12.5$ (if men are heavier than women and the weights of men are in the first column), $\tau_2 = -12.5$, and ε_{ij} depends on the actual weight measurement of the person located in the ith row and jth column. Of the myriad of factors that affect a person's weight, all but one, sex, are embraced by the last term, ε_{ij}. Of course, ordinarily we don't know the values of μ, τ_j, or ε_{ij} and therefore need to estimate them in order to achieve our ultimate goal, that of determining whether the level of the factor has an impact on Y_{ij}, the yield.

Estimation of the Parameters of the Model

We need to compute the column means to proceed with the estimation process. Our notation for the column means is $\bar{Y}_{\cdot j} = \Sigma_{i=1,R}\, Y_{ij}/R$, for the mean of column j; that is, $\bar{Y}_{\cdot 1}$, $\bar{Y}_{\cdot 2}$, $\bar{Y}_{\cdot 3}$, . . . , $\bar{Y}_{\cdot j}$, . . . , $\bar{Y}_{\cdot C}$ are the means for the first, second, third, . . . , jth, . . . , and Cth columns, respectively.

We append the column means to the data array as follows:

1	2	3		. . .	j		. . .	C
Y_{11}	Y_{12}	Y_{13}						Y_{1C}
Y_{21}	Y_{22}	Y_{23}						Y_{2C}
Y_{31}	Y_{32}	Y_{33}						Y_{3C}
					Y_{ij}			
Y_{R1}	Y_{R2}	Y_{R3}						Y_{RC}
$\bar{Y}_{\cdot 1}$	$\bar{Y}_{\cdot 2}$	$\bar{Y}_{\cdot 3}$			$\bar{Y}_{\cdot j}$			$\bar{Y}_{\cdot C}$

We use the $.j$ notation to explicitly indicate that it is the second subscript, the column subscript, that remains. That is, the act of averaging over all the rows for any specific column removes the dependence of the result on the row designation; indeed, if you were to inquire: "The mean of the first column is associated with which row?" the answer would be either "none of them" or "all of them." Correspondingly, we "dot out" the row subscript (in a later chapter, we will use $\bar{Y}_{1\cdot}$ to designate the mean of the first row—a quantity that has no useful meaning in the current study). For the first column, for example,

$$\bar{Y}_{\cdot 1} = (Y_{11} + Y_{21} + Y_{31} + \cdots + Y_{i1} + \cdots + Y_{R1})/R$$

Similarly, the average of all RC data points is called the **grand mean,** and is a function of neither row nor column (or, perhaps, all rows and columns), has the consistent notation of $\bar{Y}_{\cdot\cdot}$, and equals

$$\begin{aligned}\bar{Y}_{\cdot\cdot} = [&(Y_{11} + Y_{21} + Y_{31} + \cdots + Y_{i1} + \cdots + Y_{R1})\\ &+ (Y_{12} + Y_{22} + Y_{32} + \cdots + Y_{i2} + \cdots + Y_{R2})\end{aligned}$$

$$+ (Y_{13} + Y_{23} + Y_{33} + \cdots + Y_{i3} + \cdots + Y_{R3})$$
$$+ \cdots + (Y_{1j} + Y_{2j} + Y_{3j} + \cdots + Y_{ij} + \cdots + Y_{Rj})$$
$$+ \cdots + (Y_{1C} + Y_{2C} + Y_{3C} + \cdots + Y_{iC} + \cdots + Y_{RC})]/RC$$

An example of these and subsequent calculations appears in the section 2.2 example. The grand mean can also be computed as the mean of the column means, given our portrayal of each column as having the same number of rows, R. Thus, $\bar{Y}_{..}$ also equals

$$\bar{Y}_{..} = (\bar{Y}_{.1} + \bar{Y}_{.2} + \bar{Y}_{.3} + \cdots + \bar{Y}_{.j} + \cdots + \bar{Y}_{.C})/C$$

If the number of data points is not the same for each column, the grand mean, which always equals the arithmetic average of all the data values, can also be computed as a weighted average of the column means, with the weights being the $R_{.j}$ for each column j.

Recall our statistical model:

$$Y_{ij} = \mu + \tau_j + \varepsilon_{ij}$$

It is useful to understand that if we had infinite data (obviously, we never do), we would then know with certainty the values of the parameters of the model; for example, if one takes a sample mean of an infinite number of data values, the sample mean is then viewed as equaling the true mean. Of course, in the real world, we have (or will have, after experimentation), only a "few" data points, the number typically limited by affordability and/or time considerations.

Indeed, we use the data to form estimates of μ, τ_j, and ε_{ij}. We use the principle of least squares developed by Gauss at about the beginning of the 19th century as the criterion for determining these estimates. This is the criterion used 99.99% of the time in these situations, and one would be hard pressed to find a commercial software program that uses any other estimation criterion. In simplified form, the principle of least squares says that the optimal estimate of a parameter is the estimate that minimizes the sum of the squared differences (so-called deviations) between the actual Y_{ij} values and the "predicted values" (the latter computed by inserting the estimates into the equation). In essence, the difference is an estimate of ε; most often, this estimate is labeled e. Then, the criterion is to choose T_j (an estimate of τ_j, for each j) and M (an estimate of μ) to minimize the sum of the squared deviations.

$$e_{ij} = (Y_{ij} - M - T_j) \quad \text{and} \quad \Sigma\Sigma(e_{ij})^2 = \Sigma\Sigma(Y_{ij} - M - T_j)^2$$

where $\Sigma\Sigma$ indicates double sums, all of which are over i from 1 to R, and over j from 1 to C; the order doesn't matter.

We won't go through the derivation of the estimates here; several texts illustrate a derivation, in most cases by using calculus. In fact, you may have seen a similar derivation in the context of regression analysis in your introductory statistics course.

It can be shown, by the least-squares criterion, that

$$\bar{Y}_{..} \quad \text{estimates} \quad \mu$$

and that

$$(\bar{Y}_{.j} - \bar{Y}_{..}) \quad \text{estimates} \quad \tau_j$$

for all j. These estimates are not only the least-squares estimates but also (we would argue) commonsense estimates. After all, what is more common than to have μ, the true overall mean, estimated best by the grand mean of the data? The estimate of τ_j is also a commonsense estimate. If we were reading about the difference between the mean age of Massachusetts residents and the mean age of residents of all 50 states (that is, the United States as a whole), what would we likely read as a quantitative description? Most likely, we would read a statement such as "Massachusetts residents are, on average, 1.7 years older than the average of all U.S. residents." In essence, this 1.7 is the difference between the Massachusetts mean and the mean over all 50 states. And this is exactly what $(\bar{Y}_{.j} - \bar{Y}_{..})$ does—it takes the difference between the mean of one column (equivalent to a state in our example) and the mean of all columns (equivalent to the entire United States in our example).

If we insert our estimates into the model in lieu of the parameter values, it follows, from routine algebra, that e_{ij}, the estimate of ε_{ij}, equals

$$Y_{ij} - \bar{Y}_{.j}$$

Note that we have $e_{ij} = Y_{ij} - M - T_j$, or $e_{ij} = Y_{ij} - \bar{Y}_{..} - (\bar{Y}_{.j} - \bar{Y}_{..})$, and the above follows. We then have, when all estimates are inserted into the equation,

$$Y_{ij} = \bar{Y}_{..} + (\bar{Y}_{.j} - \bar{Y}_{..}) + (Y_{ij} - \bar{Y}_{.j})$$

This, too, may be seen as a tautology; remove the parentheses, implement the algebra, and one gets $Y_{ij} = Y_{ij}$. Nevertheless, as we shall soon see, this formulation leads to very useful results. With minor modification, we have

$$(Y_{ij} - \bar{Y}_{..}) = (\bar{Y}_{.j} - \bar{Y}_{..}) + (Y_{ij} - \bar{Y}_{.j})$$

This relationship says: "The difference between any data value and the average value of all the data is the sum of two quantities: the difference associated with the specific level of the independent variable (that is, how level j [column j] differs from the average of all levels, or columns), plus the difference between the data value and the mean of all data points at the same level of the independent variable." In our example, the difference between any sales volume value and the average over all sales volume values is equal to the sum of the difference associated with the particular level of region of residence plus the difference associated with everything else.

Sums of Squares

We can square both sides of the previous equation (obviously, if two quantities are equal, their squares are equal); this gives us:

$$(Y_{ij} - \bar{Y}_{..})^2 = (\bar{Y}_{.j} - \bar{Y}_{..})^2 + (Y_{ij} - \bar{Y}_{.j})^2 + 2(\bar{Y}_{.j} - \bar{Y}_{..})(\bar{Y}_{ij} - \bar{Y}_{.j})$$

An equation similar to the preceding exists for each data point—that is, for each *i,j* combination. There are *RC* such data points and, correspondingly, *RC* such equations. Clearly, the sum of the *RC* left-hand sides of these equations is equal to the sum of the *RC* right-hand sides. Taking these sums, at least symbolically, we get

$$\Sigma\Sigma(Y_{ij} - \bar{Y}_{..})^2 = \Sigma\Sigma(\bar{Y}_{.j} - \bar{Y}_{..})^2 + \Sigma\Sigma(Y_{ij} - \bar{Y}_{.j})^2 + 2 \cdot \Sigma\Sigma[(\bar{Y}_{.j} - \bar{Y}_{..})(Y_{ij} - \bar{Y}_{.j})] \quad (2.1)$$

where, again, all double sums are over i (from 1 to R) and j (from 1 to C). As an outgrowth of the way we selected the estimates of the parameters in our model, that is, using the principle of least squares, equation (2.1) becomes greatly simplified; writing the last term as . . . $2 \cdot \Sigma_j\{(\bar{Y}_{.j} - \bar{Y}_{..})[\Sigma_i(Y_{ij} - \bar{Y}_{.j})]\}$, we can easily show that the term in the brackets is zero for all j, and the entire cross-product (last) term thus equals zero. Furthermore, the first term following the equal sign can be reduced from a double sum to a single sum; this is because the term is, in essence, a sum of terms (R of them) that are identical. That is,

$$\begin{aligned}\Sigma\Sigma(\bar{Y}_{.j} - \bar{Y}_{..})^2 &= \Sigma_i[\Sigma_j(\bar{Y}_{.j} - \bar{Y}_{..})^2] \\ &= \Sigma_j(\bar{Y}_{.j} - \bar{Y}_{..})^2 + \Sigma_j(\bar{Y}_{.j} - \bar{Y}_{..})^2 \\ &\quad + \cdots + \Sigma_j(\bar{Y}_{.j} - \bar{Y}_{..})^2 \quad (2.2) \\ &= R[\Sigma_j(\bar{Y}_{.j} - \bar{Y}_{..})^2] \quad (2.3)\end{aligned}$$

In other words, this double sum (over i and j) of equation (2.2) can be written as R times a single sum (over j), as in (2.3). The algebra involved in reducing the double sum to a single sum is no different than if we had something like the sum from 1 to 20 of the number 7: $\Sigma 7$, from 1 to 20. Of course, this is $7 + 7 + \cdots + 7$, twenty times, or, more efficiently, $20 \cdot 7$. The result of all this apparently fortuitous simplification is the following equation, which undergirds much of what we do (again, both double sums over i [from 1 to R] and j [from 1 to C]):

$$\Sigma\Sigma(Y_{ij} - \bar{Y}_{..})^2 = R[\Sigma_j(\bar{Y}_{.j} - \bar{Y}_{..})^2] + \Sigma\Sigma(Y_{ij} - \bar{Y}_{.j})^2 \quad (2.4)$$

This says that the first term, the total sum of squares (TSS), is the sum of the second term, which is the sum of squares between columns (SSBc), plus the third term, the sum of squares within columns (SSW)[3] (alternatively called SSE, explained shortly).

TSS, the **total sum of squares,** is the sum of the squared difference between each data point and the grand mean. It would be the numerator of an estimate of the variance of the probability distribution of data points if all data points were viewed as coming from the same column or distribution (which would then have the same value of μ). It is a measure of the variability in the data under this supposition. In essence, putting aside the fact that it isn't normalized, and thus doesn't directly reflect the number of data points included in its calculation, TSS is a measure of the degree to which the data points aren't all the same.

SSBc, the **sum of squares between columns,** is the sum of the squares of the difference between each column mean and the grand mean, multiplied by R, the number of rows, and is also akin to a variance term. It's also not normalized, and it doesn't reflect the number of columns or column means whose differences from the grand mean are squared. However, SSBc is larger or smaller depending on the extent to which column means vary from one another. We might expect that, if region of residence has no influence on sales volume, the column means (average sales volume corresponding to different regions of residence) would be more similar than dissimilar. Indeed, we shall see that the size of SSBc, in relation to other quantities, is a measure of the influence of the column factor (here, region of residence) in accounting for the behavior of the dependent variable (here, sales volume).

Although the summation expression of SSBc might be called the heart of the SSBc, it is instructive to consider the intuitive role of the multiplicative term R. For a given set of column means, the summation part of the SSBc is determined. We can think of R as an amplifier; if, for example, $R = 50$ instead of $R = 5$, SSBc is 10 times larger. We would argue that this makes good sense. Think of the issue this way: Suppose that a random sample of household incomes was taken for two towns, Framingham and Natick (two towns in the western suburbs of Boston), with outliers (statistically extreme data values) not counted. If you were told that the sample means were \$5000 apart, wouldn't this same \$5000 difference suggest something very different to you depending on whether 6 or 6000 households were sampled from each town? Of course! In the former case ($n = 6$), we wouldn't be convinced at all that the \$5000 value was meaningful; one slightly aberrant household income could easily lead to the difference. In the latter case ($n = 6000$), the \$5000 difference would almost surely indicate a real difference.[4] In other words, R amplifies the sum of the squares of the differences between the column means and the grand mean, giving the SSBc a value that more meaningfully conveys the evidentiary value of the differences among the column means.

Finally, **SSW,** the **sum of squares within columns,** is a measure of the influence of factors other than the column factor on the dependent variable. After all, SSW is the sum of the squares of the differences between each data point in a column and the mean of that column. Any differences among the values within a specific column, or level of the factor under study, can have nothing to do with what the level of the factor is; all data points in a column have the same level of the factor. It seems reasonable that, if SSW is almost as big as TSS, the column factor has not explained very much about the behavior of the dependent variable; instead, factors other than the column factor dominate the reasons that the data values differ to the degree they do. Conversely, if the SSW is near zero relative to the TSS, it seems reasonable to conclude that which column the data point is in just about says it all. If we view our major task as an attempt to account for the variability in Y, partitioning it into that part associated with the level of the factor under investigation, and that part associated with other factors ("error"), then SSW is a measure of that part associated with error. This is why SSW is often called SSE, the sum of squares due to error.

2.2 Analysis of (the) Variance (ANOVA)

Traditionally, we analyze our model by use of what is called an analysis-of-variance (ANOVA) table, as shown in Table 2.1. Let's look at each of the columns of the table. The first and second columns are, by now, familiar quantities. The heading of the second column, SSQ, indicates the various sum-of-squares quantities (numerical values) due to the sources shown in the first column.

The third column heading, *df*, stands for degrees of freedom. The degrees-of-freedom number for a specific sum of squares is the appropriate value by which to divide the sum of squares to yield the fourth column, called the mean square. The reason that we need to divide by something is to make the resulting values legitimately comparable (comparing mean squares is part of our analysis, as will be seen). After all, if we were comparing whether men and women have a different mean grade point average (GPA), and we had the total of the GPAs of ten men and of six women, we would never address our inquiry simply by comparing totals (although one woman student was heard to say that if we did, perhaps the men would have a chance to come out higher). We would, of course, divide the male total by ten and the female total by six, and only then would we compare results. A mean square is conceptually the average sum of squares. That is, we noted earlier that neither the SSBc nor the SSW reflected the number of terms that went into its sum; dividing these sums of squares by the number of terms that went into the sum would then give us, in each case, the average sum of squares. However, instead of dividing by the exact number of terms going into the sum, the statistical theory behind our analysis procedure mandates that we instead divide by the number of terms that are free to vary (hence, the name **degrees of freedom**). We need to clarify what the phrase "free to vary" means.

Suppose that I am at a faculty meeting with 50 other faculty members and that we are willing to assume that the 51 of us compose a random sample of the Boston University faculty of, let's say, 1500. If I tell you the weight of each of the 50 faculty members other than myself, and I also tell you that the average weight of the 1500 Boston University faculty members is 161 pounds, can you, from that information, determine my weight? Of course, the answer is no. We could say that my weight is free to vary. It is not determined by the other information and could be virtually any value without being inconsistent with the available information. However, suppose now that I tell you the weight of each of the 50 faculty members other than myself, and I also tell you that the average weight of the 51 of us in the room is 161. Then can you determine my weight? The answer is yes. Take the average of the 51 faculty members, multiply by 51, getting the total weight of the 51 faculty, and, then subtract the 50 weights that are given; what remains is my weight. It is not free to vary but rather is completely predetermined by the given information. We can say that we have 51 data values, but that when we specify the mean of the 51 values, only 50 of the 51 values are then free to vary. The 51st is then uniquely

Table 2.1 ANOVA Table

Source of Variability	SSQ	*df*	Mean Square	
Between columns (due to region of residence)	SSBc	$C - 1$	MSBc = SSBc/$(C - 1)$	
Within columns (due to error)	SSW	$(R - 1)C$	MSW = SSW/$[(R - 1)C]$	
Total	TSS	$RC - 1$		

determined. Equivalently, the 51 values have only 50 degrees of freedom. In general, we can express this as the $(n - 1)$ rule:

> When we take each of n data values, subtract from it the mean of all n data values, square the difference, and add up all the squared differences (that is, when n data values are "squared around the mean" of those same n values), the resulting sum of squares has only $(n - 1)$ degrees of freedom associated with it.

(Where did the "missing" degree of freedom go? We can think of it as being used in the calculation of the mean.) The application of this rule stays with us throughout the entire text, in a variety of situations, applications, and experimental designs. The statistical theory dictates that, given our assumptions and goals, the proper value by which to divide each sum of squares is its degrees of freedom. An important example of these goals is that the estimates be unbiased—that is, for each estimate, the expected value of the sample value is equal to the true, unknown, value.

In the context of introductory statistics, one usually first encounters this notion when learning how to use sample data, say $x_1, x_2, \ldots, x_n$, to produce the best linear unbiased estimate of a population variance, typically called S^2 (summation over i):

$$S^2 = [\Sigma(x_i - \bar{x})^2]/(n - 1)$$

If we examine SSBc, we note that its summation component is taking each of the C column means and squaring each around the mean of those C values (the latter being the grand mean). Thus, applying the $(n - 1)$ rule, the degrees-of-freedom value used to divide SSBc is $(C - 1)$. To repeat, the statistical theory mandates this result; the $(n - 1)$ rule and the example of the professors' weight are attempts to give the reader some degree of intuitive understanding of why the statistical theory yields the results it does. There are other ways to explain the $(n - 1)$; the most notable ones deal with constraints and rank of matrices; however, the authors believe that the most intuitive way is that described above.[5]

Now consider the SSW. One way to describe its calculation is to say that first we pick a column, square each value in the column around the mean of that column, and add up the squared differences; then we do this for the next column, and eventually every column; finally, we add together the sum from each column. With R data values in each column, each column contributes $(R - 1)$ degrees of freedom; having $(R - 1)$ degrees of freedom associated with each of the C columns results in a total of $C(R - 1)$ degrees of freedom for SSW.

The degrees of freedom for the TSS is $(RC - 1)$, corresponding to RC data values squared around the mean of the RC data values. It is useful to note that the degrees of freedom in the body of the table add up to $(RC - 1)$. (As we did for the $(n - 1)$ rule, we can think of this one degree of freedom as having been used in the calculation of the grand mean to estimate μ.) The fact that the total degrees of freedom are equal to the total number of data points less one will turn out to be true for all designs, even the most complicated; indeed, in some situations we will use this fact to back into a degrees-of-freedom value that is difficult to reason using the $(n - 1)$ argument. For example, suppose that we were unable to reason what the degrees of freedom value for the SSW is. We could calculate it by knowing that the total number of degrees of freedom is $(RC - 1)$, that SSBc has $(C - 1)$ degrees of freedom, and that there are only two sources of variability, and thus the SSW must be associated with the rest of the degrees of freedom. If we take $(RC - 1)$ and subtract $(C - 1)$, we arrive at the number of degrees of freedom for the SSW of $[RC - 1 - (C - 1)] = RC - C = (R - 1)C$.

When we divide the SSBc by its degrees of freedom, we get the ***mean* square between columns (MSBc);** similarly, the SSW divided by its degrees of freedom becomes the ***mean* square within columns (MSW).**

The fifth and last column of the ANOVA table has purposely been left blank for the moment; we discuss it shortly. However, first let's look at an example to illustrate the analysis up to this point, and use it as a springboard to continue the development.

EXAMPLE 2.2 Study of Battery Lifetime

Suppose that we wish to inquire how the mean lifetime of a certain manufacturer's AA-cell battery under constant use is affected by the specific device in which it is used. It is well known that batteries of different cell sizes (such as AAA, AA, C, 9-volt) have different mean lifetimes that depend on how the battery is used—constantly, intermittently with certain patterns of usage, and so on. Indeed, usage mode partly determines which cell size is appropriate. However, an additional question is whether the same usage pattern (in this case, constant) would lead to different mean lifetimes as a function of the device in which the battery is used. The results of battery lifetime testing are necessary to convince a TV network to run an advertisement that claims superiority of one brand of battery over another, because possessing results that back up the claim reduces the network's legal liability. The testing is traditionally carried out by an independent testing agency, and the data are analyzed by an independent consultant.

Suppose that we choose a production run of AA high-current-drain alkaline batteries and decide to randomly assign three batteries (of the same brand) to each of eight test devices; all test devices have the same nominal load impedance:

1. Cell phone, brand 1
2. Cell phone, brand 2
3. Flash camera, brand 1
4. Flash camera, brand 2
5. Flash camera, brand 3
6. Flashlight, brand 1
7. Flashlight, brand 2
8. Flashlight, brand 3

Our dependent variable (yield, response, quality indicator), Y, is lifetime of the battery, measured in hours. Our independent variable (factor under study), X, is test device ("device"). The number of levels of this factor (and the number of columns), C, is eight.[6]

Because we have three data values for each device, this is a replicated experiment with the number of replicates (rows), R, equal to three. We have $RC = 24$ data points (sample values), as shown in Table 2.2. The column means (in the row at the bottom) are averaged to form the grand mean, $\bar{Y}_{..}$, which equals 5.8. The sum of squares between columns, SSBc, using equation (2.3), is

$$\begin{aligned} \text{SSBc} &= 3[(2.6 - 5.8)^2 + (4.6 - 5.8)^2 + \cdots + (7.4 - 5.8)^2] \\ &= 3[23.04] = 69.12 \end{aligned}$$

Table 2.2 **Battery Lifetime Study—Yield in Hours**

Device							
1	2	3	4	5	6	7	8
1.8	4.2	8.6	7.0	4.2	4.2	7.8	9.0
5.0	5.4	4.6	5.0	7.8	4.2	7.0	7.4
1.0	4.2	4.2	9.0	6.6	5.4	9.8	5.8
2.6	4.6	5.8	7.0	6.2	4.6	8.2	7.4

Table 2.3 **ANOVA Table—Battery Lifetime Study**

Source of Variability	SSQ	*df*	MS	
Device	69.12	7	9.87	
Error	46.72	16	2.92	
Total	115.84	23		

The sum of squares within columns, SSW, using the last term of equation (2.4), is

$$\begin{aligned}\text{SSW} &= [(1.8-2.6)^2 + (5.0-2.6)^2 + (1.0-2.6)^2] + [(4.2-4.6)^2 \\ &\quad + (5.4-4.6)^2 + (4.2-4.6)^2] + \cdots + [(9.0-7.4)^2 \\ &\quad + (7.4-7.4)^2 + (5.8-7.4)^2] \\ &= 8.96 + .96 + \cdots + 5.12 \\ &= 46.72\end{aligned}$$

The total sum of squares, TSS, as noted in equation (2.4), is the sum of these:

$$\text{TSS} = \text{SSBc} + \text{SSW} = 69.12 + 46.72 = 115.84$$

Next, we observe that our total degrees of freedom are $RC - 1 = 24 - 1 = 23$, with $C - 1 = 8 - 1 = 7$ degrees of freedom associated with device, and $(R-1)C = (2)8 = 16$ degrees of freedom associated with error. We embed these quantities in our ANOVA table as shown in Table 2.3 (again, we omit the last column—it will be filled in soon).

2.3 Forming the *F* Statistic: Logic and Derivation

The Key Fifth Column of the ANOVA Table

It can be proven that

$$E(\text{MSW}) = \sigma^2$$

where $E(\ )$ indicates the expected-value operator (which is similar to saying "on the average, with a theoretically infinite number of random samples, from each of which [in this case] MSW is calculated") and σ^2 is the (unknown) variance (square of the standard deviation) of the probability distribution of each data value, under the assumption that each data value has the same variance.

This assumption of constant variance is one of the assumptions that are often made when performing an ANOVA—more about assumptions in the next chapter. It is, perhaps, useful to elaborate a bit about this notion of an expected value; in essence, the above equation says that if we somehow repeated this entire experiment a very large number of times (with the same number of similar subjects, same levels of the same factor, and so on), we would get different values for the MSW each time, but on average, we would get whatever the value of σ^2 is. Another way to say this is that MSW is a random variable, with its true mean equal to σ^2; since this is true, MSW is said to be an **unbiased estimate** of σ^2. By definition, the expected value of a random variable is synonymous with its true mean.

It can also be proven that

$$E(\text{MSBc}) = \sigma^2 + V_{\text{col}}$$

where V_{col} is our notation for "variability due to differences in population (that is, true) column means." The actual expression we are calling V_{col} equals the following (summations are over j, from 1 to C):

$$[R/(C-1)] \cdot \Sigma(\mu_j - \mu)^2 = [R/(C-1)] \cdot \Sigma\, \tau_j^2$$

where μ_j is defined as the true mean of column j. V_{col} is, in a very natural way, a measure of the differences between the true column means. A key point is that V_{col} *equals zero if there are no differences among (true) column means, and is positive if there are such differences.*

It is important to note something implied by the E(MSBc) formula:

> There are two separate (that is, independent) reasons why the MSBc almost surely will not be calculated to be zero. One reason is that, indeed, the true column means might not be equal. The other, more subtle reason is that even if the true column means happen to be equal, routine sample error will lead to the calculated column means being unequal. After all, the column means we calculate are merely sample means ($\bar{x}$'s, in the usual notation of an introductory course in statistics), and C sample means generated from a distribution with the same true mean will, of course, never all be equal (except in the rarest of coincidences and even then only with rounding).

Again, note that you cannot know σ^2 nor V_{col}; it would take infinite data or divine guidance to know the true value of these quantities. Too bad! If we knew the value of V_{col}, it would directly answer our key question—are the true column means equal or not? Even if we knew only E(MSBc) and E(MSW), we could get the value of V_{col} by subtracting out σ^2 or by forming the ratio

$$\text{E(MSBc)/E(MSW)} = (\sigma^2 + V_{\text{col}})/\sigma^2$$

and comparing this ratio to 1. A value larger than 1 would indicate that V_{col}, a nonnegative quantity, is not zero.

Because we don't know this ratio, we do the next best thing: we examine our estimate of this quantity, MSBc/MSW, and compare it to the value 1.[7] We call this ratio F_{calc}. We use the notation of "calc" as a subscript, to indicate that it is a value calculated from the data, and to clearly differentiate the quantity from a critical value—a threshold value, usually obtained from a table, indicating a point on the abscissa of the probability density function of the ratio. The F is in honor of Sir Ronald Fisher, who is often called "the father of experimental design," and who was the originator of the ANOVA procedure we are discussing.

To review, we have

$$E(\text{MSBc}) = \sigma^2 + V_{col} \quad \text{and} \quad E(\text{MSW}) = \sigma^2$$

This suggests that if

$$\text{MSBc}/\text{MSW} > 1$$

there is some evidence that V_{col} is not zero, or, equivalently, that the level of X affects Y, or in our example, the level of device affects battery lifetime. But if

$$\text{MSBc}/\text{MSW} \leq 1$$

there is no evidence that the level of X affects Y, or that the (level of) device affects battery lifetime.

However (and this is a big however), because F_{calc} is formed from data, as are all statistical estimates, it is possible that F_{calc} could be greater than 1 even if the level of X has no effect on Y. In fact, if $V_{col} = 0$, both MSBc and MSW have the same expected value, σ^2, and we would expect $F_{calc} > 1$ about half the time (and $F_{calc} < 1$ about half the time).[8] Since F_{calc} comes out greater than 1 about half the time when X has no effect on Y, we cannot conclude that the level of X does have an effect on Y solely on the basis of $F_{calc} > 1$.

You may wonder how F_{calc} can be less than 1. This really addresses the same issue as the previous paragraph—that the values of MSBc and MSW virtually never equal their expected-value counterparts. Think of this: on average, men weigh more than women; yet, for a random sample of 10 men and 10 women, it is possible that the women outweigh the men. Similarly, on average, the value of MSBc is never less than the value of MSW; yet, for a particular set of data, MSW might exceed MSBc.

Anyway, how do we resolve the issue that the value of F_{calc} being greater than 1 or not doesn't necessarily indicate whether the level of X affects or does not affect Y, respectively? Questions such as this are resolved through the discipline of hypothesis testing. We review this topic in some detail in Chapter 3. Here we borrow from that section (calling on your recollection of these notions from introductory statistics courses) to complete the current discussion.[9]

We need to choose between two alternatives. One is that the level of the factor under study has no impact on the yield, and the other is that the level of the factor under study does influence the yield. We refer to these two alternatives as **hypotheses,** and designate them as H_0 and H_1, respectively.

H_0: level of X does not affect Y
H_1: level of X does affect Y

We can, equivalently, define H_0 and H_1 in terms of the differential effects from column to column, the τ_j, a term in our statistical model:

$$H_0: \tau_1 = \tau_2 = \tau_3 = \ldots = \tau_j = \ldots = \tau_C = 0$$
$$H_1: \text{not all } \tau_j = 0 \text{ (that is, at least one } \tau_j \text{ is not zero)}$$

Finally, we can also express our hypotheses in terms of μ_j, defined earlier as the true column means:

$$H_0: \mu_1 = \mu_2 = \ldots = \mu_j = \ldots = \mu_C$$
(that is, all column means are equal)
$$H_1: \text{not all } \mu_j \text{ are equal}$$
(at least one column mean is different from the others)

H_0 is called the **null hypothesis.** We will accept H_0 unless evidence to the contrary is overwhelming; in the latter case, we will reject H_0 and conclude that H_1, the **alternate hypothesis,** is true. By tradition, H_0 is the basis of discussion; that is, we refer to accepting H_0 or rejecting H_0. The benefit of the doubt goes to H_0 and the burden of proof is on H_1; this concept of the benefit of the doubt and the burden of proof is well analogized by the credo of criminal courtroom proceedings, in which the H_0 is that the defendant is not guilty (innocent), and the H_1 is that the defendant is guilty, and the assumption is "innocent until proven guilty." Our decision to accept or reject H_0 will be guided by the size of F_{calc}. Given, as noted earlier, that F_{calc} will be greater than 1 about half the time even if X has no effect on Y, we require F_{calc} to be much greater than 1 if we are to reject H_0 (and conclude that X affects Y); otherwise, we accept H_0, and do not conclude that X has an effect on Y; we might think of the decision to accept H_0 as a conclusion that there is insufficient evidence to reject H_0, in the same sense that a finding of not guilty in a criminal case is not an affirmation of innocence but rather a statement that there is insufficient evidence of guilt.

The quantity F_{calc}, the value of which is the function of the data on which our decision will be based, is called the test statistic. This **test statistic** F_{calc} is a random variable that we can prove has a probability distribution called the F distribution if the null hypothesis is true and the customary assumptions (discussed in Chapter 3) about the error term, ε_{ij}, are true. (We are consistent in our notation; for example, when we encounter a quantity whose probability distribution is a t curve, we call the test statistic t_{calc}.) Actually, there is a family of F distributions, indexed by two quantities; these two quantities are the degrees of freedom associated with the numerator (MSBc) and the denominator (MSW) of F_{calc}. That is, the probability distribution of F_{calc} is a bit different for each (C, R) combination. Thus, we talk about an F distribution with $(C - 1)$ and $(R - 1)C$ degrees of freedom. Because F_{calc} cannot be negative (after all, it is the ratio of two squared quantities), we are not surprised to find that the F distribution is nonzero only for nonnegative values of F; a typical F distribution is shown in Figure 2.2.

The shaded area in the tail, to the right of **c** (for **c**ritical value) is typically some small proportion (such as .05 or .01) of the entire area under the curve;

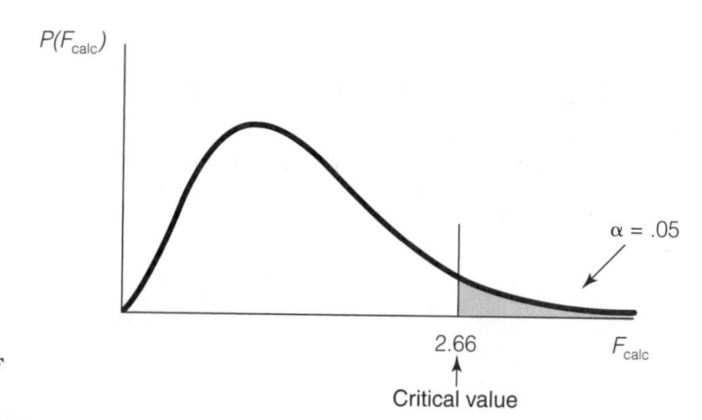

Figure 2.2 A typical *F* distribution

this entire area is, by definition, equal to 1 (the units don't really matter). Then, if the null hypothesis is true, the shaded area represents the probability that F_{calc} is larger than **c** and is designated α (the Greek letter alpha). The critical value **c** is obtained from tables of the *F* distribution because the *F* distribution is quite complex, and it would be difficult for most people analyzing experiments to derive the critical values without these tables. The tables are indexed by the two values for degrees of freedom $(C - 1)$ and $(RC - C)$, which detail the particular probability distribution of a specific F_{calc}. As in all hypothesis testing, we set up the problem presuming H_0 is true and reject H_0 only if the result obtained (here, the value of F_{calc}) is so unlikely to have occurred under that H_0-true assumption as to cast substantial doubt on the validity of H_0. This amounts to seeing if F_{calc} falls in the tail area defined by the set of values greater than **c** (the rejection [of H_0] region). In other words, our procedure is akin to a proof by contradiction: we say that if the probability of getting what we got, assuming that the level of *X* has no effect on *Y*, is smaller than α (perhaps α = .05, a 1 in 20 chance), then getting what we got for F_{calc} is too unlikely to have been by chance, and we reject that the level of *X* has no effect on *Y*. The double-negative tone of the conclusion is characteristic of the hypothesis testing procedure; however, in essence, the double negative implies a positive, and we conclude that the level of *X* actually does affect *Y*. In summary, if F_{calc} falls in the rejection region, we conclude that H_0 is false and that not all column means are equal; if F_{calc} falls in the acceptance region, we conclude that H_0 is true, and that all column means are, indeed, equal.

EXAMPLE 2.3 Battery Lifetime Example, Revisited

A portion of the *F* distribution table is shown in Table 2.4, specifically for α = .05. Tables with more extensive values of α, and for a larger selection of degrees of freedom, are in an appendix at the back of the text.

Table 2.4 **Portion of *F* Table, for Right Tail Area, α = .05**

Denominator df_2	Numerator df_1: 1	2	3	4	5	6	7	8	9
1	161.4	199.5	215.7	224.6	230.2	234.0	236.8	238.9	240.5
2	18.51	19.00	19.16	19.25	19.30	19.33	19.35	19.37	19.38
3	10.13	9.55	9.28	9.12	9.01	8.94	8.89	8.85	8.81
4	7.71	6.94	6.59	6.39	6.26	6.16	6.09	6.04	6.00
5	6.61	5.79	5.41	5.19	5.05	4.95	4.88	4.82	4.77
6	5.99	5.14	4.76	4.53	4.39	4.28	4.21	4.15	4.10
7	5.59	4.74	4.35	4.12	3.97	3.87	3.79	3.73	3.68
8	5.32	4.46	4.07	3.84	3.69	3.58	3.50	3.44	3.39
9	5.12	4.26	3.86	3.63	3.48	3.37	3.29	3.23	3.18
10	4.96	4.10	3.71	3.48	3.33	3.22	3.14	3.07	3.02
11	4.84	3.98	3.59	3.36	3.20	3.09	3.01	2.95	2.90
12	4.75	3.89	3.49	3.26	3.11	3.00	2.91	2.85	2.80
13	4.67	3.81	3.41	3.18	3.03	2.92	2.83	2.77	2.71
14	4.60	3.74	3.34	3.11	2.96	2.85	2.76	2.70	2.65
15	4.54	3.68	3.29	3.06	2.90	2.79	2.71	2.64	2.59
16	4.49	3.63	3.24	3.01	2.85	2.74	2.66	2.59	2.54
17	4.45	3.59	3.20	2.96	2.81	2.70	2.61	2.55	2.49
18	4.41	3.55	3.16	2.93	2.77	2.66	2.58	2.51	2.46
19	4.38	3.52	3.13	2.90	2.74	2.63	2.54	2.48	2.42
20	4.35	3.49	3.10	2.87	2.71	2.60	2.51	2.45	2.39
21	4.32	3.47	3.07	2.84	2.68	2.57	2.49	2.42	2.37
22	4.30	3.44	3.05	2.82	2.66	2.55	2.46	2.40	2.34
23	4.28	3.42	3.03	2.80	2.64	2.53	2.44	2.37	2.32
24	4.26	3.40	3.01	2.78	2.62	2.51	2.42	2.36	2.30
25	4.24	3.39	2.99	2.76	2.60	2.49	2.40	2.34	2.28
26	4.23	3.37	2.98	2.74	2.59	2.47	2.39	2.32	2.27
27	4.21	3.35	2.96	2.73	2.57	2.46	2.37	2.31	2.25
28	4.20	3.34	2.95	2.71	2.56	2.45	2.36	2.29	2.24
29	4.18	3.33	2.93	2.70	2.55	2.43	2.35	2.28	2.22
30	4.17	3.32	2.92	2.69	2.53	2.42	2.33	2.27	2.21
40	4.08	3.23	2.84	2.61	2.45	2.34	2.25	2.18	2.12
60	4.00	3.15	2.76	2.53	2.37	2.25	2.17	2.10	2.04
120	3.92	3.07	2.68	2.45	2.29	2.17	2.09	2.02	1.96
∞	3.84	3.00	2.60	2.37	2.21	2.10	2.01	1.94	1.88

Source: M. Merrington and C. M. Thompson (1943), "Tables of Percentage Points of the *F* Distribution." *Biometrika*, 33, p. 73. Reprinted with permission of Oxford University Press.

Note that, as indicated earlier, we need two degrees-of-freedom values to properly make use of the *F* tables. The first, $(C - 1)$, associated with the numerator of F_{calc} (that is, MSBc), is shown across the top of the table, and is often labeled simply "numerator *df*," or "df_1." The second, $(R - 1)C$, associated with the denominator of F_{calc} (that is, MSW), is shown along the left side of the table, and is often simply labeled "denominator *df*," or "df_2." The vast majority of texts that include *F* tables have adopted the convention that the table has numerator *df* indexed across the top, and denominator *df* indexed going down the left-hand column (or, on occasion, the right-hand column for a right-side page). Remember that the *F* distribution is used here (at least for

Table 2.5 **ANOVA Table—Battery Lifetime Study**

Source of Variability	SSQ	*df*	MS	F_{calc}
Device	69.12	7	9.87	3.38
Error	46.72	16	2.92	
Total	115.84	23		

the moment) under the presumption that H_0 is true—that the level of the column factor does not influence yield. Though nobody does this in practice, we could designate the probability distribution of F as $P(F_{calc}/H_0)$ to emphasize the point.

From the F table, we see that the critical value **c** equals 2.66 when $\alpha = .05$, numerator $df = 7$, and denominator $df = 16$. We repeat our earlier ANOVA table as Table 2.5 with F_{calc}, the key fifth column, now filled in with the value of 3.38 = MSBc/MSW = 9.87/2.92. Figure 2.3 is a picture of the appropriate F distribution for Table 2.5, with F_{calc} and **c** labeled. As shown in the Figure 2.3 plot, $F_{calc} = 3.38$ falls within the rejection region (for H_0). Again, $F_{calc} >$ **c** is interpreted as saying "It is too improbable that, were H_0 true, we would get an F_{calc} value as large as we did. Therefore, the H_0 premise (hypothesis) is rejected; the results we observed are too inconsistent with H_0." In our example, rejection of H_0 is equivalent to concluding that the type of device does affect battery lifetime. When H_0 is rejected, we often refer to the result as statistically significant.

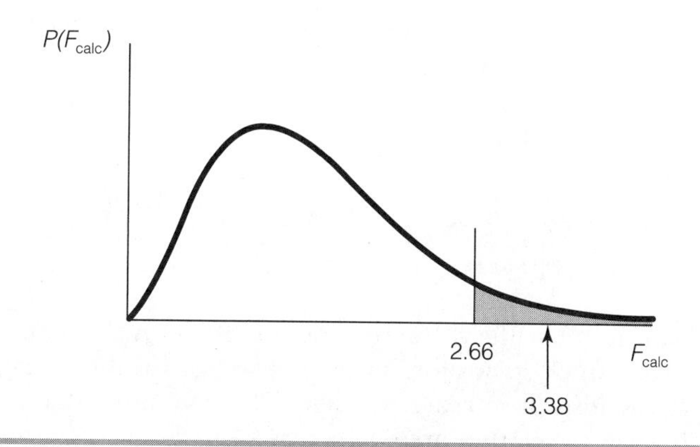

Figure 2.3 ***F* distribution for Table 2.5**

EXAMPLE 2.4 Battery Lifetime Example Using Excel Software

Several software programs perform one-factor ANOVA. Some of these are designed as statistical software programs, whereas others are not primarily for statistical analysis but perform some of the more frequently encountered statistical techniques. Excel is in this latter category; we illustrate its use in analyzing the battery lifetime data from the previous section.

After entering the data as illustrated in the spreadsheet in Table 2.6, we used the "Anova: Single-Factor" command from the "Analysis Tools" section. The output is below the eight-column, three-replicate data input. Table 2.6 shows how the **count** (number of replicates per column), the **sum** of the data points in a column, the column **average** (mean), and the column **variance** (the value of s^2 for the column) are provided, followed by the ANOVA table. The first five columns of the table have exactly the same information, in the same order, as described in the previous section. This is not true for all software packages. Some change the order of columns two (SS, which is the same as SSQ in Tables 2.1, 2.3, and 2.5) and three (df). Others present the information in a somewhat different fashion, but in a form from which all the above information can be determined. For example, one software program gives the degrees of freedom, the F value, the square root of the MSW, and the percentages of the TSS that are the SSBc and SSW; from this information, we can derive all the values in the first five columns of Table 2.5.

Table 2.6 output adds two additional columns to the ANOVA table in Table 2.5. First, it adds the p value, which, in a problem such as this, is the area on the F curve to the right of the calculated F_{calc} (here, 3.38). In other words, the area to the right of 3.38 on the F curve with (7, 16) degrees of freedom corresponds to a probability of .02064. This .02064 speaks to the "degree of evidence against H_0," where a smaller value represents stronger evidence; the meaning and usefulness of p value is discussed in more depth in section 3.4. The other added column is labeled F_{crit} and represents **c**, the critical value, that is, the appropriate F table value with $\alpha = .05$; this value is 2.66. Of course, once we observe that the p value (the area to the right of F_{calc}) is less than .05, we know that the value 3.38 is in the critical region (rejection region) for $\alpha = .05$.

EXAMPLE 2.5 Battery Lifetime Example Using SPSS Software

Now let's use the software program SPSS for this same battery life example. SPSS, which stands for Statistical Package for the Social Sciences, is designed as a statistical software package. The data are entered somewhat differently than in Excel; they are entered as 24 rows and 2 columns. One column contains the 24 values of the dependent variable, and the other column contains

Table 2.6 Excel Format: ANOVA Table for Battery Lifetime Study

1.8	4.2	8.6	7	4.2	4.2	7.8	9
5	5.4	4.6	5	7.8	4.2	7	7.4
1	4.2	4.2	9	6.6	5.4	9.8	5.8
	Anova: Single-Factor						
	Summary						
	Groups	*Count*	*Sum*	*Average*	*Variance*		
	Column 1	3	7.8	2.6	4.48		
	Column 2	3	13.8	4.6	0.48		
	Column 3	3	17.4	5.8	5.92		
	Column 4	3	21	7	4		
	Column 5	3	18.6	6.2	3.36		
	Column 6	3	13.8	4.6	0.48		
	Column 7	3	24.6	8.2	2.08		
	Column 8	3	22.2	7.4	2.56		
	ANOVA						
	Source of Variation						
		SS	*df*	*MS*	*F*	*P-value*	*F crit*
	Between Groups	69.12	7	9.87429	3.3816	0.02064	2.657
	Within Groups	46.72	16	2.92			
	Total	115.8	23				

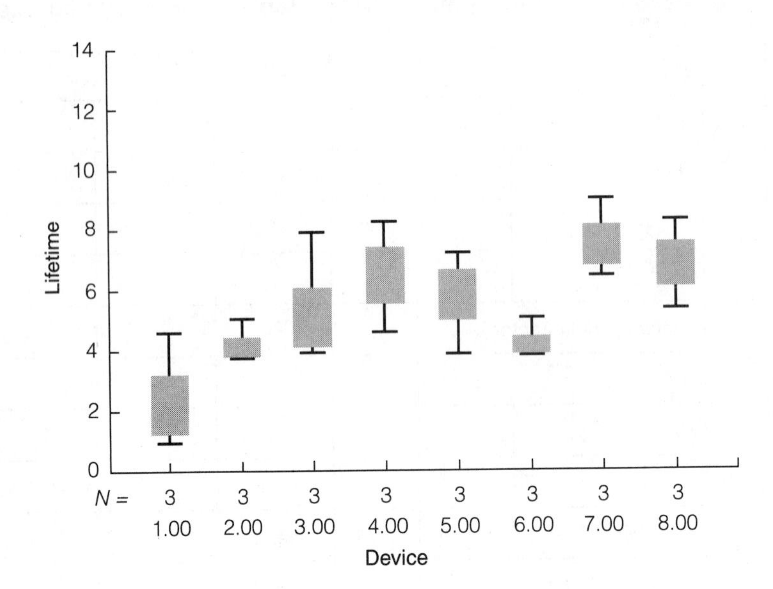

Figure 2.4 Box plot for battery life study

the corresponding level of the factor under study for each of these values—that is, three 1s, three 2s, . . . , three 8s.

Figure 2.4 shows a box plot for the data. A box plot's usefulness is much greater when there are large numbers of data values per column. Even here, however, with three data points per column, it graphically indicates the range of values within each column.

Table 2.7 **SPSS Format: ANOVA Table for Battery Lifetime Study**

- - - - - ONEWAY - - - - -

Variable Lifetime

By Variable Device

Analysis of Variance

Source	D.F.	Sum of Squares	Mean Squares	F Ratio	F Prob.
Between Groups	7	69.1200	9.8743	3.3816	0206
Within Groups	16	46.7200	2.9200		
Total	23	115.8400			

Table 2.7 shows the ANOVA table in SPSS format. We can see that the SPSS output reverses the order for the SSQ and *df* columns compared to the Excel output. SPSS also provides the *p* value, though it uses the phrase "F Prob." instead. However, the output is more or less the same for both packages.

EXAMPLE 2.6 A Larger-Scale Example: Customer Satisfaction Study Using JMP Software

The Merrimack Valley Pediatric Clinic (MVPC) conducted a customer satisfaction study at its four locations: Amesbury, Andover, and Methuen in Massachusetts, and Salem in southern New Hampshire. A series of questions were asked, and a respondent's "overall level of satisfaction" (using MVPC's terminology) was computed by adding together the numerical responses to the various questions. The response to each question was 1, 2, 3, 4, or 5, corresponding to, respectively, "very unsatisfied," "moderately unsatisfied," "neither unsatisfied nor satisfied," "moderately satisfied," and "very satisfied." In our discussion, we ignore the possibility that responses can be treated as an interval scale.

There were 16 questions with the possibility of a 5 rating on each, so the minimum score total was 16 and the maximum score total was 80. (For proprietary reasons, we cannot provide the specific questions.)

Marion Earle, MVPC's medical director, wanted to know (among other things) if there were differences in the average level of satisfaction among customers in four locations. Data from a random sample of 30 responders from each of the four locations are provided in Table 2.8, and in the data sets as Example—Chapter 2.

We use the JMP statistical software package to provide the analysis in Figure 2.5. As can be seen in the table part of the figure, the F_{calc} (called "F Ratio" by JMP) is quite large, 205.29, with a *p* value (called "Prob > F" by JMP) of <.0001, indicating that at any practical significance level, we reject the hypothesis that there is no difference among mean satisfaction levels for the four locations, in favor of there being differences among the four. In addition to the ANOVA table, the output provides the mean and standard error of the mean (the pooled standard deviation estimate, 2.93, which equals the square root of the mean square error, 8.60, divided by the square root of 30, the "sample size" per column) for each column. Also provided are the $\boldsymbol{R}^2$, which is the SSBc (5296.425) divided by the TSS (6293.9917), and the adjusted R^2:

$$\text{Adjusted } R^2 = 1 - [(1 - R^2)(n - 1)/(n - C)]$$

where $n = RC$, the total number of data values, and as before, R and C are the number of rows and columns, respectively.

Table 2.8 **Data from MVPC Satisfaction Study**

Amesbury	Andover	Methuen	Salem
66	55	56	64
66	50	56	70
66	51	57	62
67	47	58	64
70	57	61	66
64	48	54	62
71	52	62	67
66	50	57	60
71	48	61	68
67	50	58	68
63	48	54	66
60	49	51	66
66	52	57	61
70	48	60	63
69	48	59	67
66	48	56	67
70	51	61	70
65	49	55	62
71	46	62	62
63	51	53	68
69	54	59	70
67	54	58	62
64	49	54	63
68	55	58	65
65	47	55	68
67	47	58	68
65	53	55	64
70	51	60	65
68	50	58	69
73	54	64	62

Above the ANOVA table in Figure 2.5, JMP presents a "means diamonds" figure. The horizontal center line in the graph is the grand mean (60.09); the center line for each diamond is the mean at that level (here, city), and the top and bottom of the diamonds, vertically, represent a 95% confidence interval for the mean. The shorter lines near the vertices represent what JMP calls overlap lines, lines that are at distances of .707 of the distance of 95% confidence limits from the mean. For sample sizes per level that are the same, seeing if the top overlap line of one level and the bottom overlap line of another level indeed overlap or not determines whether a *t* test for differences in mean with significance level of .05 would accept (if they do overlap) or reject (if they don't overlap) the null hypothesis that those two means are really the same.

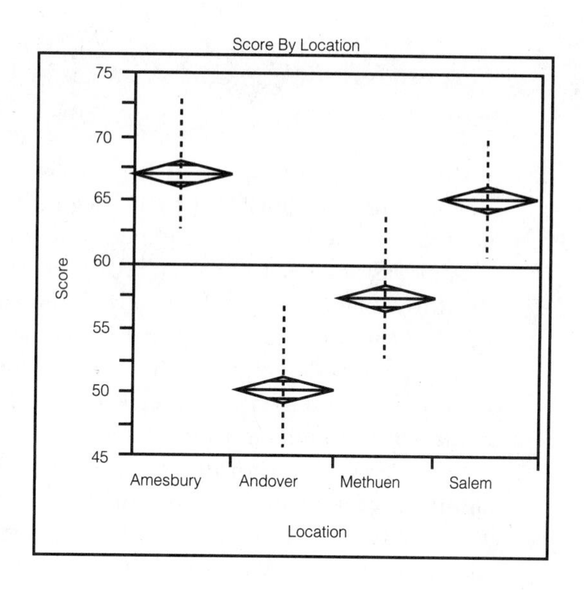

Oneway Anova
Summary of Fit

RSquare	0.841505
RSquare Adj	0.837406
Root Mean Square Error	2.932527
Mean of Response	60.09167
Observations (or Sum Wgts)	120

Analysis of Variance

Source	DF	Sum of Squares	Mean Square	F Ratio
Model	3	5296.4250	1765.47	205.2947
Error	116	997.5667	8.60	Prob>F
C Total	119	6293.9917	52.89	<.0001

Means for Oneway Anova

Level	Number	Mean	Std Error
Amesbury	30	67.1000	0.53540
Andover	30	50.4000	0.53540
Methuen	30	57.5667	0.53540
Salem	30	65.3000	0.53540

Std Error uses a pooled estimate of error variance

Figure 2.5 JMP format: Graphic and ANOVA analysis

EXAMPLE 2.7 Corporate Environmental Behavior at Clean Air Electric Co., Revisited

The experiment described at the beginning of the chapter was actually conducted. On the basis of the experimental results and other industry-wide indicators, as of early 1999, Clean Air Electric did expand its activities to two of the three states that had already deregulated the electric power industry. (One of the three states, Massachusetts, has deregulated its electric power industry in a way that makes it difficult for any company not from Massachusetts to enter its market.) A year later, Clean Air Electric had well over a million and a quarter customers in these states.

The factor "corporate environmental behavior," although not the only significant factor in the experiment, was indeed significant with a very low p value. The level of publicized corporate environmental behavior clearly did have an impact on the self-reported intent to use Clean Air Electric. The detailed results for the factor, level by level, cannot be made public at this time.

Interestingly, based on specific location within one of the states, some customers were able to switch to Clean Air Electric and simultaneously reduce their cost for electric power. It is not yet clear to Clean Air Electric whether the environmental behavior has had a material impact on their customers' choice to switch. Other customers who have switched are now knowingly paying more for their electric power; it seems clear that they were affected by the stated corporate environmental behavior. In anticipation of other states' deregulating, as well as for further promotion in the current states of operation, further experimentation is under consideration.

The marketing strategies developed from the results of this experiment are given significant credit for the success of Clean Air Electric so far.

2.4 A Comment

We have completed the basics of the analysis of variance for a study inquiring about the impact on the dependent variable of one factor. In the next chapter we continue with several additional topics that connect to the core of this chapter.

Exercises

1. Suppose that we have the data in Table 2EX.1, with four replicates of sales for each of three levels of one factor, treatment. Perform an ANOVA to test whether there is sufficient evidence of differences in sales due to the level of the treatment. Use $\alpha = .05$.

Table 2EX.1 Sales by Treatment

Treatment		
1	2	3
6	6	11
3	5	10
8	4	8
3	9	11

2. Having completed exercise 1, you are now told that the actual data available include eight replicates, as in Table 2EX.2. A bureaucrat noticed that the last four replicates were duplicates of the first four replicates; thus, he thought, why not give only the top half of the data to you, the consultant, thus saving time and money. After all, he reasoned, the second half of the data doesn't add or subtract from the story contained by the data. Perform an ANOVA on these data to test whether there is sufficient evidence that differences in sales are due to the level of the treatment. Again, use $\alpha = .05$.

Table 2EX.2 Complete Sales Data

Treatment		
1	2	3
6	6	11
3	5	10
8	4	8
3	9	11
6	6	11
3	5	10
8	4	8
3	9	11

3. Compare your results for exercise 1 with the results for exercise 2, and discuss.

4. An industry can be dominated by one of five manufacturing process technologies:
   A: Project
   B: Job shop
   C: Batch
   D: Mass production
   E: Continuous process
   It has been established by research conducted by Professor Peter Ward of Ohio State for his doctoral dissertation at the School of Management, Boston University, that certain industry average "staffing ratios" (that is, 100 times the ratio of the number of workers in a particular occupation, such as functional manager, to the number of production workers) are not the same for industries having different dominant technologies.

   Suppose that we randomly choose four industries having each of the dominant technologies (20 industries in total), and record each industry's staffing ratio for occupation K (one particular occupation), as done in Table 2EX.4. Using a significance

Table 2EX.4 Staffing Ratios by Dominant Technology

A	B	C	D	E
1.7	1.7	2.4	1.8	3.1
1.2	1.9	1.2	2.2	2.9
0.9	0.9	1.6	2.0	2.4
0.6	0.9	1.0	1.4	2.4

level of .05, and studying the staffing ratio for occupation K, is there sufficient evidence of a difference due to dominant technology? What is the p value of the test?

5. Suppose that we have data on the weight loss of 100 people, each person assigned to one specific diet, each diet having assigned to it the same number of people. In performing an ANOVA, the analysts arrived at Table 2EX.5, which is incomplete; complete the table.

Table 2EX.5 ANOVA for Diet and Weight Loss Study

Source of Variability	SSQ	*df*	MSQ	F_{calc}
Diet	—	3	—	15
Error	—	—	600	
Total	—	—		

6. Consider the data in Table 2EX.6, representing four levels (literally!) of lodging on a cruise ship, and a random sample of six passengers from each level. The yield is an assessment of the amount of motion felt during cruising (scaled from 1 to 30). Is there sufficient evidence from which to conclude that the level of the room on the cruise ship affects perception of the degree of motion? Perform an F test with a significance level of .05. What is the p value of the test?

Table 2EX.6 Motion Assessment by Ship Level

1	2	3	4
16	16	28	24
22	25	17	28
14	14	27	17
8	14	20	16
18	17	23	22
8	14	23	25

7. Suppose that an animal hospital wished to determine whether the cost of a service call for a dog differs by species of dog. The hospital examined its records for three months for four different dog species and arrived at the data in Table 2EX.7 based on cost per visit ($). At a significance level of .05, is there evidence of a difference in cost per visit for the four dog species? Find the p value of the test.

Table 2EX.7 Statistics for Veterinary Costs by Dog Species

	Species			
Statistic	**1**	**2**	**3**	**4**
Mean	38.12	29.72	33.40	36.15
Std. dev.	9.18	10.42	11.34	9.36
Sample size	23	35	17	29

8. The computer science department of a university wishes to test whether there are differences among three programming texts with respect to how long it takes a student completing the text to write a program in C++ language. Eight students from an introductory class were randomly assigned each of the texts, asked to complete the steps outlined in the text, and then given a program to write. The results, in minutes, are shown in Table 2EX.8. Conduct a one-factor ANOVA to determine if the texts are equally effective, and report the p value.

Table 2EX.8 Programming Time (Minutes)

Text 1	Text 2	Text 3
10	14	12
9	12	8
12	14	10
13	13	12
15	15	13
12	15	10
13	14	14
12	16	11

9. The data in Table 2EX.9 represent the cost per bottle ($) of a certain international brand of bottled water in five supermarkets in each of three different contiguous cities on the east coast of southern Florida. At a significance level of .05, is there evidence of a difference in price of the brand of bottled water among the three cities? What is the p value of the test?

Table 2EX.9 Supermarket Prices for Bottled Water ($)

Boynton Beach	Delray Beach	Boca Raton
1.19	1.29	1.44
1.24	1.29	1.51
1.09	1.29	1.59
1.14	1.25	1.39
1.19	1.19	1.49

10. There is a common belief that the price of many products is higher in Boca Raton than in neighboring cities, simply because those who live in Boca Raton have a higher average income than in the neighboring cities, and many of them will pay the higher prices for convenience. Based on the data in Table 2EX.9, at $\alpha = .05$, is there a significant difference between prices of the brand of bottled water between Boynton Beach and Boca Raton? Between Delray Beach and Boca Raton? Between Boynton Beach and Delray Beach?

11. A rental car company wondered whether car size—subcompact, compact, midsize, or full size—makes any difference in how many days a customer rents a car. A random sample of four rentals was chosen for each size car, and the number of days before the car was returned was recorded in Table 2EX.11. Perform an ANOVA to test to see if there is evidence at $\alpha = .01$ of a difference in duration of rentals among the four different sizes of cars. What is the p value of the test?

Table 2EX.11 Days per Rental by Car Size

Subcompact	Compact	Midsize	Full Size
7	3	4	4
3	3	3	5
3	3	1	3
7	1	3	7

12. One of the complaints mentioned by patients at the four offices of the Merrimack Valley Pediatric Clinic (MVPC) related to the time they had to wait before a receptionist was available to welcome them, ask them about the purpose of their visit, and give them an estimate of how soon a physician or nurse practitioner would be able to see them. The waiting times for 45 patients at each of the four offices were unobtrusively measured with a stop watch. The data, in minutes, are in Table 2EX.12 (and on the data disk, titled Problem 2.12). Perform an ANOVA to test whether there is a difference

Table 2EX.12 Minutes of Waiting Time by Office

Amesbury	Andover	Methuen	Salem	Amesbury	Andover	Methuen	Salem
7.71	5.79	4.24	9.59	3.58	5.78	3.98	11.98
4.88	2.83	8.43	13.19	6.43	4.85	4.98	5.61
5.59	9.48	3.24	10.40	5.59	7.44	7.09	10.06
3.38	3.75	6.66	10.42	4.89	3.74	7.37	10.77
8.36	6.03	5.89	12.80	6.83	3.73	8.62	10.68
4.06	4.83	3.67	9.92	4.05	4.32	4.11	7.84
5.77	3.81	6.28	10.05	4.05	4.85	4.40	14.15
4.87	5.06	4.40	11.03	4.49	6.93	7.52	11.15
4.11	3.52	5.82	12.25	4.89	2.57	8.62	11.53
5.05	4.50	4.37	10.93	6.45	7.37	4.11	11.61
3.89	6.37	7.45	9.72	3.17	4.31	4.81	7.17
4.62	3.39	6.09	9.50	6.78	7.18	5.77	10.08
6.03	3.51	9.66	10.47	4.48	4.73	7.40	8.87
3.79	3.41	5.34	9.00	6.64	2.28	7.70	12.66
3.88	5.68	9.05	4.64	4.80	0.34	5.22	10.48
3.81	4.40	4.06	13.15	2.96	4.18	5.79	10.34
5.51	2.30	5.48	12.06	1.50	6.31	7.98	12.67
4.55	5.87	6.34	12.72	4.39	5.80	4.10	10.68
2.98	7.61	4.44	11.29	5.98	4.06	5.90	13.46
5.65	7.85	6.90	8.41	5.60	6.73	8.85	8.59
6.96	4.17	3.08	13.07				
7.14	8.23	7.48	9.59				
4.38	2.88	7.64	12.71				
7.42	3.10	6.66	11.84				
3.41	6.91	6.65	12.82				

in average waiting time for a receptionist among the four MVPC offices. What is the p value of the test?

13. Members of the golf league at Eastern Electric are looking for a new golf course; the course they've used for years has been sold to developers of a retirement community. A search team has gathered the data in Table 2EX.13 (also in the data sets, titled Problem 2.13) on four local courses; for each course, they have the most recent scores for players like those in the Eastern Electric golf league. Perform an ANOVA to determine whether there is a significant difference in average score among the four local golf courses. Find the p value of the test.

Table 2EX.13 Golf Scores on Four Courses

Near Corners	Meadow Brook	Birch Briar	Fountainbleau	Near Corners	Meadow Brook	Birch Briar	Fountainbleau
115	99	107	135	108	95	108	142
106	101	111	144	106	100	118	136
108	98	114	131	100	106	111	116

Table 2EX.13 (continued)

Near Corners	Meadow Brook	Birch Briar	Fountainbleau	Near Corners	Meadow Brook	Birch Briar	Fountainbleau
101	99	107	130	107	97	116	138
117	90	109	139	102	98	110	127
103	96	119	133	106	106	106	141
109	100	117	137	102	94	111	142
106	93	116	139	110	96	111	142
103	89	117	138	106	99	120	144
106	99	108	128	116	96	115	138
102	103	111	136	104	97	114	136
105	102	115	145	114	98	112	148
109	102	110	142	101	100	113	138
102	97	121	122	105	99	114	137
102	104	114	132	107	100	115	129
102	104	113	133	102	95	109	140
108	93	101	139	108	100	107	137
104	96	114	130	98	90	114	140
99	90	114	140	110	97	120	142
108	94	116	125	110	101	121	132
113	102	111	123	108	102	110	135
113	99	113	141	108	96	119	136
104	100	114	141	101	99	116	137
114	94	115	137	103	105	116	129
101	92	113	142	109	105	116	148
101	97	112	135	110	90	120	134
111	97	114	138	113	97	112	135
108	98	114	129	101	99	117	138
106	93	115	136	102	96	111	138
112	102	113	142	110	95	117	135
103	100	118	121	113	99	113	145
103	100	112	126	101	93	113	129
104	99	110	137	103	97	115	140
106	95	112	133	105	114	112	134
111	100	121	126	110	92	104	137
100	104	114	130	110	99	120	152
112	104	113	132	104	104	120	137
104	93	108	135	105	97	119	139
111	93	118	134	111	99	117	133
105	105	117	134	101	92	113	127
99	94	124	128	106	102	108	135
94	100	118	131	113	99	115	128
104	99	117	142	103	97	114	128
109	98	114	132	99	109	121	126
108	97	113	127	104	98	119	143
104	87	110	132	101	95	118	139
110	107	113	138	94	103	113	133
101	102	117	138	102	104	116	144
112	98	120	130	105	107	115	134
99	106	119	133	112	106	120	130

14. Consider the data in Table 2EX.14, which represent the amount of life insurance (in $1000s) for a random selection of seven state senators from each of three states: California, Kansas, and Connecticut. All of the senators are male, married, with two or three children, and between the ages of 40 and 45. Are there significant differences due to state? Use $\alpha = .05$.

Table 2EX.14 Life Insurance ($1000s)

State		
1	2	3
90	80	165
200	140	160
225	150	140
100	140	160
170	150	175
300	300	155
250	280	180

15. Suppose that a symphony orchestra has tried three approaches, A, B, and C, to solicit funds from 30 generous local sponsors, 10 sponsors per approach. The dollar amounts of the donations that resulted are in Table 2EX.15. The column means are listed in the bottom row. For convenience when we revisit this example at the end of the next chapter, we rank-order the results in each column in descending order in presenting the table. Use an F test at $\alpha = .05$ to determine whether there are differences in amount of charitable donations due to solicitation approach.

Table 2EX.15 Funds Raised

Approach		
A	B	C
3500	3200	2800
3000	3200	2000
3000	2500	1600
2750	2500	1600
2500	2200	1200
2300	2000	1200
2000	1750	1200
1500	1500	800
1000	1200	500
500	1200	200
2205	2125	1310

16. The battery-lifetime study was repeated with new and more extensive data; the results are in Table 2EX.16. (Recall that the devices represent two brands of cell phone, three brands of flash camera, and three brands of flashlight.) Develop the ANOVA table. What do you conclude at $\alpha = .05$ about the influence of the device on mean battery lifetime? (Data are in the data sets as Problem 2.16.)

17. One of the authors recently taught two sections of the same course, called Quantitative Methods, in the same semester. This course was a core MBA course covering the basics of introductory statistics, ranging from probability, through discrete and continuous distributions, confidence intervals, hypothesis testing, and extensive model-building techniques, including multiple regression and stepwise regression. One class was taught on Tuesday evenings, the other Wednesday evenings (each class of three hours was held once a week for 14 weeks, plus a final exam week).

The distribution (in alphabetical order) of the final numerical grades (prior to translating them into letter grades) was tabulated by evening, status (part-time/full-time), and gender. The results for the 55 students are in Table 2EX.17 (and in the data sets as Problem 2.17). Does mean grade differ by evening? Use $\alpha = .05$.

18. Using the data in Table 2EX.17, does mean grade differ by status? Use $\alpha = .05$.

19. Using the data in Table 2EX.17, does mean grade differ by gender? Use $\alpha = .05$.

20. When you tested differences in grades by gender (exercise 19), you may have noticed that the number of females who are part-

Table 2EX.16	**Battery Lifetime Study by Device—Yield in Hours**						

Cell 1	Cell 2	Fl. Cam. 1	Fl. Cam. 2	Fl. Cam. 3	Flash. 1	Flash. 2	Flash. 3
4.74	5.39	4.92	8.67	3.47	4.93	8.40	4.31
3.24	5.21	6.46	8.90	4.20	5.07	7.48	4.80
2.09	4.55	6.17	6.62	6.52	6.11	11.06	6.62
4.06	5.85	5.76	11.35	3.32	5.40	8.25	5.56
6.20	4.45	3.43	7.80	9.03	3.78	7.18	7.99
3.63	4.19	0.51	4.57	8.92	4.59	8.84	9.61
4.00	4.07	9.42	3.91	8.77	3.25	7.45	8.16
2.94	4.36	5.27	9.31	3.12	4.48	6.54	8.82
1.19	4.99	7.25	8.66	9.24	5.09	8.12	10.91
0.59	4.59	4.27	4.56	9.42	4.73	9.58	7.71
1.54	4.38	6.08	6.28	7.20	2.99	7.56	6.13
1.54	5.22	2.82	7.88	4.58	5.25	5.79	7.37
1.72	5.51	1.08	8.20	3.21	5.38	8.91	3.94
5.91	4.68	4.51	7.37	7.35	5.02	7.31	6.91
2.72	4.29	8.81	5.42	3.90	4.05	8.61	7.32
5.11	4.64	7.66	5.11	5.73	5.76	9.25	4.29
3.97	4.75	4.65	6.22	8.27	3.00	7.78	11.02
1.87	5.23	5.96	5.71	6.35	4.33	9.35	9.71
3.18	4.27	8.23	7.99	7.09	4.34	7.95	5.17
3.08	3.62	6.08	5.37	3.97	6.01	8.59	6.20
1.72	4.80	3.86	7.93	6.73	5.13	8.41	4.54
1.71	3.96	7.02	3.25	6.58	5.14	8.44	7.25
1.08	5.13	1.01	6.44	5.50	5.32	10.23	8.87
5.03	4.89	9.69	4.69	4.24	4.90	10.16	6.27
1.50	3.99	5.05	9.01	7.60	5.47	8.61	7.83
1.69	4.45	2.79	7.05	7.25	3.80	9.33	6.49
4.17	4.09	5.11	8.60	6.09	3.71	8.89	8.06
6.28	3.78	4.90	13.33	6.75	5.05	8.74	9.45
2.96	5.06	8.01	7.47	7.11	6.08	7.47	5.24
1.47	4.76	2.57	6.72	8.98	3.11	10.51	7.95
5.65	4.44	5.06	5.88	3.67	5.08	12.06	6.75
1.67	4.78	3.64	11.86	6.94	4.46	6.37	10.88
5.72	3.88	7.66	6.91	5.22	4.76	7.90	9.40
1.52	4.64	2.15	6.00	5.66	4.39	12.03	8.66
3.73	3.74	3.90	6.70	2.02	4.77	6.81	6.36
6.75	5.31	9.43	8.36	7.32	4.09	9.58	6.89
1.69	4.48	8.36	10.37	7.28	5.12	8.59	9.61
1.93	4.21	6.56	6.13	5.46	4.56	9.29	6.02
6.08	4.89	7.98	11.02	5.48	3.59	10.42	5.19
3.51	4.74	5.50	5.65	2.83	4.48	7.77	6.13
2.28	5.30	10.32	6.16	5.02	3.13	9.10	7.97
1.34	4.25	8.33	8.93	3.13	5.91	7.89	5.12
5.54	5.12	10.48	7.89	5.14	5.45	7.88	6.39
2.57	3.44	4.98	5.91	7.71	5.50	6.37	7.32
4.38	4.17	3.09	6.80	1.61	4.58	4.94	10.50
1.46	5.57	6.38	4.77	5.79	4.59	8.51	6.09
1.48	4.71	6.20	6.32	5.17	4.74	6.26	7.86
4.59	3.74	2.28	4.51	8.44	3.71	7.88	8.26
2.19	3.69	8.69	10.59	5.86	3.82	6.60	8.00
1.82	4.37	4.95	5.45	8.73	4.15	9.30	6.43

(continued)

Table 2EX.16 (continued)

Cell 1	Cell 2	Fl. Cam. 1	Fl. Cam. 2	Fl. Cam. 3	Flash. 1	Flash. 2	Flash. 3
1.73	4.80	5.94	9.44	5.76	3.97	9.85	10.86
1.58	5.09	2.24	7.43	8.17	4.67	10.19	6.84
1.26	5.03	9.26	8.97	8.89	6.34	8.55	9.02
5.06	4.98	4.18	10.12	6.39	4.27	5.38	8.30
2.18	4.96	2.97	7.21	2.49	3.44	8.85	8.24
0.65	3.74	8.09	7.00	4.84	5.11	7.38	5.08
2.16	5.26	4.89	8.71	7.34	4.69	7.92	10.65
2.14	4.04	8.37	8.00	5.80	3.46	7.03	8.37
4.44	4.97	7.25	5.11	4.07	4.44	7.14	5.11
4.37	4.59	0.51	7.93	5.76	5.65	9.55	5.64
6.04	4.74	5.95	5.10	4.52	4.65	8.87	9.08
5.34	3.76	5.15	7.76	4.49	4.76	7.22	8.25
1.63	4.94	6.78	8.92	7.00	5.84	8.89	6.30
3.49	5.18	6.83	7.76	8.58	4.58	5.69	7.79
3.27	4.97	5.73	2.86	6.33	5.45	7.56	7.20
1.27	4.02	3.92	10.09	2.62	5.73	8.08	8.38
6.03	4.14	9.09	8.38	7.97	2.99	7.68	9.25
0.54	4.05	11.71	6.59	7.45	5.16	8.66	7.67
2.22	4.13	7.63	10.80	9.36	5.01	6.96	7.18
3.95	5.27	1.86	6.17	5.91	4.87	11.21	6.39
6.59	3.07	7.61	4.18	4.93	4.42	7.93	10.21
6.45	4.71	6.18	3.18	4.65	4.13	8.59	6.09
4.48	5.03	1.59	8.10	6.57	5.52	8.45	10.04
5.95	4.25	4.00	8.91	6.98	5.23	7.51	8.09
2.61	3.88	7.25	4.76	8.19	4.28	6.64	5.51
1.73	4.95	0.48	5.06	4.11	4.22	7.67	6.81
6.12	4.32	6.27	7.15	6.03	4.40	9.23	7.96
2.55	5.37	0.09	6.96	7.43	6.36	9.74	5.28
1.86	4.04	6.80	8.87	3.88	4.18	8.05	7.87
3.79	5.06	7.65	10.46	7.53	4.43	10.45	7.37
1.03	4.29	3.62	5.56	5.94	5.04	9.80	9.96
5.21	3.76	7.64	7.68	6.55	4.41	11.85	5.79
3.66	4.09	6.30	5.44	8.84	5.86	9.42	8.34
2.08	5.08	3.75	8.17	6.05	5.42	9.21	8.58
4.74	4.40	3.35	6.39	4.64	5.31	8.46	8.67
1.55	5.03	10.74	6.27	7.32	3.93	7.73	8.11
0.33	2.88	5.39	2.58	5.77	4.69	6.10	5.97
2.59	4.35	3.85	8.22	7.30	3.34	6.63	8.58
2.39	4.10	6.47	8.92	5.92	5.72	7.40	10.13
2.37	5.46	6.89	8.94	5.84	4.14	8.63	8.58
1.78	3.62	2.80	7.28	3.23	4.84	8.09	9.75
1.07	5.14	1.20	9.61	5.78	4.56	9.79	7.63
2.08	3.31	3.08	5.32	7.37	3.76	7.66	6.76
2.24	4.27	2.34	5.33	3.25	4.16	8.58	4.95
6.54	4.62	7.14	7.15	3.16	5.27	9.61	6.14
6.51	4.54	5.36	7.42	7.13	5.41	9.39	6.60
3.03	6.07	6.70	4.98	8.09	3.07	8.59	7.81
1.10	5.95	6.37	7.61	7.30	3.66	7.66	5.74
2.60	4.64	5.55	8.59	5.88	4.94	9.18	5.07
0.66	4.41	5.95	8.04	6.58	4.07	5.90	6.70

Table 2EX.17 **Course Grade, Fall Semester**

Student	Evening	Status	Gender	Grade	Student	Evening	Status	Gender	Grade
1	Tuesday	Part-time	Male	61.39	31	Wednesday	Part-time	Female	54.80
2	Tuesday	Part-time	Male	81.47	32	Wednesday	Part-time	Male	54.30
3	Tuesday	Part-time	Male	65.59	33	Wednesday	Full-time	Female	77.80
4	Tuesday	Full-time	Female	70.30	34	Wednesday	Full-time	Female	67.30
5	Tuesday	Part-time	Male	85.20	35	Wednesday	Full-time	Female	72.20
6	Tuesday	Part-time	Female	63.70	36	Wednesday	Part-time	Male	62.51
7	Tuesday	Part-time	Female	51.94	37	Wednesday	Full-time	Female	81.90
8	Tuesday	Part-time	Male	68.74	38	Wednesday	Part-time	Female	80.90
9	Tuesday	Part-time	Female	69.28	39	Wednesday	Part-time	Female	39.00
10	Tuesday	Full-time	Male	66.90	40	Wednesday	Part-time	Male	65.85
11	Tuesday	Full-time	Male	64.30	41	Wednesday	Full-time	Female	52.58
12	Tuesday	Full-time	Male	58.88	42	Wednesday	Full-time	Male	86.30
13	Tuesday	Part-time	Male	86.09	43	Wednesday	Part-time	Female	30.56
14	Tuesday	Full-time	Female	74.40	44	Wednesday	Part-time	Male	75.29
15	Tuesday	Part-time	Female	56.70	45	Wednesday	Part-time	Male	71.11
16	Wednesday	Full-time	Female	61.80	46	Wednesday	Part-time	Female	56.04
17	Wednesday	Part-time	Male	62.08	47	Wednesday	Part-time	Male	71.10
18	Wednesday	Part-time	Male	83.80	48	Wednesday	Part-time	Male	63.60
19	Wednesday	Full-time	Female	76.10	49	Wednesday	Part-time	Male	47.90
20	Wednesday	Part-time	Male	60.59	50	Wednesday	Full-time	Male	83.30
21	Wednesday	Full-time	Female	71.70	51	Wednesday	Part-time	Male	74.60
22	Wednesday	Full-time	Female	83.50	52	Wednesday	Part-time	Male	59.68
23	Wednesday	Part-time	Male	50.15	53	Wednesday	Part-time	Female	50.51
24	Wednesday	Part-time	Male	88.90	54	Wednesday	Full-time	Male	87.30
25	Wednesday	Part-time	Female	53.08	55	Wednesday	Part-time	Male	70.59
26	Wednesday	Part-time	Male	66.05					
27	Wednesday	Full-time	Female	62.49					
28	Wednesday	Part-time	Male	66.24					
29	Wednesday	Part-time	Female	61.72					
30	Wednesday	Part-time	Male	72.49					

time versus full-time is not the same proportion as that for males. Does this potentially affect your conclusions for exercise 18 or 19?

Notes

1. The word *level* is traditionally used to denote the value, or amount, or category of the independent variable or factor under study, to emphasize two issues: (1) The factor can be quantitative/numerical, in which case the word *value* would likely be appropriate, or it can be nominal (e.g., male/female, or supplier A/supplier B/supplier C), in which case the word *category* would likely be appropriate. *Level* caters to both cases. (2) The analysis we perform, at least at the initial stage, always treats the variable as if it is in cate-

gories. Of course, any numerical variable can be represented as categorical: income, for example, can be represented as high, medium, or low.

2. Many different notational schemes are available, and notation is not completely consistent from one text/field/topic to another. We believe that using i for the row and j for the column is a natural, reader-friendly choice, and likewise for the choice of C for the number of columns and R for the number of rows. Naturally, when we go beyond just rows and columns, we will need to expand the notation (for example, if we wanted to investigate the impact of two factors, say region of residence and year of first purchase, with replication at each combination of levels of the two factors, we would need three indices). However, we believe that this choice of notation offers the wisest trade-off between being user-friendly (especially here at the initial stage of the text) and allowing an extrapolation of notation that remains consistent with the principles of the current notation.

3. It would certainly seem that a notation of SSWc would be more consistent with SSBc, at least in this chapter. However, the former notation is virtually never used in English language texts. The authors suspect that this is because of the British ancestry of the field of experimental design, and the sensitivity to WC as "water closet." In subsequent chapters, the "within" sum of squares will not always be "in columns," and the possible inconsistency becomes moot.

4. Of course, along with the sample sizes and difference in sample means, the standard deviation estimates for each town's data need to be considered, along with a significance level, and so on; this description simply attempts to appeal to intuition.

5. Virtually all theoretical results in the field of statistics have some intuitive reasoning behind them; it remains for the instructor to convey it to the students.

6. One could argue that this study really has two factors—one the actual test device and the other the brand of the device. However, from another view, one can validly say that there are eight treatments of one factor. What is sacrificed in this latter view is the ability to separate the variability associated with the actual device from the variability associated with the brand of the device. We view the study as a one-factor study so that it is appropriate for this chapter. The two-factor viewpoint is illustrated in later chapters.

7. One may say, "Why not examine (MSBc − MSW) and compare it to the value 0? Isn't this conceptually just as good as comparing the ratio to 1?" The answer is a qualified yes. To have the ratio be 1 or different from 1 is equivalent to having the difference be 0 or different from 0. However, since MSBc and MSW are random variables, and do not exactly equal their respective parameter counterparts, $(\sigma^2 + V_{col})$ and σ^2, as we shall discuss, we will need to know the probability distribution of the quantity examined. The distribution of the difference between MSBc and MSW depends critically on scale—in essence the value of σ^2, something we don't know. Examining the ratio avoids this problem—the ratio is a dimensionless quantity! Its probability distribution is complex but can be determined with known information (R, C, and so on). Hence we always study the ratio.

8. It is not exactly half the time for each, because, although the numerator and denominator of F_{calc} both have the same expected value, their ratio does not have an expected value of 1; the expected value of F_{calc} in our current discussion is $(RC - C)/(RC - C - 2) > 1$, although the result is only slightly more than 1 in most real-world cases. Also, the probability distribution of F_{calc} is not symmetric.

9. It often happens, in an exposition such as this, that the best order of presentation is a function of the level of knowledge of the reader; background material required by one may be unnecessary for another. The flow of presentation is, of course, influenced by how these disparate needs are addressed. At this point, we present just enough of the hypothesis-testing background to allow us to continue with the analysis. Some readers may find it advantageous to first read section 3.4 and then return to the current section.

CHAPTER 3

Some Further Issues in One-Factor Designs and ANOVA

3.1 Introduction

We need to consider several important collateral issues that complement our discussion in Chapter 2. We first examine the standard assumptions typically made about the probability distribution of the ε's in our statistical model. Next, we discuss a nonparametric test that is appropriate if the assumption of normality, one of the standard assumptions, is seriously violated. We then review hypothesis testing, a technique that is an essential part of the ANOVA and that we heavily rely on throughout the text. This leads us to a discussion of the notion of statistical power and its determination in an ANOVA. Finally, we find a confidence interval for the true mean of a column and for the difference between two true column means.

3.2 Basic Assumptions of ANOVA

Certain assumptions underlie the valid use of the F test to perform an ANOVA, as well as some other tests we encounter in the next chapter. The actual statement of assumptions depends on whether our experiment corresponds with what is called a "fixed" model or with what is called a "random" model. Because the F test described in Chapter 2 is identical for either model, we defer some discussion of the distinction between the two models to Chapter 6, when we introduce two-factor designs. For designs with two or more factors, the appropriate F_{calc} is often different for the two models.

However, we'll now consider a basic description of a fixed model and a random model.

A **fixed model** applies to cases in which there is inherent interest in the specific levels of the factor(s) under study, and there is no direct interest in extrapolating results to other levels. Indeed, inference will be limited to the actual levels of the factor that appear in the experiment. This would be the case if we were testing three specific promotional campaigns, or four specific treatments of an uncertain asset situation on a balance sheet. A **random model** applies to cases in which we test randomly selected levels of a factor, where these levels are from a population of such levels, and inference is to be made concerning the entire population of levels. An example would be the testing of six randomly selected post offices to see if post offices in general differ on some dimension (the Y_{ij}); another example would be the testing of whether there are differences in sales territories by randomly selecting five of them as the levels of the factor "territory."

For a fixed model, those well-versed in regression analysis will find the assumptions familiar; the same so-called standard assumptions are common to both techniques. We ascribe no special meaning to the order in which we list them.

Recall the statistical model $Y_{ij} = \mu + \tau_j + \varepsilon_{ij}$. We assume the following three statements:

1. The ε_{ij} are independent random variables for all i, j.

This means that each error term, ε_{ij}, is independent of each other error term. Note that this assumption, as well as the assumptions to follow, pertain to the error term. In essence, we assume that although each column may have a different true mean (indeed, what we wish to determine is whether or not this is true), knowing the deviation of any one data value from its particular true column mean sheds no light on the deviation of any other data point from its particular true column mean. If this assumption is violated, it is often because of the correlation between error terms of data values from different time periods. If the correlation is between error terms of data values of successive time periods, it is referred to as first-order autocorrelation; if between error terms two periods apart, it is referred to as second-order autocorrelation, and so on.

2. Each ε_{ij} is normally distributed; with no loss of generality, we can assume that $E(\varepsilon_{ij}) = 0$ (that is, the true mean of ε_{ij} is zero).

This is equivalent to saying that if we consider all the data values in a specific column, they would (theoretically, or if we had an infinite number of them) be distributed according to a normal distribution, with a mean equal to whatever is the true mean of that column.

3. Each ε_{ij} has the same (albeit unknown) variance, σ^2_ε.

This says that the normal distribution of each respective column, though perhaps differing in mean, has the same variance. This assumption is often referred to as the *assumption of constant variance,* and sometimes the *assumption of homoscedasticity.*[1]

If the random model applies, the key difference is essentially that the τ_j values are random variables, as opposed to the fixed model, where the τ_j values are unknown constants. We would state the assumptions as follows:

1. The ε_{ij} are independent random variables for all i, j (same as assumption 1 above).
2. Each ε_{ij} is normally distributed with a constant variance (same as assumptions 2 and 3 above).
3. a. The τ_j values are independent random variables having a normal distribution with a constant variance.
   b. The τ_j and ε_{ij} are independent random variables.

When these assumptions obtain, the estimates of the grand and column means, $\bar{Y}_{..}$ and $\bar{Y}_{.j}$, are maximum-likelihood estimates (this is a good thing—a property complementing the unbiasedness property noted earlier) and, perhaps more importantly, the conventional F test we have introduced (and t-test we make use of in the next chapter) are valid for the hypothesis testing we undertake.

It is not likely that all of the assumptions above are completely true in any particular study. This is especially so for the assumption of constant variance. However, the last two assumptions (2, 3a, 3b) are said to be **robust;** that is, a moderate (a term never precisely quantified) violation of the assumption is likely not to have a material effect on the resulting α value (the significance level) or the β value, the probability of Type II error (that is, the chance of obtaining results that suggest accepting that there is no effect of the factor, when in fact there actually is an effect). We discuss the probability of reaching incorrect conclusions—rejecting H_0 when it is actually true (Type I error), and accepting H_0 when it is actually false (Type II error)—a bit later in this chapter.

The first assumption, that of independence of the error terms, is not especially robust, and hence can seriously affect the significance level and the probability of Type II error. As noted above, the other two assumptions, those of normality and constant variance of the error terms, are considered robust. The degree to which these assumptions are robust is not generally quantified in a rigorous way; however, a number of researchers have investigated these issues. We report on some of these studies, to give the reader a feel for the topic.

The robustness of the normality assumption partly depends on what the departure from normality primarily involves: a skewness or a kurtosis that is different from a normal curve. Skewness is a measure of the extent to which distribution is not symmetric; the normal curve, of course, is symmetric. Kurtosis for a random variable, X, is defined as the fourth central moment divided by the square of the variance, or $(E\{[X - E(X)]^4\})/(E\{[X - E(X)]^2\})^2$; in essence, kurtosis is a dimensionless measure of the degree to which the curve "tails off." One extreme would be a rectangle, the other extreme would be toward a distribution with thicker and thicker tails. A kurtosis departure from the normal curve is considered more serious than nonzero skewness. The effect is slight on α; α can actually be a bit smaller or a bit larger, depending on the way in which the kurtosis deviates from that of a normal curve. The same is true for the probability of Type II error, though the latter may be affected more seriously if the sample size per column is small. Remember that all of this discussion is based on what we have referred to as a "moderate departure," though as noted earlier, the term is not precisely defined. Scheffé, in his text *The Analysis of Variance* (New York, Wiley, 1959), presents an example in which the skewness (defined as the third central moment, divided by the cube of the standard deviation) of the error terms is two (for normality it is zero), and a nominal α of .05 resulted in an actual significance level of .17. Too many results would be found to be significant if there is really no effect of the factor; most people would regard this difference between .05 and .17 as, indeed, material.

The impact of nonconstant variance depends, to a degree, on whether the sample size per column is equal for each column; that is, whether R is a constant or is R_j for column j, the R's not all the same. In the former case, the impact of nonconstant variance on α is quite minimal. In the latter case, with unequal sample size per column, the impact can be more serious if the variances corresponding to the columns with smaller sample sizes are also the larger variances. In the battery lifetime example in Chapter 2, the sample variances for the eight columns are listed as part of the Excel output in Table 2.6. Note that the largest is 5.92 and the smallest is 0.48, a ratio of about 12.3 to 1. This ratio might seem to indicate a more than moderate violation of the equal-variance assumption, but remember that there are only three data values per column; thus, each column does have the same value of R, and perhaps more important, each sample variance is a not especially reliable estimate of its column's true variance. We can test for equality among the eight (true) column variances using the Hartley test (described, for example, in Ott, L., *An Introduction to Statistical Methods and Data Analysis,* 2nd edition, p. 340). The Hartley test specifically bases its conclusions on the ratio of the highest sample variance to the lowest sample variance (in this case, the ratio is the 12.3 value). The 12.3 ratio was nowhere near high enough at $\alpha = .05$ to reject the null hypothesis of equality of variances; the critical ratio value was over 400, a high critical value that reflects the unreliability of variance estimates based on only three data values.

Conventional wisdom indicates that the effects of nonnormality and nonconstant variance are additive and not multiplicative. This provides some further comfort.

We end this discussion about the assumptions by noting that there are ways to test for the validity of each assumption. Furthermore, if a serious violation of an assumption is found, remedial actions can be undertaken to deal with the problem. These remedies are, generally, either to transform the data (for example, replace each Y by the log of Y) to try to eliminate the problem, or to incorporate a more onerous model. We leave discussion of these tests and remedies to other sources. Another alternative is to simply avoid the whole issue of the probability distribution associated with the data. We discuss this option next.

3.3 Kruskal-Wallis Test

One way to avoid the distributional aspect of the standard assumptions (that is, the assumption of normality) is to perform what is called a nonparametric test. An alternate, perhaps more appropriate, name is a distribution-free test. One distribution-free test that is analogous to the one-factor ANOVA F test was developed by Kruskal and Wallis and takes their name: the Kruskal-Wallis test. However, there is a drawback to its use (otherwise, why not always use the Kruskal-Wallis test?); the F test is more powerful than the Kruskal-Wallis test. Power, as defined in section 3.5, equals $(1 - \beta)$, that is, one minus the probability of Type II error. Thus power is the probability of rejecting H_0 when it indeed should be rejected. So the probability of drawing the correct conclusion is higher for the F test, everything else being equal and when the assumptions are valid (or close to valid—see earlier remarks about robustness), than it is for the Kruskal-Wallis test. The key to the Kruskal-Wallis test, as is true for the majority of distribution-free tests, is that the data are converted to ranks. The test assumes that the distributions of the data in each column are continuous and have the same shape, except perhaps for the mean.

EXAMPLE 3.1 Battery Lifetime Study with the Kruskal-Wallis Test

We illustrate the technique of the Kruskal-Wallis test using the battery lifetime example from Chapter 2. Table 3.1 reiterates the data in Table 2.2; the last row shows the column means. The use of the Kruskal-Wallis test could be motivated by the fact that the sample variances are somewhat different from column to column, although not significantly different, and by the fact that each is based on only three data values.

Table 3.1 Battery Lifetime Study—Yield in Hours

Device							
1	2	3	4	5	6	7	8
1.8	4.2	8.6	7.0	4.2	4.2	7.8	9.0
5.0	5.4	4.6	5.0	7.8	4.2	7.0	7.4
1.0	4.2	4.2	9.0	6.6	5.4	9.8	5.8
2.6	4.6	5.8	7.0	6.2	4.6	8.2	7.4

The question, as in Chapter 2, is whether the different devices yield different mean battery lifetimes. We formulate this as a hypothesis-testing problem:

H_0: Device does not affect battery lifetime (technically: if a randomly generated value from device 1 is X_1, from device 2 is X_2, from device 3 is X_3, and so on, then $P(X_1 > X_2) = P(X_2 > X_3) = P(X_3 > X_1) = \ldots = .5$; that is, a battery lifetime from any device has an equal chance of being lower or higher than a battery lifetime from any other device).

H_1: Device does affect battery lifetime (or, technically, not all of the probabilities stated in the null hypothesis equal .5).

The Kruskal-Wallis test proceeds as follows. First, we rank the data in descending order, *as if all the data points were from one large column.* For example, the largest value in any column is 9.8, so this value from column seven gets the highest rank of 24 (given that there are 24 data values in total). The next highest value is 9; however, two data values equal 9—a tie; therefore, the ranks of 23 and 22 are split or averaged, each becoming a rank of 22.5. Next comes the value of 8.6, which gets a rank of 21; next is the value of 7.8, of which there are two, each getting the rank of 19.5. The process continues until the lowest value gets assigned a rank of 1 (unless it is tied with other values).

In Table 3.2 we replace each data value with its rank. The quantities in the last two rows are the sum of the rank values in that column (T's) and the number of data points in that column (n's). Now, instead of using the actual data points to form the test statistic, we use the ranks. We might expect, under the null hypothesis, that the high, medium, and low ranks would be uniformly distributed over the columns, and thus that the T's would be close to one another—if the column factor didn't matter. An indication to the contrary would be seen to indicate that device affects battery lifetime.

The Kruskal-Wallis test statistic, which equals zero when the T's are the same for each column, equals (summation over j, j indexing columns)

$$H = \{12/[N(N + 1)]\} \Sigma (T_j^2/n_j) - 3(N + 1)$$

Table 3.2 Battery Lifetime Study—Rank Orders

Device							
1	2	3	4	5	6	7	8
2	5.5	21	16.5	5.5	5.5	19.5	22.5
10.5	12.5	9	10.5	19.5	5.5	16.5	18
1	5.5	5.5	22.5	15	12.5	24	14
13.5	23.5	35.5	49.5	40	23.5	60	54.5
3	3	3	3	3	3	3	3

where

n_j = number of data values in jth column
N = total number of data points, equal to the sum of the n_j over j
K = number of columns (levels)
T_j = sum of ranks of the data in jth column

For our example,

$$H = \{12/[24(25)]\}(13.5^2/3 + 23.5^2/3 + \cdots + 54.5^2/3) - 3(25)$$
$$= 12.78$$

However, there is one extra step: a correction to H for the number of ties. This corrected H, H_c, equals

$$H_c = H/[1 - \Sigma(t^3 - t)/(N^3 - N)]$$

where, for each tie, t = the number of tied observations. In our data, we have six ties, one of six data values (the ranks of 5.5), and five of two data values (ranks of 22.5, 19.5, 16.5, 12.5, and 10.5). The correction factor is usually negligible, and in fact $H_c = 13.01$, not much different than $H = 12.78$.

Presuming H_0 is true, the test statistic H (or H_c) has a distribution that is well approximated by a chi-square (χ^2) distribution with $df = K - 1$. The χ^2 distribution looks similar to the F distribution (both have a range of zero to infinity, and are skewed to the right). In fact, it can be shown that for any given value of α, any percentile point of a χ^2 distribution with $K - 1$ degrees of freedom, divided by $K - 1$, is equal to the percentile point of the corresponding F distribution, with numerator degrees of freedom equal to $K - 1$, and with denominator degrees of freedom equal to infinity.[2] In our battery lifetime problem, with eight columns, $K - 1 = 7$. A plot of a χ^2 distribution for $df = 7$, $\alpha = .05$, and our test statistic value of $H_c = \chi^2_{\text{calc}} = 13.01$, is shown in Figure 3.1. The critical value is $\mathbf{c} = 14.07$; χ^2 tables appear in an appendix at the end of the text. $H_c = 13.01$ falls in the acceptance region, though close to the critical value. We cannot (quite) conclude that device affects battery lifetime.

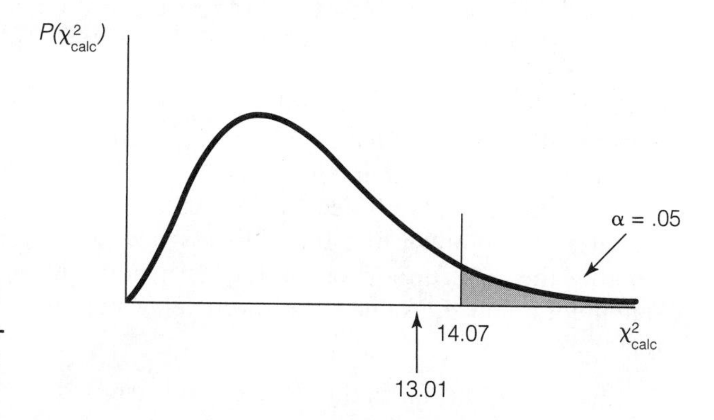

Figure 3.1 χ^2 **distribution for battery lifetime study**

Interestingly, the overall result changed! The p value is about .02 using the F test and about .06 using the Kruskal-Wallis test; because the α value used was .05, the results are on opposite sides of the respective critical values. However, given that $\alpha = .05$ is arbitrary (though very traditional), one could debate how meaningful the difference in results really is—it's similar to comparing 94% confidence to 98% confidence in a result.

3.4 Review of Hypothesis Testing

When analyzing our data to determine whether the factor under study has an effect on the yield—the dependent variable—the discipline we use is called hypothesis testing; it is sometimes referred to as significance testing. It recognizes that we do not have an infinite amount of data and therefore need to use statistical inference, where we "make inference" about one or more parameters based on a set of data. Typically in ANOVA, we are both estimating the value of each τ (and μ), and testing whether we should believe that the τ's are equal to zero or not; we do this by computing F_{calc} from the data and comparing its value to a critical value. In this section, we elaborate on the logic of the discipline of hypothesis testing. If we had to single out one portion of the world of statistics that could be labeled The Statistical Analysis Process, it would be the thought process and logic of hypothesis testing. It is a relevant concept for situations we encounter routinely, every day—even though in most of these situations we don't formally collect data and manipulate numerical results.

The essence of hypothesis testing is accepting or rejecting some contention called, not surprisingly, a hypothesis. The formulation is structured in a way such that we choose between two mutually exclusive and collectively exhaustive hypotheses.[3]

EXAMPLE 3.2 Internal Revenue Service Watchdog

Let's consider a simple example. Suppose that the U.S. Internal Revenue Service (IRS) suggests that a new version of the 1040 form takes, on the average, 160 minutes to complete, and we, an IRS watchdog agency, wish to investigate the claim. We collect data on the time it takes a random sample of n taxpayers to complete the form. In essence, we are interested in deciding whether the data support or discredit a hypothesis: here, the hypothesis is a statement about the value of μ, the true average time it takes to fill out the form. We state:

H_0: $\mu = 160$ (that is, the IRS's claim is true)

versus

H_1: $\mu \neq 160$ (that is, the IRS's claim is *not* true)

By tradition, we call H_0 the null hypothesis and H_1 the alternate hypothesis. We must decide whether to accept or reject H_0. (Also by tradition, we always talk about H_0, though of course whatever we do to H_0, we in essence do the opposite to H_1.) How are we to decide? Here, we decide by examining the average of the n data values (often called $\overline{X}$ in introductory courses), and considering how close or far away it is from the alleged value of 160.

We start with the presumption that the null hypothesis is true; that is, the null hypothesis gets the benefit of the doubt. Indeed, we decide which statement we label as the null hypothesis and which we label as the alternate hypothesis, specifically depending on which side should get the benefit of the doubt and which thus has the burden of proof. This usually results in the null hypothesis being the status quo, or the hypothesis that historically has been viewed as true. The analogy to a criminal court proceeding is very appropriate and useful, though in that setting, there is no doubt which side gets the benefit of the doubt (the not-guilty side, of course). In other words, H_0 will be accepted unless there is substantial evidence to the contrary. In the criminal courts, H_0 is the presumption of innocence; it is rejected only if the evidence (data) is judged to indicate guilt beyond a reasonable doubt (that is, the evidence against innocence is substantial).

What would common sense suggest about choosing between accepting H_0 and rejecting H_0? If $\overline{X}$ is close to 160, and thus consistent with a true value of 160, then accept H_0; otherwise, reject H_0. Of course, one needs to clarify the definition of "close to." Our basic procedure follows this commonsense suggestion. Here are the steps to be followed:

1. Assume, to start, that H_0 is true.
2. Find the probability that, if H_0 is true, we would get a value of the test statistic, $\overline{X}$, as far from 160 as we indeed got.
3. If, under the stated presumption, the probability of getting the discrepancy we got is *not especially small,* we view the resulting $\overline{X}$ as "close" to 160, and accept H_0; if this probability is *quite small* we view the resulting $\overline{X}$ as inconsistent with ("not close" to) a true value of 160, and thus we reject H_0.

Of course, we now must define the dividing line between "not especially small" and "quite small." It turns out that the experimenter can choose the dividing line any place he or she wants. In fact, this dividing line is precisely what we have called the "significance level," denoted by α. The traditional value of α is .05, though, as stated above, the experimenter can choose it to be any value desired. In the vast majority of real-world cases it is chosen to be either .01, .05, or .10. (Sometimes one doesn't explicitly choose a value of α but rather examines the results after the fact and considers the p value, the "after-the-fact α," to compare to .05 or another value. Whether choosing a significance level at the beginning or examining the p value later, the salient features of the hypothesis-testing procedure are maintained.)

Specific analysis for IRS example Continuing with a detailed look at our IRS example, we need to know the probability distribution of $\overline{X}$ given that H_0 is true, or given that the true mean, μ, equals 160. In general, the standard deviation of the distribution may or may not be known (in this example, as in the large majority of examples, it likely would not be). However, for simplicity, we shall assume that it is known,[4] and that $\sigma = 50$. Supposing a sample size n of 100, we can appeal to the central limit theorem and safely assume that the probability distribution of $\overline{X}$ is very well approximated by a Gaussian (or normal or bell-shaped) distribution with mean $\mu = 160$ and $\sigma(\overline{X}) = \sigma/\sqrt{n} = 50/10 = 5$. Also, we will specify an α value of .05. Now, we find a range of values, in this case symmetric around the (alleged) center of 160, that contains an area of .95 (that is, $1 - \alpha$); this is called the *acceptance region*.[5] The area outside the range of values is the *critical (or rejection) region*. In Figure 3.2 we show the probability curve, upper and lower limits of the acceptance region, and the shaded critical region. The limits are found by computing $160 \pm 1.96(5) = (150.2 \text{ to } 169.8)$, where 1.96 is the 97.5% cumulative point on the standard normal (Z) curve.

If, for example, we found our sample mean was $\overline{X} = 165$, we would reason that the true mean might well be 160, but that due to statistical fluctuation (that is, error: the varying effects of factors not controlled, and perhaps not even overtly recognized as factors), our sample mean was a bit higher than the true (population) mean. After all, even if $\mu = 160$, we don't expect $\overline{X}$ to come out to exactly 160! On the other hand, if our sample mean had instead been 175, we would not be so understanding; we would conclude that a value so far from 160 ought not be attributed to statistical fluctuation. Why? Because if μ truly equals 160, the probability that we would get a value of $\overline{X}$ that is as far away as 175 (in either direction) is simply too low. Indeed, "too low" is defined as less than α (or $\alpha/2$ in each tail). We know that this probability is less than α because the $\overline{X}$ is in the critical region. We would conclude that the more appropriate explanation is that the population mean is, in fact, not 160 but probably higher, and that 175 is perhaps not inconsistent with the actual population mean. Could we be wrong? Yes! More about this later.

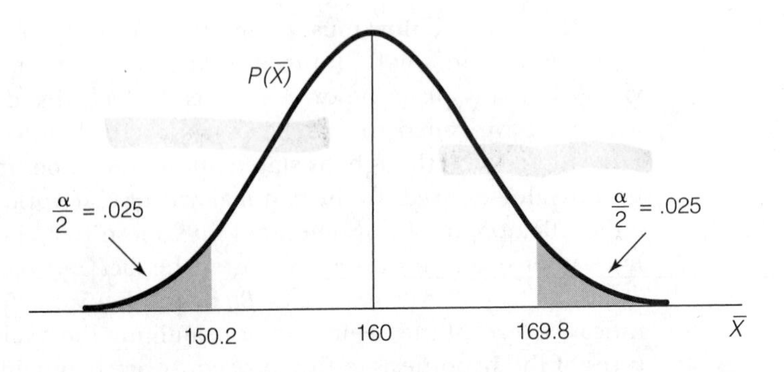

Figure 3.2
Acceptance and critical regions for two-sided hypothesis test

We have portrayed this analysis as determining whether 160 is a correct depiction of the average time it takes to fill out the new form. The analysis, of course, recognizes that we don't insist that $\overline{X}$ come out at exactly 160 in order to be considered supportive of the hypothesis of a true mean of 160.

One could have portrayed this problem a bit differently. Suppose that the group was not interested in whether the 160 was an accurate value per se but in whether the IRS was *understating* the true time it takes to fill out the form. Then the issue would not be whether $\mu = 160$ or not, but whether μ was actually greater than 160. We would then formulate the two hypotheses as follows:

H_0: $\mu \leq 160$ (that is, the IRS-claimed mean time is *not* understated)

versus

H_1: $\mu > 160$ (that is, the IRS-claimed mean time *is* understated)

These hypotheses suggest a one-tailed critical region; common sense says that we would reject H_0, in favor of H_1, only if $\overline{X}$ comes out appropriately *higher* than 160. No sensible reasoning process says that $\overline{X}$ can be so low as to push us toward H_1. We thus perform a so-called one-tailed test, as pictured in Figure 3.3. In the figure, the critical value 168.25 is calculated from $160 + 1.65(5) = 168.25$, where 1.65 is the 95% cumulative point on the standard normal (Z) curve.

Notice that although α still equals .05, all of this quantity is allocated to the upper tail. The critical value (there's only one) is 168.25; any value of $\overline{X}$ below this value falls in the acceptance region and indicates acceptance of H_0. A value of $\overline{X}$ above 168.25 falls in the critical region and causes us to reject H_0. One-sided hypothesis tests may, of course, be either upper-tailed or lower-tailed; for a lower-tailed test, the analysis proceeds in a similar, albeit mirror-image, manner.

As we have seen, the logic of the F test in ANOVA suggests using the one-tailed (*upper*-tailed) test, as we have just done. The same is true of the χ^2 test we performed when conducting a Kruskal-Wallis test. However, in subsequent chapters we shall encounter some two-sided tests; this is especially true in Chapter 4.

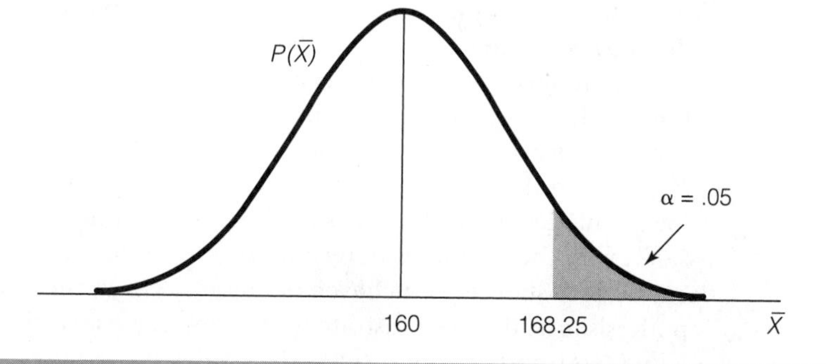

Figure 3.3
Acceptance and critical regions for one-sided hypothesis test

p Value

The ANOVA results for the battery life example as presented by Excel in Table 2.6 refer to a quantity called the *p* value. This quantity was also part of the output of the SPSS presentation, as well as the JMP presentation of the MVPC problem, although labeled by a different name. Indeed, it is a quantity that is part of every software package that performs hypothesis testing (whether an *F* test, a *t* test, or any other test).

Just what is the ***p* value** of a test? One way to describe it would be the weight of evidence against a null hypothesis. Simply picking a significance level, say $\alpha = .05$, and noting whether the data indicate acceptance or rejection of H_0 lacks a precision of sorts. Think of an *F* curve with a critical/rejection region to the right of the critical value; a result stated as "reject" doesn't distinguish between an F_{calc} that is just a tad to the right of the critical value and one that is far into the critical region, possibly orders of magnitude larger than the critical value. Or consider the example in Figure 3.3 for testing the hypotheses

H_0: $\mu \leq 160$ (that is, the IRS-claimed mean time is *not* understated)

versus

H_1: $\mu > 160$ (that is, the IRS-claimed mean time *is* understated)

The critical value for the test is 168.25. Consider two different results, $\overline{X} = 168.5$ and $\overline{X} = 200$. The former value just makes it into the critical region, whereas the latter value is just about off the page! In the real world, although both of these results say, "Reject the null hypothesis at the 5% significance level," they would not be viewed as equivalent results. In the former case, the evidence against H_0 is just sufficient to reject; in the latter case it's much greater than that required to reject. The determination of a *p* value is one way to quantify what one might call the degree to which the null hypothesis should be rejected. In Figure 3.3, the *p* value is the area to the right of the $\overline{X}$ value that obtains. If $\overline{X} = 168.5$, the area to the right of $\overline{X}$ (which we would instantly know is less than .05, since $168.5 > 168.25$) is .0446, corresponding to 168.5 being 1.7 standard deviations of the mean $[\sigma(\overline{X}) = 5]$ above 160. If $\overline{X} = 200$, the *p* value is zero to many decimal places, as 200 is eight standard deviations of the mean above 160.

To be more specific, we can define the *p* value as the highest (preset) α for which we would still accept H_0. For a one-sided upper-tailed test, the *p* value is the area to the right of the test statistic (on an *F* curve, to the right of F_{calc}; on an $\overline{X}$ curve, the area to the right of $\overline{X}$, and so on); for a one-sided lower-tailed test, the *p* value is the area to the left of the test statistic; for a two-sided test, the *p* value is determined by doubling the area to the left or right of the test statistic, whichever of these areas is smaller. The majority of hypothesis tests that we illustrate in this text are *F* tests that are one-sided upper-tailed tests, and as noted above, the *p* value is then the area to the right of F_{calc}.

Type I and Type II Errors

Will we always get a value of $\overline{X}$ that falls in the acceptance region when H_0 is true? Of course not. Sometimes $\overline{X}$ is higher than μ, sometimes lower, and occasionally $\overline{X}$ is a lot higher or a lot lower than μ—far enough away to cause us to reject H_0—even though H_0 is true. How often? Indeed, α is precisely the probability that we reject H_0 when H_0 is true. If you look back at either of the hypothesis-testing figures, Figures 3.2 or 3.3, you can see that the curve is centered at 160 (H_0, or at the limit of the range of H_0 values), and the shaded-in critical region has an area of precisely α. In fact, the value of α was an input to determining the critical value. The error of rejecting an H_0 when it is true is called a Type I error. We can make the probability of a Type I error as small as we wish. If, going back to the two-tailed example, we had picked an acceptance region of 140 to 180, we would have had $\alpha = .00006$—small by most standards.

Why don't we make α vanishingly small? Because there's another error, called a Type II error, which becomes more probable as α decreases. This, of course, is the "other side of the coin," the error we make when H_0 is false, but we accept it as true. The probability of a Type II error is called β; that is, $\beta = P(\text{accept } H_0/H_0 \text{ false})$. In our two-tailed test earlier, $\beta = P(\text{accept that } \mu = 160/\text{in actuality } \mu \neq 160)$. As we'll see, to actually quantify β, we need to specify exactly how $\mu \neq 160$.

It may be useful to consider the following table, which indicates the four possibilities that exist for any hypothesis-testing situation. The columns of the table represent the truth, over which we have no control (and which we don't know; if we did know, why would we be testing?); the rows represent our conclusion, based on the data we observe.

	H_0 True	H_0 False
We Accept H_0	Correct $1 - \alpha$	Type II Error β
We Reject H_0	Type I Error α	Correct $1 - \beta$

Holding the sample size constant, α and β trade off—that is, as one value gets larger, the other gets smaller. The optimal choices for α and β depend, in part, on the consequences of making each type of error. When trying to decide the guilt or innocence of a person charged with a crime, society has judged (and the authors agree) that an α error is more costly—that is, sending an innocent person to jail is more costly to society than letting a guilty person go free. In many other cases, it's the β error that is more costly; for example, for routine medical screening (where the null hypothesis is that the person does not have the disease), it's usually judged more costly to conclude a person is disease free when he actually has the disease, compared with concluding that he has the disease when he actually doesn't. (In the latter case, the error will often be discovered by further medical testing later, anyway.) Hence, we can't generalize about which error is more costly. Is it more costly to conclude that the factor has an effect when it really doesn't (an α error)? Or to conclude that the factor has no effect, when it actually does have an effect? It's situation-specific!

As we have seen, we generally preset α, and as noted, often at .05. One reason for this, as we'll see, is that β depends on the real value of μ, and as we just noted above, we don't know the true value of μ! Let's look back at the one-tailed test we considered in Figure 3.3:

H_0: $\mu \leq 160$ (that is, the IRS-claimed mean time is *not* understated)

versus

H_1: $\mu > 160$ (that is, the IRS-claimed mean time *is* understated)

For β, we then have

$$\begin{aligned}\beta &= P(\text{accept } H_0/H_0 \text{ false})\\ &= P(\overline{X} < 168.25/\mu > 160)\end{aligned}$$

However, we have a problem: $\mu > 160$ lacks specificity. We must consider a specific value of μ in order to have a definite value at which to center our normal curve and determine the area under 168.25 (that is, in the acceptance region—while H_0 is actually false). In essence, what we are saying is that we have a different value of β for each value of μ (as long as the μ considered is part of H_1; otherwise there is no Type II error).

Going back to our one-tailed example of Figure 3.3 where the critical value was 168.25, with $\alpha = .05$, $\sigma = 50$, and $n = 100$, we can illustrate β for a true mean of $\mu = 180$ minutes, as in Figure 3.4, and for a true mean of $\mu = 185$, as in Figure 3.6. These values are simply examples; perhaps the true $\mu = 186.334$, or 60π, or anything else.

We find the area below 168.25 by transforming the curve to a Z curve and using the Z table; the results are in Figure 3.5.

We computed the -2.35 value by a routine transformation to the Z curve:

$$(168.25 - 180)/(50/\sqrt{100}) = -11.75/5 = -2.35$$

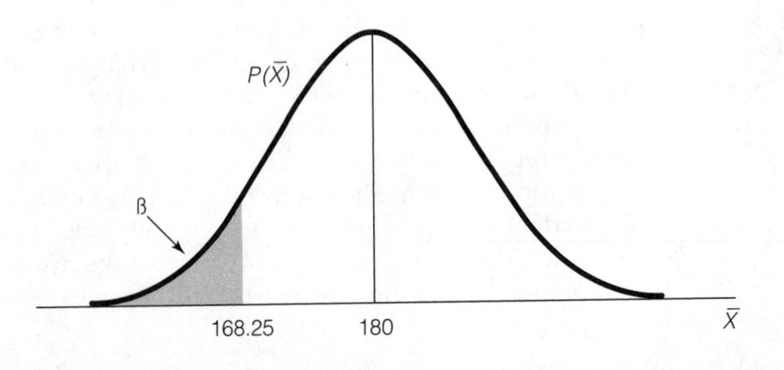

Figure 3.4 β if the true value of μ is 180

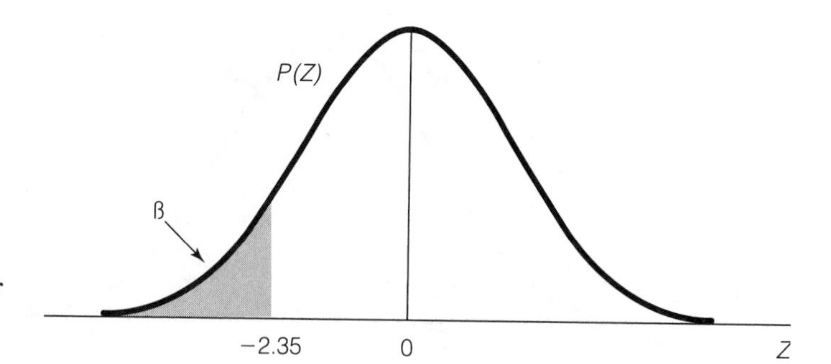

Figure 3.5 Z **value for 168.25 when the true value of** μ **is 180**

Routine use of a Z table (see appendix to the book) reveals that $\beta = .0094$.

Figure 3.6 shows β for the value $\mu = 185$. We computed the -3.35 value in Figure 3.7 as follows:

$$(168.25 - 185)/(50/\sqrt{100}) = -16.75/5 = -3.35$$

And $\beta = .0006$.

Note that as the separation between the mean under H_0 and the assumed true mean under H_1 increases, β decreases. This is to be expected because discrimination between the two conditions becomes easier as they are increasingly different. One can graph β versus the (assumed) true μ (for example, the values 180, 185, other values between and beyond those two values, and so on); this is called an operating characteristic curve, or OC curve.

Back to ANOVA

Recall that in ANOVA,

H_0: $\mu_1 = \mu_2 = \ldots = \mu_j = \ldots = \mu_C$; all column means are equal
H_1: Not all column means are equal

We saw in Chapter 2 that the test statistic, $F_{\text{calc}} = \text{MSB}_c/\text{MSW}$, has an F distribution with degrees of freedom $(C - 1, RC - C)$. α is the probability that we

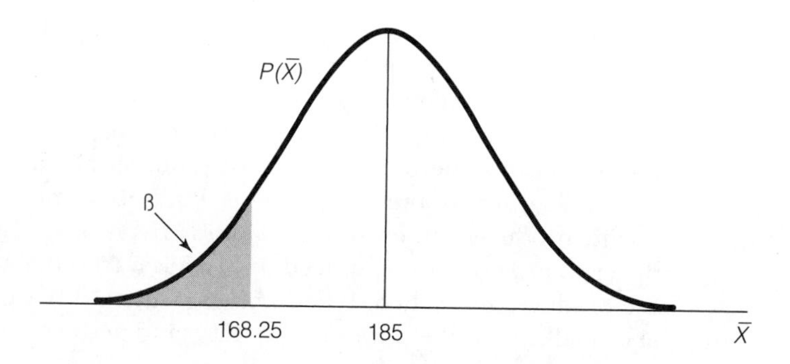

Figure 3.6 β **if the true value of** μ **is 185**

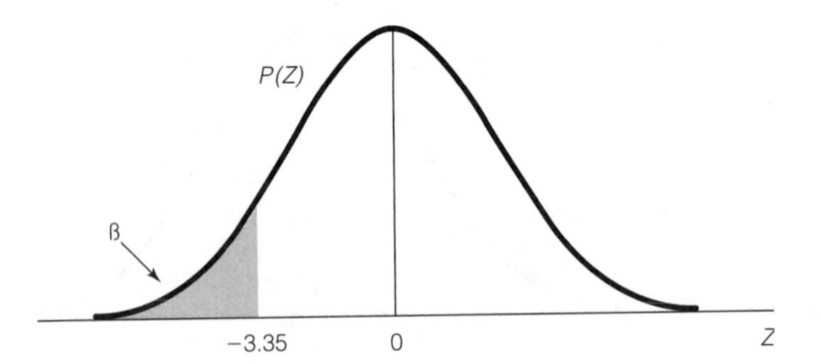

Figure 3.7 Z value for 168.25 when the true value of μ is 185

reject the contention that the factor has no effect, when it actually does have no effect. The critical value, **c**, is determined from an F table, such that $P(F_{\text{calc}} > \mathbf{c}) = \alpha$. What is β?

$$\beta = P(F_{\text{calc}} < \mathbf{c}/\text{not all } \mu_j \text{ are equal—that is, given that the level of } X \text{ does affect } Y)$$

But the condition "not all μ_j are equal" is less easy to make specific. It would be extremely rare in the real world to know with any confidence the precise way in which "not all μ_j are equal," if indeed they are not all equal.[6] Thus, determining β is usually meaningful only under the most general of assumptions about the μ values (for example, if they are assumed to be some specific uniform distance apart from one another). We consider this determination in the next section. Always keep in mind that just because we cannot easily conceive with confidence a specific value with which to determine β it does not change the fact that β error exists, or the fact that, as noted earlier, it trades off against α.

3.5 Power

Often, instead of considering the β of a hypothesis test, we speak of the power of a hypothesis test. Whereas β is the probability of *accepting* H_0 when H_0 is false, the **power** of a test is merely the probability of (correctly) *rejecting* H_0 when H_0 is false. That is, power $= 1 - \beta$. Switching the focus from β to power is simply a matter of working with a quantity in which higher is better instead of one in which lower is better. In some fields it is traditional to work with β. An example is quality control, in which it is customary to talk about α and β as producer's risk and consumer's risk, respectively. (These terms arise from the notion that the producer is hurt economically by rejecting good-quality products, whereas the consumer is hurt economically by accepting bad-quality products.) Interestingly, *consumer risk* is often more costly to the producer than *producer risk;* indeed, it is seldom true that a Type II error leaves the producer unscathed. When working with ANOVA problems, it is customary to talk about the power of the test being performed.

It can be shown that, for a one-factor ANOVA,

$$\text{Power} = f(\alpha, \nu_1, \nu_2, \text{and } \phi)$$

where

α = significance level
ν_1 = *df* of numerator of F_{calc}, $C - 1$
ν_2 = *df* of denominator of F_{calc}, $RC - C$
ϕ = noncentrality parameter, a measure of how different the μ's are from one another; specifically

$$\phi = (1/\sigma)\sqrt{[\Sigma R_j(\mu_j - \mu)^2]/C}$$

where summation is over j, j indexing columns. Note that ϕ includes the sample size by virtue of its dependence on R and C. Of course, if R is constant, it can be factored out and placed in front of the summation sign.

Although we have indicated what quantities affect power, we have not indicated explicitly the nature of the functional relationship; it's quite complex. The probability distribution of F_{calc}, given that H_0 is false, is called the noncentral F distribution and is not the same as the "regular" F distribution, which is appropriate when H_0 is true, that we have already used. We can infer something qualitative about power from what we already know. All other things being equal,

- Power should increase with increasing α (corresponding to the tradeoff between α and β).
- Power should increase with lower σ (corresponding to the increased ability to discriminate between two alternative μ's when the curves' centers are more standard deviations apart).
- Power should increase with increased R (corresponding to the standard deviation of each column mean being smaller with a larger R).
- Power should decrease with increased C (corresponding to an increased number of columns being equivalent to levels of a factor that are closer together).

Obviously, an important aspect of ANOVA is the attempt to discriminate between H_0 and H_1. As we've noted, the smaller the difference between them, the more likely we are to make an error. Because our approach usually starts by fixing α, and α is independent of the specifics of the alternate hypothesis, the difference between H_0 and H_1 is a nonissue for α. Yet it's a driving force for power. The issue is further complicated because, as also indicated earlier, the study of β (or power) requires that prior to running the experiment we have knowledge of some values that are what we wish to discover from the experiment (for example, σ, μ_j's, μ). Hence, in practice, we need to make assumptions about these quantities. Often, we make assumptions in terms of multiples of σ, which can avoid an explicit input value for σ. Consider the following example.

Suppose we have a one-factor ANOVA with $\alpha = .05$, $R = 10$, $C = 3$. Suppose further that we (arbitrarily, but not totally unrealistically) decide to calculate power assuming that the μ's are one standard deviation apart (for example, $\mu_1 = \mu_2 - \sigma$, μ_2, and $\mu_3 = \mu_2 + \sigma$). Then, noting that the mean of the three means equals μ_2, the noncentrality parameter is

$$\phi = (1/\sigma)\sqrt{10[(-\sigma)^2 + 0 + \sigma^2]/3} = \sqrt{20/3} = 2.58$$

and

$$\nu_1 = C - 1 = 2$$
$$\nu_2 = (R - 1)C = 27$$

and we already specified that $\alpha = .05$.

We now are able to refer to what are called power tables (Figures 3.8, 3.9, and 3.10).[7] Using the values above, we find that power approximately equals .976. The sequence of steps is as follows:

1. Find the table for the appropriate value of ν_1 (here, equal to 2—the first one shown).
2. Find which horizontal-axis measure of ϕ is appropriate (it depends on the value of α, with the tables including .05 and .01), and determine the appropriate point on the horizontal axis (here, indicated on the $\nu_1 = 2$ table by a thick dot at 2.58).
3. Find which set of curves corresponds with the value of α (here, it's the set on the left, labeled .05).
4. Find among the appropriate set of curves the one corresponding to the value of ν_2, $RC - C$ (here, equal to 27, although given that they are so close together, we used the nearest listed value, 30).
5. Find the intersection of that curve with the horizontal axis value identified in step 2 (here, indicated by a thick square).
6. Read the value of the power along the left vertical scale (here, we see a value somewhere between .97 and .98, approximately .976).

Power Considerations in Determination of Required Sample Size

We can examine the issue of power from a slightly different perspective. Rather than determine what power we have for a given set of input values, it is often more useful to specify the desired power ahead of time (along with the number of levels of the factor to be included, C, and the desired value of α), and find out how large our sample size (in terms of the number of replicates per column) must be to achieve the desired power. We view the number of columns, C, as an input value that was previously determined based on other, presumably important, considerations; however, one can instead find the required number of replicates for varying values of C, and then decide on the value of C. Of course, as for any determination involving β or power, one must specify a degree to which the μ's are not equal, analogous to the ϕ of the previous section.

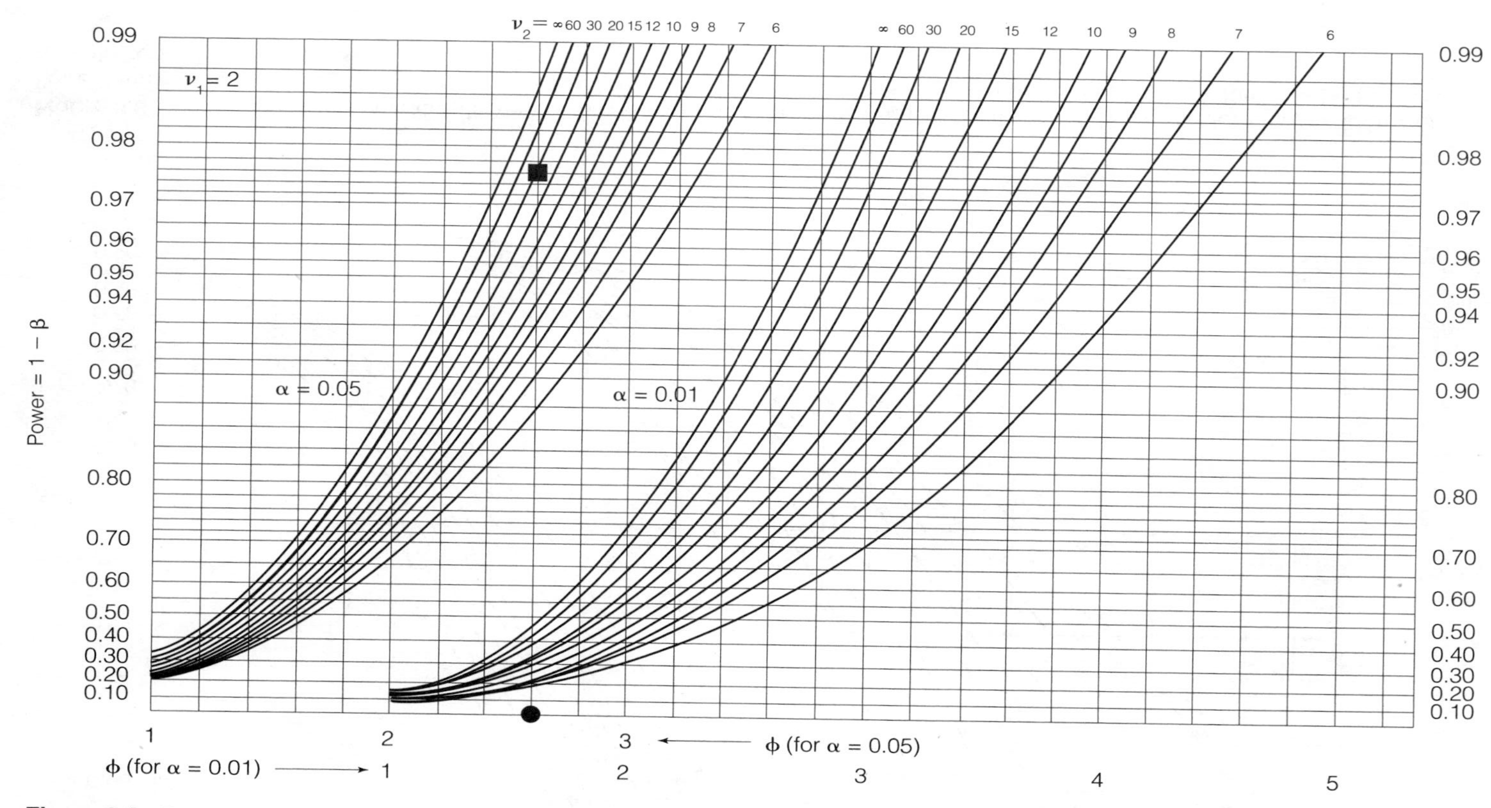

Figure 3.8 Power table for numerator degrees of freedom, ν_1, equal to 2. *Source:* E. S. Pearson and H. O. Hartley (1951), "Charts of Power Function for Analysis of Variance Tests, Derived from the Non-Central *F*-Test." *Biometrika,* 38, pp. 112–130. Reprinted with permission of Oxford University Press.

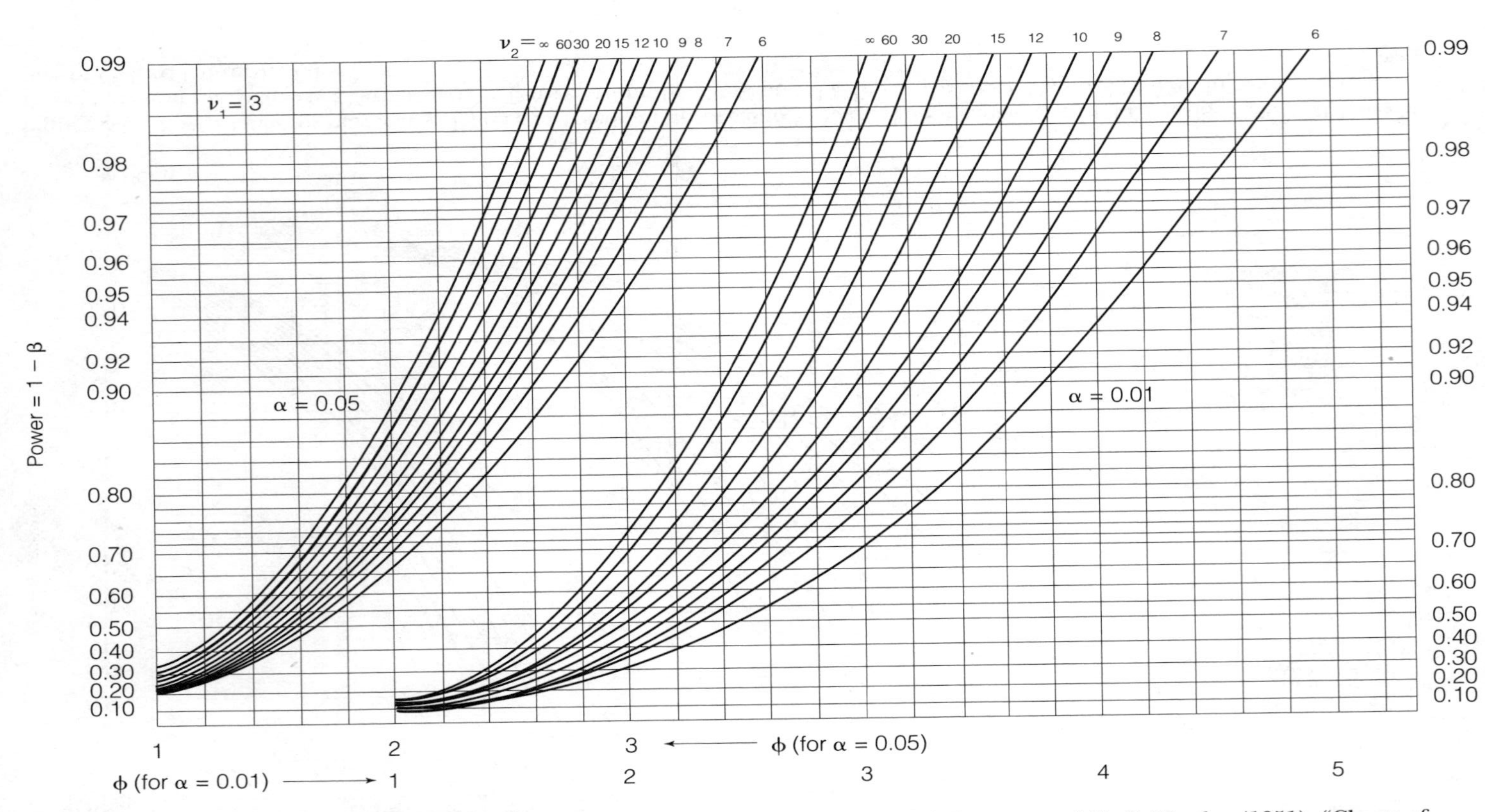

Figure 3.9 Power table for numerator degrees of freedom, ν_1, equal to 3. *Source:* E. S. Pearson and H. O. Hartley (1951), "Charts of Power Function for Analysis of Variance Tests, Derived from the Non-Central *F*-Test." *Biometrika*, 38, pp. 112–130. Reprinted with permission of Oxford University Press.

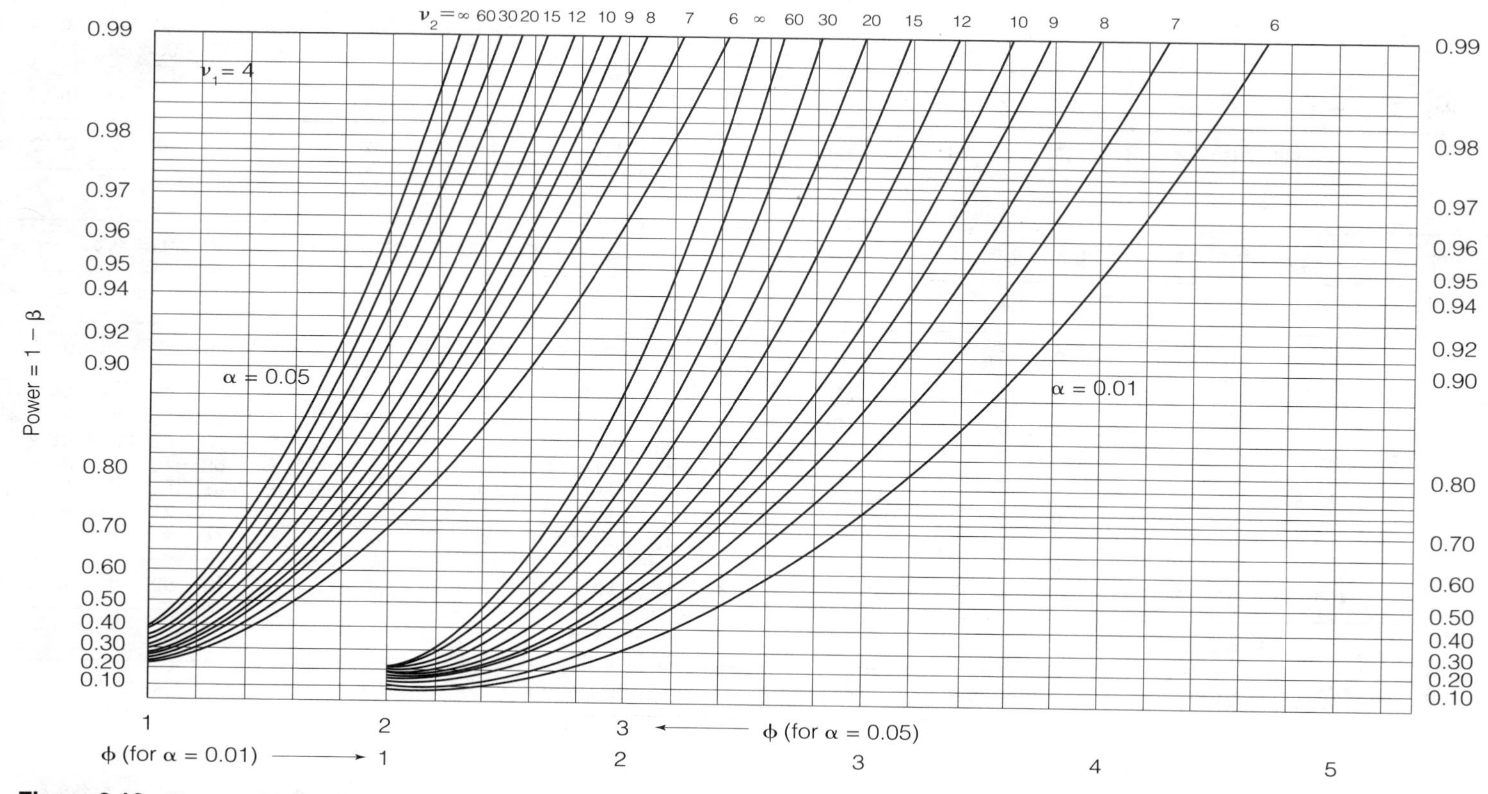

Figure 3.10 **Power table for numerator degrees of freedom, ν_1, equal to 4.** *Source:* E. S. Pearson and H. O. Hartley (1951), "Charts of Power Function for Analysis of Variance Tests, Derived from the Non-Central *F*-Test." *Biometrika,* 38, pp. 112–130. Reprinted with permission of Oxford University Press.

Table 3.3 Power Tables for Sample Size

Power $1 - \beta = .70$

	$\Delta/\sigma = 1.0$				$\Delta/\sigma = 1.25$				$\Delta/\sigma = 1.50$				$\Delta/\sigma = 1.75$				$\Delta/\sigma = 2.0$				$\Delta/\sigma = 2.5$				$\Delta/\sigma = 3.0$			
	α				α				α				α				α				α				α			
C	.2	.1	.05	.01	.2	.1	.05	.01	.2	.1	.05	.01	.2	.1	.05	.01	.2	.1	.05	.01	.2	.1	.05	.01	.2	.1	.05	.01
2	7	11	14	21	5	7	9	15	4	6	7	11	3	4	6	9	3	4	5	7	2	3	4	5	2	3	3	5
3	9	13	17	25	6	9	11	17	5	7	8	12	4	5	7	10	3	4	5	8	3	3	4	6	2	3	3	5
4	11	15	19	28	7	10	13	19	5	7	9	13	4	6	7	10	4	5	6	8	3	4	4	6	2	3	4	5
5	12	17	21	30	8	11	14	20	6	8	10	14	5	6	8	11	4	5	6	9	3	4	5	6	3	3	4	5
6	13	18	22	32	9	12	15	21	6	9	11	15	5	7	8	12	4	5	7	9	3	4	5	7	3	3	4	5
7	14	19	24	34	9	13	16	22	7	9	11	16	5	7	9	12	4	6	7	10	3	4	5	7	3	3	4	5
8	15	20	25	35	10	13	16	23	7	10	12	17	6	7	9	13	5	6	7	10	3	4	5	7	3	3	4	5
9	15	21	26	37	10	14	17	24	7	10	12	17	6	8	9	13	5	6	8	10	3	4	5	7	3	4	4	6
10	16	22	27	38	11	14	18	25	8	10	13	18	6	8	10	14	5	6	8	11	4	5	6	7	3	4	4	6

Power $1 - \beta = .80$

	$\Delta/\sigma = 1.0$				$\Delta/\sigma = 1.25$				$\Delta/\sigma = 1.50$				$\Delta/\sigma = 1.75$				$\Delta/\sigma = 2.0$				$\Delta/\sigma = 2.5$				$\Delta/\sigma = 3.0$			
	α				α				α				α				α				α				α			
C	.2	.1	.05	.01	.2	.1	.05	.01	.2	.1	.05	.01	.2	.1	.05	.01	.2	.1	.05	.01	.2	.1	.05	.01	.2	.1	.05	.01
2	10	14	17	26	7	9	12	17	5	7	9	13	4	5	7	10	3	4	6	8	3	3	4	6	2	3	4	5
3	12	17	21	30	8	11	14	20	6	8	10	14	5	6	8	11	4	5	6	9	3	4	5	7	3	3	4	5
4	14	19	23	33	9	13	15	22	7	9	11	16	5	7	9	12	4	6	7	10	3	4	5	7	3	3	4	5
5	16	21	25	35	10	14	17	23	8	10	12	17	6	8	9	13	5	6	7	10	4	4	5	7	3	4	4	6
6	17	22	27	38	11	15	18	25	8	11	13	18	6	8	10	13	5	7	8	11	4	5	6	8	3	4	4	6
7	18	24	29	39	12	16	19	26	9	11	14	18	7	9	10	14	5	7	8	11	4	5	6	8	3	4	5	6
8	19	25	30	41	12	16	20	27	9	12	14	19	7	9	11	15	6	7	9	12	4	5	6	8	3	4	5	6
9	20	26	31	43	13	17	21	28	9	12	15	20	7	9	11	15	6	7	9	12	4	5	6	8	3	4	5	6
10	21	27	33	44	14	18	21	29	10	13	15	21	8	10	12	16	6	8	9	12	4	5	6	8	3	4	5	6

Power 1 − β = .90

	Δ/σ = 1.0				Δ/σ = 1.25				Δ/σ = 1.50				Δ/σ = 1.75				Δ/σ = 2.0				Δ/σ = 2.5				Δ/σ = 3.0			
	α				α				α				α				α				α				α			
C	.2	.1	.05	.01	.2	.1	.05	.01	.2	.1	.05	.01	.2	.1	.05	.01	.2	.1	.05	.01	.2	.1	.05	.01	.2	.1	.05	.01
2	14	18	23	32	9	12	15	21	7	9	11	15	5	7	8	12	4	6	7	10	3	4	5	7	3	3	4	6
3	17	22	27	37	11	15	18	24	8	11	13	18	6	8	10	13	5	7	(8)	11	4	5	6	8	3	4	5	6
4	20	25	30	40	13	16	20	27	9	12	14	19	7	9	11	15	6	7	9	12	4	5	6	8	3	4	5	6
5	21	27	32	43	14	18	21	28	10	13	15	20	8	10	12	15	6	8	9	12	4	5	6	9	4	4	5	7
6	22	29	34	46	15	19	23	30	11	14	16	21	8	10	12	16	7	8	10	13	5	6	7	9	4	4	5	7
7	24	31	36	48	16	20	24	31	11	14	17	22	9	11	13	17	7	9	10	13	5	6	7	9	4	5	5	7
8	26	32	38	50	17	21	25	33	12	15	18	23	9	11	13	17	7	9	11	14	5	6	7	9	4	5	6	7
9	27	33	40	52	17	22	26	34	13	16	18	24	9	12	14	18	8	9	11	14	5	6	8	10	4	5	6	7
10	28	35	41	54	18	23	27	35	13	16	19	25	10	12	14	19	8	10	11	15	5	7	8	10	4	5	6	7

Power 1 − β = .95

	Δ/σ = 1.0				Δ/σ = 1.25				Δ/σ = 1.50				Δ/σ = 1.75				Δ/σ = 2.0				Δ/σ = 2.5				Δ/σ = 3.0			
	α				α				α				α				α				α				α			
C	.2	.1	.05	.01	.2	.1	.05	.01	.2	.1	.05	.01	.2	.1	.05	.01	.2	.1	.05	.01	.2	.1	.05	.01	.2	.1	.05	.01
2	18	23	27	38	12	15	18	25	9	11	13	18	7	8	10	14	5	7	8	11	4	5	6	8	3	4	5	6
3	22	27	32	43	14	18	21	29	10	13	15	20	8	10	12	16	6	8	9	12	5	6	7	9	4	4	5	7
4	25	30	36	47	16	20	23	31	12	14	17	22	9	11	13	17	7	9	10	13	5	6	7	9	4	5	5	7
5	27	33	39	51	18	22	25	33	13	15	18	23	10	12	14	18	8	9	11	14	5	6	7	10	4	5	6	7
6	29	35	41	53	19	23	27	35	13	16	19	25	10	12	14	19	8	10	11	15	6	7	8	10	4	5	6	8
7	30	37	43	56	20	24	28	36	14	17	20	26	11	13	15	19	8	10	12	15	6	7	8	10	4	5	6	8
8	32	39	45	58	21	25	29	38	15	18	21	27	11	14	16	20	9	11	12	16	6	7	8	11	5	5	6	8
9	33	40	47	60	22	26	30	39	15	19	22	28	12	14	16	21	9	11	13	16	6	8	9	11	5	6	6	8
10	34	42	48	62	22	27	31	40	16	19	22	29	12	15	17	21	9	11	13	17	6	8	9	11	5	6	7	8

Source: T. I. Bratcher, M. A. Moran, and W. J. Zimmer, "Tables of Sample Size in Analysis of Variance," *Journal of Quality Technology,* 1970, pp. 156–164. Reprinted with permission of the American Society for Quality.

The customary formulation for quantifying the degree to which the μ's are not equal is through the quantity Δ/σ, where

$$\Delta = \text{the range of the } \mu\text{'s} = \max(\mu_j) - \min(\mu_j)$$

In essence, the table we are about to describe assumes that the μ's are uniformly spread between the maximum value and the minimum value. Because we are determining the required sample size before we have collected any data, we typically do not have an estimate of σ^2, such as the MSW (or mean square error, MSE, as it is often called). Thus we usually input a value of Δ/σ, as opposed to separate values of Δ and σ. A popular choice is $\Delta/\sigma = 2$ (that is, the range of values of the true column means is two standard deviations).

We determine the required sample size (again, in terms of replicates per column) using the sample size tables (Table 3.3). The sequence of steps is as follows:

1. Find the portion of the table with the desired power; the table provided includes powers of .70, .80, .90, and .95.
2. Within that portion of the table, find the section with the desired Δ/σ ratio.
3. Find the column with the desired value of α.
4. Find the row for the appropriate number of columns, C.
5. Match the row found in step 4 with the column found in step 3, and read the value at their intersection; this is the value of R.

For example, if we wish the following values:

$$\begin{aligned} 1 - \beta &= .9 \\ \Delta/\sigma &= 2 \\ \alpha &= .05 \\ C &= 3 \end{aligned}$$

then $R = 8$ (circled in the table). Note also that we can now go to the section of power equal to .80, and for the same α, we see that $R = 8$ also provides 80% power to detect a Δ/σ of 1.75; we can go to the section of power equal to .70, and note that, again for $\alpha = .05$, $R = 8$ provides power of 70% to detect a Δ/σ of 1.50. Indeed, $R = 8$ has 95% power of detecting a Δ/σ partway between 2.0 and 2.5.

3.6 Confidence Intervals

In this section we present the procedure for finding a confidence interval for (1) the true mean of a column (or level) and (2) the true difference between two column means. We assume that the standard assumptions described earlier in this chapter hold true. In general, with a normally distributed sample mean, $\bar{X}$, and with a known value for the standard deviation, σ, a $100(1 - \alpha)\%$ confidence interval for the true μ is formed by taking $\bar{X} \pm e$, with

$$e = z_{1-\alpha/2}(\sigma/\sqrt{n}) \tag{3.1}$$

where $z_{1-\alpha/2}$ is the $100(1 - \alpha/2)\%$ cumulative value of the standard normal curve, and n is the number of data values in that column, also known as R. For example, $z_{1-\alpha/2}$ equals 1.96 for 95% confidence. However, in performing an ANOVA, we do not know the true standard deviation (although by assumption it is a constant for each data value). When a standard deviation is unknown (which is most of the time in real-world data analysis), the z is replaced by a t in equation (3.1), and the true standard deviation is replaced by our estimate of the standard deviation, s. This gives us

$$e = t_{1-\alpha/2}(s/\sqrt{n}) \tag{3.2}$$

where $t_{1-\alpha/2}$ is the $100(1 - \alpha/2)\%$ cumulative value of the Student t curve, with the number of degrees of freedom that corresponds with the degrees of freedom used to estimate the sample standard deviation, s. With one column of n data points, the number of degrees of freedom is $(n - 1)$. However, in ANOVA, our estimate of the standard deviation is the square root of the MSW (or of the mean square error), and is based on pooling variability from each of the columns, resulting, indeed, in $(RC - C)$ degrees of freedom (that is, the degrees of freedom of the error term).

EXAMPLE 3.3 Confidence Interval for Clinic Satisfaction Study

To find a confidence interval for a particular column mean, we simply apply the formula in (3.2). We can demonstrate this using the data from the Merrimack Valley Pediatric Clinic (MVPC) satisfaction study in Chapter 2. Recall that the study asked 30 respondents at each of the clinic's four locations to rate satisfaction as earlier described. The ANOVA results, along with the mean of each column, were presented as JMP output in Figure 2.5 and are repeated in Table 3.4.

Table 3.4 **JMP Output: ANOVA for Clinic Satisfaction Study**

Analysis of Variance

Source	DF	Sum of Squares	Mean Square	F Ratio
Model	3	5296.4250	1765.47	205.2947
Error	116	997.5667	8.60	Prob>F
C Total	119	6293.9917	52.89	<.0001

Means for Oneway Anova

Level	Number	Mean	Std Error
Amesbury	30	67.1000	0.53540
Andover	30	50.4000	0.53540
Methuen	30	57.5667	0.53540
Salem	30	65.3000	0.53540

Std Error uses a pooled estimate of error variance

Suppose that we want to find a 95% confidence interval for the true mean of the Amesbury site (column one, as the data were set up in Table 2.8). The column mean is 67.1. The standard deviation estimate is the square root of 8.60 = 2.933. With 116 degrees of freedom (see output above) the value of $t_{1-\alpha/2}$ is equal to about 1.98, and with $n = R$ per column of 30, we have[8]

$$\begin{aligned} e &= t_{1-\alpha/2}(s/\sqrt{n}) \\ &= 1.98(2.933/\sqrt{30}) \\ &= 1.06 \end{aligned}$$

Thus, our 95% confidence interval is

$$\overline{X} \pm e \quad \text{or} \quad 67.10 \pm 1.06 \quad \text{or} \quad 66.04 \text{ to } 68.16$$

What if we want to find a confidence interval for the true *difference* between two column means? Then, with the two columns of interest labeled 1 and 2, our confidence interval centers at the difference in the column means, $\overline{X}_1 - \overline{X}_2$, and is

$$(\overline{X}_1 - \overline{X}_2) \pm e$$

However, now, recognizing that under the standard assumption of independence among data values, the standard deviation estimate of $\overline{X}_1 - \overline{X}_2$ is

$$\sqrt{s^2/n_1 + s^2/n_2} = s\sqrt{1/n_1 + 1/n_2}$$

where again, $s = \sqrt{\text{MSW}}$, we have

$$e = t_{1-\alpha/2}(s\sqrt{1/n_1 + 1/n_2}) \tag{3.3}$$

or, if $n_1 = n_2 = R$,

$$e = t_{1-\alpha/2}(s\sqrt{2/R}) \tag{3.4}$$

In the MVPC example above, let's find a 95% confidence interval for the difference in means between Methuen and Andover. The difference in column means is (57.5667 − 50.4000) = 7.1667. Plugging into equation (3.4) and noting that the degrees of freedom is again 116, we get

$$\begin{aligned} e &= 1.98(2.933\sqrt{2/30}) \\ &= 1.499 \end{aligned}$$

Thus, our 95% confidence interval for the true difference in mean satisfaction score between Methuen and Andover is

$$(\overline{X}_1 - \overline{X}_2) \pm e \quad \text{or} \quad 7.1667 \pm 1.499 \quad \text{or} \quad 5.6677 \text{ to } 8.6657$$

Exercises

1. Consider the data in Table 3EX.1 (a repeat of Table 2EX.6), representing four levels (literally!) of lodging on a cruise ship, and a random sample of six passengers from each level. The yield is an assessment of the amount of motion felt during cruising (scaled from 1 to 30). Is there sufficient evidence from which to conclude that the level

of the room on the cruise ship affects perception of the degree of motion? Analyze using the Kruskal-Wallis test with $\alpha = .05$. What is the p value of the test?

Table 3EX.1 **Motion Assessment by Ship Level**

1	2	3	4
16	16	28	24
22	25	17	28
14	14	27	17
8	14	20	16
18	17	23	22
8	14	23	25

2. Consider the Table 3EX.2 data, which represent the amount of life insurance (in $1000s) for a random selection of seven state senators from each of three states: California, Kansas, and Connecticut. All of the senators are male, married, with two or three children, and between the ages of 40 and 45. Because it appears that there could be major differences in variability from column to column, it was decided that a Kruskal-Wallis test would be performed to inquire whether amounts of life insurance differed by state/part of the country, at least with respect to state senators with these demographics. Conduct this test at $\alpha = .05$.

Table 3EX.2 **Life Insurance ($1000s)**

State		
1	2	3
90	80	165
200	140	160
225	150	140
100	140	160
170	150	175
300	300	155
250	280	180

3. Repeat exercise 2 using a conventional F test. Do your conclusions differ? Discuss.
4. Suppose that we are conducting a one-factor ANOVA and have four levels of the factor under study, six replicates at each of the four levels, and desire a significance level of .01. At a value of ϕ of 2.5, what power would your F test have?
5. What is the gain in power in exercise 4 if we increase the sample size to nine replicates per column? What is the loss in power if we reduce the number of replicates to three replicates per column? What does this suggest about the way power varies with number of replicates per column?
6. Consider exercise 4 again. Does the power increase more if we increase the number of replicates per column to nine, or if we change the significance level to .05?
7. Consider the situation with one factor under study at four levels. If we desire a significance level of .01, and insist that the power of the F test be .80 with a Δ/σ value of 2, what is the number of replicates needed at each level of the factor?
8. If we have performed an ANOVA for a one-factor design with four columns and ten replicates per column, and found MSB_c to be 100, and MSW to be 25, what is the estimate of $\Sigma(\mu_j - \mu)^2$?
9. Suppose that a symphony orchestra has tried three approaches, A, B, and C, to solicit funds from 30 generous local sponsors, 10 sponsors per approach. The dollar amounts of the donations that resulted are in Table 3EX.9. The column means are listed in the bottom row. For convenience, we rank-order the results in each column in descending order in the table. Use the Kruskal-Wallis test at $\alpha = .05$ to determine whether there are differences in solicitation approach with respect to amount of charitable donations.

Table 3EX.9 Funds Raised

Approach		
A	B	C
3500	3200	2800
3000	3200	2000
3000	2500	1600
2750	2500	1600
2500	2200	1200
2300	2000	1200
2000	1750	1200
1500	1500	800
1000	1200	500
500	1200	200
2205	2125	1310

10. a. For each of the columns in Table 3EX.9, find a 95% confidence interval for the true column mean. Assume all of the standard assumptions hold.
    b. Are there any values that are included in all three of these confidence intervals?
    c. Discuss the implications of the answer to part b.

Notes

1. We note the word *homoscedasticity* solely to prepare readers for it, should they see it in other texts or treatises on the subject. It means "constant variance," or something close to that, in Greek. It is sometimes spelled *homoskedasticity*.

2. In fact, there are close relationships among the F distribution, the chi-square distribution (χ^2), the Student t distribution (t), and the standard normal, Z distribution; here we relate each to the F distribution:

$$\chi^2(df_1)/df_1 = F(df_1, \infty)$$
$$t^2(df_2) = F(1, df_2)$$
$$Z^2 = F(1, \infty)$$

3. Although not necessarily the case from a mathematical perspective, the two hypotheses are collectively exhaustive for all practical purposes, and certainly in the applications we undertake.

4. In most cases the standard deviation is not known. (Indeed, we have encountered a known standard deviation only when (1) the process being studied had a standard deviation that historically has remained constant, and the issue was whether the process was properly calibrated or was off-center, or (2) the quantity being tested is a proportion, in which case the standard deviation is treated as if known, as a direct function of the hypothesized value of the proportion.) However, the assumption of known standard deviation is useful in this presentation. Our goal at this point does not directly include distinctions between the Z and the t; that changes in Chapter 4, where the Student t distribution is discussed.

5. Notice that in this example, logic suggests a critical (rejection) region that is two-sided (more formally called two-tailed, and the test is said to be a two-tailed test). After all, common sense says that we should reject H_0: $\mu = 160$ if the $\overline{X}$ is *either* too small (that is, a lot below 160) *or* too large (that is, a lot above 160). Because α is whatever it is (here, .05) in total, it must be split between the upper and lower tails; it is traditional, when there are two tails, to split the area equally between the tails.

6. Obviously, we don't know for sure the exact values of the μ's, or we would not have a need to test. However, in very rare cases we are not certain whether the μ's are equal or not, but we do know what they are if they are not all equal.

7. It may be possible to obtain these values, and those of the next section, via software. However, there is insight to be gained through seeing the entire tables in print.

8. Usually, when degrees of freedom for the t distribution exceed 30, we simply use the corresponding z value, which here is 1.96. However, for 120 degrees of freedom, the value of the t is 1.98. For 116 degrees of freedom, it is very close to 1.98.

CHAPTER 4

Multiple-Comparison Testing

EXAMPLE 4.1 The Qualities of a Superior Motel

A relatively low-priced motel chain was interested in inquiring about certain factors that might play an important role in the consumers' choice of motel. As mentioned in Chapter 1, the managers knew that certain factors, such as location, couldn't be altered (at least in the short run), and that price played a significant role but, for most locations, needed to be considered as already set by market forces for the specific location. The chain was interested in exploring the impact on customer satisfaction and choice of motel of a set of factors involving the availability and quality of food and beverages, entertainment, and business amenities. The company believed that the impact of these factors would be relatively uniform across locations.

Two of the factors that the company believed would be very important, and their levels, are listed in Figure 4.1.

The sample of respondents was separated into four segments. Two of the segments were frequent users of the motel chain (based on the proportion of total hotel or motel use involving the sponsoring chain), split into business users and leisure users. The other two segments were infrequent or nonusers of the motel chain, split the same way. The key dependent variable (among other, more intermediary variables, such as attitude toward the chain) was the respondent's estimate of the number of nights he or she would stay at the motel chain during the next 12 months. We return to this example at the end of the chapter.

1. Breakfast (at no extra charge)
   - None available
   - Continental breakfast buffet—fruit juices, coffee, milk, fresh fruit, bagels, doughnuts
   - Enhanced breakfast buffet—add some hot items, such as waffles and pancakes, *that the patron makes himself*
   - Enhanced breakfast buffet—add some hot items, such as waffles and pancakes, *with a "chef" who makes them for the patron*
   - Enhanced breakfast buffet—add some hot items, such as waffles and pancakes, *that the patron makes himself,* **and also pastry (dough from a company like Sara Lee) freshly baked on premises**
   - Enhanced breakfast buffet—add some hot items, such as waffles and pancakes, *with a "chef" who makes them for the patron,* **and also pastry (dough from a company like Sara Lee) freshly baked on premises**
2. Entertainment
   - Three local channels and **five** of the more popular cable stations, plus pay-per-view movies
   - Three local channels and **fifteen** of the more popular cable stations, plus pay-per-view movies
   - Three local channels and fifteen of the more popular cable stations, plus pay-per-view movies **and Nintendo games**
   - Three local channels and fifteen of the more popular cable stations, plus pay-per-view movies **and VCR**
   - Three local channels and fifteen of the more popular cable stations, plus pay-per-view **movies and both Nintendo games and VCR**

Figure 4.1 **Some factors and levels for motel study.**

4.1 Logic of Multiple-Comparison Testing

In Chapter 2, we saw how to determine whether the level of some factor affects a dependent variable of interest. After we decided on the dependent variable and the factor to study, the procedure was primarily objective and quantitative (with some necessary judgments, such as evaluating the validity of the assumptions and choosing the value of α). If we conclude that the factor under study does have an impact on the dependent variable, we would certainly not want to stop our analysis there. After all, if the F test led to rejection of H_0: $\mu_1 = \mu_2 = \mu_3 = \cdots = \mu_C$ in favor of H_1: not all column means are equal, we would still have no indication of the way in which they are not equal—only that they're not all the same! If $C = 4$, are all four μ's different, or are three the same and one different from those three? If one is different from the rest, which one? We will demonstrate procedures to answer these types of more detailed inquiries. These procedures are referred to as **multiple-comparison tests** and, nicely enough, use only the data used in the already completed F test.

EXAMPLE 4.2 Multiple-Comparison Testing for Battery Lifetime

Having learned that the device in which the AA-cell battery is used affects the battery's average lifetime, for example, we would likely want to know more detail. Is it whether it's a cell phone, flash camera, or flashlight that makes the difference? Or is it the brand of cell phone, flash camera, or flashlight that seems to be key? Is the average lifetime higher for a particular brand of flash camera than for that same brand of flashlight? Or is it a case of all eight devices simply having different average battery lifetimes? Multiple-comparison tests extend the analysis from the aggregate impact of levels of the factor (for example, does device affect average battery lifetime for AA batteries?) to the detailed differences among subsets of levels of the factor (for example, how does one subset of devices vary from another subset of devices with respect to impact on average battery lifetime?).

Several different multiple comparison tests are available. The choice among them depends on a number of considerations, but the primary one is: what questions do the experimenters wish to answer, and how did they arrive at them? Sometimes, the experimenters state, in advance of seeing the data, a particular set of comparisons they wish to explore (a comparison, typically, is a linear combination of the means); these are called **a priori** or **planned comparisons.** Other times, the experimenters wish to routinely compare each column mean (that is, each level of the factor) with each other column mean (each other level of the factor); these are called **pairwise comparisons.** Yet other times, the experimenters wait until after examining the data to decide which comparisons look interesting for testing; these after-the-fact comparisons are called **post hoc, a posteriori,** or **exploratory comparisons.** (Traditionally, pairwise comparisons are also viewed as being post hoc.)

However, even after concluding which of the above scenarios is relevant, more than one test is usually eligible for selection. The next decision that the experimenters may have to decide concerns the probability of Type I error, α, in terms of both magnitude and "philosophy."

4.2 Type I Errors in Multiple-Comparison Testing

We already know that if we reject a null hypothesis when it is true, we are committing what is called a Type I error. In our discussions, this false rejection means concluding that there are differences among the levels of a factor when, in truth, there are no such differences. However, in multiple-comparison testing, it is not automatic what the term Type I error probability, when set equal to .05, or .01, or any other value, means.

If we test three hypotheses, each involving independent test statistics (for example, separately testing the value of three different column means), and each having a probability of Type I error, α, of .05, what is the probability that, if the three null hypotheses are all true, we reject *at least one* of them? Given the independence among the test statistics, the answer is

$$P(\text{at least one Type I error}) = \mathbf{a} = 1 - (.95)^3 = .143$$

(This calculation uses the so-called multiplication rule for independent events, and then the complementary events rule of elementary probability theory.) The quantity **a** (boldfaced to distinguish it from the routine indefinite article) is called the **experimentwise error rate.** It represents the probability that one or more Type I errors are committed in an experiment (hence the name experimentwise). Type I error probability *per comparison* is, in a sense, replaced with Type I error probability *per experiment.* For six independent tests, we would have an experimentwise error rate of .265; for ten independent tests, .401. Although it is universal to use the Greek letter α to represent the probability of making a Type I error, it makes sense to the authors to use some other symbol to represent the probability of making *at least one* Type I error; we have adopted the bold letter **a** to represent the experimentwise error rate. The notation in this area is not consistent from text to text. Some texts simply use α for both errors; others use α and **a** as we do; yet others differentiate between the two error rates in other ways—such as $\alpha_{\text{individual}}$ and $\alpha_{\text{experimentwise}}$.

In many of the applications we shall pursue, the determination of the experimentwise error from the knowledge of the probability of Type I error of a single comparison (a hypothesis test) is not as easily accomplished as by the formula above. This is because, most often, the test statistics of the hypothesis tests (that is, comparisons) we wish to include in our analysis are not independent. Consider the following example of three hypotheses to be tested:

TEST 1	TEST 2	TEST 3
H_0: $\mu_1 = \mu_2$	H_0: $\mu_1 = \mu_3$	H_0: $\mu_2 = \mu_3$
H_1: $\mu_1 \neq \mu_2$	H_1: $\mu_1 \neq \mu_3$	H_1: $\mu_2 \neq \mu_3$

Intuitively, if we were told that H_0 was accepted for test 1, and that H_0 was rejected for test 2, isn't it more likely than it was before we had any knowledge about test 1 and test 2 that we will reject H_0 for test 3? Indeed, the test statistics for the three tests are *not* independent. When the test statistics are not independent, it is not straightforward to determine the experimentwise error rate if each individual test has a marginal error rate of α; likewise, it is not straightforward to determine the α of each individual test given an experimentwise error rate of $\mathbf{a}$. What we can say is that if the k tests are not independent, $\mathbf{a} \leq 1-(1-\alpha)^k$.

A decision has to be made which to control (that is, specify): the traditional Type I error, which we could call the "comparison error rate," or the experimentwise error rate. Typically, we specify one of the two error rates, and the other error rate is not directly addressed. Most of the multiple-comparison tests control the experimentwise error; however, one notable test we introduce deals solely with the comparison error rate.

EXAMPLE 4.3 Broker Study

For our discussion of multiple-comparison tests we use a common example, which we now introduce, to illustrate each of them. An experiment was conducted by a financial services firm (let's call the firm American Financial Services, AFS) to determine if brokers they use to execute trades differ with respect to their ability to provide an equity (for simplicity, say stock) purchase for AFS at a low buying price per share.[1] To measure cost, AFS used, for each trade of a particular stock, an index, Y, that is clearly "the higher the better":

$$Y = 1000(A - P)/A$$

where

P = per share price paid for the stock
A = average of high price and low price per share, for the day

The study segmented trades by size (as a percent of all shares traded that day in that stock). Other types of segments, such as industry sector of the stock, and liquidity of the stock (measured by *average* volume of that stock per day), were considered but not used in this experiment; they may be used in future experiments. The current data are for trades involving 0–20% of the amount of that stock traded during that day. Five brokers were in the study, and six trades were randomly assigned to each broker.

With six trades per broker, we have a replicated (R = 6) one-factor study, with that factor (broker) at five levels (C = 5). A summary of the data (rounded to nearest integer) is in Table 4.1. For example, the value 11 in Table 4.1 (boldfaced to make it easy to find) was derived by taking a price per share (P) of 67.25, with high and low for the day of 69.25 and 66.75, respectively, for an average (A) of 68.00; then, $1000(68.00 - 67.25)/68.00 = 11.03$, which, rounded to the nearest integer, equals 11.

Table 4.1 **AFS Broker Study**

	BROKER				
Trade	**1**	**2**	**3**	**4**	**5**
1	12	7	8	21	24
2	3	17	1	10	13
3	5	13	7	15	14
4	−1	**11**	4	12	18
5	12	7	3	20	14
6	5	17	7	6	19
Column mean	6	12	5	14	17

The ANOVA table appears as Table 4.2. Recall that with five columns, the numerator $df = C - 1 = 4$ and the denominator $df = C(R - 1) = 5 \cdot 5 = 25$. $F_{calc} = 7.56$ is compared to the threshold value, which is $\mathbf{c} = 2.76$ for $\alpha = .05$ and $df = 4, 25$ ($\mathbf{c} = 4.18$ for $\alpha = .01$). We reject H_0 (that there is no difference between column means) and conclude that there *are* differences among the brokers examined with respect to the index of purchase price per share. Now we delve further into the analysis.

Table 4.2 **ANOVA Table for Broker Study**

Source of Variability	SSQ	*df*	MS	F_{calc}
Broker	640.8	4	160.2	7.56
Error	530	25	21.2	
Total	1170.8	29		

We noted earlier that there are three basic categories of multiple comparison tests: a priori (planned) comparisons, pairwise comparisons, and post hoc exploratory comparisons. We treat the first category, a priori comparisons, in Chapter 5.[2] This separate treatment is useful in illustrating the different mind-set involved when using planned comparisons, which are often based at least partly on theoretical considerations. In this chapter we discuss pairwise comparisons and post hoc exploratory comparisons.

4.3 Pairwise Comparisons

Several multiple-comparison tests have as their basic procedure the comparison of all pairs of column means. Pairwise comparison tests are likely the most frequently used type of multiple-comparison tests. We discuss four of the more popular of these tests in detail, and briefly mention two others. For a variety of reasons, not necessarily identical for each test discussed, all of these tests should be used only when the original F test has indicated the rejection of H_0. Indeed, one can reasonably argue that if the original F test indicates that we cannot reject equality of all of the column means, what more is there to explore?

Fisher's Least Significant Difference Test

The first method we discuss, devised by R. A. Fisher and called Fisher's least significant difference (LSD) test, essentially involves performing a series of pairwise t tests, each with a specified value of α. In isolation, the equality of each pair of means is tested with a Type I error rate of α. But the different tests are clearly not independent, and there is no simple formula to derive the experimentwise error rate—the latter is not a focal point, and is "whatever it is." There is one slight modification to just doing a series of unconnected t tests: the unknown population standard deviation is estimated for all tests by the same value, the square root of MSW. This modification is supported by the assumption of constant variance, mentioned in Chapter 3, and provides us with a larger number of degrees of freedom for the error term. This, in turn, results in each test having greater power $(1 - \beta)$, relative to using the respective error estimate derived from only the two specific columns being tested.[3]

Recall that in any hypothesis test, we establish an acceptance region for H_0 and accept H_0 if the appropriate test statistic falls within that region. If it falls in the critical region, we reject H_0. Here, for each pair of columns, i and j, the test statistic is the difference between the column means, $\overline{Y}_i - \overline{Y}_j$. If this difference is small, we conclude that the true means are equal (that is, their true difference is really zero, $\mu_i = \mu_j$, and any difference in the observed column means is just that due to statistical fluctuation). If the difference is not small, we conclude that the two levels of the factor being tested produce different true means, that is, $\mu_i \neq \mu_j$. We apply this method to our AFS broker study soon.

Figure 4.2 illustrates the distribution for $(\overline{Y}_i - \overline{Y}_j)$. It is a linear transformation of a t distribution, centering at zero, but with variance equal to the variance of a t distribution with $(RC - C)$ degrees of freedom, multiplied by the variance estimate of $(\overline{Y}_i - \overline{Y}_j)$, which is $\sqrt{\text{MSW}}\sqrt{1/n_i + 1/n_j}$, as discussed below. The acceptance region falls between the lower and upper critical values, C_l and C_u, respectively. The variance estimate for a data value is the MSW. (In this context, it may be useful to note that the MSW is equivalent to the

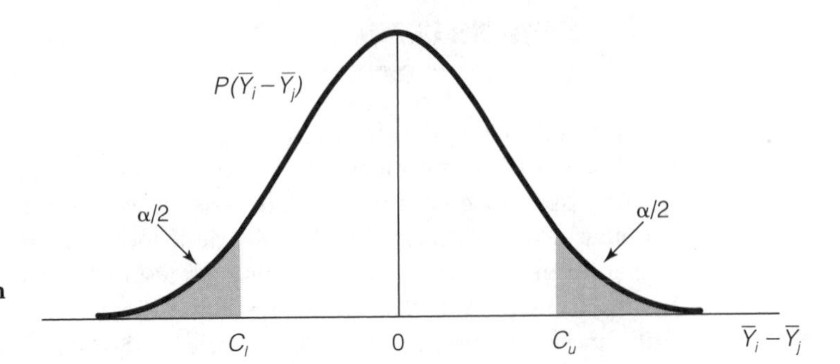

Figure 4.2 Distribution and acceptance region for $\overline{Y}_i - \overline{Y}_j$.

"pooled variance," s_p^2, a term you might recall from a basic statistics course.) Then the acceptance region, AR, for $(\overline{Y}_i - \overline{Y}_j)$ is

$$0 \pm t_{1-\alpha/2}\sqrt{\text{MSW}}\sqrt{1/n_i + 1/n_j} \tag{4.1}$$

where $t_{1-\alpha/2}$ is the abscissa (that is, horizontal axis) value from the *t* table with the appropriate degrees of freedom, and a left tail cumulative area of $(1 - \alpha/2)$; if $\alpha = .05$, the value of $(1 - \alpha/2)$ is .975, reflecting that the .05 is split evenly between the two tails. The number of replicates in column *i* is denoted as n_i; similarly for n_j. In our AFS broker example, n_i and n_j are both equal to $R = 6$. Note the plus-or-minus sign: when the plus sign is applied, we get C_u in Figure 4.2, and when the minus sign is applied we get C_l.

To find a confidence interval for the difference between two of the column means, we use equation (3.3), which takes the difference in the $\overline{Y}$'s and, essentially, adds to and subtracts from that difference the same quantity as added to and subtracted from zero in (4.1):

$$t_{1-\alpha/2}\sqrt{\text{MSW}}\sqrt{1/n_i + 1/n_j}$$

In (3.3), $\sqrt{\text{MSW}}$ was written as *s*, representing the general case of the square root of the pooled variance estimate.

In our example, MSW = 21.2, $df = 25$, and, as noted, the number of replicates in each column is $n_i = n_j = R = 6$. With $\alpha = .05$, the acceptance region becomes

$$0 \pm 2.060\sqrt{21.2}\sqrt{1/6 + 1/6} \quad \text{or} \quad 0 \pm 5.48$$

The *t* value of 2.060 comes directly from a *t* table with 25 degrees of freedom (see Table 4.3). The value, 5.48, is called Fisher's **least significant difference (LSD).** Indeed, it is accurately named, for this is the smallest difference between two column means (in the AFS broker example) that would indicate a significant result—the rejection of the hypothesis of the equality of the two true means. With the same number of data points, *R*, in each column, the LSD formula reduces to

$$\text{LSD} = t_{1-\alpha/2}\sqrt{2\text{MSW}/R}$$

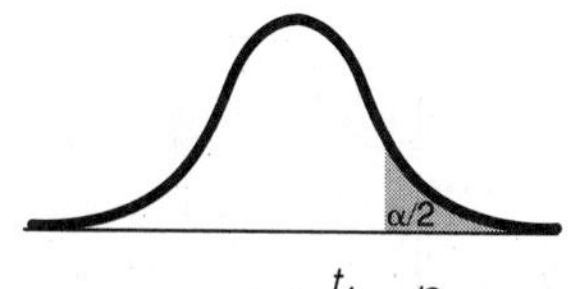

Table 4.3 *t* Table

df	$\alpha/2 = .10$	$\alpha/2 = .05$	$\alpha/2 = .025$	$\alpha/2 = .010$	$\alpha/2 = .005$
1	3.078	6.314	12.706	31.821	63.657
2	1.886	2.920	4.303	6.965	9.925
3	1.638	2.353	3.182	4.541	5.841
4	1.333	2.132	2.776	3.747	4.604
5	1.476	2.015	2.571	3.365	4.032
6	1.440	1.943	2.447	3.143	3.707
7	1.415	1.895	2.365	2.998	3.499
8	1.397	1.860	2.306	2.896	3.355
9	1.383	1.833	2.262	2.821	3.250
10	1.372	1.812	2.228	2.764	3.169
11	1.363	1.796	2.201	2.718	3.106
12	1.356	1.782	2.179	2.681	3.055
13	1.350	1.771	2.160	2.650	3.012
14	1.345	1.761	2.145	2.624	2.977
15	1.341	1.753	2.131	2.602	2.947
16	1.337	1.746	2.120	2.583	2.921
17	1.333	1.740	2.110	2.567	2.898
18	1.330	1.734	2.101	2.552	2.878
19	1.328	1.729	2.093	2.539	2.861
20	1.325	1.725	2.086	2.528	2.845
21	1.323	1.721	2.080	2.518	2.831
22	1.321	1.717	2.074	2.508	2.819
23	1.319	1.714	2.069	2.500	2.807
24	1.318	1.711	2.064	2.492	2.797
25	1.316	1.708	2.060	2.485	2.787
26	1.315	1.706	2.056	2.479	2.779
27	1.314	1.703	2.052	2.473	2.771
28	1.313	1.701	2.048	2.467	2.763
29	1.311	1.699	2.045	2.462	2.756
30	1.310	1.697	2.042	2.457	2.750
40	1.303	1.684	2.021	2.423	2.704
60	1.296	1.671	2.000	2.390	2.660
120	1.289	1.658	1.980	2.358	2.617
∞	1.282	1.645	1.960	2.326	2.576

Source: E. S. Pearson and H. O. Hartley (1966), "Probability Points of the *t* Distribution with ν Degrees of Freedom," *Biometrika Tables for Statisticians,* vol. 1, 3d ed. Reprinted with permission of Oxford University Press.

Now, to complete the use of Fisher's LSD method, we compare each pairwise difference to the LSD. In Figure 4.3a, we show a box plot for the data; in Figure 4.3b, we show a plot of the means for the same data. However, the LSD process is facilitated by first arranging the column means in ascending order:

Column:	3	1	2	4	5
Mean:	5	6	12	14	17

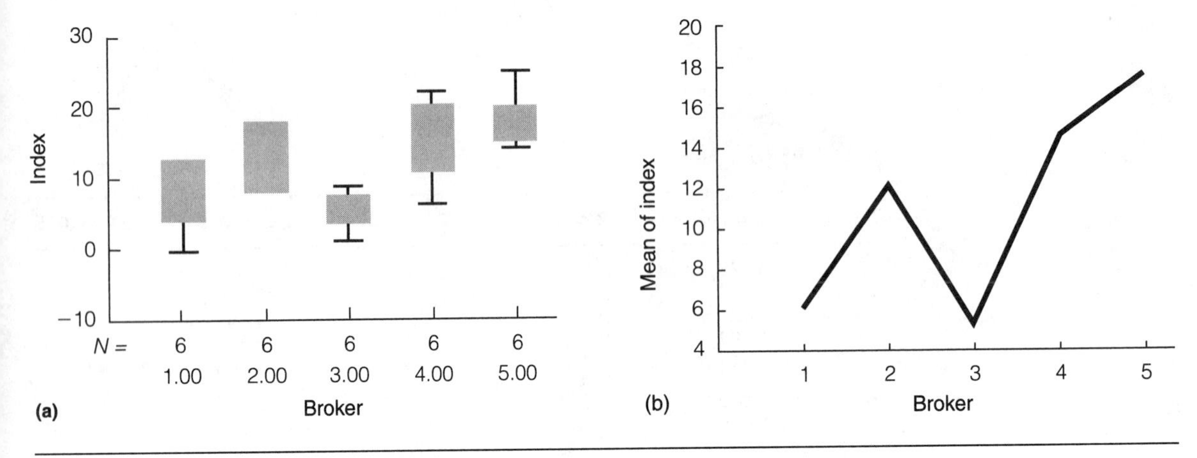

Figure 4.3 Plots for AFS broker study: (a) box plot, (b) mean plot.

Next, we examine the column means and identify any differences in *adjacent* column means that exceed our LSD of 5.48, marking them with a vertical bar between them. (Note that we carry the column identity through the analysis—just another example of "practicing safe statistics.") The only adjacent column means with a difference greater than 5.48 are the 6 and 12:

3	1	|	2	4	5
5	6	|	12	14	17

In this notation, if there are any vertical bars, all column means to one side of the vertical bar are significantly different from all column means to the other side of the vertical bar.

Next, we compare with the LSD the difference in column means for each pair of means within each subset:

Columns Compared	Difference in Means	Difference < or > LSD
3 vs. 1	*	<
2 vs. 4	*	<
2 vs. 5	5	<
4 vs. 5	*	<

*Adjacent columns; comparison already done

Note that the preceding table has an asterisk for adjacent comparisons; this is to reflect that if they were significantly different, the fact would have been noted at the earlier stage when adjacent column means were first examined.

Now to draw the conclusions: We would say that brokers 1 and 3 are equivalent (their difference is not statistically significant, and, hence, we cannot reject their equality) with respect to buying price index, Y; similarly for brokers in the second subset (brokers 2, 4, and 5). However, the two subsets of columns (brokers) *are* concluded to be different. We can show this diagrammatically:

Conclusion: (3 1) (2 4 5)

We read this as follows: "Means three and one are equal, but are different from (smaller than) means two, four, and five, these latter three means being equal."[4]

Sadly, it must be said that things are not always so clean. Suppose that the mean for column five were a bit larger—18 instead of 17. The analysis would look as follows:

3	1	2	4	5
5	6	12	14	18

Conclusion: (3 1) (2 (4) 5)

Columns Compared	Difference in Means	Difference vs. LSD
3 vs. 1	*	<
2 vs. 4	*	<
2 vs. 5	6	>
4 vs. 5	*	<

The preceding table appears to foster an inconsistency: "column mean two is equal to column mean four, column mean four is equal to column mean five, but column mean two is different from column mean five. With multiple comparisons, we cannot impute transitivity. Think of the example of a series of streets that are next to each other; we might describe each pair of adjacent streets as near each other, but the two farthest apart might well *not* be described as near each other. The conclusions literally read: "Column mean three and column mean one are equal, but different from (smaller than) column means two, four, and five. Column means two and four are equal, and columns four and five are equal, but column mean two is different from (smaller than) column mean five." However, as a practical matter, we would conclude that column means three and one are equal and different from (smaller than) the other three column means; column mean two is different from (smaller than) column mean five; we do not get a clear indication concerning the whereabouts of column mean four, except that it's between column means two and five, inclusive.

As we indicated earlier, Fisher's least significant difference test starts with a fixed value for Type I error probability (for each *comparison*). In this case, with $\alpha = .05$, the probability of at least one false rejection, given that all H_0's are true, would be an experimentwise error rate of .401 if the comparisons were independent; of course, they're not, and the true experimentwise error rate is not straightforward to obtain.

EXAMPLE 4.4 Using Excel to Perform a *t* Test and Fisher's LSD

The Excel software package does not routinely include multiple-comparison testing of any kind. However, it does include the *t* test for a single pair of columns (and in theory, one could do this *t* test for each pair of columns). We

Table 4.4 **Excel Output: *t* Test for AFS Broker Study**

12	7		
3	17		
5	13		
−1	11		
12	7		
5	17		
	t-Test: Two-Sample Assuming Equal Variances		
		Variable 1	*Variable 2*
	Mean	6	12
	Variance	26.4	20.4
	Observations	6	6
	Pooled Variance	23.4	
	Hypothesized Mean Difference	0	
	df	10	
	t	−2.1483446	
	P(T<=t) one-tail	0.02861202	
	t Critical one-tail	1.81246151	
	P(T<=t) two-tail	0.05722404	
	t Critical two-tail	2.22813924	

illustrate the use of this *t* test in Excel solely because of its likelihood of availability to virtually all readers; add-ins produced by several companies do include most of the techniques discussed later in the chapter, and this section may help with the application of these add-ins. We use SPSS later to illustrate all of the techniques in the chapter.

The *t* test for columns one and two is illustrated in Table 4.4; we choose to illustrate the *t* test assuming equal variances, as noted in the body of the computer output. As we mentioned earlier, this is in keeping with the assumption made in Fisher's LSD analysis. Note that the pooled variance equals 23.4, a value not equal to the MSW used earlier in Fisher's LSD of 21.2; indeed, the pooled variance will be different for each pair tested. Commensurate with this fact is that the degrees-of-freedom value of the variance estimate is only 10 (see Table 4.4), instead of 25 as in the earlier LSD analysis; correspondingly, a lower value of power would result if a series of unconnected t tests were to be conducted.

The output in Table 4.4 indicates that at $\alpha = .05$, we *accept* the equality of column means one and two (though it's close—the p value of .057 is only slightly above .05). The output doesn't literally determine an LSD for the difference in column means, but it does the algebraic equivalent of finding a t_{calc} of -2.148 (simply called "t" in Table 4.4), equal to the difference in column means divided by the square root of (the pooled variance times the quantity $[1/n_1 + 1/n_2]$), and comparing it with the "t critical" (our $t_{1-\alpha/2}$) value of 2.228. It properly uses the value for a two-tailed test, which implies another "t critical" of -2.228. This result would seem to contradict our conclusion from the previous LSD analysis that column means one and two were different.

It is instructive to consider why the conclusions here are different. One reason is that the single t test uses a slightly higher variance estimate of 23.4 instead of the 21.2 of the earlier LSD analysis (the latter is the average variance estimate for the *five* columns, not simply the average of the two columns being tested). However, a second reason is that with only 10 degrees of freedom, a larger difference in observed column means is necessary to conclude that there is a real difference, retaining the same probability of Type I error; this is related to the fact that the single t test has less power—if there is a real difference, this test is less likely to reveal it. Finally, as noted earlier, the accept/reject decision was a close one.

This concludes our examination of Fisher's LSD test (although we revisit its illustration later in the chapter in the SPSS section). It is the one pairwise comparison procedure that directly considers the Type I error as a *per comparison* error rate. The remaining pairwise comparison (and exploratory comparison) procedures focus on the experimentwise error rate.

Tukey's Honestly Significant Difference Test

The honestly significant difference (HSD) test, devised by J. W. Tukey, focuses on the experimentwise error rate, **a.** Except for the value against which we compare the difference in column means to determine significance, the HSD test works in a manner similar to that of the LSD test in the previous section. That there is any difference at all results from the focus on the experimentwise error rate instead of on the comparison error rate. There are two different, equivalent approaches to applying the HSD test. One resembles the approach used in the LSD discussion, and simply replaces $t_{1-\alpha/2}$ in the formula for LSD,

$$\text{LSD} = t_{1-\alpha/2}\sqrt{2\text{MSW}/R}$$

by a special Tukey t value, which could be notated as $\text{tuk}_{1-\mathbf{a}/2}$. Its derivation could be described and a table of these values, as a function of **a,** provided. This would give us the HSD value of

$$\text{HSD} = \text{tuk}_{1-\mathbf{a}/2}\sqrt{2\text{MSW}/R}$$

The table of $\text{tuk}_{1-\mathbf{a}/2}$ values would be those that control **a** at the desired value. Although this approach seems to the authors to have great pedagogical virtue, especially after having introduced the LSD technique, the vast majority of discussions of the HSD form the same numerical value of HSD as above, but by viewing it as

$$\text{HSD} = q_{1-\alpha/2}\sqrt{\text{MSW}/R}$$

where $q_{1-\alpha/2} = (\text{tuk}_{1-\mathbf{a}/2})\cdot\sqrt{2}$. (The traditional notation for indexing q uses the symbol α in place of **a,** but this α is meant to represent the **experimentwise** error rate.) In other words, instead of simply replacing $t_{1-\alpha/2}$ by $\text{tuk}_{1-\mathbf{a}/2}$, we replace $(t_{1-\alpha/2})\sqrt{2}$ by $q_{1-\alpha/2}$. In a sense, we can say that the table value of q includes the multiplier $\sqrt{2}$.

To repeat, the HSD procedure works the same way as does the LSD procedure (including the rank-order step and so on—we illustrate it below), except that the differences between pairs of column means are compared to HSD, not LSD, and

$$\text{HSD} = q_{1-\alpha/2}\sqrt{\text{MSW}/R}$$

A few words about $q_{1-\alpha/2}$ are in order. The difference between the largest and smallest of a set of column means, that is, the *range* of the column means, under the null hypothesis assumption that all true column means are equal, depends on how many column means there are, as well as on the standard deviation of the data. The "how many column means under study" question comes into the discussion in a natural way; if we sample two people, the range of heights of the two is their height difference. If we add another person to the mix, the range of heights can only stay the same or get larger. Surely, we'd expect the range in heights of 100 people (maximum height minus minimum height) to be much larger than the range of heights of only two randomly chosen people. This concept is related to the primary ideas in the field called order statistics. The random variable, q, is defined, with $\bar{Y}$'s as column means and R as the number of replicates per column, as

$$q = (\bar{Y}_{\text{largest}} - \bar{Y}_{\text{smallest}}) \;/\; \sqrt{\text{MSW}/R}$$

The probability distribution of q is called the Studentized range distribution. The percentiles of the distribution of q, $q_{1-\alpha/2}$ being the $(1 - \alpha/2)$th percentile, depend on how many columns are under study and, similar to the t distribution, how many degrees of freedom, $C(R-1)$, are associated with the MSW. Therefore, sometimes, $q_{1-\alpha/2}$ is written as $q(C, df)_{1-\alpha/2}$, and

$$\text{HSD} = q(C, df)_{1-\alpha/2}\sqrt{\text{MSW}/R}$$

Table 4.5 is a Studentized range table for various values of C, df, and α.

Table 4.5 **Studentized Range Table**

Error df	a	Number of Treatment Means 2	3	4	5	6	7	8	9	10	11
5	.05	3.64	4.60	5.22	5.67	6.03	6.33	6.58	6.80	6.99	7.17
	.01	5.70	6.98	7.80	8.42	8.91	9.32	9.67	9.97	10.24	10.48
6	.05	3.46	4.34	4.90	5.30	5.63	5.90	6.12	6.32	6.49	6.65
	.01	5.24	6.33	7.03	7.56	7.97	8.32	8.61	8.87	9.10	9.30
7	.05	3.34	4.16	4.68	5.06	5.36	5.61	5.82	6.00	6.16	6.30
	.01	4.95	5.92	6.54	7.01	7.37	7.68	7.94	8.17	8.37	8.55
8	.05	3.26	4.04	4.53	4.89	5.17	5.40	5.60	5.77	5.92	6.05
	.01	4.75	5.64	6.20	6.62	6.96	7.24	7.47	7.68	7.86	8.03
9	.05	3.20	3.95	4.41	4.76	5.02	5.24	5.43	5.59	5.74	5.87
	.01	4.60	5.43	5.96	6.35	6.66	6.91	7.13	7.33	7.49	7.65
10	.05	3.15	3.88	4.33	4.65	4.91	5.12	5.30	5.46	5.60	5.72
	.01	4.48	5.27	5.77	6.14	6.43	6.67	6.87	7.05	7.21	7.36
11	.05	3.11	3.82	4.26	4.57	4.82	5.03	5.30	5.35	5.49	5.61
	.01	4.39	5.15	5.62	5.97	6.25	6.48	6.67	6.84	6.99	7.13
12	.05	3.08	3.77	4.20	4.52	4.75	4.95	5.12	5.27	5.39	5.51
	.01	4.32	5.05	5.50	5.84	6.10	6.32	6.51	6.67	6.81	6.94
13	.05	3.06	3.73	4.15	4.45	4.69	4.88	5.05	5.19	5.32	5.43
	.01	4.26	4.96	5.40	5.73	5.98	6.19	6.37	6.53	6.67	6.79
14	.05	3.03	3.70	4.11	4.41	4.64	4.83	4.99	5.13	5.25	5.36
	.01	4.21	4.89	5.32	5.63	5.88	6.08	6.26	6.41	6.54	6.66
15	.05	3.01	3.67	4.08	4.37	4.59	4.78	4.94	5.08	5.20	5.31
	.01	4.17	4.84	5.25	5.56	5.80	5.99	6.16	6.31	6.44	6.55
16	.05	3.00	3.65	4.05	4.33	4.56	4.74	4.90	5.03	5.15	5.26
	.01	4.13	4.79	5.19	5.49	5.72	5.92	6.08	6.22	6.35	6.46
17	.05	2.98	3.63	4.02	4.30	4.52	4.70	4.86	4.99	5.11	5.21
	.01	4.10	4.74	5.14	5.43	5.66	5.85	6.01	6.15	6.27	6.38
18	.05	2.97	3.61	4.00	4.28	4.49	4.67	4.82	4.96	5.07	5.17
	.01	4.07	4.70	5.09	5.38	5.60	5.79	5.94	6.08	6.20	6.31
19	.05	2.96	3.59	3.98	4.25	4.47	4.65	4.79	4.92	5.04	5.14
	.01	4.05	4.67	5.05	5.33	5.55	5.73	5.89	6.02	6.14	6.25
20	.05	2.95	3.58	3.96	4.23	4.45	4.62	4.77	4.90	5.01	5.11
	.01	4.02	4.64	5.02	5.29	5.51	5.69	5.84	5.97	6.09	6.19
24	.05	2.92	3.53	3.90	4.17	4.37	4.54	4.68	4.81	3.92	5.01
	.01	3.96	4.55	4.91	5.17	5.37	5.54	5.69	5.81	5.92	6.02
30	.05	2.89	3.49	3.85	4.10	4.30	4.46	4.60	4.72	4.82	4.92
	.01	3.89	4.45	4.80	5.05	5.24	5.40	5.54	5.65	5.76	5.85
40	.05	2.86	3.44	3.79	4.04	4.23	4.39	4.52	4.63	4.73	4.82
	.01	3.82	4.37	4.70	4.93	5.11	5.26	5.39	5.50	5.60	5.69
60	.05	2.83	3.40	3.74	3.98	4.16	4.31	4.44	4.55	4.65	4.73
	.01	3.76	4.28	4.59	4.82	4.99	5.13	5.25	5.36	5.45	5.53
120	.05	2.80	3.36	3.68	3.92	4.10	4.24	4.36	4.47	4.56	4.64
	.01	3.70	4.20	4.50	4.71	4.87	5.01	5.12	5.21	5.30	5.37
∞	.05	2.77	3.31	3.63	3.86	4.03	4.17	4.29	4.39	4.47	4.55
	.01	3.64	4.12	4.40	4.60	4.76	4.88	4.99	5.08	5.16	5.23

Source: E. S. Pearson and H. O. Hartley (1966), *Biometrika Tables for Statisticians*, vol. 1, 3d ed. Reprinted with permission of Oxford University Press.

The HSD test is very conservative in that it determines the HSD for comparison with the difference in the two most extreme column means ($\overline{Y}_{\text{largest}} - \overline{Y}_{\text{smallest}}$) and uses it as the benchmark for *all* pairwise differences. Hence, it is possible that a difference other than that of the two extremes, but which is nearly as large as the difference in the two extremes, may not be judged as contradicting H_0, as it might be judged under a less conservative test in which the critical difference to reject H_0 is set at a more appropriate value.

EXAMPLE 4.5 Tukey's HSD Test for the Broker Study

We now illustrate the HSD test for our AFS broker study example. We begin the same way we began for the LSD test, by writing the column means in ascending order.

Column:	3	1	2	4	5
Mean:	5	6	12	14	17

However, now we must compute the HSD; note that $C = 5$, $df = 25$, MSW = 21.2, $R = 6$, and suppose we choose **a** = .05 (recall that **a** is called α in the q expression):

$$\begin{aligned} \text{HSD} &= q(C, df)_{1-\alpha/2}\sqrt{\text{MSW}/R} \\ &= (4.16)\sqrt{21.2/6} \\ &= 7.82 \end{aligned}$$

(The value 4.16 is linearly interpolated from the $df = 24$ value, 4.17, and the $df = 30$ value, 4.10.)

Next, as we did for the LSD, we examine the column means and identify any differences in *adjacent* column means that exceed our HSD of 7.82. If we find any, we note this fact with a vertical bar. In this case, no adjacent column means have a difference greater than 7.82.

Thus we need to compare each of the 10 differences in column means with the HSD, as shown in Table 4.6. What is the story that Table 4.6 tells? If we ignore the column-two mean for the moment, we conclude that column means three and one are the same, but different from (smaller than) column means four and five, which are themselves the same; that is,

Conclusion: (3 1) (4 5)

However, column mean two is "the same as column means three and one," but also "the same as column means four and five." (As pointed out in the end-of-chapter note 4, this type of phrasing is more concise than it is elegant.) We have an inconsistency of the type expressed earlier. In a consulting capacity, we would tell a client that "column means three and one cannot be said to be different, column means four and five cannot be said to be different, but the former two can be said to be different from (smaller than) the latter two; we

Table 4.6 **Tukey's HSD Test**

Columns Compared	Difference in Means	Difference vs. 7.82	Reject Equality
3 vs. 1	*	<	
3 vs. 2	7	<	
3 vs. 4	9	>	Yes
3 vs. 5	12	>	Yes
1 vs. 2	*	<	
1 vs. 4	8	>	Yes
1 vs. 5	11	>	Yes
2 vs. 4	*	<	
2 vs. 5	5	<	
4 vs. 5	*	<	

*Adjacent columns; comparison already done

cannot determine the role of column mean two." We would express this thought diagrammatically as

It should be noted that one potential reason for obtaining different results from those of the Fisher's LSD analysis is that the per comparison error rates are very different. In the immediately preceding Tukey HSD analysis, with 10 comparisons and **a** = .05, each *comparison* error rate is, relatively speaking, very small; if they were independent comparisons, which they're not, each would be only .0051. The per comparison error rate for the Fisher's LSD analysis was $\alpha = .05$, about 10 times larger.

As we mentioned earlier, Tukey's HSD test uses for all comparisons the HSD based on the difference in the two extreme means. It is, therefore, considered by many to be too conservative, yielding fewer significant results than might be warranted (that is, it is said to be *negatively biased*). For this reason, the Newman-Keuls test, which we cover next, is considered by many to be preferable to Tukey's HSD test.

Newman-Keuls Test with Example

The Newman-Keuls test[5] is an alternative to Tukey's HSD test. It is similar to the HSD test in that it uses the same Studentized range distribution table. However, it differs from the HSD test in that it does not use the same value of $q_{1-\alpha/2} = q(C, df)_{1-\alpha/2}$ for all comparisons, but uses a value, $q(s, df)_{1-\alpha/2}$, that takes into account how many "steps," s, separate the two means being compared after the means are rank-ordered. If, as in our AFS broker study example,

$C = 5$, then when comparing the two extreme column means, $s = 5$. If we label our rank-ordered means in ascending order as means I, II, III, IV, and V (for the moment, irrespective of original column label), we then use $q(5, df)_{1-\alpha/2}$ when comparing means I and V (five steps apart, inclusive), $q(4, df)_{1-\alpha/2}$ when comparing means I and IV or means II and V (four steps apart, inclusive), $q(3, df)_{1-\alpha/2}$ when comparing means I and III or means II and IV or means III and V (three steps apart, inclusive), and lastly, $q(2, df)_{1-\alpha/2}$ (two steps apart, inclusive) when comparing the sets of adjacent means.

Thus, the format for performing the Newman-Keuls test is very much like that of the HSD procedure, except that we compare different differences to different benchmarks. In our example, with $C = 5$, $df = 25$, and again choosing **a** =.05, we have (again, with interpolation):

$q(5, df)_{1-\alpha/2} = 4.16$ (same as in the HSD analysis)

$q(4, df)_{1-\alpha/2} = 3.89$

$q(3, df)_{1-\alpha/2} = 3.52$

$q(2, df)_{1-\alpha/2} = 2.91$

With MSW = 21.2 and $R = 6$, we multiply each of these q values by $\sqrt{21.2/6} = 1.880$, as we did in the HSD analysis, getting the NKD (our notation, standing for Newman-Keuls difference) values of

$s = 5$, NKD = 7.82

$s = 4$, NKD = 7.31

$s = 3$, NKD = 6.62

$s = 2$, NKD = 5.47

Table 4.7 shows the Newman-Keuls test results for the same rank-ordered column means as earlier, repeated here for convenience:

Column:	3	1	2	4	5
Mean:	5	6	12	14	17

We observe that the Newman-Keuls test finds two results that are different from those of the HSD test: column means three and two, as well as column means one and two, are now concluded to be different, whereas the HSD test did not conclude that they were different. In some sense, it is no surprise that the Newman-Keuls test indicated more significant comparisons than Tukey's HSD test. As noted earlier, each benchmark used in the Newman-Keuls test is less than or equal to its counterpart comparison when performing Tukey's HSD test; hence, the Newman-Keuls test will always yield an equal or greater number of significant comparisons.

The Newman-Keuls results happen to come out much cleaner than the Tukey HSD results, and now happen to equal the results yielded by the earlier Fisher's LSD test:

Conclusion: (3 1) (2 4 5)

Table 4.7 Newman-Keuls Test

Columns Compared	Difference in Means	Difference vs. NKD	Reject Equality
3 vs. 1	1	< (5.47)	
3 vs. 2	7	> (6.62)	Yes
3 vs. 4	9	> (7.31)	Yes
3 vs. 5	12	> (7.82)	Yes
1 vs. 2	6	> (5.47)	Yes
1 vs. 4	8	> (6.62)	Yes
1 vs. 5	11	> (7.31)	Yes
2 vs. 4	2	< (5.47)	
2 vs. 5	5	< (6.62)	
4 vs. 5	3	< (5.47)	

That is, column means three and one are the same, and are different from (smaller than) column means two, four, and five, the latter three column means being the same.

The Newman-Keuls test has one minor caveat that needs to be mentioned. For the procedure to actually control the experimentwise error rate at the stated value, it is necessary that any time a comparison with a certain number of steps apart, *s*, is found to be nonsignificant, all comparisons with a smaller value of *s*, within these bounds, are automatically judged to be nonsignificant. In other words, no significant differences can be concluded within the bounds of an already-judged difference that is nonsignificant.

Two Other Tests Comparing All Pairs of Column Means

We briefly mention two other pairwise comparison tests that are similar to the three tests we have examined in detail. Tukey, perhaps in reaction to the Newman-Keuls test, has suggested a middle ground between his HSD test and the Newman-Keuls test, using the average of the HSD and the NKD as the benchmark to establish significance. This is called Tukey's wholly significant difference (WSD). In our AFS broker study example, we have, at **a** = .05,

s	HSD	NKD	WSD
5	7.82	7.82	**7.82**
4	7.82	7.31	**7.57**
3	7.82	6.62	**7.22**
2	7.82	5.47	**6.64**

In fact, using the WSD in this example happens to produce the same results as the HSD test.

D. B. Duncan argues that both the Tukey and Newman-Keuls tests tend to use **a** = .05, and that, therefore, the individual comparison error rate is quite small, smaller than traditional (indeed, we commented on this α earlier). Duncan suggests that the value of **a** should be

$$\mathbf{a} = 1 - (1 - \alpha)^{s-1}$$

where s is the number of steps, as defined in the Newman-Keuls test. You may recall that the above formula correctly relates **a** to α only when the tests are all independent of one another; Duncan, of course, knew that the tests to which his technique would normally be applied are not independent, but nevertheless suggested this choice. The procedures followed in Duncan's test mirror those of the Newman-Keuls test, especially the proceeding to a comparison with a lower value of s only when finding a significant result. What singles out Duncan's test is that it requires values of the Studentized range tables that are not the usual ones of $\mathbf{a} = .05$ or .01, but are values of **a** such as .098 ($\alpha = .05$, $s = 3$), .143 ($\alpha = .05$, $s = 4$), and .185 ($\alpha = .05$, $s = 5$). Tables for use with Duncan's test can be found in D. B. Duncan, "Multiple Range and Multiple F Tests," *Biometrics,* 1955, 11: 1–42.

The choice of which of the all-pairwise comparisons tests to use is, of course, up to the individual experimenter. However, the Newman-Keuls test appears to be recommended by textbooks and other sources more than any other single choice.

Dunnett's Test

Let's consider one further pairwise comparison test. What distinguishes it from the others of this section is that it does not test *all* pairs (or potentially all pairs)—only a specific subset of pairs. Devised by C. W. Dunnett, this procedure considers one of the columns as a "control," or "control group," and addresses only the comparison of each other column mean to this control column mean. For example, if there are five columns, Dunnett's test conducts four comparisons, not the ten comparisons that would ensue were all pairwise comparisons to be performed. Naturally, with fewer comparisons, and a different pattern of interdependency among those comparisons, the same experimentwise error rate, **a,** in Dunnett's test yields a set of table value benchmarks that are different from the other tests that perform all pairwise comparisons.

Dunnett's test is conducted similarly to Fisher's LSD test, in that the Dunnett difference (Dut-D, our notation) can be considered as simply replacing the $t_{1-\alpha/2}$ in the LSD formula,

$$\text{LSD} = t_{1-\alpha/2}\sqrt{2\text{MSW}/R}$$

with $\text{Dut}_{1-\mathbf{a}/2}$:

$$\text{Dut-D} = \text{Dut}_{1-\mathbf{a}/2}\sqrt{2\text{MSW}/R}$$

where the $\text{Dut}_{1-\mathbf{a}/2}$ value is taken from Table 4.8.

Table 4.8 Dunnett's Table, $a = .05$

	One Control and Number of Treatments								
df	1 & 1	1 & 2	1 & 3	1 & 4	1 & 5	1 & 6	1 & 7	1 & 8	1 & 9
5	2.57	3.03	3.29	3.48	3.62	3.73	3.82	3.90	3.97
6	2.45	2.86	3.10	3.26	3.39	3.49	3.57	3.64	3.71
7	2.36	2.75	2.97	3.12	3.24	3.33	3.41	3.47	3.53
8	2.31	2.67	2.88	3.02	3.13	3.22	3.29	3.35	3.41
9	2.26	2.61	2.81	2.95	3.05	3.14	3.20	3.26	3.32
10	2.23	2.57	2.76	2.89	2.99	3.07	3.14	3.19	3.24
11	2.20	2.53	2.72	2.84	2.94	3.02	3.08	3.14	3.19
12	2.18	2.50	2.68	2.81	2.90	2.98	3.04	3.09	3.14
13	2.16	2.48	2.65	2.78	2.87	2.94	3.00	3.06	3.10
14	2.14	2.46	2.63	2.75	2.84	2.91	2.97	3.02	3.07
15	2.13	2.44	2.61	2.73	2.82	2.89	2.93	3.00	3.04
16	2.12	2.42	2.59	2.71	2.80	2.87	2.92	2.97	3.02
17	2.11	2.41	2.58	2.69	2.78	2.85	2.90	2.95	3.00
18	2.10	2.40	2.56	2.68	2.76	2.83	2.89	2.94	2.98
19	2.09	2.39	2.55	2.66	2.75	2.81	2.87	2.92	2.96
20	2.09	2.38	2.34	2.63	2.73	2.80	2.80	2.90	2.95
24	2.06	2.35	2.31	2.61	2.70	2.76	2.81	2.86	2.90
30	2.04	2.32	2.47	2.58	2.66	2.72	2.77	2.82	2.86
40	2.02	2.29	2.44	2.54	2.62	2.68	2.73	2.77	2.81
60	2.00	2.27	2.41	2.51	2.58	2.64	2.69	2.73	2.77
120	1.98	2.24	2.38	2.47	2.55	2.60	2.65	2.69	2.73
∞	1.96	2.21	2.35	2.44	2.51	2.57	2.61	2.65	2.69

Source: C. W. Dunnett, "A Multiple Comparison Procedure for Comparing Several Treatments with a Control." *Journal of the American Statistical Association,* 50, 1995, pp. 1096–1121. Reprinted with permission.

EXAMPLE 4.6 Dunnett's Test for the Broker Study

We illustrate Dunnett's test using our AFS broker study example, assuming broker one (that is, column one) is the control. This would correspond with a situation in which broker one is the default broker, used the majority of the time, and a switch will be made only if there is overwhelming evidence that another broker is superior. For an experimentwise error rate of .05, $df = 25$, and four columns (brokers, or in Dunnett's terms, "treatments") in addition to the control, we have $\text{Dut}_{1-\mathbf{a}/2} = 2.61$. This gives us

$$\begin{aligned}\text{Dut-D} &= \text{Dut}_{1-\mathbf{a}/2}\sqrt{2\text{MSW}/R}\\ &= 2.61\sqrt{2 \cdot 21.2/6}\\ &= 6.94\end{aligned}$$

The results are in Table 4.9, with the columns and means repeated here for convenience. Note that there is no need to rank-order the column means.

Table 4.9 Dunnett's Test for AFS Broker Study

Columns Compared	Difference in Means	Difference vs. 6.94	Reject Equality
1 vs. 2	6	<	
1 vs. 3	1	<	
1 vs. 4	8	>	Yes
1 vs. 5	11	>	Yes

Column:	1	2	3	4	5
Mean:	6	12	5	14	17

We conclude that brokers 2 and 3 are not significantly different from the "control broker"—broker 1 (although, in the case of broker 2, it is very close), but that brokers 4 and 5 are, indeed, significantly different from (larger than) the control broker.

4.4 Post Hoc Exploratory Comparisons— The Scheffé Test

Sometimes the data themselves suggest comparisons. That is, the experimenters had no anticipation at all that a specific set of column means should be compared with another specific set of column means, but after examining the results, they notice something that appears to be an identifiable pattern of differences, and decide to see if the differences are "statistically significant." For example, suppose that, in the AFS broker study example discussed in this chapter, the experimenters, just by chance, noticed that broker 1 was the only female broker of the five, and had nearly the lowest mean of the five, and had nearly twice as high a standard deviation as the other low-scoring broker. Further suppose that as a consequence of this realization, the experimenters decide to test whether column mean one differs from the average of the other four column means.

What is the potential statistical problem with this test? The problem is that the decision to conduct the test was **post hoc,** that is, based at least partly on a finding from the data "after the fact." Presumably the experimenters had no intention to test the impact of gender. Likely, the issue was of interest only because the female column mean was overtly smaller than all but one other column mean, and the data in the column had a higher standard deviation than that for the other low-scoring broker. Obviously, comparisons determined to be interesting after the fact often reflect surprisingly large differences. Testing one of these interesting differences as if it were a planned test is capitalizing too much on a chance occurrence; the probability of Type I error is greatly magnified above its nominal value. This kind of "data snooping," combined with the fact that an infinite number of comparisons can be made among a set of column means (when we do not limit the comparisons to pairwise comparisons, and include all comparisons of one linear combination of the column means to another linear combination of the column means), requires that we do something to acknowledge (and control) the Type I error potential.[6]

A real-world example of an analysis tainted by data snooping is an article one of the authors saw in a newspaper that had the headline, "Incidence of certain diseases later in life depends what month you're born in." The article went on to report the results of a hypothesis test (believed to have been a χ^2 goodness of fit test—the type of test wasn't perfectly clear from the article) that showed a statistically significant relationship between birth month and incidence of disease for two diseases: cancer and schizophrenia. Not believing the conclusions in general, and noting that the two diseases put forth were somewhat far apart along the spectrum of ill health, the author investigated further (by reading the article in detail and trying to infer what analysis was actually performed). His conclusion was that the experimenter tested for a relation between each of about 250 different diseases and/or combinations of diseases and birth month, finding a strong relationship for these two diseases in particular; for all 250 tests, the experimenter used the routine critical value *with an individual comparison error rate of .01!* Is there any wonder that the author of the article found two tests/diseases highly significant?

H. Scheffé has developed a test for controlling the experimentwise error rate at whatever value is desired, in a situation in which all of the infinite possible comparisons among the column means (not just pairs of column means) take place, and regardless of how these comparisons are chosen. The Scheffé test can be based on the F distribution or the t distribution, both of which were discussed earlier. We illustrate it using the F distribution, since that is how it is usually introduced.

Carrying Out the Test

Scheffé's procedure tests the following hypotheses:

$$H_0\text{: } L = 0$$

$$H_1\text{: } L \neq 0$$

where L is, in a sense, *any* linear combination of the true column means. Of course, a pairwise comparison can be written in this form; for example, to compare column means one and two when there are five columns:

$$L = 1(\mu_1) + -1(\mu_2) + 0(\mu_3) + 0(\mu_4) + 0(\mu_5)$$

The above L is simply $\mu_1 - \mu_2$. The general case, when there are C column means, is

$$L = a_1(\mu_1) + a_2(\mu_2) + \cdots + a_C(\mu_C)$$

The test statistic is

$$L' = a_1(\overline{Y}_1) + a_2(\overline{Y}_2) + \cdots + a_C(\overline{Y}_C)$$

The critical value that the absolute value of L' must exceed in order to reject H_0 (and conclude that the difference expressed by L is not equal to zero) depends, in part, on the values of the a_j in the L' expression. This is because the variance of L' can be shown to equal (summation over j from 1 to C).

$$(\sigma^2/R)\ \Sigma(a_j^2)$$

Of course, σ^2 is unknown, and the above expression assumes that the variance is the same for each data point. Dividing σ^2 by R reflects the well-known fact that the variance of a sample mean (of independent data values) is the variance of an individual data value, divided by the number of data values. The summation of the squares of the a_j values is basically derived from the matter of scale (that is, if all the a_j values were doubled, the variance would quadruple, and the standard deviation would double).

As expected, we estimate σ^2 by MSW. The critical value for the absolute value of L' is

$$\sqrt{(\text{MSW}/R) \cdot \Sigma(a_j^2) \cdot (C-1) \cdot F_{1-\mathbf{a}}(df_1, df_2)}$$

where $F_{1-\mathbf{a}}(df_1, df_2)$ stands for the $(1-\mathbf{a})$th percentile of the F distribution (the value of $\mathbf{a}$ is the experimentwise error desired), with numerator $df_1 = C-1$, and denominator $df_2 = C(R-1)$. The degrees of freedom are the same as the degrees of freedom for the original F test. The $(C-1)$ multiplicative factor enters the above formula for the critical value of L' for a subtle reason. The MSB_c has $C-1$ degrees of freedom; the maximum L'^2 value, divided by $(\text{MSW}/R) \cdot \Sigma(a_j^2)$, is equal to the MSB_c, but that can occur only when the particular L' explains all of the variability expressed in the SSB_c. Because the SSB_c is divided by $(C-1)$ degrees of freedom to produce the MSB_c, whereas a comparison has only one degree of freedom, the maximum L'^2 divided by $(\text{MSW}/R) \cdot \Sigma(a_j^2)$ requires a critical F value that is $(C-1)$ times the original $F_{1-\mathbf{a}}(df_1, df_2)$ value.

EXAMPLE 4.7 Scheffé Test for Post Hoc Study

We now apply the Scheffé test to the hypothetical situation mentioned in the opening paragraph of this section: it was noticed that column mean one corresponded to the only female broker, and was nearly the lowest value of all the column means, and the experimenters decided to test whether column mean one was statistically different from the average of the other four column means (that consisted of male brokers).

The test of whether column mean one equals the average of the other columns means or not is essentially taking a comparison:

$$L' = \overline{Y}_1 - (1/4)[\overline{Y}_2 + \overline{Y}_3 + \overline{Y}_4 + \overline{Y}_5]$$

or

$$L' = 1(\overline{Y}_1) - 1/4(\overline{Y}_2) - 1/4(\overline{Y}_3) - 1/4(\overline{Y}_4) - 1/4(\overline{Y}_5)$$

For the column means (in the order 1, 2, 3, 4, 5) of 6, 12, 5, 14, 17, we calculate an absolute value of L' of $|6 - .25(12 + 5 + 14 + 17)| = |6 - 12| = 6$.

And,

$$\Sigma(a_j^2) = 1 + (-1/4)^2 + (-1/4)^2 + (-1/4)^2 + (-1/4)^2$$
$$= 1.25$$

With MSW = 21.2, $R = 6$, $C = 5$, and choosing $\mathbf{a} = .05$, which yields $F_{1-\mathbf{a}}(df_1, df_2) = F_{.95}(4, 25) = 2.76$, we calculate the critical value for L' to be

$$\sqrt{(21.2/6) \cdot 1.25 \cdot 4 \cdot 2.76} = 6.98$$

Because L' equals 6, we accept H_0, and cannot reject the hypothesis that the true column one mean equals the average of the other four true column means.

Discussion of Scheffé Test

If the original F test accepts H_0 (that all the column means are equal), there can never be a comparison for which the Scheffé test rejects H_0 (that the true value of the comparison equals zero). If the original F test rejects its H_0, then there exists at least one comparison for which the Scheffé test would reject its H_0. Of course, in this latter case, the experimenter may not actually choose to test a particular comparison for which the Scheffé test rejects H_0.

Because the Scheffé test goes out of its way to cover all possible comparisons while still retaining an experimentwise error rate of **a,** it has relatively low power. It should not be used for pairwise comparisons (since the other tests presented are more powerful), nor when there are planned comparisons.

The notion of testing whether a linear combination of column means is zero or not is an important one. Indeed, this notion is a good introduction to the next chapter, in which we consider orthogonal contrasts; these are sets of linear combinations of the column means with special properties.

EXAMPLE 4.8 SPSS Examples

In this section, we use the computer software package SPSS to illustrate all of the techniques discussed in this chapter. The data of our AFS broker study are entered into SPSS, and as noted in Chapter 2, in a format different from that of Excel. SPSS, as a *statistical* software package, recognizes that there are two

Table 4.10 AFS Data Input for SPSS

Score	Broker	Score	Broker	Score	Broker	Score	Broker
12.00	1.00	13.00	2.00	3.00	3.00	24.00	5.00
3.00	1.00	11.00	2.00	7.00	3.00	13.00	5.00
5.00	1.00	7.00	2.00	21.00	4.00	14.00	5.00
−1.00	1.00	17.00	2.00	10.00	4.00	18.00	5.00
12.00	1.00	8.00	3.00	15.00	4.00	14.00	5.00
5.00	1.00	1.00	3.00	12.00	4.00	19.00	5.00
7.00	2.00	7.00	3.00	20.00	4.00		
17.00	2.00	4.00	3.00	6.00	4.00		

Table 4.11 SPSS Output: ANOVA for AFS Broker Study

Variable Score
By Variable Broker

Analysis of Variance

Source	D.F.	Sum of Squares	Mean Squares	F Ratio	F Prob.
Between Groups	4	640.8000	160.2000	7.5566	.0004
Within Groups	25	530.0000	21.2000		
Total	29	1170.8000			

variables in this study, a dependent variable and one independent variable; Excel requires the data to be entered as it appears for analysis—five columns of six data values each. In SPSS, the user indicates which variable is the dependent and which is the independent. The variables can also be given their contextual names, as shown in Table 4.10, which lists the data for SPSS. (Table 4.10 is represented as eight columns solely to save space; in SPSS, all the data would be in two long columns.)

The ANOVA results, in SPSS output format, are in Table 4.11.

Fisher's LSD We request Fisher's LSD test; the first part of the output is in Table 4.12. Before we consider the rest of the output, note that if we compute (using the SPSS operator SQRT)

$$3.2558 \cdot 2.91 \cdot \text{SQRT}(1/6 + 1/6)$$

we get (with minor rounding error) the 5.48 value determined earlier in the chapter. The output in Table 4.13 indicates the results. The form of the results

Table 4.12 **SPSS for Macintosh Output: Fisher's LSD Test for Broker Study**

```
  Variable Score
 By Variable Broker

Multiple Range Tests: LSD test with significance level .05

The difference between two means is significant if
        MEAN(J)-MEAN(I) > = 3.2558 * RANGE * SQRT (1/N(I) + 1/N(J))
        with the following value(s) for RANGE: 2.91
```

Table 4.13 **SPSS for Macintosh Output: Fisher's LSD Test, Continued**

```
(*) Indicates significant differences which are shown in the lower triangle

                        G  G  G  G  G
                        r  r  r  r  r
                        p  p  p  p  p

                        3  1  2  4  5

Mean      Broker

 5.000    Grp 3
 6.000    Grp 1
12.0000   Grp 2         *  *
14.0000   Grp 4         *  *
17.0000   Grp 5         *  *

Subset 1
Group     Grp 3     Grp 1
Mean      5.0000    6.0000
- - - - - - - - - - - - - - -
Subset 2
Group     Grp 2     Grp 4     Grp 5
Mean      12.0000   14.0000   17.0000
- - - - - - - - - - - - - - - - - - - -
```

is to provide a matrix that has an asterisk in each place where the columns are judged to be different. Here, the results indicate two groups, as found earlier:

(3, 1)   (2, 4, 5)

In the table, "G r p" stands for the word "group"; note also that the output rank-orders the column means.

Different versions of SPSS yield different formats for the output, although the substance of the results is the same. The format in Tables 4.12 and 4.13

Table 4.14 SPSS 8.0 for Windows Output: Fisher's LSD Test

	(1)	(2)	(3)	(4)	(5)	(6)	(7)
LSD	1	2	−6.00*	2.658	.033	−11.47	−.53
		3	1.00	2.658	.710	−4.47	6.47
		4	−8.00*	2.658	.006	−13.47	−2.53
		5	−11.00*	2.658	.000	−16.47	−5.53
	2	1	6.00*	2.658	.033	.53	11.47
		3	7.00*	2.658	.014	1.53	12.47
		4	−2.00	2.658	.459	−7.47	3.47
		5	−5.00	2.658	.072	−10.47	.47
	3	1	−1.00	2.658	.710	−6.47	4.47
		2	−7.00*	2.658	.014	−12.47	−1.53
		4	−9.00*	2.658	.002	−14.47	−3.53
		5	−12.00*	2.658	.000	−17.47	−6.53
	4	1	8.00*	2.658	.006	2.53	13.47
		2	2.00	2.658	.459	−3.47	7.47
		3	9.00*	2.658	.002	3.53	14.47
		5	−3.00	2.658	.270	−8.47	2.47
	5	1	11.00*	2.658	.000	5.53	16.47
		2	5.00	2.658	.072	−.47	10.47
		3	12.00*	2.658	.000	6.53	17.47
		4	3.00	2.658	.270	−2.47	8.47

was generated from SPSS 5.0 on the Macintosh; the format in Table 4.14 is from SPSS 8.0 for Windows. Table 4.14 column one is "Broker i," which is compared with column two, "Broker j." Column three is the actual difference in means; column four is the standard error of the difference; column five is the p value of the difference (with an * next to a column three value with p value less than .05); columns six and seven are the lower and upper 95% confidence limits for the difference, respectively. Note that all SPSS output is boxed, and in Table 4.14 and some later examples of output, column headings are added above the box for use in our discussion.

Tukey's HSD We present the results of Tukey's HSD analysis in Table 4.15; the output format is basically the same as for the LSD analysis. The calculation of

$$3.2558 \cdot 4.15 \cdot \text{SQRT}(1/6 + 1/6)$$

(see top of Table 4.15) gives us the value of 7.82 (with minor rounding error) that we found earlier in the chapter. Of course, the results are also duplicated; with no asterisks appearing at the intersection of the Grp 2 row and any of the columns, we get the grouping result

$$(3, 1, 2) \qquad (2, 4, 5)$$

Table 4.15 SPSS for Macintosh Output: Tukey's HSD Test

```
  Variable Score
 By Variable Broker

Multiple Range Tests: Tukey-HSD test with significance level .05

The difference between two means is significant if
       MEAN(J)-MEAN(I) > = 3.2558 * RANGE * SQRT (1/N(I) + 1/N(J))
       with the following value(s) for RANGE: 4.15

(*) Indicates significant differences which are shown in the lower triangle

                        G  G  G  G  G
                        r  r  r  r  r
                        p  p  p  p  p

                        3  1  2  4  5
Mean        Broker

 5.000      Grp 3
 6.000      Grp 1
12.0000     Grp 2
14.0000     Grp 4       *  *
17.0000     Grp 5       *  *

Subset 1
Group       Grp 3      Grp 1      Grp 2
Mean         5.0000     6.0000    12.0000
- - - - - - - - - - - - - - -
Subset 2
Group       Grp 2      Grp 4      Grp 5
Mean        12.0000    14.0000    17.0000
- - - - - - - - - - - - - - - - - - - -
```

Table 4.16 shows the same results as Table 4.15 but in the SPSS 8.0 for Windows format, with the same column meanings as those in Table 4.14.

For Tukey's HSD, the SPSS output also provides another format, called "homogeneous subsets," shown in Table 4.17. (SPSS doesn't provide this format for Fisher's LSD.) The "Sig." values at the bottom of each subset (.094 for subset 1, .353 for subset 2) refer to the p values when testing whether the means in that subset are equal; hence, all of these p values are above .05, or they wouldn't be in that same subset.

Newman-Keuls test Next we demonstrate the Newman-Keuls test. As mentioned in note 5 at the end of the chapter, some texts (and the SPSS output) call it by the name Student-Newman-Keuls test. Similar to the other tests, if we compute

$$3.2558 \cdot \text{RANGE} \cdot \text{SQRT}(1/6 + 1/6)$$

Table 4.16 SPSS 8.0 for Windows Format: Tukey's HSD Test

	(1)	(2)	(3)	(4)	(5)	(6)	(7)
Tukey HSD	1	2	−6.00*	2.658	.192	−13.81	1.81
		3	1.00	2.658	.995	−6.81	8.81
		4	−8.00*	2.658	.043	−15.81	−.19
		5	−11.00*	2.658	.003	−18.81	−3.19
	2	1	6.00	2.658	.192	−1.81	13.81
		3	7.00	2.658	.094	−.81	14.81
		4	−2.00	2.658	.942	−9.81	5.81
		5	−5.00	2.658	.353	−12.81	2.81
	3	1	−1.00	2.658	.995	−8.81	6.81
		2	−7.00	2.658	.094	−14.81	.81
		4	−9.00*	2.658	.018	−16.81	−1.19
		5	−12.00*	2.658	.001	−19.81	−4.19
	4	1	8.00*	2.658	.043	.19	15.81
		2	2.00	2.658	.942	−5.81	9.81
		3	9.00*	2.658	.018	1.19	16.81
		5	−3.00	2.658	.790	−10.81	4.81
	5	1	11.00*	2.658	.003	3.19	18.81
		2	5.00	2.658	.353	−2.81	12.81
		3	12.00*	2.658	.001	4.19	19.81
		4	3.00	2.658	.790	−4.81	10.81

Table 4.17 SPSS Output: Homogeneous Subsets Format

	Broker	N	Subset 1	Subset 2	Subset 3
Tukey HSD	3	6	5.00		
	1	6	6.00		
	2	6	12.00	12.00	
	4	6		14.00	
	5	6		17.00	
	Sig.		.094	.353	

using the differing values for RANGE depending on the step, as shown in Table 4.18, we obtain the values arrived at earlier in the chapter.

Likewise, our test results duplicate those found earlier, which are the same groupings as those in the LSD analysis in Table 4.13, repeated in Table 4.19. In SPSS 8.0 for Windows, the only output format provided is that of the homogeneous subsets, shown in Table 4.20.

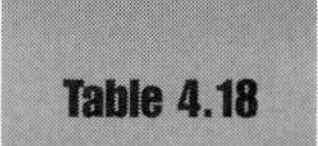

Table 4.18 SPSS for Macintosh Output: Student-Newman-Keuls Test

```
  Variable Score
 By Variable Broker

Multiple Range Tests: Student-Newman-Keuls test with significance level .05

The difference between two means is significant if
       MEAN(J)-MEAN(I) > = 3.2558 * RANGE * SQRT (1/N(I) + 1/N(J))
       with the following value(s) for RANGE:
Step         2       3       4       5
RANGE     2.91    3.52    3.89    4.15
```

Table 4.19 SPSS for Macintosh Output: Student-Newman-Keuls Test, Continued

```
(*) Indicates significant differences which are shown in the lower triangle

                        G  G  G  G  G
                        r  r  r  r  r
                        p  p  p  p  p

                        3  1  2  4  5

Mean       Broker

 5.000     Grp 3
 6.000     Grp 1
12.0000    Grp 2        *  *
14.0000    Grp 4        *  *
17.0000    Grp 5        *  *

Subset 1
Group      Grp 3      Grp 1
Mean        5.0000     6.0000
- - - - - - - - - - - - - - -
Subset 2
Group      Grp 2      Grp 4      Grp 5
Mean       12.0000    14.0000    17.0000
- - - - - - - - - - - - - - - - - - - -
```

SPSS includes Tukey's WSD and Duncan's test, two other tests we briefly discussed without full analysis. The SPSS output, shown here only in homogeneous subset form, is in Table 4.21.

Dunnett's test We illustrated Dunnett's test, assuming that broker 1 was the "control" broker, in section 4.3. Table 4.22 shows the SPSS output for this same test.

Scheffé test We discussed the Scheffé test for the hypothesis pair

H_0: $L = 0$

H_1: $L \neq 0$

Table 4.20 **SPSS 8.0 for Windows Output: Student-Newman-Keuls Test**

	Broker	N	Subset 1	Subset 2	Subset 3
Student-Newman-Keuls (S-N-K)	3	6	5.00		
	1	6	6.00		
	2	6		12.00	
	4	6		14.00	
	5	6		17.00	
	Sig.		.710	.165	

Table 4.21 **SPSS Output: Tukey's WSD and Duncan's Test**

Tukey B (Tukey's-b)	3	6	5.00	
	1	6	6.00	
	2	6	12.00	12.00
	4	6		14.00
	5	6		17.00
Duncan	3	6	5.00	
	1	6	6.00	
	2	6		12.00
	4	6		14.00
	5	6		17.00
	Sig.		.710	.086

Table 4.22 **SPSS Output: Dunnett's Test**

Multiple Comparisons
Dependent Variable: SCORE
Dunnett t (2-sided)

(I) BROKER	(J) BROKER	Mean Difference (I-J)	Std. Error	Sig.	95% Confidence Interval Lower Bound	Upper Bound
2	1	6.00	2.658	.103	−.93	12.93
3	1	−1.00	2.658	.987	−7.93	5.93
4	1	8.00*	2.658	.020	1.07	14.93
5	1	11.00*	2.658	.001	4.07	17.93

*The mean difference is significant at the .05 level.

Table 4.23 SPSS Output: Step 2 for Scheffé Test

Contrast Coefficients

	BROKER				
Contrast	1	2	3	4	5
	1	−.25	−.25	−.25	−.25

where L is any linear combination of the means. We illustrated it for the following contrast in particular:

$$L' = 1(\overline{Y}_1) - 1/4(\overline{Y}_2) - 1/4(\overline{Y}_3) - 1/4(\overline{Y}_4) - 1/4(\overline{Y}_5)$$

SPSS output directly provides the Scheffé test only for pairwise comparisons. However, we noted earlier that we would not recommend the use of the Scheffé test for this purpose. One cannot test a nonpairwise contrast "directly" with SPSS by use of Scheffé's test. However, one can ask for the Scheffé test results for pairwise comparisons, just to get the critical value; then, one can specify the coefficients of the contrast (Table 4.23), and the contrast's value will be computed (here, 6). These results, along with the formulas in section 4.4, can be used to piece together a Scheffé test.

EXAMPLE 4.9 A Larger-Scale Example: Customer Satisfaction Study

We revisit the Merrimack Valley Pediatric Clinic (MVPC) example of Chapter 2, in which we had customer satisfaction data for its four locations (Amesbury, Andover, Methuen, and Salem), and wished to inquire, among other things, whether the average level of satisfaction differed by location. For details about the scale, questions, and so on, see Example 2.3. For convenience, we repeat the data in Table 4.24; we have 30 responders from each location. (In the data sets see Example—Chapter 2.)

We also repeat in Figure 4.4 the JMP software output of the means diamonds figure, as described in Chapter 2 for Figure 2.5. Finally, we repeat the ANOVA table from JMP as Table 4.25, and list the column means. Now we illustrate the multiple-comparison tests of this chapter for the MVPC data. We use the SPSS software because JMP is somewhat limited in its coverage of the multiple-comparison tests discussed in the chapter, having only Fisher's LSD (called "Student's t statistics for all combinations of group means"), Tukey's HSD, and Dunnett's test. For simplicity, we begin with the four cities already

Table 4.24 Data from MVPC Satisfaction Study

Amesbury	Andover	Methuen	Salem
66	55	56	64
66	50	56	70
66	51	57	62
67	47	58	64
70	57	61	66
64	48	54	62
71	52	62	67
66	50	57	60
71	48	61	68
67	50	58	68
63	48	54	66
60	49	51	66
66	52	57	61
70	48	60	63
69	48	59	67
66	48	56	67
70	51	61	70
65	49	55	62
71	46	62	62
63	51	53	68
69	54	59	70
67	54	58	62
64	49	54	63
68	55	58	65
65	47	55	68
67	47	58	68
65	53	55	64
70	51	60	65
68	50	58	69
73	54	64	62

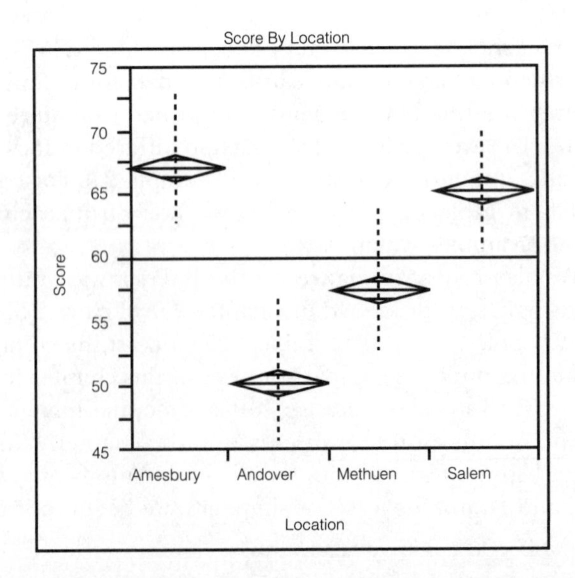

Figure 4.4 JMP output: means diamonds for clinic satisfaction study.

Table 4.25 **JMP Output: ANOVA for Clinic Satisfaction Study**

Analysis of Variance

Source	DF	Sum of Squares	Mean Square	F Ratio
Model	3	5296.4250	1765.47	205.2947
Error	116	997.5667	8.60	Prob>F
C Total	119	6293.9917	52.89	<.0001

	Mean
Amesbury	67.1000
Andover	50.4000
Methuen	57.5667
Salem	65.3000

arranged in ascending order (only because the table of means in Table 4.25 was generated by JMP, whereas the following are the results SPSS provides):

LOCATION	MEAN
Andover	50.4000
Methuen	57.5667
Salem	65.3000
Amesbury	67.1000

Fisher's LSD In essence, Fisher's LSD test (Table 4.26) with $\alpha = .05$ indicates that all four groups are different from one another; the LSD value is 1.498, and differences between each pair of means exceeds 1.498.

Table 4.26 **SPSS Output: Fisher's LSD Test for Clinic Study**

```
   Variable SCORE
By Variable LOCATION

Multiple Range Tests: LSD test with significance level .05

The difference between two means is significant if
        MEAN(J)-MEAN(I) > = 2.0736 * RANGE * SQRT (1/N(I) + 1/N(J))
        with the following value(s) for RANGE: 2.80

(*) Indicates significant differences which are shown in the lower triangle

Mean       LOCATION
                      1  2  3  4

50.4000    Grp 1
57.5667    Grp 2      *
65.3000    Grp 3      *  *
67.1000    Grp 4      *  *  *
```

Tukey's HSD Tukey's HSD test (Table 4.27) says "significance level .050"; of course, this means the experimentwise **a** = .05. The individual test α is much lower than .05. The test indicates that location 1 (Andover) has a different mean from the others, location 2 (Methuen) has a different mean from the others, and locations 3 and 4 (Salem and Amesbury) have means that cannot be said to differ. The HSD value is 1.975, and differences between each pair of means exceed 1.975, except those of Salem and Methuen. Of course, the different result from that of Fisher's LSD is caused primarily by the much lower α per comparison in the HSD test.

Newman-Keuls test The Newman-Keuls results (Table 4.28) duplicate the Fisher's LSD results. When Tukey's HSD is adjusted for the number of steps between the two means being compared, Salem and Amesbury are judged as different. In the HSD test, the critical value was 1.975; in the Newman-Keuls test, the critical value for *adjacent* means in rank-order (step = 2 for adjacent means) is 1.510; the difference in means for Salem and Amesbury is 1.800.

Dunnett's test To perform Dunnett's test (Table 4.29) we assume that location 4, Amesbury, is the "control" location. Obviously, we can perform Dunnett's test with any of the locations as the control location. All three "treatment" locations differ from the control location.

Scheffé test With respect to Scheffé's test, as noted earlier, SPSS gives us the critical value, but not directly the contrast value, if the contrast is other than a pairwise comparison. (As noted earlier, we would not recommend use of the

Table 4.27 **SPSS Output: Tukey's HSD Test for Clinic Study**

```
  Variable SCORE
By Variable LOCATION

Multiple Range Tests: Tukey-HSD test with significance level .050

The difference between two means is significant if
        MEAN(J)-MEAN(I) > = 2.0736 * RANGE * SQRT (1/N(I) + 1/N(J))
        with the following value(s) for RANGE: 3.69

(*) Indicates significant differences which are shown in the lower triangle

Mean        LOCATION
                     1 2 3 4

50.4000     Grp 1
57.5667     Grp 2    *
65.3000     Grp 3    * *
67.1000     Grp 4    * *
```

Table 4.28 **SPSS Output: Newman-Keuls Test for Clinic Study**

```
   Variable SCORE
By Variable LOCATION

Multiple Range Tests: Student-Newman-Keuls test with significance level .050

The difference between two means is significant if
        MEAN(J)-MEAN(I) > = 2.0736 * RANGE * SQRT (1/N(I) + 1/N(J))
        with the following value(s) for RANGE:

Step       2      3      4
RANGE   2.82   3.36   3.69

(*) Indicates significant differences which are shown in the lower triangle

Mean        LOCATION
                     1  2  3  4

50.4000     Grp 1
57.5667     Grp 2    *
65.3000     Grp 3    *  *
67.1000     Grp 4    *  *  *
```

Table 4.29 **SPSS Output: Dunnett's Test for Clinic Study**

Multiple Comparisons
Dependent Variable: SCORE
Dunnett's (2-sided)

(I) LOCATION	(J) LOCATION	Mean Difference (I-J)	Std. Error	Sig.	95% Confidence Interval Lower Bound	95% Confidence Interval Upper Bound
1	4	−16.70*	0.757	.000	−18.20	−15.20
2	4	−09.53*	0.757	.000	−11.03	−08.03
3	4	−01.80*	0.757	.020	−03.30	−00.30

*The mean difference is significant at the .05 level.

Scheffé test for pairwise comparisons.) Table 4.30 shows part of the output SPSS gives when requested to perform Scheffé's test for this MVPC data. Further output is provided with respect to pairwise comparisons. However, we can compute Scheffé's critical value by calculating it in accordance with the output's information:

$$2.0736 \cdot 4.01 \cdot \text{SQRT}(1/30 + 1/30) = 2.15$$

Table 4.30 **SPSS Output: Scheffé's Test for Clinic Study**

```
    Variable SCORE
By Variable LOCATION

Multiple Range Tests: Scheffé test with significance level .05

The difference between two means is significant if
        MEAN(J)-MEAN(I) > = 2.0736 * RANGE * SQRT (1/N(I) + 1/N(J))
        with the following value(s) for RANGE: 4.01
```

Then, using only the column means and critical value, we can compute any contrast value desired. The subject of the next chapter is, indeed, analysis of sets of comparisons made up of nonpairwise contrasts.

EXAMPLE 4.10 The Qualities of a Superior Motel, Revisited

A major experiment, including the two key factors as noted at the beginning of the chapter, was conducted. Of the two factors listed, "breakfast" was highly significant, but "entertainment" was not significant. However, as noted earlier, what was key in each case was not solely to determine whether the level of the factor "mattered," but more specifically to identify the pattern of differences, if any, because different costs apply to implementing different levels of each factor. Indeed, it was conceivable that a level of one of the factors could engender a higher estimated demand than another level, but not exceed that other level by enough to be economically justified.

For the factor "breakfast," Figure 4.5 shows the mean for each level for the infrequent users or (up to now) nonusers in terms of estimated number of nights of stay at the chain over the next 12 months.

Using a Newman-Keuls test with **a** = .05, levels 1, 2, and 3 were not significantly different from one another, and were all different from (lower than) levels 4, 5, and 6. Levels 4 and 5 were not different from one another, but level 6 was different from (higher than) all of the others. It appeared to motel management that even though it was truly difficult to interpret the actual means with any precision, never mind an exact economic meaning, the addition of a "chef" (in reality, simply a somewhat-experienced cook who dresses in white with a chef's hat) was worth implementing. Motel management interpreted the data to suggest that it was economically advisable to also add the Sara Lee-like pastry. Both were implemented as of early 1999.

Mean	Level	Breakfast (at no extra charge)
3.5	1	• None available
3.8	2	• Continental breakfast buffet—fruit juices, coffee, milk, fresh fruit, bagels, doughnuts
4.0	3	• Enhanced breakfast buffet—add some hot items, such as waffles and pancakes, *that the patron makes himself*
5.2	4	• Enhanced breakfast buffet—add some hot items, such as waffles and pancakes, *with a "chef" who makes them for the patron*
5.4	5	• Enhanced breakfast buffet—add some hot items, such as waffles and pancakes, *that the patron makes himself,* **and also pastry (dough from a company like Sara Lee) freshly baked on premises**
6.4	6	• Enhanced breakfast buffet—add some hot items, such as waffles and pancakes, *with a "chef" who makes them for the patron,* **and also pastry (dough from a company like Sara Lee) freshly baked on premises**

Figure 4.5 Mean for levels of the factor "breakfast."

As mentioned, for entertainment, the results were not significant. Motel management thought that this was because the motel chain was used primarily by travelers on their way from one place to another, most often staying one night at a time. For those travelers, breakfast is still important (folks are still going to eat!), but they often don't spend sufficient time in the room to make the entertainment issue a significant one. Of course, this insight came after seeing the results. If true, and if it had been recognized before the experiment was conducted, it might have suggested not including the factor in the experiment.

Exercises

1. Consider the data in Table 4EX.1 on sales results for three levels of the factor "treatment." (The data were presented in exercise 2 in Chapter 2 and are repeated here for convenience.) Perform Fisher's LSD test for all pairwise comparisons. Use $\alpha = .05$.

Table 4EX.1 **Sales Data**

Treatment		
1	2	3
6	6	11
3	5	10
8	4	8
3	9	11
6	6	11
3	5	10
8	4	8
3	9	11

2. For the Table 4EX.1 data, perform Tukey's HSD test for all pairwise comparisons. Use an experimentwise error rate of **a** = .05.
3. For the Table 4EX.1 data, perform the Newman-Keuls test for all pairwise comparisons. Use an experimentwise error rate of **a** = .05.
4. For the Table 4EX.1 data, perform Dunnett's test, assuming that column one is the control group. Use an experimentwise error rate of **a** = .05.
5. For the Table 4EX.1 data, perform Scheffé's test to test whether or not the mean of column two is equal to the average of column means one and three.
6. Consider exercise 4 in Chapter 2, in which we noted the five different manufacturing process technologies that can be dominant:

   A: Project

   B: Job shop

   C: Batch

   D: Mass production

   E: Continuous process

   Using the data for that exercise (repeated for convenience in Table 4EX.6), perform a Newman-Keuls test at **a** = .05. Explain your conclusions in practical terms.

Table 4EX.6 **Staffing Ratios by Dominant Technology**

A	B	C	D	E
1.7	1.7	2.4	1.8	3.1
1.2	1.9	1.2	2.2	2.9
0.9	0.9	1.6	2.0	2.4
0.6	0.9	1.0	1.4	2.4

7. Consider the Table 4EX.7 data (repeated from exercise 6 of Chapter 2), representing four levels of lodging on a cruise ship. The yield is an assessment of the amount of motion felt during cruising (scaled from 1 to 30). Perform Tukey's HSD test on these data at **a** = .05. Explain your conclusions in practical terms.

Table 4EX.7 **Motion Assessment by Ship Level**

1	2	3	4
16	16	28	24
22	25	17	28
14	14	27	17
8	14	20	16
18	17	23	22
8	14	23	25

8. For Table 4EX.7, perform a Newman-Keuls test on the data at **a** = .05. Explain your conclusions in practical terms.
9. In exercises 7 and 8, do the two tests yield exactly the same practical conclusions? Explain.
10. Consider the data in Table 4EX.10 (repeated from exercise 14 of Chapter 2), which represents the amount of life insurance (in $1000s) carried by a random selection of seven state senators from each of three states: California, Kansas, and Connecticut. Perform Fisher's LSD test for all three pairwise comparisons.

Table 4EX.10 Life Insurance ($1000s)

State		
1	2	3
90	80	165
200	140	160
225	150	140
100	140	160
170	150	175
300	300	155
250	280	180

11. For Table 4EX.10, perform Tukey's HSD test for all three pairwise comparisons. Use an experimentwise error rate of **a** = .05.
12. For Table 4EX.10, perform Dunnett's test assuming that column one is the control; repeat Dunnett's test assuming column two is the control; repeat Dunnett's test assuming column three is the control. Are your results consistent? How do the Dunnett results, combined, relate to the HSD test results? Use an experimentwise error rate of **a** = .05.
13. Among the brokers in the example used throughout the chapter, suppose that broker five differed from the others in some identifiable way (such as years of experience). Using Scheffé's test, would you accept or reject the hypothesis that the mean for broker five equals the average of the means for brokers two, three, and four (that is, considering only the male brokers)? Use **a** = .05.
14. Tables 4EX.14a and b show the data and ANOVA results for the AA-cell battery problem of Chapter 2. Perform a Tukey's HSD test on the eight column means. Use **a** = .05.
15. For Table 4EX.14a, perform a Newman-Keuls test on the eight column means. Use **a** = .05.
16. Do the results for exercise 14 differ from the results for exercise 15?
17. For Table 4EX.14a, perform a Dunnett's test with column 3 as the control column. Use **a** = .05.
18. For Table 4EX.14a, use Scheffé's test to test whether the average true column mean for the columns with the four *lowest* observed column means is equal (or not) to the average of the true column means for the four columns with the four *highest* observed column means. Use **a** = .05. (The wording of this question itself illustrates well the conceptual issues involved in post hoc comparisons.)
19. Using the data from Table 2EX.12 in Chapter 2 (in the data sets as Problem 2.12), which consist of the waiting times of patients at four offices of the Merrimack Valley Pediatric Clinic (MVPC), perform Fisher's LSD test for the four offices, using $\alpha = .05$.
20. Using the same MVPC waiting time data as for exercise 19, perform Tukey's HSD test on the four columns, using **a** = .05.

Table 4EX.14a Battery Lifetime Study—Yield in Hours

Device							
1	2	3	4	5	6	7	8
1.8	4.2	8.6	7.0	4.2	4.2	7.8	9.0
5.0	5.4	4.6	5.0	7.8	4.2	7.0	7.4
1.0	4.2	4.2	9.0	6.6	5.4	9.8	5.8
2.6	4.6	5.8	7.0	6.2	4.6	8.2	7.4

Table 4EX.14b ANOVA Table for Battery Lifetime Study

Source of Variability	SSQ	*df*	MS	F_{calc}
Device	69.12	7	9.87	3.38
Error	46.72	16	2.92	
Total	115.84	23		

21. Repeat exercise 20, using the Newman-Keuls test.
22. Repeat exercise 20 using Dunnett's test and assuming that Andover is the "control group."
23. Using the data from exercise 13 of Chapter 2 (in the data sets as Problem 2.13), which consist of the golf scores at four local golf courses for golfers similar to those of Eastern Electric, perform Fisher's LSD test for the four courses, using $\alpha = .05$.
24. Using the same golf-score data as for exercise 23, perform Tukey's HSD test on the four columns, using **a** = .05.
25. Repeat exercise 24, using the Newman-Keuls test.
26. Repeat exercise 24, using Dunnett's test and assuming that the Meadow Brook golf course is the "control group."

Notes

1. Given the large volume of virtually every trade, the issue of how much would be paid in commissions was essentially irrelevant. The issue was, indeed, the buying price.

2. A priori planned comparisons may be non-orthogonal. The title of Chapter 5 (Orthogonality, Orthogonal Decomposition, and Their Role in Modern Experimental Design) is not meant to imply otherwise.

3. Several tests are available for testing the different levels of the factor for the equality of variances. We provided a reference for these tests in Chapter 3.

4. This quote is not elegantly said. To be very technical, we should say, "We cannot reject the contention that true column means three and one are equal; we also cannot reject that contention for column means two, four, and five. However, we can reject the contention that each column mean in the first subset is equal to each column mean in the second subset." The authors view the issue as a choice between clarity and elegance, and prefer the clarity.

5. Some of the tests discussed in this chapter go by more than one name. For example, Fisher's LSD test is often called, simply, the LSD test. The Newman-Keuls test is sometimes called the Student-Newman-Keuls test.

6. Some authors emphasize that the contrasts examined under this post hoc thought process must be ones that are generated from examining the data "after the fact"—that is, they must be the result of "data snooping." One could debate whether the female/male issue is in that category; however, we take the view that the key issue generating interest in the contrast *is* from the data: that the column one mean is nearly the lowest, and the data composing the mean seem to have a somewhat higher amount of variability, whereas we would never have considered gender if the data didn't stand out.

CHAPTER 5

Orthogonality, Orthogonal Decomposition, and Their Role in Modern Experimental Design

EXAMPLE 5.1 Planning Travel Packages at Joyful Voyages, Inc.

With the cruise ship industry experiencing enormous growth—consumer demand growing at an increasing rate and new entrants into the market making the entire marketing process more competitive—Joyful Voyages, Inc.,[1] decided to run a set of experiments to help determine terms of a travel package that would enhance both demand and profit. The cost structure of cruises is, in some sense, a "step function." Once the cruise is "a go," that is, is sufficiently full so that the ship will indeed sail, profit margins are extremely high on filling otherwise empty berths; at least 80% of a typical price paid for the cruise will be profit. The food costs are minimal, and the primary incremental costs are land costs (for example, the per person entry fee to a museum, included in the cruise price).

Joyful Voyages was interested in testing various factors: the "offer" (discussed in more detail below), different positionings, itinerary length, and cobranding with strategic allies (for example, a well-known Japanese company for a trip to Japan). Price (per person) was considered part of the offer. It was clear that, at least to begin with, each destination had to be viewed as a "separate" study. If results were similar for several destinations, then generalizations might be considered. The first experiment was carried out for a specific destination in the South Pacific, encompassing a series of biweekly sailings during a 12-week period. Joyful Voyages was most interested in the offer factor, as listed in Figure 5.1.

1. Price of $2989—no other provisions (as is customary and routine, full payment is due four weeks prior to the sailing)
2. Price of $3489—no other provisions (as is customary and routine, full payment is due four weeks prior to the sailing)
3. Price of $2989—ability to **finance the payment** over three months in three equal payments, one payment before the trip takes place
4. Price of $3489—ability to **finance the payment** over three months in three equal payments, one payment before the trip takes place
5. Price of $2989—**flexibility to wait until last minute** (two weeks before the sailing, on a first-come, first-served basis) **to decide on the date of the trip,** out of the six trips in the 12-week period
6. Price of $3489—**flexibility to wait until last minute** (two weeks before the sailing, on a first-come, first-served basis) **to decide on the date of the trip,** out of the six trips in the 12-week period
7. Price of $2989—opportunity to **pay at least three months in advance** ("early booking") and **get a 15% discount**
8. Price of $3489—opportunity to **pay at least three months in advance** ("early booking") and **get a 15% discount**

Boldface type is solely for ease of reader identification of the difference among the levels.

Figure 5.1 Levels of the offer factor.

Each of the eight offers was sent to 12,500 different (randomly selected) potential responders (that is, 100,000 responders in total). All potential responders came from Joyful Voyage's house list of people having taken at least one voyage with Joyful. The number of responses to each offer was recorded. We return to this example at the end of the chapter.

5.1 Introduction

In Chapter 2, we saw how to investigate whether or not one factor influences some dependent variable. Our approach was to partition the total sum of squares (TSS), the variability in the original data, into two components—the sum of squares between columns (SSB_c), attributable to the factor under study, and the sum of squares within a column (SSW), the variability not explained by the factor under study, and instead explained by "everything else." Finally, these quantities were combined with the appropriate degrees of freedom in order to assess statistical significance. We were able to accept or reject the null hypothesis that all column means are equal (or, correspondingly, reject or accept that the factor under study has an impact on yield). In Chapter 4, we discussed multiple-comparison techniques for asking more detailed questions about the factor under study; for example, if not all column means are equal, how do they differ? We now present a more sophisticated, flexible, and potent way to analyze (or "decompose") the impact of a factor on the yield, not limited to pairwise comparisons.

The procedure amounts to partitioning SSB_c, the variability attributed to the factor represented by the C columns, into $C - 1$ "orthogonal" (we define this word in more detail later—for now, consider the word to mean "independent") components. Recall that the B_c in SSB_c relates to the variability "between columns." The reference to columns is our way of referring to various values, or levels, of the factor under study. There is no inherent meaning to the notion of columns; indeed, in Chapter 2 we could just as well have written the data horizontally and called the different levels "rows."

Each of the $C - 1$ orthogonal components represents a one-degree-of-freedom test statistic (such as F_{calc}), which can be viewed as addressing one particular question (that is, testing one particular hypothesis) about the way in which the factor under investigation affects the yield. In terms of the battery example of Chapter 2, where average battery lifetime was analyzed for eight different devices, we might decompose the differences among the eight devices into seven subsources: for example, the differences between cell phone, flash camera, and flashlight, the differences between brands of cell phone, the differences among brands of flash camera, the difference between one brand of flash camera and that same brand of flashlight, and so forth. Each of the seven questions is formulated as one of seven[2] orthogonal inquiries into the data's message. We revisit this battery lifetime example in an exercise at the end of the chapter.

5.2 Forming an Orthogonal Matrix

Asking questions of the data to break down the variability into subsources is equivalent, as we shall see below, to constructing linear combinations of the column means (that is, $\bar{Y}_{\cdot j}$) and testing, in each case, whether its "true value counterpart" is zero or not. These linear combinations are sometimes called "orthogonal contrasts." We form each linear combination by multiplying each column mean by some constant and the resulting products are then summed. We can show this with the following notation. For simplicity, we assume for the moment that we have four columns (that is, $j = 1, \ldots, 4$ and $C = 4$).

$$Z_1 = a_{11}\, Y_1 + a_{12}\, Y_2 + a_{13}\, Y_3 + a_{14}\, Y_4$$

where Y_j represents the column mean for the jth column and Z_1 is the first linear combination to be evaluated.[3]

Similarly, for our second linear combination,

$$Z_2 = a_{21}\, Y_1 + a_{22}\, Y_2 + a_{23}\, Y_3 + a_{24}\, Y_4$$

Finally,

$$Z_3 = a_{31} Y_1 + a_{32} Y_2 + a_{33} Y_3 + a_{34} Y_4$$

Although we have set the stage for the development, we have not said how these a_{ij} coefficients are determined, nor have we yet explained the significance of the term "orthogonal." We do so now.

We can organize these coefficients into a table, or more accurately, into an array known as a matrix. (The fact that these coefficients are arranged in a matrix facilitates the process of orthogonal decomposition when we use a spreadsheet.) It is always true that the a_{ij} coefficient matrix has C columns and $C - 1$ rows,[4] where C still designates the number of levels of the factor under study, and $C - 1$ still indicates the number of degrees of freedom associated with the column sum of squares, SSB_c. Each row of the matrix corresponds to one of the Z's; indeed, the first subscript (i in a_{ij}) designates both the ith Z and the ith row. The number of Z's and the number of rows in the coefficient matrix logically equals the number of degrees of freedom available. Similarly, the columns of the matrix and the columns (representing levels) of the original data set are, in essence, the same; that is, the second subscript (j in a_{ij}) designates both the jth column in the coefficient matrix and the jth column in the original data set. Hence, the number of columns in the coefficient matrix and the number of columns in the original data set must be equal as well. Here is the coefficient matrix:

COEFFICIENT MATRIX

$$\begin{matrix} a_{11} & a_{12} & a_{13} & a_{14} \\ a_{21} & a_{22} & a_{23} & a_{24} \\ a_{31} & a_{32} & a_{33} & a_{34} \end{matrix}$$

If every pair of different rows of the matrix satisfies the condition that their inner product (also called the dot product) is zero, then the rows are said to be **orthogonal.** The terms **inner product** and **dot product** may sound complex, but they are just simple arithmetic operations: multiply, for a given two rows, the coefficients, column by column, and then sum the products. For example, for rows one and two, the inner product is

$$a_{11} \cdot a_{21} + a_{12} \cdot a_{22} + a_{13} \cdot a_{23} + a_{14} \cdot a_{24}$$

In general, for a four-column analysis, the condition that the inner product equals zero can be expressed as

$$\Sigma_{j=1,4}(a_{i_1 j} \cdot a_{i_2 j}) = 0$$

where i_1 and i_2 designate two different rows. Mathematically, this condition of having the inner product equal to zero can be proven to indicate that the two Z's (that is, rows), and the inquiries they represent, are uncorrelated. This lack of correlation, combined with the normality assumption noted in Chapter 3, indicates that the two Z's are statistically independent—a very desirable condition, discussed in the previous chapter, that enables us to make inferences not otherwise available; we also revisit this issue later in the chapter.

If, furthermore, the inner product of each row *with itself* (or equivalently, and perhaps more easily visualized, the sum of the squares of the coefficients in each row) equals 1, the rows (and the matrix) are said to be **orthonormal.** These terms, orthogonal and orthonormal, are commonly used in the areas of science and engineering, although they are not often used in areas of social science and management. The literature on the subject matter of this chapter uses these terms, so we do also. In essence, an orthonormal table or matrix is simply an orthogonal table or matrix that is scaled or normalized to 1 or 100%, as are many numerical results we encounter routinely. For example, for row one, the condition is

$$a_{11} \cdot a_{11} + a_{12} \cdot a_{12} + a_{13} \cdot a_{13} + a_{14} \cdot a_{14} = 1$$

or, equivalently,

$$a_{11}^2 + a_{12}^2 + a_{13}^2 + a_{14}^2 = 1$$

The making of an orthogonal matrix into an orthonormal matrix is simply a matter of dividing all coefficients, originally chosen by contextual considerations—the "how" of which we shall soon see—by whatever appropriate number, or scale factor, makes the sum of squared coefficients equal to one.

Finally, we add a third condition to what *we* shall call an orthonormal matrix: that the sum of the coefficients of each row (without squaring!) equals zero; for example, for row one:

$$a_{11} + a_{12} + a_{13} + a_{14} = 0$$

This condition ensures that the question being addressed by the Z (the linear combination represented by the a_{ij}'s of that row) is arithmetically sensible. For example, to compare the *total* of five column means with the *total* of three other column means makes no logical, arithmetic, or any other kind of sense! This lack of sense would manifest itself in the resulting coefficients: five $+1$s and only three -1s for a total of $+2$, not zero (perhaps divided by the square root of eight, if we include the condition that the matrix be orthonormal). Note that we *can* compare five column means with three other column means (if it's contextually useful to do so), but before we adjust to make the system orthonormal we first give each column mean in the group of five a coefficient of $1/5$ and give each of the other three column means a coefficient of $-1/3$; this is sensible, and the coefficients now sum to zero.

With these three conditions on the coefficients, we have a set of questions (Z's) that allows independent assessment of three (in general, $C - 1$) different hypotheses concerning the division of the variability accounted for by the original column factor. That is, the SSB_c can be partitioned (decomposed) by these orthogonal contrasts into $C - 1$ components.[5] Each Z (each query of the data) is independent[6] of every other Z; thus, the probabilities of Type I and Type II errors (α and β) in the ensuing hypothesis tests are independent and stand alone. (Recall the complexities noted in the previous chapter in this regard.) In addition, *the sum of the $C - 1$ separate sums of squares (SSQ) precisely equals the SSB_c.* This is very helpful in interpreting the data's message—we can now take ratios of these sum-of-squares quantities (SSQs) in a meaningful way. For example, in the earlier average battery lifetime example, we might find that "70% of the variability in lifetime among batteries used in the eight different devices can be attributed to differences between flash camera usage [averaged over brands of flash cameras] and flashlight usage [averaged over the three brands of flashlight]" or that "differences in average battery lifetime among the three brands of flash camera explain 2.5 times as much variability as do differences in average battery lifetime among the two brands of cell phone."

An example of an orthonormal matrix (that is, a $C - 1$ by C matrix of coefficients that satisfies the three conditions) is the following 3×4 matrix:

ORTHONORMAL MATRIX

1/2	1/2	−1/2	−1/2
1/2	−1/2	1/2	−1/2
1/2	−1/2	−1/2	1/2

Note the following three characteristics of an **orthonormal matrix:** each row has coefficients that sum to zero, each row has squared coefficients that sum to one, and the inner product of each pair of rows equals zero. As a final reminder: if we relaxed the condition that the sum of the coefficients, when first squared, equals one, we could fill the cells of the table above with ones instead of halves [negative ones instead of negative halves], and we'd have an *orthogonal* matrix, but one that isn't *orthonormal.* However, scaling the coefficients to achieve an orthonormal matrix has useful benefits, as we stated earlier and will subsequently illustrate.

EXAMPLE 5.2 Calculation Example

We'll now go through an example demonstrating how the sum of squares of the individual components equals the original SSB_c of the factor composing the columns. Suppose we have four column means—Y_1, Y_2, Y_3, and Y_4, as follows:

Y_1	Y_2	Y_3	Y_4
6	4	1	−3

We calculate the grand mean and SSB_c as follows:

$$\text{Grand mean} = \bar{Y} = (6 + 4 + 1 - 3)/4 = 2$$
$$SSB_c = R[(6-2)^2 + (4-2)^2 + (1-2)^2 + (-3-2)^2] = 46R$$

where R is the number of rows (replications per column). Suppose that we use the orthonormal matrix above.

Z_1 is computed by calculating a linear combination of the column means, using the first row of the coefficient matrix as the weights. Equivalently, we

Table 5.1 **Calculation of Z_i from Y_j**

	Y_1	Y_2	Y_3	Y_4		
	6	**4**	**1**	**−3**	Z	Z^2
$Z_1 =$	1/2	1/2	−1/2	−1/2	6	36
$Z_2 =$	1/2	−1/2	1/2	−1/2	3	9
$Z_3 =$	1/2	−1/2	−1/2	1/2	−1	1
					$SSB_c = 46R$	

can say that Z_1 is calculated by taking the inner product of the first row of the coefficient matrix and the row of column means. Either way, we get

$$\begin{aligned} Z_1 &= (1/2)(6) + (1/2)(4) + (-1/2)(1) + (-1/2)(-3) \\ &= 3 + 2 - 0.5 + 1.5 \\ &= 6 \end{aligned}$$

Z_2 and Z_3 are calculated in a similar fashion to be 3 and −1, respectively.

Note that $Z_1^2 + Z_2^2 + Z_3^2 = 6^2 + 3^2 + (-1)^2 = 36 + 9 + 1 = 46$. And, as we saw above, $SSB_c = 46R$.

A summary of the calculations is in Table 5.1.

EXAMPLE 5.3 Portfolio Rating

It may not be perfectly clear yet how the above helps us. As it turns out, the column means above have been obtained from a one-factor ANOVA experiment in which four different individual retirement account (IRA) options—that is, portfolios—were each offered to a test group of four different people (16 people in all) in order to determine client preference. Each person rated the one portfolio presented to him or her on a scale of −10 to 10.[7] Three attributes of these investment portfolios are identified—"aggressiveness" (degree of risk), "balance" (mixture of debt and equity), and "environmental responsibility" (overt commitment to conduct business in an environmentally responsible way). The four fund prospectuses have been designed to promise either a high degree or a low degree of each of these attributes. The characteristics of the four portfolios (pf-1 through pf-4) are shown in Table 5.2.

The raw data are as follows, with the column means in the last row:

pf-1	pf-2	pf-3	pf-4
5	5	2	−4
7	4	0	−5
8	2	3	−1
4	5	−1	−2
6	4	1	−3

Note that the column means are the same as those we used in the previous calculation example in Table 5.1. When the data are analyzed as a one-factor design with $C = 4$ (four columns) and $R = 4$ (four rows, or replicates per column), the ANOVA table is as shown in Table 5.3.

At $\alpha = .05$, $F(3, 12) = 3.49$ and the F test thus indicates that client ratings are not equal for all portfolios offered (that is, the factor "portfolio" is significant; $p < .01$). Yet this result by itself does not provide all of the information that might be useful. By partitioning the results further, into three orthogonal components, Z_1, Z_2, and Z_3, we can break up the SSQ associated with portfolio (that is, the SSB_c, which equals 184) into separate SSQs for the three attributes, thus determining *which* investment attributes are driving the differences among the ratings. We use the same orthonormal matrix as used in the previous section (repeated for convenience):

	pf-1	pf-2	pf-3	pf-4
$Z_1 =$	1/2	1/2	−1/2	−1/2
$Z_2 =$	1/2	−1/2	1/2	−1/2
$Z_3 =$	1/2	−1/2	−1/2	1/2

Z_1, by multiplying the first and second column means by $+1/2$, and the third and fourth column means by $-1/2$ (see the orthonormal matrix in the table above), is calculating the difference between the average rating for the two highly

Table 5.2 **Characteristics of Four IRA Portfolios**

Feature	pf-1	pf-2	pf-3	pf-4
Aggressiveness	High	High	Low	Low
Balance	High	Low	High	Low
Environmental responsibility	High	Low	Low	High

Table 5.3 **One-Way ANOVA Table**

Source of Variability	SSQ	*df*	MSQ	F_{calc}
Portfolio	184	3	61.33	20.44
Error	36	12	3	
Total	220	15		

aggressive choices (the first two portfolios; see Table 5.2) and the average rating of the two less-aggressive choices (the last two portfolios). Thus the difference between Z_1 and zero is an indication of how much of the differences in ratings among the four portfolios can be placed on the shoulder of the *degree of aggressiveness* of the portfolio. How different Z_2 is from zero is an indication of how much of the differences in ratings among the four portfolios can be placed on the shoulder of the *balance* of a portfolio. Z_2 is formed as the difference between the average rating of the more balanced portfolios (1 and 3) and the less-balanced portfolios (2 and 4). Finally, how different Z_3 is from zero is an indication of how much of the differences in ratings among the four portfolios can be placed on the shoulder of the *environmental responsibility* of a portfolio. It is formed as the difference between the average rating of the more environmentally responsible portfolios (1 and 4) and the less environmentally responsible portfolios (2 and 3). (Note also that limiting the detailed analysis to pairwise comparisons would not fully enable clarification of the potential subsources of variability.)

We can now form an augmented ANOVA table (Table 5.4), statistically testing these three attributes, or subfactors, explicitly to see whether a different level of the attribute truly has an impact on portfolio rating. The SSB_c is 184 with three degrees of freedom (see Table 5.3). The 184 can be partitioned (orthogonally decomposed) into three orthogonal components. If we look back at Table 5.1 for the three Z^2 values of 36, 9, and 1, respectively, and multiply each of these by $R = 4$ (the number of replicates), we get 144, 36, and 4, and $144 + 36 + 4 = 184$; getting a total of exactly 184 is no coincidence! We knew that would be the result because the coefficient matrix was indeed orthonormal (and not merely orthogonal). We'll soon discuss how this helps us.

It can be mathematically proven that each of these SSQ components, formed in the manner described, can be interpreted as a sum-of-squares value that is properly associated with one degree of freedom, and that can also legitimately serve as the numerator of an F_{calc}, and all that this implies.[8]

At $\alpha = .05$, $F(1, 12) = 4.75$, and the degree of aggressiveness and the degree of balance of the portfolio are each significant at .05 ($p < .001$ for both), but the degree of environmental responsibility is not ($p > .20$). Furthermore, differences in portfolios that are aggressive from those that are not agressive

Table 5.4 Augmented ANOVA Table

Source of Variability	SSQ	df	MSQ	F_{calc}
Portfolio	184	3	61.3	20.4
Aggressiveness	144	1	144	48.0
Balance	36	1	36	12.0
Env. resp.	4	1	4	1.33
Error	36	12	3	
Total	220	15		

are estimated to explain four times as much variability in ratings among the four portfolios as differences in the portfolios' balance (that is, 144 is 4 times 36). Indeed, differences in described aggressiveness of the portfolios are estimated to explain 100(144/184) = 78% of the differences in ratings among the four portfolios.

EXAMPLE 5.4 Drug Comparison

Consider another example, in which we evaluate the efficacy of two brands of aspirin and one brand of non-aspirin-based pain killer (let's simply call it "Non-A"), along with a placebo[9] for reference.

Note that the need for experimental control (here a placebo) extends beyond the testing of drugs. In a famous experiment held at Hawthorn Works, an AT&T Western Electric factory in Chicago, Illinois, in the 1940s, industrial engineers were testing to see if the light level affected production. Engineers increased the amount of light reaching each work location and observed an increase in production output. Having completed the experiment, they restored the light level to its previous value, only to discover another increase in production output. It was subsequently determined that it was the experimental activity, and not the level of light, that was affecting production. This phenomenon, against which every experimenter must guard, is known as the Hawthorn effect.

For the drug study, Table 5.5 shows the raw data and experimental results including column means. Each drug was administered to $R = 8$ different peo-

Table 5.5 **Drug Efficacy Data and Column Means**

Placebo	Aspirin 1	Aspirin 2	Non-A
4	9	8	10
5	6	7	8
7	2	7	12
3	5	8	10
5	7	6	9
9	6	3	14
6	6	7	6
1	7	10	11
$Y_1 = 5$	$Y_2 = 6$	$Y_3 = 7$	$Y_4 = 10$

ple. Y_{ij} was the self-reported degree of improvement (reduction of pain) in the patient's headache, on a scale of 1 (virtually no reduction) to 15 (virtually no pain remaining). Table 5.6 is the associated one-way ANOVA table.

We conclude that at $\alpha = .05$, with $F(3, 28) = 2.95$, "drug" is significant ($p < .01$). But again, this "macro" statement reveals only a small portion of the story. We want to understand just what about the four drugs is driving the differences in patient improvement. Suppose that we decide to decompose the differences in improvement among the four drugs into the following three components; with respect to degree of patient improvement, we ask the following questions:

1. Are the "real" drugs, overall, different from the placebo? (P′ versus P)
2. Are the aspirin drugs different from one another? (A1 versus A2)
3. Are aspirin drugs different from the Non-A? (A versus Non-A)

We need to construct our questions in the form of linear combinations of the column means. For the first question, how would you form a linear com-

Table 5.6 **One-Way ANOVA Table**

Source of Variability	SSQ	*df*	MSQ	F_{calc}
Drug	112	3	37.33	7.47
Error	140	28	5	
Total	252	31		

bination to compare placebo to nonplacebo? Your first thought might be to say, "Compare the column average of the placebo to the average of the other three column averages," and that is correct. But what does *compare* really mean here? It means to subtract one from the other and examine the difference! The direction of subtraction doesn't matter (as long as the later interpretation is consistent with the chosen direction). In terms of coefficients in a linear combination, this means to take 1/3 of the sum of the column means of A1, A2, and Non-A, and subtract from it (or it from) the mean of the one placebo column. In essence, this gives a coefficient of −1 to the placebo column mean, and coefficients of 1/3 to each of the other three column means. Though these coefficients don't *yet* conform to an orthonormal row (that is, the sum of the squares of −1, 1/3, 1/3, and 1/3 is not equal to one), it is noteworthy that the unsquared coefficients do sum to zero. We see these coefficients in the first coefficient row of Table 5.7. It might be instructive to contrast what this first row is doing with the goal of the Dunnett test of the previous chapter. This first row is comparing the *mean* of the last three columns ("treatments") to the mean of the first column, the placebo. The Dunnett test compares each of the other column means, *individually,* to the mean of the control column.

By a similar logic, the second row looks at the difference between the degree of improvement obtained with one brand of aspirin and that obtained with the other brand of aspirin. The third row analyzes the difference between the average degree of improvement from the two brands of aspirin and that from the nonaspirin. In each case, if the absolute value of the difference is large (that is, "very nonzero"), we infer that the corresponding columns, or combinations of columns, are not the same with respect to degree of improvement—the differences are statistically significant.

We observe that these rows each add to zero (that is, each calculation of Z is arithmetically sensible) and that they are orthogonal (the inner product of any two rows equals zero). They might not have come out orthogonal. The authors purposely chose a set of questions so that they would be; however, these seem to be intuitively appealing questions. It could have been otherwise. For example, if question one were to compare the placebo to only the aspirin drugs, and question two were to compare the placebo to the nonaspirin drug, the two questions would not be orthogonal; the inner product of (−1, 1/2, 1/2, 0) and (−1, 0, 0, 1) is not zero; it's one. It should always

Table 5.7 **Linear Combinations for Questions 1–3 for Drug Study**

	P	A1	A2	Non-A
P vs. P′	−1	1/3	1/3	1/3
A1 vs. A2	0	−1	1	0
A vs. Non-A	0	−1/2	−1/2	1

be remembered that the choice of questions that make contextual sense is not the purview of the statistician, it's the purview of (in this case) the pharmacist. This issue is revisited in exercise 14 at the end of the chapter.

The rows of coefficients in Table 5.7 are not, however, orthonormal: the inner product of each row with itself (or equivalently, the sum of the squares of the coefficients in a given row) does not equal 1. We can divide the coefficients in any given row by any nonzero constant without losing the first two properties (orthogonality and unsquared coefficients summing to zero). If we calculate the sum of the squares of the coefficients in a row and then divide each coefficient of that row by the square root of that exact value, it will guarantee that we have scaled the row such that it does have a sum of squared coefficients that equals 1. We illustrate this as follows:

First row: $\sqrt{(-1)^2 + (1/3)^2 + (1/3)^2 + (1/3)^2} = \sqrt{12/9} = 1.1547$

Second row: $\sqrt{(0)^2 + (-1)^2 + (1)^2 + (0)^2} = \sqrt{2} = 1.4142$

Third row: $\sqrt{(0)^2 + (-1/2)^2 + (-1/2)^2 + (1)^2} = \sqrt{3/2} = 1.2247$

When we divide each coefficient in row one by 1.1547, each coefficient in row two by 1.4142, and each coefficient in row three by 1.2247, we obtain the following orthonormal matrix (due to rounding, a row may sum to a tiny nonzero amount):

ORTHONORMAL MATRIX

−.8660	.2887	.2887	.2887
0	−.7071	.7071	0
0	−.4082	−.4082	.8165

As before, we can embed the orthonormal matrix into the template for calculating the Z_i from the Y_j as shown in Table 5.8.

Now we determine whether or not the estimates of the indicated differences are statistically significant. Table 5.9 shows the augmented ANOVA

Table 5.8 **Calculation of Z_i from Y_j**

	Y_1	Y_2	Y_3	Y_4		
	5	6	7	10	Z	Z^2
Z_1 =	−.8660	.2887	.2887	.2887	2.309	5.33
Z_2 =	0	−.7071	.7071	0	.707	.50
Z_3 =	0	−.4082	−.4082	.8165	2.858	8.17
					SSB_c = 14R	

Table 5.9 **Augmented ANOVA Table**

Source of Variability	SSQ		df		MSQ		F_{calc}	
Drug	112		3		37.33		7.47	
P vs. P′		42.64		1		42.64		8.53
A1 vs. A2		4.00		1		4.00		.80
A vs. Non-A		65.36		1		65.36		13.07
Error	140		28		5			
Total	252		31					

table; each of the subitems in the SSQ column represents a Z^2 value in Table 5.8 multiplied by $R = 8$.

We conclude, at $\alpha = .05$, with $F(1, 28) = 4.20$, that with respect to degree of improvement of patients' condition, there is no difference between the two aspirin brands ($p > .20$), but that the group of "real" drugs is different from the placebo, and that the aspirin drugs are different from the nonaspirin drug.

At this point, we can make some managerially meaningful observations; for example, 38.1% [= 100(42.64/112)] of the variability among the four drugs is estimated to be due to the difference between the real drugs and the placebo, and 58.4% can be attributed to the difference between aspirin brands and the nonaspirin. Note that even though the intra-aspirin brand difference cannot, beyond a reasonable doubt, be said to be nonzero, we still allocate 3.6% [= 100(4/112)] to that source. This is a matter of convention, for 3.6% is the best estimate, even though we are not convinced that it is different from zero. Furthermore, due to the statistical independence of the values associated with each source of variability, we can say that, of the "within-real-drug differences" (that is, ignoring differences concerning the placebo versus the real drugs), virtually all of the variability, 94.2% of it [= 100(65.36/ 69.36)], is explained by the difference between the aspirin brands and the nonaspirin.

EXAMPLE 5.5 Amended Drug Comparison

We can use this numerical example again to further illustrate the process of formulating the equations for orthogonal decomposition. Suppose these same

Table 5.10 Linear Combinations for Amended Drug Study

	A1	A2	Non-A1	Non-A2
A1 vs. A2	−1	1	0	0
Non-A1 vs. Non-A2	0	0	−1	1
A vs. Non-A	−1/2	−1/2	1/2	1/2

data were from a different experiment—one in which the four columns consisted of two aspirin brands and two types of nonaspirin medications, as follows:

ASPIRIN 1	ASPIRIN 2	NON-A1	NON-A2
$Y_1 = 5$	$Y_2 = 6$	$Y_3 = 7$	$Y_4 = 10$

Since the column means and the error have not changed, and we still have eight rows, we know that $F_{\text{calc}} = 7.47$ and not all columns are the same—the null hypothesis of no difference among the columns is rejected at $\alpha = .05$. The questions we would ask about the data are different, however. Suppose that we would like to test if the two aspirin brands differ from one another, if the two nonaspirin types are different from one another, and finally, if the aspirin brands on average are different from the nonaspirin types on average. The not-yet-normalized coefficients appear in Table 5.10. The first row measures the difference between the mean of the two aspirin brands. Similarly, the second row measures the difference between the mean of the two nonaspirin types. Finally, the difference between aspirin brands and nonaspirin types is determined by taking the difference between the average of the nonaspirin types (columns three and four) and the average of the aspirin brands (columns one and two). Could anything else be as intuitively sensible?

Note, once again, that the rows are orthogonal and each row adds to zero. It remains for us to normalize the rows. We determine the scale factors as before—for each row, we calculate the square root of the sums of the squares of the coefficients in each row. Convince yourself that the orthonormal matrix in Table 5.11 results.

Table 5.11 Orthonormal Matrix

	A1	A2	Non-A1	Non-A2
A1 vs. A2	−.707	.707	0	0
Non-A1 vs. Non-A2	0	0	−.707	.707
A vs. Non-A	−.5	−.5	.5	.5

Table 5.12 Calculation of Z_i from Y_j

	Y_1	Y_2	Y_3	Y_4		
	5	6	7	10	Z	Z^2
$Z_1 =$	−.707	.707	0	0	.707	.5
$Z_2 =$	0	0	−.707	.707	2.121	4.5
$Z_3 =$	−.5	−.5	.5	.5	3.000	9.0
					$SSB_c = 14R$	

As before, we embed this matrix into our calculation table, Table 5.12. Once again, we turn to the augmented ANOVA table, shown in Table 5.13, to determine which differences are significant. We find, at $\alpha = .05$, with $F(1, 28) = 4.20$, that there is no difference between the aspirin brands ($p > .20$); there is, however, a difference between the nonaspirin types ($p < .02$) and a difference between aspirin brands and nonaspirin types ($p < .001$). Furthermore, the variability explained by the difference between aspirin brands and nonaspirin types is twice as much as that explained by the difference between nonaspirin types.

Table 5.13 Augmented ANOVA Table

Source of Variability	SSQ	df	MSQ	F_{calc}
Drug	112	3	37.33	7.47
A1 vs. A2	4	1	4	.80
Non-A1 vs. Non-A2	36	1	36	7.20
A vs. Non-A	72	1	72	14.40
Error	140	28	5	
Total	252	31		

EXAMPLE 5.6 Food Additives Study Using SPSS

We use a tasty example to illustrate the use of SPSS when using orthogonal contrasts. Here we study the effect of five different food additives, A, B, C, D, and E, on the taste of a certain vanilla pound cake (a pastry). "Taste" was measured averaging a set of responses about different aspects of taste; each of five questions used a five-point Likert scale, ranging from (1) "strongly disagree" to (5) "strongly agree," and each question was worded so that a 5 indicated the most favorable rating concerning the cake's taste. The study was sponsored by the federal government and conducted in a facility that routinely considers, among other issues, food and nutrition for the U.S. military forces. The data have been coded to mask the actual values while retaining the essential elements of the message. Recall that having *five* levels of the factor allows the investigation of, and decomposition into, as many as *four* sources of variability. We summarize the means for each additive as follows:

Additive:	A	B	C	D	E
Mean:	5	28	30	19	27

The experiment was run with replication $R = 20$; that is, 20 pound cakes were baked for each treatment, one cake per responder, 100 responders in total.[10] (It seemed that one could never visit this facility, say for an unrelated consulting project, without getting recruited for a taste test of something!) Table 5.14 shows the one-way ANOVA table.

Running the analysis using SPSS yields the same basic result, shown in Table 5.15. For $\alpha = .05$, $F(4, 95) \approx 2.5$, we conclude that our result is significant ($p < .0001$), so not all additives are the same with respect to taste.

We now set out to see *how* they differ. Additives B and *C* were supplied by supplier U, and additives D and E were supplied by supplier V. Additive A was actually *no additive*. The actual set of questions asked was the following set of orthogonal questions:

Table 5.14 One-Way ANOVA Table

Source of Variability	SSQ	*df*	MSQ	F_{calc}
Cake	8456	4	2114	235.1
Error	854.3	95	8.99	
Total	9310.3	99		

Table 5.15 SPSS Output: ANOVA Table for Pound Cake Study

Variable TASTE
By Variable TREATMENT

Analysis of Variance

Source	D.F.	Sum of Squares	Mean Squares	F Ratio	F Prob.
Between Groups	4	8456.0000	2114.0000	235.0951	.0000
Within Groups	95	854.2500	8.9921		
Total	99	9310.2500			

Z_1: Difference between cake without an additive and cake with an additive
Z_2: Difference between additives from the two different suppliers (U versus V)
Z_3: Difference between the two additives from U
Z_4: Difference between the two additives from V

As before, we first set up the linear combinations that will calculate appropriate indicators of these differences (Table 5.16), momentarily ignoring the normalization. (Convince yourself that these linear combinations will, in fact, calculate appropriate estimates of the four differences defined above.) Note, once again, that the rows of this matrix are orthogonal and that each row sums to zero. We scale the rows as before. For example, the first row has a sum of squares of the coefficients of

$$1^2 + (-1/4)^2 + (-1/4)^2 + (-1/4)^2 + (-1/4)^2 = 1.25$$

Dividing each coefficient in the first row by the square root of 1.25, approximately 1.118, and similarly by 1.0 for row 2 (by chance, the sum of the squared coefficients is already 1.0), and by .707 for rows 3 and 4, we get the following orthonormal matrix:

ORTHONORMAL MATRIX

.8944	−.2236	−.2236	−.2236	−.2236
0	.500	.500	−.500	−.500
0	.707	−.707	0	0
0	0	0	.707	−.707

Table 5.16 Linear Combinations for Questions Z_1–Z_4

	1	2	3	4	5
Z_1	1	−1/4	−1/4	−1/4	−1/4
Z_2	0	1/2	1/2	−1/2	−1/2
Z_3	0	1	−1	0	0
Z_4	0	0	0	1	−1

To use SPSS to enter the contrast values, we click on "Contrasts" and enter the four sets (rows) of coefficients in the above orthonormal matrix; the output in Table 5.17 verifies that these were the coefficients used. Embedding the matrix into the template used for calculation, we have the information in Table 5.18. As before, we write out the augmented ANOVA table, shown in Table 5.19.

Again using $\alpha = .05$, with $F(1, 95) \approx 3.9$, all four differences ("contrasts") are concluded to be nonzero. For the first, second, and fourth contrasts, $p <$ .001; for the contrast examining differences between the additives of U, $p <$ .05. Differences between suppliers account for 8.5% of the variability [100(720/8456)], and differences between additives from supplier V amount to 7.6% of the variability. The most prominent effect (83.4% of the variability explained) is due to the difference between having an additive and not having an additive.

Unfortunately, SPSS does not provide the augmented ANOVA table in one fell swoop. What it does provide, in the notation of this chapter, are the values of the Z's and of the t_{calc}'s, which are the respective square roots of the F_{calc}'s in the augmented ANOVA table, as well as the p values. It does this under the banner of an analysis that uses the mean square error in the original ANOVA table as the basis of the standard error for all contrasts (as opposed to using a variance estimate based only on the particular columns, weighted according to the contrast's coefficients). The SPSS results are in Table 5.20. Note that the Value column in Table 5.20 consists of the same set of values as the Z's in Table 5.18, and the numbers in the T Value column are each the square root of the corresponding F_{calc} value in Table 5.19, the augmented ANOVA table (some numbers are off a tiny bit due to rounding).

Not having the key to how the data are coded, we cannot report precisely how much better the cakes with additives taste compared with cakes without additive (it *was* an increase, because the transformation of the data was a positive linear one). However, the government agency that sponsored this experiment does know the data coding and, hence, how taste varies with the various food additives. The agency is now in a position to examine the other relevant issues (such as cost

Table 5.17 SPSS Output: Orthonormal Matrix

Contrast Coefficient Matrix

	Grp 1	Grp 2	Grp 3	Grp 4	Grp 5
Contrast 1	.8944	−.2236	−.2236	−.2236	−.2236
Contrast 2	0	.5000	.5000	−.5000	−.5000
Contrast 3	0	.7071	−.7071	0	0
Contrast 4	0	0	0	.7071	−.7071

Table 5.18 Calculation of Z_i from Y_j

	Y_1	Y_2	Y_3	Y_4	Y_5		
	5	**28**	**30**	**19**	**27**	Z	Z^2
$Z_1 =$	.8944	−.2236	−.2236	−.2236	−.2236	−18.78	352.8
$Z_2 =$	0	.500	.500	−.500	−.500	6.00	36.0
$Z_3 =$	0	.707	−.707	0	0	1.41	2.0
$Z_4 =$	0	0	0	.707	−.707	5.66	32.0
						$SSB_c = 422.8R$ $= 8456$	

Table 5.19 Augmented ANOVA Table

Source of Variability	SSQ	*df*	MSQ	F_{calc}
Cake	8456	4	2114	235.1
Std. vs. additive	7056	1	7056	784.90
Supplier (U vs. V)	720	1	720	80.09
Additives from U	40	1	72	4.45
Additives from V	640	1	640	71.19
Error	854.3	95	8.99	
Total	9310.3	99		

Table 5.20 SPSS Output

Pooled Variance Estimate

	Value	S. Error	T Value	D.F.	T Prob.
Contrast 1	−18.7829	.6705	−28.012	95.0	.000
Contrast 2	6.0000	.6705	8.948	95.0	.000
Contrast 3	−1.4141	.6705	−2.109	95.0	.038
Contrast 4	−5.6573	.6705	−8.436	95.0	.000

Table 5.21 Linear Combinations for New Questions Z_1–Z_4

	1	2	3	4	5
Z_1	1	−1/4	−1/4	−1/4	−1/4
Z_2	0	1/2	−1/2	1/2	−1/2
Z_3	0	1	0	−1	0
Z_4	0	0	1	0	−1

of additives and process changeover costs) that are necessary to combine with the experimental results to make the correct managerial decisions.

The questions embodied in Z_1 through Z_4 above are representative of the questions that could be asked. We might ask: are there other questions that could have been asked? Yes! By way of example, consider the following:

Z_1: Standard cake versus cake with an additive
Z_2: Difference between additives using chemical composition P for sweetener (B and D) and additives using composition Q (C and E)
Z_3: Difference between the two suppliers with respect to additive using composition P
Z_4: Difference between the two suppliers with respect to additive using composition Q

Table 5.21 shows the linear combinations for these questions. Its rows, too, are orthogonal, each adding to zero. They can be normalized and we can repeat the process of orthogonal decomposition. It might appear that by repeating the process of orthogonal decomposition, we can ask more questions than allowed by the degrees-of-freedom constraint. Such is not the case, however. Only (at most) four (that is, $C-1$) independent questions (linear combinations) can be asked. Were we to continue the process immediately above, we would generate four other questions that are *internally* orthogonal, but which are not orthogonal to the first set of four questions; even for bakery goods, there's no free lunch!

Finally, as noted earlier, we need not ask four questions just because the number of degrees of freedom allows four. We could ask, say, only two questions. That portion of SSB_c remaining simply corresponds to "other sources of variability not investigated," and would have two degrees of freedom.

EXAMPLE 5.7 Planning Travel Packages at Joyful Voyages, Inc., Revisited

The dependent variable for the study described at the beginning of the chapter was the number of bookings (out of the same number of names mailed for each

Table 5.22 **Levels of the Price and Incentive Factors**

	Level							
Factor	**1**	**2**	**3**	**4**	**5**	**6**	**7**	**8**
Price	Low	High	Low	High	Low	High	Low	High
Incentive	N	N	Fi	Fi	Fl	Fl	D	D

Note: Low price = $2989, high price = $3489. N = none, Fi = finance, Fl = flexibility, D = discount

offer), although the number of inquiries was also recorded. The anticipated general level of response (that is, bookings), based on experience with this and similar locations, and for similar offers, was about 0.4%. Other promotional vehicles that contributed to the final number of bookings were also used.

A one-factor ANOVA revealed that, indeed, there were significant differences among responses to the different offers. Then an orthogonal breakdown of the sum of squares associated with "offer" was conducted. Once the concept of an orthogonal breakdown was explained to the company executives, they indicated that they wanted to study four specific "propositions" (that was their word—to us, they would be *contrasts,* or *questions*). Recall that there is no need to have a full set of contrasts—seven contrasts in this case, given the eight columns—indeed, we have only four.

We repeat the eight levels of the factor in Table 5.22 for convenience, using an abbreviated form.

Contrast 1 examined overall differences between the low price ($2989) and the high price ($3489). Its set of coefficients (before the normalizing), was as follows:

Level:	1	2	3	4	5	6	7	8
Contrast 1:	−1	1	−1	1	−1	1	−1	1

Contrast 2 compared the levels whose incentive didn't focus on price (none and flexibility), versus those that did focus on price (finance and discount). Contrast 3 examined whether the flexibility offers truly differed from no incentive at all. Contrast 4 compared the two types of price-focused incentives.

	Level							
	1	**2**	**3**	**4**	**5**	**6**	**7**	**8**
Contrast 2	1	1	−1	−1	1	1	−1	−1
Contrast 3	1	1	0	0	−1	−1	0	0
Contrast 4	0	0	1	1	0	0	−1	−1

The analyses of the contrasts yielded the following results: Contrast 1 was significant ($p < .01$), and negative. That's no big surprise—the lower price did generate significantly more responses; of course, whether the lower price would generate more *profit* was a different question. Contrast 2 was also significant ($p < .01$) and negative; the offers with monetary inducements generated more responses. Contrast 3 was not significant (or was "marginally" significant, $p = .08$); apparently, flexibility didn't help response much. The contrast 4 result caused the most discussion among the company executives, because nobody believed that it could be not significant. However, the result was not such a mystery when it was discovered that even the executives had split opinions: before the results were known, three of the six executives present had been convinced that the "finance the payments" offers would outbook the price-discount offers, and the other three executives had been convinced that the price-discount offers would generate the most bookings. However, based on the result of contrast 2, both trios claimed to have had the right insight! (We didn't comment.)

Exercises

1. Suppose that we have data on the response to various analgesics, in terms of the reduction of headache pain; we have two replicates per treatment, as shown in Table 5EX.1. Analyze as a one-factor design to test for differences among analgesics. Use $\alpha = .05$.

Table 5EX.1 Headache Pain Reduction

Aspirin (Brand 1)	Aspirin (Brand 2)	Nonaspirin (Brand 1)	Nonaspirin (Brand 2)	Placebo
8	12	7	7	5
7	10	6	5	4

2. In exercise 1, design a set of orthogonal contrasts to compare the following:
   a. Placebo versus nonplacebo
   b. Aspirin (both brands) versus nonaspirin brand 2
   c. Nonaspirin brand 1 versus aspirin (both brands) and nonaspirin brand 2
   d. Aspirin brand 1 versus aspirin brand 2

3. For the orthogonal contrasts in exercise 2, test for significant differences in each of the four comparisons. Use $\alpha = .05$.

4. A journal (*Decision Sciences*) article reported the results of an experiment conducted at a series of graduate business schools, dealing with the job satisfaction of junior faculty members as a function of the leadership style of the dean. Essentially, each dean was classified into *one* of the following leadership styles:
   a. Charismatic
   b. Autocratic
   c. Democratic
   d. Laissez-faire

   Job satisfaction was measured for junior faculty at the various business schools that participated in the study. The key question was whether job satisfaction varied by decanal leadership style. However, the authors also professed interest in three other subissues:

   I. Is there a difference in job satisfaction between some kind of leadership (charismatic, autocratic, democratic) versus no leadership (laissez-faire)?

II. Is there a difference in job satisfaction between democratic leadership and nonparticipative (charismatic, autocratic) leadership?

III. Is there a difference in job satisfaction between charismatic leadership and autocratic leadership?

Form an appropriate orthogonal matrix that can be used to decompose the variability associated with leadership style into components that allow the experimenter to address the three questions above.

5. For Example 5.6, the vanilla pound cake study, generate four questions that are internally orthogonal and different from those in the example. Demonstrate that these four new questions are *not* orthogonal to the original set of four questions.

6. Suppose that we have a one-factor ANOVA, with three columns, in which each level of the factor is the amount of shelf space given to a product, and the dependent variable is sales (in dollars—a numerical value). With three columns we can, of course, break up the sum of squares between columns into two sources. What two sources (that is, what orthogonal contrasts) might make sense in this situation? (We cover this real-world situation explicitly in Chapter 12.)

7. Revisit exercise 2.12, specifically the data in Table 2EX.12 for the waiting time of patients at four offices of the Merrimack Valley Pediatric Clinic (MVPC). Recall that the data are in the data sets as Problem 2.12. Consider the following three questions:
   a. Do the more recently opened offices, Amesbury and Andover, differ in waiting time from the less recently opened offices, Methuen and Salem?
   b. Do the larger offices, Amesbury and Methuen, differ in waiting time from the smaller offices, Andover and Salem?
   c. Do the offices that are in the two less affluent cities, Amesbury and Salem, differ in waiting time from the offices in the two more affluent cities, Andover and Methuen?

   Are these questions orthogonal?

8. If you conclude that the answer to exercise 7 is yes, set up an orthonormal table, preparing to analyze the questions.

9. Continuing with exercise 8, test each contrast. Which question of the three is associated with the largest source of variability?

10. In exercise 7, can we replace question c with another question, with the result that, with the new third question, the three questions are orthogonal?

11. Revisit exercise 2.13, specifically the data in Table 2EX.13, which deal with scores of golfers at four local golf courses that have clientele similar to the golfers at Eastern Electric. The four golf courses are named Near Corners, Meadow Brook, Birch Briar, and Fountainbleau. Recall that the data are on the data disk as Problem 2.13. Suppose that further data about the golf courses reveal that Near Corners has relatively few sand traps, relatively few water hazards, and relatively few par-3's; that Meadow Brook has relatively few sand traps, relatively many water hazards, and relatively many par-3's; that Birch Briar has relatively many sand traps, relatively few water hazards, and relatively many par-3's; and that Fountainbleau has relatively many sand traps, relatively many water hazards, and relatively few par-3's. Determine three questions to orthogonally decompose the sum of squares due to golf course.

12. Form the orthonormal matrix for the three questions you composed for exercise 11.

13. Test each contrast at $\alpha = .05$.

14. Discuss the trade-off between decomposing the sum of squares due to the factor into orthogonal questions, versus decomposing them into questions that are perhaps more desirable, but not orthogonal. Is there a way to accomplish both goals?

Table 5EX.15 Battery Lifetime Study by Device

Cell Phone 1	Cell Phone 2	Flash Cam. 1	Flash Cam. 2	Flash Cam. 3	Flashlight 1	Flashlight 2	Flashlight 3
Domestic	Foreign	Domestic	Foreign	Foreign	Domestic	Foreign	Foreign
1.8	4.2	8.6	7	4.2	4.2	7.8	9
5	5.4	4.6	5	7.8	4.2	7	7.4
1	4.2	4.2	9	6.6	5.4	9.8	5.8

Note: Results are in hours.

15. The data for the battery example from Chapter 2, along with an indication of whether the test devices were made within the United States or elsewhere, are presented in Table 5EX.15. Questions of interest are as follows (all with respect to average battery lifetime): Is the battery lifetime for the cell phone made by the domestic manufacturer different from the battery lifetime for the cell phone made by the foreign manufacturer? Is the battery lifetime of the flash camera brand made by the domestic manufacturer different from the battery lifetime of the flash camera brands made by foreign manufacturers? Is the battery lifetime of the flashlight brand made by the domestic manufacturer different from the battery lifetime of the flashlight brands made by foreign manufacturers? Set up an orthogonal matrix to study these three questions.
16. In exercise 15, suppose we have the following two additional questions of interest: Is there a difference in battery lifetime between flash cameras and flashlights? Is there a difference in battery lifetime between cell phones and the other devices? Add these two inquiries as rows in the orthogonal matrix constructed in exercise 15.
17. Add two (more) orthogonal questions as rows to the five-row orthogonal matrix constructed in exercise 16.
18. Consider again the five-row orthogonal matrix of exercise 16. Would an inquiry comparing the battery lifetime of cell phones to the battery lifetime of flashlights be a sixth orthogonal row of this matrix?

Notes

1. A fictitious name. However, the example is real, with only a few changes in levels of factors, results, and locations, in order not to reveal the identity of the company.

2. For a situation with C levels of a factor, we have referred to $C - 1$ orthogonal questions into which the SSB_c can be partitioned; in this example, with eight levels, we refer to seven orthogonal questions. In fact, we do not need to specify a full seven questions—we can specify any number *up to seven*—the remaining variability is simply labeled as "other differences." We note this again later in the chapter.

3. Where the context makes it clear, we replace $\bar{Y}_{\cdot j}$ with the less cumbersome Y_j.

4. Again, the same exception applies as in note 2: the matrix may be specifically chosen to have fewer than $C - 1$ rows, along with a catchall row describing "other differences."

5. Those who studied (and remember) their high-school physics may recognize the similarity of

orthogonal decomposition to the resolution of vectors, loosely defined as a quantity with a magnitude and a direction, into components in the "X, Y, and Z directions." Typical examples include force and velocity. **V** (the vector, let's say) is decomposed along three unit vectors, one each in the X, Y, and Z directions, respectively. Unit vectors have a magnitude of one, and are perpendicular; in this instance, perpendicular and orthogonal are synonyms. The result of the vector decomposition is a magnitude in the X direction, a magnitude in the Y direction, and a magnitude in the Z direction, such that the sum of squares of these magnitudes is equal to the square of the magnitude of **V.** The orthonormal rows of the coefficient matrix are, in fact, orthogonal unit vectors; the calculation of the Z_i's is the resolution of the effect under study into components along these unit vectors.

6. If two events, A and B, are independent, knowledge of the occurrence of one of the events sheds no light on the occurrence of the other event. If A and B are independent, $P(\text{A and B}) = P(\text{A}) \cdot P(\text{B})$.

7. There would be some major advantages to designing this experiment so that each of the four (or 16) people evaluate each of the four portfolios. However, "people" would then be a second factor, and we have not yet covered designs with two factors. If each person evaluated each of the four portfolios, the design would be called a "repeated-measures" or "within-subject" design.

8. In essence, one can prove that each component sum of squares follows a chi-square distribution with one degree of freedom, and is independent of the error sum of squares. After all, an F distribution is derived as the ratio of two independent chi-square random variables, each divided by its respective degrees of freedom.

9. A placebo is an inactive substance used for control. Placebos are frequently used in studies of drugs to account for the tendency of patients to perceive an improvement in symptoms merely due to the ingestion of "medicine." Double-blind studies, in which the identity of the real drug and the placebo are hidden from both the patient and the experimenter, mitigate against the unintentional giving of cues that otherwise might undermine the integrity of the experiment. Such studies have revealed, for example, that yohimbine hydrochloride, a drug long prescribed for temporary male impotence, performs no better than a placebo. (R. Berkow, *Merck Manual of Medical Information,* Rahway, N.J., Merck, 1997)

10. Normal procedure in a bakery would likely not be to randomly bake 100 cakes individually. However, this is done in the government facility, since the sole purpose of the baking is for testing, not for actual feeding.

CHAPTER 6

Two-Factor Cross-Classification Designs

EXAMPLE 6.1 Planning Travel Packages at Joyful Voyages, Inc.: A Second Look

Examples 5.1 and 5.7 introduced Joyful Voyages, Inc., a company seeking to use experimentation to take full advantage of the increasing demand for cruises to various locations. The company wanted to investigate how the number of bookings for a particular voyage varied by the parameters of the offer. The offer had two core components—one was the base price, and the other was the "incentive" (a traditional direct-mail term for this type of factor)—to further entice the prospect to actually book a cruise. The price factor had two levels and the incentive factor had four levels, including that of no incentive. In Chapter 5 we examined the eight combinations of price and incentive, viewing the problem as a one-factor design with eight levels, as seen in Figure 5.1.

However, the experiment can be viewed and designed as a two-factor design, or two-factor experiment, where the two factors are as shown in Figure 6.1. The same eight offers (combinations of the two factors) are considered, and the same data are used for analysis. However, the analysis will be different, explicitly capturing the *interaction* between the two factors—the factor effects acting nonadditively. That is, we will now be able to address the question of whether the appeal of each incentive differs or not depending on the price, or conversely, whether or not the price effect differs by the type of incentive. We return to this example, and the specific issue of interaction, at the end of the chapter.

Factor 1—Price

Levels:

1. $2989 per person
2. $3489 per person

Factor 2—Incentive

Levels:

1. None—payment due in full four weeks before the sailing
2. Finance payment—over a period of three months, first payment due one month before the sailing
3. Flexibility—ability to keep the dates of the sailing optional (on a first-come, first served basis) until two weeks before the sailing; six trips embark over a 12-week period
4. Discount—opportunity to pay three months in advance, and receive a 15% discount off the price.

Figure 6.1 Levels of price and incentive

6.1 Introduction to Studying Two Factors

Chapter 2 introduced one-factor designs—experiments designed to determine whether the level of a factor, the (one) independent variable, affects the value of some quantity of interest, the dependent variable. By way of example, we considered whether device/usage influences battery life. We expanded on this initial analysis by introducing multiple-comparison testing and orthogonal breakdowns of sums of squares.

Now suppose that we want to determine whether battery life varies by battery brand (a different issue from the brand of the device in which the battery is used). How should we proceed? We could design another one-factor experiment and note its results along with the results of the first experiment, which considered the impact of the test device. Indeed, this sequential process takes place frequently in practice. Unfortunately, for a whole host of reasons, it's usually an unwise process. A major reason is that the conclusions may be incorrect. For example, maybe one brand of battery is superior when used in one device, but another brand is superior in a different device. In other words, device and brand of battery may be synergistic (or interactive) in their effect on battery lifetime.[1] A prudent experimental design would allow for such interaction; an experiment that doesn't, such as the one-factor-at-a-time sequence mentioned above, may easily yield results that are suboptimal—that is, for two factors, A and B, the optimal level of factor A, without regard to the level of factor B, may be A^0 (A "optimal"); the optimal level of factor B, without regard to factor A, may be B^0; but the optimal combination of levels of factors A and B may not be (A^0, B^0). We discuss the critical issue of interaction at length in a later section.

There are other reasons that the one-factor-at-a-time process is not a wise choice. Indeed, even if we knew that there were no interaction effect between the two factors under study, studying the two factors one at a time would result in less reliable estimates of effects given the same total number of data values. And if we study one factor while holding the other factor constant, we may not be able to generalize the results. This issue is elaborated upon in Chapter 9.

We now develop the preferred procedure for examining the possible effect of two cross-classified factors on some quantity of interest. Two factors are said to be "crossed" or **cross-classified** when each level of one factor is in combination with each level of the other factor.[2] (In Chapter 7, we discuss two-factor experiments in which the factors are not crossed but "nested.")

6.2 Designs with Replication

EXAMPLE 6.2 Battery Brand as a Second Factor

Continuing with our study of battery lifetime, the dependent variable, we now use two independent variables (factors): battery brand and the device in which the battery is used. Suppose we study three devices and four brands, and decide to run each combination of levels of factors for two batteries; that is, each combination of levels of factors, or "treatment combination," is replicated twice. (In the real world, battery testing often involves 16 batteries, sometimes more, for each combination of device and brand. However, to make the arithmetic easier to follow so we can focus on the concepts, we use only two batteries per cell in this first example; some other examples in the chapter use a much larger data set.)

Our model development assumes that, as in this example, each treatment combination has the same number of replicates, n. If all treatment combinations do not include the same number of replicates, we use the notation n_{ij} for the number of replicates of treatment combination (i, j); only minor changes in algebra are then required. We present the data in an array, as in Table 6.1. The data are in hours; instead of "devices/usages" we use only the word "device," since these data are for one type of usage (continuous running of the battery) in three different devices. Having signed confidentiality agreements, we do not reveal which brand is which column, and simply note that the brands are four of the following (in alphabetical order): Duracell, Eveready, Kodak, Panasonic, RadioShack, and Ray-O-Vac. For example, we read Table 6.1 as follows: when we ran the experiment (and observed battery

Table 6.1 **Battery Lifetime in Hours by Brand and Device**

	Brand			
Device	**1**	**2**	**3**	**4**
1	17.9, 18.1	17.8, 17.8	18.1, 18.2	17.8, 17.9
2	18.2, 18.0	18.0, 18.3	18.4, 18.1	18.1, 18.5
3	18.0, 17.8	17.8, 18.0	18.1, 18.3	18.1, 17.9

life of two batteries), at the combination of device 1 and brand 1, battery life values were 17.9 hours and 18.1 hours, respectively, for the two replicates.

The first row of experimental results in Table 6.1 corresponds to all trials of the experiment for which the device in which the battery was inserted was held at device 1 ("level 1"). Remember, the different levels may or may not correspond to metric values (here, they don't); either way they are treated simply as "categories." (We discussed this point at length in note 1 at the end of Chapter 2.) Similarly, the first column of experimental results corresponds to all trials of the experiment using brand/level 1. The cell located at the intersection of this row and column contains all experimental outcomes for which device was held at level 1 and, simultaneously, brand was held at level 1; that is, a cell is the location for a treatment combination. As we noted earlier, this array is known as a cross-classification (or cross-tabulation). Each entry may be said to be at a "crossing" of the two factors; hence the name, "two-factor cross-classification design."

Table 6.1 has three rows ($R = 3$), four columns ($C = 4$), and two replicates ($n = 2$) for each cell or treatment combination. Remember, each cell corresponds to a specific and unique combination of the levels of the row and column factors. This arrangement, which has the same number of replicates for each cell, is an orthogonal design: if you're told which row a data value is in, it gives you no clue as to which column it's in, and vice versa.[3] This is said to be a "balanced" design.

Of course, a design need not have the same number of replicates per cell; however, the analysis is simpler and the experiment more efficient and easier to interpret if it does. (Alternatively, as we'll see later in the book, there are carefully planned, systematic designs in which the number of replicates is not the same for each cell—for example, a specific half of the cells have the same number of replicates, n, and a specific half of the cells are not run at all.) If an array is not balanced, we often try to make it balanced. For example, if all but one cell has five replicates, and that one has four replicates, this might be because the missing data value had been lost or never taken, for whatever reason. We might create a fifth replicate for that cell by taking the average of the other four replicates of that cell, and account for having done so by dropping one degree of freedom from the error term. Or, if there were an extra data value in one cell, we could randomly choose one to drop. Although this might be considered "throwing away good data," in practice the facilitation of analysis and interpretation is often considered more important than the added reliability from one additional data value.

The Model

As in Chapter 2, we begin with a statistical model. Every data point can be written

$$Y_{ijk} = \mu + \rho_i + \tau_j + I_{ij} + \epsilon_{ijk}$$

where

$i = 1, 2, 3, \ldots, R$; that is, i indexes rows
$j = 1, 2, 3, \ldots, C$; j indexes columns
$k = 1, 2, 3, \ldots, n$; k indexes replication per cell
Y_{ijk} = the data point that corresponds with the kth replicate in the cell at the intersection of the ith row and jth column; for example, $Y_{231} = 18.4$ in Table 6.1
μ = the grand mean
ρ_i = the difference between the ith row mean and the grand mean
τ_j = the difference between the jth column mean and the grand mean
I_{ij} = a measure of the interaction associated with the ith row and the jth column (more about this later)
ϵ_{ijk} = the "error" or "noise" in the data value (i, j, k)—the difference between the data value and the true mean of that cell

The terms τ_j and ϵ_{ijk} are the same quantities as they were in Chapter 2, where we considered one-factor designs, except that the error term has a necessary additional subscript. ρ_i is for rows what τ_j is for columns, and is symmetric in the role it plays. As noted, I_{ij} represents the degree of interaction associated with the ith row and jth column, and will be elaborated on soon. In general, we have n observations per cell, and RC cells, corresponding to a total of (nRC) data values, and $(nRC - 1)$ degrees of freedom.

Parameter Estimates

Our estimates of the parameters of this model follow a development similar to that in Chapter 2. Replacing each parameter in the model by its corresponding estimate, we find

$$Y_{ijk} = \bar{Y}_{\cdots} + (\bar{Y}_{i\cdot\cdot} - \bar{Y}_{\cdots}) + (\bar{Y}_{\cdot j\cdot} - \bar{Y}_{\cdots}) + (\bar{Y}_{ij\cdot} - \bar{Y}_{i\cdot\cdot} - \bar{Y}_{\cdot j\cdot} + \bar{Y}_{\cdots}) + (Y_{ijk} - \bar{Y}_{ij\cdot})$$

where

$\bar{Y}_{\cdots}$ is the grand mean (mean of all the data)
$\bar{Y}_{i\cdot\cdot}$ is the mean of row i
$\bar{Y}_{\cdot j\cdot}$ is the mean of column j
$\bar{Y}_{ij\cdot}$ is the mean of *cell* $[i, j]$

All the terms in the above equation are similar to those of the corresponding one-factor-design equation, and are intuitive, with the notable exception of the term $(\bar{Y}_{ij\cdot} - \bar{Y}_{i\cdot\cdot} - \bar{Y}_{\cdot j\cdot} + \bar{Y}_{\cdots})$; let's consider this term in detail. In doing so, recall that we're examining an equation for the value of one specific data value—the *k*th replicate in the *i*th row and the *j*th column (that is, in the *ij*th cell).

We may write the term $(\bar{Y}_{ij\cdot} - \bar{Y}_{i\cdot\cdot} - \bar{Y}_{\cdot j\cdot} + \bar{Y}_{\cdots})$ as

$$(\bar{Y}_{ij\cdot} - \bar{Y}_{\cdots}) - (\bar{Y}_{i\cdot\cdot} - \bar{Y}_{\cdots}) - (\bar{Y}_{\cdot j\cdot} - \bar{Y}_{\cdots})$$

where

$(\bar{Y}_{ij\cdot} - \bar{Y}_{\cdots})$ is the difference between the cell mean and the grand mean
$(\bar{Y}_{i\cdot\cdot} - \bar{Y}_{\cdots})$ is the difference between the row mean and the grand mean
$(\bar{Y}_{\cdot j\cdot} - \bar{Y}_{\cdots})$ is the difference between the column mean and the grand mean

Thus, $(\bar{Y}_{ij\cdot} - \bar{Y}_{\cdots}) - (\bar{Y}_{i\cdot\cdot} - \bar{Y}_{\cdots}) - (\bar{Y}_{\cdot j\cdot} - \bar{Y}_{\cdots})$ may be viewed as

The degree to which a cell mean differs from the grand mean

minus

an adjustment for the data value's "row membership"

minus

an adjustment for the data value's "column membership."

Assume for a moment, *without loss of generality,* that there is no error in any of the $\bar{Y}$ values. This is unrealistic, but can certainly be envisioned—just imagine that there are an infinite number of replicates, so that the calculated mean of each cell, $\bar{Y}_{ij\cdot}$, exactly equals the true mean of that cell, $\mu_{ij\cdot}$. Given this assumption (which is made simply to elucidate the discussion), how can (or why might)

$$(\bar{Y}_{ij\cdot} - \bar{Y}_{\cdots}) - (\bar{Y}_{i\cdot\cdot} - \bar{Y}_{\cdots}) - (\bar{Y}_{\cdot j\cdot} - \bar{Y}_{\cdots})$$

be nonzero? After all, we have (by assumption) no error, and have explicitly taken the row effect and column effect into account. The answer, as alluded to in the introductory paragraph, is *interaction.* The next section defines and illustrates this important concept.

Interaction

Suppose we have two factors, A and B, each at two levels, high (H) and low (L), and (again for clarity of discussion and no loss of generality) an infinite amount of replication (and hence no error in the cell means). Further suppose that the cell means are as follows:

	B_L	B_H
A_L	5	8
A_H	10	?

Suppose we start from row 1 and column 1, (A_L, B_L), where the yield is 5. As we change the level of factor A ($A_L \rightarrow A_H$), holding the level of factor B constant (at B_L), the yield increases by 5 (= 10 − 5). If we hold the level of factor A constant (at A_L) and change the level of factor B ($B_L \rightarrow B_H$), the yield increases by 3 (= 8 − 5). What happens when the level of both factors, A and B, changes ($A_L \rightarrow A_H$ *and* $B_L \rightarrow B_H$)? There are three possibilities. If the yield is 13 (increasing by 8, precisely the sum of 3 and 5), there is no interaction. If the yield is greater than 13 (increasing by more than 8, the sum of 3 and 5), we have positive interaction. If the yield is less than 13 (increasing by less than 8, the sum of 3 and 5, or not increasing at all), we have negative interaction. To summarize:

If (A_H, B_H) = 13, there is no interaction.
If (A_H, B_H) > 13, there is positive interaction.
If (A_H, B_H) < 13, there is negative interaction.

This suggests one practical way to describe interaction, at least with respect to its sign:

Interaction = degree of difference from the sum of the separate effects

In many real-world situations this working definition of interaction has proved useful to the authors. For example, in Chapter 11 we consider a case with a dependent variable of sales in a supermarket. Suppose that we consider two factors of the study, with two levels each. One is the amount of shelf space a product gets (say, apples), and the other is whether the product is promoted (highlighted in signs at the entrance to the supermarket). If the normal amount of shelf space is doubled and there is no promotion, sales increase 12%. If the product is promoted and shelf space is normal, sales increase 8%. In practice, management considered it important to determine how sales would be affected if shelf space were doubled *and* the product were promoted. Would sales increase by 20% (the sum of 12% and 8%), more than 20% (showing positive interaction), or less than 20% (showing negative interaction)? The answer had profound implications concerning whether each of the limited resources (shelf space with respect to capacity, promotion with respect to budget) should be given to the same product (if there's positive interaction) or to different products (if there's negative interaction).

There is another way to conceptualize interaction. Suppose we have the following cell means:

	B_L	B_H
A_L	5	8
A_H	10	17

Holding the level of factor B constant at B_L, what happens as $A_L \rightarrow A_H$?
Answer: the yield increases by 5 (= 10 − 5).
Holding the level of factor B constant at B_H, what happens as $A_L \rightarrow A_H$?
Answer: the yield increases by 9 (= 17 − 8).

This suggests another useful, practical way to describe interaction:

> If the effect of one factor varies depending on the level of another factor, there is **interaction** between the two factors.

Of course, we can examine the interaction between A and B by holding the level of factor A constant and varying the level of factor B (instead of the reverse, above).

At A_L, $B_L \rightarrow B_H$ increases yield by 3 (= 8 − 5).

At A_H, $B_L \rightarrow B_H$ increases yield by 7 (= 17 − 10).

It is no coincidence that the magnitude of the interaction is the same:

$$(9 - 5) = (7 - 3) = 4$$

This second working definition of interaction has also proved useful to the authors in many real-world situations. For example, as in a problem discussed later in the chapter, suppose that one factor is gender (M, F), the other two-level factor is brand name, and the dependent variable is "purchase intent." In both this example and the later one, changing from one brand name to the other has one effect for males and a very different effect for females. To say that gender and brand name "interact," as illustrated by the fact that the brand name effect differs from one gender to the other, is fairly simple. It's a lot clearer than describing the interaction the first way: As you go from male to female at the same brand name, the change in purchase intent is this; as you go from brand name 1 to brand name 2 for the same gender, the change in purchase intent is that; when you change both brand name and gender you get a change that is different from the sum of the aforementioned two changes. What?

Each way of viewing interaction is useful in different scenarios. Sometimes both ways are practical. Either way the numerical value of the interaction is the same, regardless of which is the more useful way to view it.

We'll say much more about interaction effects in subsequent chapters. At this point, merely note that the performance of two separate one-factor experiments (as opposed to a cross-classification of the two factors) does not allow the examination of interaction effects. Prudent experimentation must allow for the possibility of interaction unless there is ample reason to believe that there is none (an issue also discussed in later chapters). The default presumption, in the absence of evidence to the contrary, should generally be that interaction may be present.

A useful way to depict interaction between two factors is with an interaction plot. For example,

	B_L	B_H
A_L	5	8
A_H	10	17

could be represented by Figure 6.2. In an interaction plot, only when the lines are linear (which is unavoidable when there are only two levels of a factor)

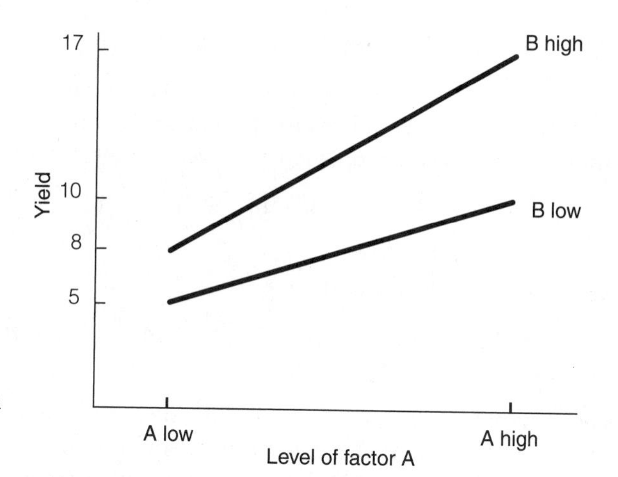

Figure 6.2 **Interaction plot**

and parallel is there no interaction. In Figure 6.2, the lines are not parallel, confirming that there is nonzero interaction, as calculated earlier. If there is interaction, the interpretation of main effects become problematic. In this example, going from A low to A high gives an average increase in yield of 7; however, it's really either 5 or 9, depending on the level of B. Thinking that the increase is 7 for both levels of B might not be too big a distortion; however, when the lines cross, the distortion is usually more serious.

For example, suppose that the numbers were

	B_L	B_H
A_L	5	8
A_H	10	7

Now, the main effect of A is 2; however, it's 5 when B is low, and -1 when B is high. This is more of a distortion. For B, it's "worse"; the main effect of B is zero, but it's either $+3$ or -3 depending on the level of A. The interaction plot in Figure 6.3 illustrates a crossover.

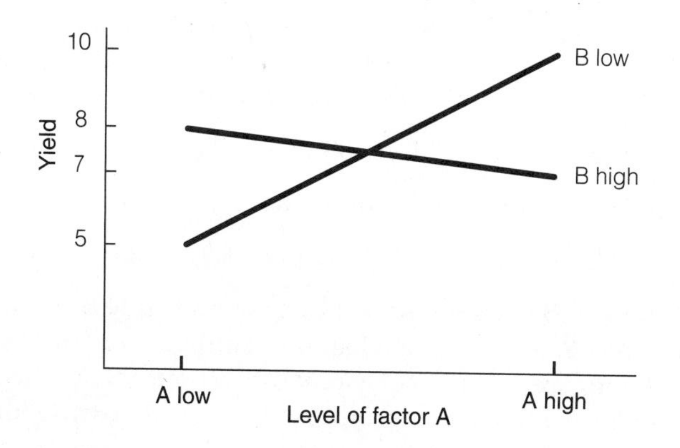

Figure 6.3 **Crossover interaction plot**

Back to the Statistical Model: Sums of Squares

We now develop the sums-of-squares relationships, as in Chapter 2, for our two-factor design. Recall from the earlier section, Parameter Estimates, the equation in which we replaced each parameter by its best estimate:

$$Y_{ijk} = \bar{Y}_{...} + (\bar{Y}_{i..} - \bar{Y}_{...}) + (\bar{Y}_{.j.} - \bar{Y}_{...}) + (\bar{Y}_{ij.} - \bar{Y}_{i..} - \bar{Y}_{.j.} + \bar{Y}_{...}) + (Y_{ijk} - \bar{Y}_{ij.})$$

We rearrange terms slightly to get

$$Y_{ijk} - \bar{Y}_{...} = (\bar{Y}_{i..} - \bar{Y}_{...}) + (\bar{Y}_{.j.} - \bar{Y}_{...}) + (\bar{Y}_{ij.} - \bar{Y}_{i..} - \bar{Y}_{.j.} + \bar{Y}_{...}) + (Y_{ijk} - \bar{Y}_{ij.})$$

As in Chapter 2, we can square both sides of this equation (if two quantities are equal, their squares are, of course, equal), and recalling that we can write this equation for every combination of i, j, and k (row, column, and replicate), we can add all (nRC) such squared equations. Again, as in Chapter 2, all cross-product terms cancel, and we get (sums over i are from 1 to R, sums over j are from 1 to C, and sums over k are from 1 to n):

$$\begin{aligned}\Sigma_i\Sigma_j\Sigma_k (Y_{ijk} - \bar{Y}_{...})^2 &= \Sigma_i\Sigma_j\Sigma_k (\bar{Y}_{i..} - \bar{Y}_{...})^2 + \Sigma_i\Sigma_j\Sigma_k (\bar{Y}_{.j.} - \bar{Y}_{...})^2 \\ &+ \Sigma_i\Sigma_j\Sigma_k (\bar{Y}_{ij.} - \bar{Y}_{i..} - \bar{Y}_{.j.} + \bar{Y}_{...})^2 \\ &+ \Sigma_i\Sigma_j\Sigma_k (Y_{ijk} - \bar{Y}_{ij.})^2\end{aligned}$$

Once again, observe that $\Sigma_i\Sigma_j\Sigma_k (\bar{Y}_{i..} - \bar{Y}_{...})^2$ does not depend on the indices j and k; we can factor the sum as follows:

$$\begin{aligned}\Sigma_i\Sigma_j\Sigma_k (\bar{Y}_{i..} - \bar{Y}_{...})^2 &= \Sigma_j\Sigma_k [\Sigma_i (\bar{Y}_{i..} - \bar{Y}_{...})^2] \\ &= nC[\Sigma_i (\bar{Y}_{i..} - \bar{Y}_{...})^2]\end{aligned}$$

Similar reasoning holds for the second and third terms to the right of the equal sign, and we arrive at

$$\begin{aligned}\Sigma_i\Sigma_j\Sigma_k (Y_{ijk} - \bar{Y}_{...})^2 &= nC\,\Sigma_i (\bar{Y}_{i..} - \bar{Y}_{...})^2 + nR\,\Sigma_j (\bar{Y}_{.j.} - \bar{Y}_{...})^2 \\ &+ n\,\Sigma_i\Sigma_j (\bar{Y}_{ij.} - \bar{Y}_{i..} - \bar{Y}_{.j.} + \bar{Y}_{...})^2 \\ &+ \Sigma_i\Sigma_j\Sigma_k (Y_{ijk} - \bar{Y}_{ij.})^2\end{aligned}$$

Note that without replication the last term equals zero since, for every value of (i, j), $\bar{Y}_{ij.} = Y_{ijk}$; that is, with no replication, a cell mean equals the individual data value. We'll come back to this observation later.

We may write the above equation symbolically as

$$\text{TSS} = \text{SSB}_r + \text{SSB}_c + \text{SSI}_{r,c} + \text{SSW}$$

and allocate the degrees of freedom as follows:

$$nRC - 1 = (R - 1) + (C - 1) + (R - 1)(C - 1) + RC(n - 1)$$

Note that SSB_c is the same basic quantity as SSB_c in our one-factor design in Chapter 2. SSB_r is the analogous quantity for the row factor; after all, rows and columns are, in essence, symmetric in their role in the analysis. In $\text{SSI}_{r,c}$, the I stands for interaction; the subscripts are, in a sense, unnecessary, since what else could be interacting except the two factors under study? However, we adopt the subscripted notation to indicate that, in a study with three or more factors, the subscripts would distinguish which term stands for which set of factors interacting.

Consider the degrees-of-freedom allocation. The terms $(nRC - 1)$ for the total, $(C - 1)$ for columns, and its analog $(R - 1)$ for rows, follow directly from the $(n - 1)$ rule discussed in Chapter 2. The degrees-of-freedom value for an interaction term is always the product of the degrees of freedom of the factors interacting. This may not be obvious, and we don't want to spend a lot of time formally proving this, but consider the following 3×4 table. If you know the value of the cell means for the cells with X's in them, along with all row means and column means, the cell means of the remaining cells (those without X's inside) are uniquely determined.

X	X	X	
X	X	X	

There are X's in six of the cells, corresponding to $(3 - 1)(4 - 1) = 2 \cdot 3 = 6$ degrees of freedom. This logic generalizes to $(R - 1)(C - 1)$ cells necessary to be known, along with all but one of the row means and column means, for all of the remaining cell means to be determined; hence, we have $(R - 1)(C - 1)$ degrees of freedom for the interaction sum of squares, a sum of squares that includes each cell and uses the row and column means in its calculation.

The number of degrees of freedom for the error term, SSW, follows directly from the $(n - 1)$ rule; each cell, with n data values, contributes $(n - 1)$ degrees of freedom, and we have RC cells, for a total of $RC(n - 1)$ degrees of freedom.

EXAMPLE 6.3 Numerical Example

Calculating sums of squares Now let's calculate these sums-of-squares quantities with our data for the battery lifetime example. We'll consider in detail the mechanics that the authors have found helpful.

Across the left and the top of the array in Table 6.2 are the level designations of the row and column factors, respectively, for our battery example; we'll need these later when we examine how the rows and columns differ (if they do). Within each cell, we enter the cell mean and put it in parentheses (or circle it by hand) to distinguish it from the yields themselves. (Another example of practicing safe statistics!) The row and column means are at the

Table 6.2 Battery Lifetime in Hours by Brand and Device

	Brand				
Device	1	2	3	4	$\bar{Y}_{i..}$
1	17.9, 18.1 (18.0)	17.8, 17.8 (17.8)	18.1, 18.2 (18.15)	17.8, 17.9 (17.85)	17.95
2	18.2, 18.0 (18.1)	18.0, 18.3 (18.15)	18.4, 18.1 (18.25)	18.1, 18.5 (18.3)	18.20
3	18.0, 17.8 (17.9)	17.8, 18.0 (17.9)	18.1, 18.3 (18.2)	18.1, 17.9 (18.0)	18.00
$\bar{Y}_{.j.}$	18.00	17.95	18.20	18.05	18.05

Note: Cell means are in parentheses, column means are in the bottom row, and row means are in the last column.

right and bottom of the array, respectively, and the grand mean is shown as the lower-rightmost entry. Check to verify that the average of the row means equals the average of the column means. Next, calculate the sums of squares.

$$\begin{aligned} SSB_r &= (2)(4)[(17.95 - 18.05)^2 + (18.20 - 18.05)^2 + (18.00 - 18.05)^2] \\ &= 8[.01 + .0225 + .0025] \\ &= .28 \end{aligned}$$

$$\begin{aligned} SSB_c &= (2)(3)[(18.00 - 18.05)^2 + (17.95 - 18.05)^2 + (18.2 - 18.05)^2 \\ &\quad + (18.05 - 18.05)^2] \\ &= 6[.0025 + .001 + .0225 + 0] \\ &= .21 \end{aligned}$$

$$\begin{aligned} SSI_{r,c} &= 2[(18.0 - 17.95 - 18.00 + 18.05)^2 \\ &\quad + (17.8 - 17.95 - 17.95 + 18.05)^2 \\ &\quad + \cdots + (18.0 - 18.0 - 18.05 + 18.05)^2] \\ &= 2[.055] \\ &= .11 \end{aligned}$$

$$\begin{aligned} SSW &= [(17.9 - 18.0)^2 + (18.1 - 18.0)^2 + (17.8 - 17.8)^2 \\ &\quad + \cdots + (17.9 - 18.0)^2] \\ &= .30 \end{aligned}$$

and as a consequence,

$$\begin{aligned} TSS &= .28 + .21 + .11 + .30 \\ &= .90 \end{aligned}$$

If doing the calculations by hand, it may be simpler to independently compute SSB_c, SSB_r, SSW, and TSS, and then derive $SSI_{r,c}$ by subtraction.

The analysis of variance We are now prepared to embark on our analysis of variance to determine if row or column factors or interaction effects are statistically significant. Our ANOVA table is in Table 6.3.

Table 6.3 **ANOVA Table: Two-Factor Study of Battery Life**

Source of Variability	SSQ	*df*	MS	F_{calc}
Rows	.28	2	.14	5.6
Columns	.21	3	.07	2.8
Interaction	.11	6	.0183	.73
Error	.30	12	.025	
Total	.90	23		

We compare F_{calc} to the appropriate critical value, **c**, taken from the F tables in the Statistical Tables appendix. For the row factor, we test these hypotheses:

H_0: All row means are equal

H_1: Not all row means are equal

In Chapter 2 we saw that

$$\text{MSB}_c = \text{SSB}_c/(C-1) \qquad \text{and} \qquad E(\text{MSB}_c) = \sigma^2 + V_c$$

where $E(\ )$ is the expected value function and V_c = variability due to differences in true column means. In the same way, we have

$$\text{MSB}_r = \text{SSB}_r/(R-1) \qquad \text{and} \qquad E(\text{MSB}_r) = \sigma^2 + V_r$$

where V_r = variability due to differences in true row means. Similarly for interaction, we have

$$\text{MSI} = \text{SSI}/[(R-1)(C-1)] \qquad \text{and} \qquad E(\text{MSI}) = \sigma^2 + V_{\text{int}}$$

where $V_{\text{int}} = n\,\Sigma_i\Sigma_j\, I_{ij}^2/[(R-1)(C-1)]$.

With $\alpha = .05$, and for $df = (2, 12)$, we have $\mathbf{c} = 3.89$. Thus, $F_{calc} = 5.6 > 3.89$, and we reject H_0 ($p < .05$); the row factor is significant, meaning that we conclude that not all true row means are equal.

For the column factor, we test these hypotheses:

H_0: All column means are equal

H_1: Not all column means are equal

With $\alpha = .05$, and for $df = (3, 12)$, we have $\mathbf{c} = 3.49$. Thus, $F_{calc} = 2.8 < 3.49$ and we accept H_0 ($p > .05$); the column factor is not significant, meaning that we cannot reject that all true column means are equal.

Finally, for interaction, we test these hypotheses:

H_0: There is no interaction between row and column factors

H_1: There is interaction between row and column factors

Table 6.4 **ANOVA with Pooled Error and Interaction**

Source of Variability	SSQ	*df*	MS	F_{calc}
Rows	.28	2	.14	6.15
Columns	.21	3	.07	3.07
Error	.41 (.11 + .30)	18 (6 + 12)	.0228	
Total	.90	23		

With $\alpha = .05$, and for $df = (6, 12)$, we have $\mathbf{c} = 3.00$. Thus, $F_{calc} = .73 < 3.00$, $p > .05$, and we accept H_0; the interaction effect between the two factors is not significant—we cannot reject that there is no interaction between the two factors. As an aside, whenever an F_{calc} is less than one, we really don't have to bother to look up the value of **c** in the F table; for any practical value of α, no F table value is under one.

Note that in our example, the mean square for interaction, MSI, is less than MSW. We concluded that there is no interaction by comparing $F_{calc} = .73$ to $\mathbf{c} = 3.00$. Since

$$E(\text{MSI}) = \sigma^2 + V_{int} \qquad \text{and} \qquad E(\text{MSW}) = \sigma^2$$

and V_{int} cannot be negative, some experimenters take the view that this constitutes strong evidence that V_{int} is zero. Correspondingly, they then argue that

$$E(\text{MSI}) = \sigma^2 + 0 = \sigma^2$$

meaning that MSI and MSW are both estimating the same quantity—σ^2. Hence, they say, one should pool the two estimates for use as the denominator of F_{calc} when testing for significance of the row and column factors. (Everyone would agree with the latter part of the discussion—if two quantities are known to estimate the same quantity, they should be pooled; it's the premise that they estimate the same quantity that not all experimenters agree with.)

The pooling leads to a modified ANOVA table, as in Table 6.4. Pooling the two estimates is algebraically equivalent to adding the sums of squares, and adding the degrees of freedom. In each case, both F_{calc} and **c** change, the latter due to the increase in the error *df*. For the row factor, $F_{calc} = 6.15 > \mathbf{c} = 3.55$ ($p < .05$) and we continue to reject H_0; for the column factor, $F_{calc} = 3.07 < \mathbf{c} = 3.16$ ($p > .05$) and we still accept H_0 (but it's a much closer call).

EXAMPLE 6.4 Analysis Using Excel

We now perform the analysis of the previous example using Excel software. As mentioned in Chapter 2, we do not recommend Excel as the software of choice for experimental design and ANOVA analyses, but we include examples using it because Excel is so commonly available, and may be the only software a reader is able to use.

The input data are shown in Table 6.5. Note certain facts about their form for Excel. The first *two* rows of the input data are data from the *first* level of the row-factor, the next two rows correspond to the second level of the row-factor, and so on. That's how replication is dealt with in Excel: a dialog box asks the user to specify how many rows of replicates correspond to each row-factor level. Specifying 2 tells the software that the six rows of data represent three levels of two replicates each, and not, for example, two levels of three replicates each or six levels of one replicate each.

Also, Excel insists that the complete first (literal) row be column labels, and that the first (literal) column have one or more row labels; thus, one needs to highlight a 7×5 table like that in Table 6.5 in order to get Excel to read the problem as a six-row, four-column set of data—hence, the X and a, b, c, d in Table 6.5. If Excel is to be a user's software of choice, perhaps a software "add-in"[4] to Excel should be purchased.

Table 6.6 shows the output, which gives various descriptive statistics (count, total, mean, variance) before providing the ANOVA table.

Table 6.5 Excel Format: Input Data

	a	b	c	d
X	17.9	17.8	18.1	17.8
	18.1	17.8	18.2	17.9
	18.2	18.0	18.4	18.1
	18.0	18.3	18.1	18.5
	18.0	17.8	18.1	18.1
	17.8	18.0	18.3	17.9

Table 6.6 **Excel Output: ANOVA analysis**

Anova: Two-Factor with Replication

Summary	a	b	c	d	Total
X					
Count	2	2	2	2	8
Sum	36	35.6	36.3	35.7	143.6
Average	18	17.8	18.15	17.85	17.95
Variance	.02	0	.005	.005	.0257
Count	2	2	2	2	8
Sum	36.2	36.3	36.5	36.6	145.6
Average	18.1	18.15	18.25	18.3	18.2
Variance	.02	.045	.045	.08	.0343
Count	2	2	2	2	8
Sum	35.8	35.8	36.4	36.0	144.0
Average	17.9	17.9	18.2	18.0	18.0
Variance	.02	.02	.02	.02	.0286
Total					
Count	6	6	6	6	
Sum	108	107.7	109.2	108.3	
Average	18	17.95	18.2	18.05	
Variance	.02	.039	.016	.063	
ANOVA					

Source of Variation	*SS*	*df*	*MS*	*F*	*P-value*	*F crit*
Sample	.28	2	.14	5.6	0.019	3.885
Columns	.21	3	.07	2.8	0.085	3.490
Interaction	.11	6	.0183	0.73	0.632	2.996
Within	.30	12	.025			
Total	.90	23				

EXAMPLE 6.5 Analysis Using SPSS

We now perform the analysis of the same example using SPSS software. Table 6.7 shows the input data; note that SPSS uses a very different input format from that of Excel. We list all of the data in column one, and in columns two and three we note to which row and column, respectively, each data point belongs. SPSS provides a dialog box that asks which column/variable is the Y (the dependent variable) and which columns/variables are the factors (the independent variables).

Table 6.8 shows the SPSS output. The format of the ANOVA table is a bit different from the one we constructed in Table 6.3 and from that of Excel; as we noted in Chapter 2, each software package presents the output somewhat differently. Of course, the basic information is the same. In Table 6.8, the SPSS ANOVA table not only provides the standard information for rows (Device) and columns (Brand) but also pools them into what it calls "main effects." (Each factor's effect, by itself, is called a "main effect"; pooling the main effects means adding, for each factor, the sums of squares and the degrees of freedom, which yields a mean square and an F_{calc} value.) If we wonder whether the main effects taken as a whole are significant (although this is

Table 6.7 SPSS Format: Input Data

Time	Device	Brand	Time	Device	Brand
17.9	1.00	1.00	18.1	1.00	3.00
18.1	1.00	1.00	18.2	1.00	3.00
18.2	2.00	1.00	18.4	2.00	3.00
18.0	2.00	1.00	18.1	2.00	3.00
18.0	3.00	1.00	18.1	3.00	3.00
17.8	3.00	1.00	18.3	3.00	3.00
17.8	1.00	2.00	17.8	1.00	4.00
17.8	1.00	2.00	17.9	1.00	4.00
18.0	2.00	2.00	18.1	2.00	4.00
18.3	2.00	2.00	18.5	2.00	4.00
17.8	3.00	2.00	18.1	3.00	4.00
18.0	3.00	2.00	17.9	3.00	4.00

Table 6.8 **SPSS Output: ANOVA Table**

*** ANALYSIS OF VARIANCE ***

Time
by Device
Brand

Source of Variation	Sum of Squares	DF	Mean Squares	F	Sig of F
Main Effects	.49000	5	.09800	3.920	.024
Device	.28000	2	.14000	5.600	.019
Brand	.21000	3	.07000	2.800	.085
2-Way Interactions	.11000	6	.01833	.733	.633
Device Brand	.11000	6	.01833	.733	.633
Explained	.60000	11	.05455	2.182	.098
Residual	.30000	12	.02500		
Total	.90000	23	.03913		

24 cases were processed.
0 cases (.0 pct) were missing.

often not a useful question), we would use the F_{calc} of 3.92 and the p value (here called "sig of F") of .024. For interaction, SPSS does the same thing, providing a row that pools all the interaction terms; here there is only one term, so the pooling consists of only the one term (with SSI = .11, $df = 6$, and so on). The output then pools all main and interaction effects, calling it "explained" (SSQ = .60, $df = 11$, and so on).

Noticing that the F_{calc} value for the interaction is under 1.0, we might decide that we want to pool its sum of squares and degrees of freedom with those of the error term (called the "residual" by SPSS) and then recompute the F values for the two main effects. We could instruct SPSS to do this pooling by going into "options" and indicating that only main effects are to be "in the model." The new ANOVA is shown in Table 6.9.

Table 6.9 **SPSS Output: ANOVA with Pooled Error and Interaction**

*** ANALYSIS OF VARIANCE ***

Source of Variation	Sum of Squares	DF	Mean Squares	F	Sig of F
Main Effects	.490	5	.098	4.302	.009
Device	.280	2	.140	6.146	.009
Brand	.210	3	.070	3.073	.054
Explained	.490	5	.098	4.302	.009
Residual	.410	18	.023		
Total	.900	23	.039		

EXAMPLE 6.6 A Larger Example: First United Federal Bank of Boston

M. L. Naughton, the vice president of operations at the First United Federal Bank of Boston, wanted to investigate the benefit of various types of teller training. Tellers may attend a formal two-week teller-training course, depending on its availability at the time of their hiring, or they may have some number of weeks of one-on-one training working with an experienced teller at a branch office, or they may have some combination of both. Branch managers usually augment formal course work with one-on-one training before allowing a new employee to work unaided with bank customers.

There are many measures of a teller's performance; one of the more important is the magnitude of the monthly gross overage and/or shortage for that teller. The bank cares about the error irrespective of the sign—either the customer or the bank is the loser in an erred transaction, and neither is acceptable to the bank.

The bank staff has assembled data on 100 new tellers, as seen in Table 6.10. For each teller, the entry in the Course column indicates the teller did (yes) or did not (no) take the two-week course. The column headed Weeks 1:1 indicates the *number of weeks* of one-on-one training provided for that teller. The entry in the Error column indicates the larger of the sum of overages and the sum of shortages for that teller. What was of prime interest is whether either (or both) of the training options affect performance, and whether there is (nonzero) interaction between the two. The data are also available in the data sets, labeled as Bank Example in Chapter 6.

There were two levels of the factor "formal two-week teller-training course": took it (yes) or not (no). And there were five levels of the factor "number of weeks of one-on-one training": two weeks, four weeks, six weeks, eight weeks, or ten weeks. For each of the resulting ten treatment combinations, there were ten tellers/"replicates."

Using the JMP software, we performed a two-factor cross-classification ANOVA. Figure 6.4 represents the "means diamonds" plot, as described for JMP in Chapter 2. Note that this dependent variable, called "Error," is one for which *lower is better.* In Figure 6.4, the data for "course = no" has an interesting feature: it appears to be bimodal. This is because the Table 6.10 data for "Course = no" and "Course = yes" include all the levels of one-on-one time; the bimodality will turn out to be a manifestation of interaction between the two factors.

Figure 6.5 shows the means diamonds plot for the factor "time (weeks) of one-on-one training." The ANOVA results are in Table 6.11.

It can be seen that both factors, "Course" and "Weeks 1:1," and their interaction are highly significant. From the diamonds plots, we can see that Course = yes is lower (better) than Course = no, and at least pictorially, it appears that Wks 1:1 = 4 is lower than Wks 1:1 = 2, and that Wks 1:1 = 6, 8, and 10 are about equally effective, and lower than Wks 1:1 = 4. Indeed, Fisher's LSD at $\alpha = .05$ and Tukey's HSD at $\mathbf{a} = .05$ both verify this.

Table 6.10 Bank Teller Performance by Training Type

Teller	Course	Weeks 1:1	Error	Teller	Course	Weeks 1:1	Error
1	No	2	$66.52	51	No	2	$59.24
2	No	4	$22.10	52	No	4	$22.22
3	No	6	$18.48	53	No	6	$10.36
4	No	8	$8.08	54	No	8	$14.23
5	No	10	$22.69	55	No	10	$7.30
6	Yes	2	$11.88	56	Yes	2	$14.52
7	Yes	4	$15.07	57	Yes	4	$8.21
8	Yes	6	$10.57	58	Yes	6	$14.82
9	Yes	8	$7.53	59	Yes	8	$10.73
10	Yes	10	$10.16	60	Yes	10	$19.64
11	No	2	$53.81	61	No	2	$55.77
12	No	4	$21.25	62	No	4	$30.53
13	No	6	$19.93	63	No	6	$11.39
14	No	8	$9.48	64	No	8	$12.13
15	No	10	$7.78	65	No	10	$12.40
16	Yes	2	$11.02	66	Yes	2	$10.89
17	Yes	4	$14.20	67	Yes	4	$14.87
18	Yes	6	$9.50	68	Yes	6	$3.32
19	Yes	8	$3.75	69	Yes	8	$14.40
20	Yes	10	$12.16	70	Yes	10	$14.31
21	No	2	$64.03	71	No	2	$59.24
22	No	4	$29.63	72	No	4	$24.20
23	No	6	$14.42	73	No	6	$11.19
24	No	8	$21.57	74	No	8	$10.81
25	No	10	$6.21	75	No	10	$14.37
26	Yes	2	$10.26	76	Yes	2	$18.60
27	Yes	4	$17.27	77	Yes	4	$19.40
28	Yes	6	$12.96	78	Yes	6	$6.03
29	Yes	8	$10.12	79	Yes	8	$6.30
30	Yes	10	$16.10	80	Yes	10	$13.99
31	No	2	$54.35	81	No	2	$64.39
32	No	4	$19.32	82	No	4	$17.53
33	No	6	$14.80	83	No	6	$13.36
34	No	8	$13.12	84	No	8	$12.89
35	No	10	$16.33	85	No	10	$15.18
36	Yes	2	$8.92	86	Yes	2	$19.48
37	Yes	4	$18.42	87	Yes	4	$10.46
38	Yes	6	$9.28	88	Yes	6	$10.44
39	Yes	8	$15.95	89	Yes	8	$15.71
40	Yes	10	$9.32	90	Yes	10	$5.01
41	No	2	$50.70	91	No	2	$57.24
42	No	4	$10.84	92	No	4	$30.00
43	No	6	$14.45	93	No	6	$13.40
44	No	8	$16.77	94	No	8	$6.94
45	No	10	$13.50	95	No	10	$9.24
46	Yes	2	$12.65	96	Yes	2	$10.24
47	Yes	4	$16.83	97	Yes	4	$0.08
48	Yes	6	$6.38	98	Yes	6	$6.87
49	Yes	8	$16.89	99	Yes	8	$9.62
50	Yes	10	$2.74	100	Yes	10	$16.49

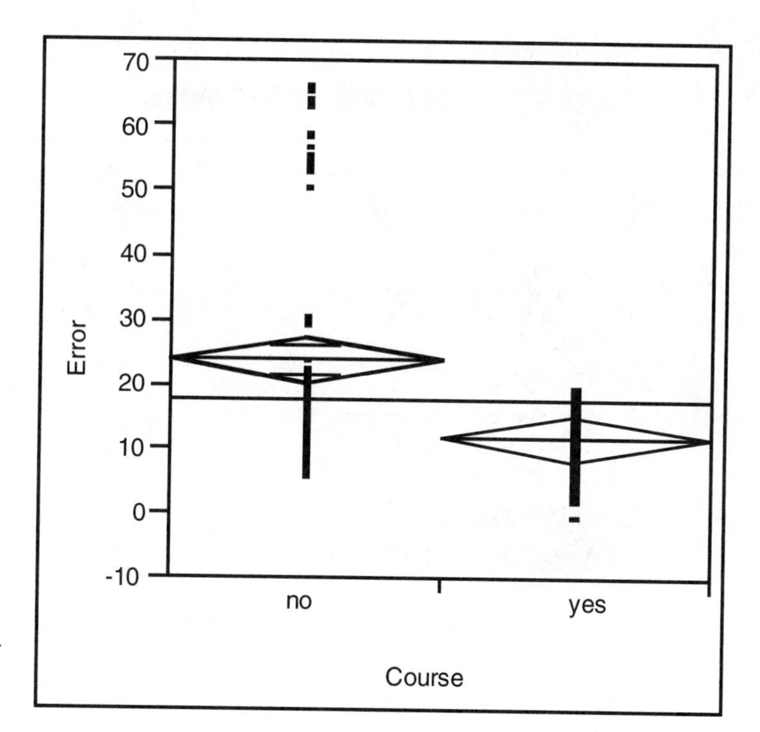

Figure 6.4 Means diamonds plot for "formal course"

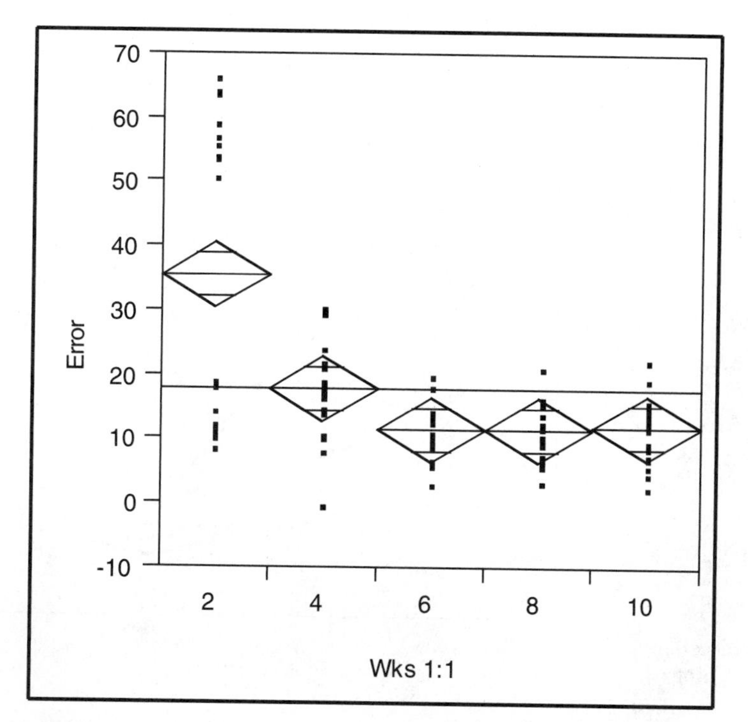

Figure 6.5 Means diamonds plot for "one-on-one training time"

Table 6.11 **JMP Output: ANOVA Table**

Analysis of Variance

Source	DF	Sum of Squares	Mean Squares	F Ratio	Prob > F
Model	9	19505.594	2167.29	96.1145	<.0001
Course	1	3860.758	3860.76	171.216	<.0001
Wks 1:1	4	8494.477	2123.62	94.178	<.0001
Course*Wks 1:1	4	7150.359	1787.59	79.276	<.0001
Error	90	2029.412	22.55		
Total	99	21535.006			

Table 6.12 **Cell Means**

	Took Course	
Wks 1:1	**0**	**1**
2	58.53	12.85
4	22.76	13.48
6	14.18	9.02
8	12.60	11.10
10	12.50	11.99

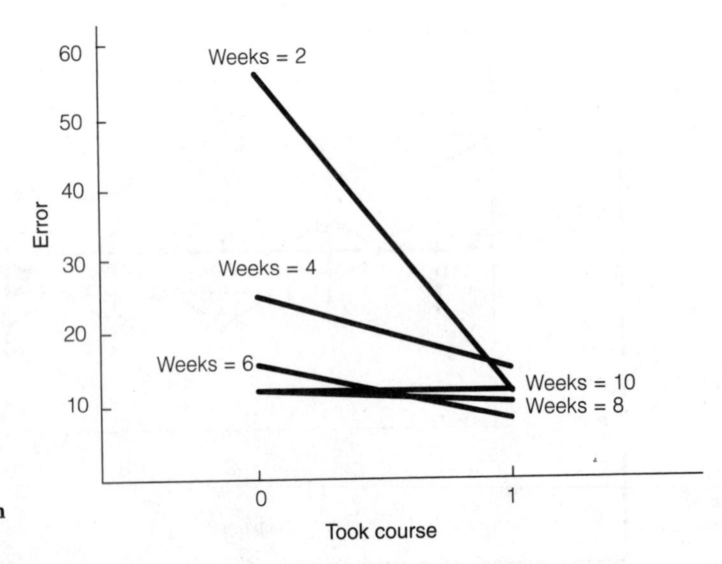

Figure 6.6 Interaction plot for teller training

Table 6.12 shows the cell means. We can see from the table the direction of the interaction. For Wks 1:1 = 2, the benefit of taking the formal course is quite high—a reduction in error from \$58.53 to \$12.85. For Wks 1:1 = 4, the benefit is not quite as large, but still relatively large. For Wks 1:1 = 6, there's a noticeable benefit, but it's somewhat lower than for the previous levels of Wks 1:1; for Wks 1:1 = 8 or 10, there's not much benefit, if any at all. Or, equivalently, the benefit of increasing the weeks of one-on-one training is quite strong (at least up to six weeks) if there is no formal course, but is not very evident if the teller does take the formal course.

This can be represented by an interaction plot of Error vs. "Took Course" for each "Wks 1:1" treatment (Figure 6.6).

6.3 Fixed Levels versus Random Levels

Thus far we have assumed (without having said so) that each factor has what are called "fixed" levels. Basically, a factor has **fixed levels** if the levels of the factor implemented in the experiment are chosen by the experimenter, and the levels of the factor in the experiment are the only levels about which inferences are to be made. The definition is not always so clean, however. Should the term apply in a case in which the factor is continuous and some specific subset is chosen systematically (say, five equally spaced values between the practical minimum and maximum levels of the factor)? Except on a mathematical level, which is not our focus, not all authors agree on a specific definition of "fixed-level" factor, but the definition above is appropriate for all practical considerations.

A factor with fixed levels is called a "fixed-level factor." A classic example of a fixed-level factor would be the factor "sex" (male, female).

The opposite of fixed levels is random levels. A reasonable working definition of a factor having **random levels** is that the levels of the factor implemented in the experiment are randomly selected (that is, they are a random sample) from a larger number of possibilities. Some authors specify that this larger set must be infinite, but others specifically include cases where it is not infinite. A classic example of a random-level factor is the amount of rainfall, where the levels in the experiment are those that "just happened" to occur.

If each factor in a design is a fixed-level factor, we have a "fixed model"; if each factor in a design is a random-level factor, we have a "random model." Of course, we can have a "mixed model"—a design in which some factors are fixed-level factors and others are random-level factors. For a two-factor study, the subject of this chapter, an example might be having one factor, "type of fertilizer," under our control, and another factor, "amount of rainfall," not under our control.

You may wonder why this matters. The answer is that it affects the analysis with regard to how the F_{calc}'s should be formed; hence, the conclusions *could* change. The F_{calc}'s are formed for different models as a function of the different expected mean squares, listed in Table 6.13. You may recognize the expected mean squares for the fixed model as those we stated earlier in the chapter.

The significance of Table 6.13 is that it guides the selection of the denominator in the calculation of F_{calc} as a function of the model. Consider the fixed-model column of Table 6.13; note that to inquire if there are nonzero row effects, we form $F_{calc} = MSB_r/MSW$. Why this ratio? Because this is the ratio that "isolates" V_r; that is, the numerator has exactly the same terms as the denominator, except for an additional nonnegative term, V_r. Only a ratio that isolates V_r can test for row effects; for example, if we formed the (silly) ratio of MSB_r divided by MSB_c, what would it definitively tell us? If it is close to one, both V_r and V_c could be nonzero but nearly equal, or they could be exactly zero: there's no way to distinguish between the two cases! Yet, distinguishing whether V_r is nonzero or zero is exactly what we're trying to do. By examining Table 6.13 for the random model, we can see that the ratios to isolate V_r and V_c are different than for the fixed model. For the mixed model, some of the appropriate ratios are different from either of the other two columns.

Why do we get different expected mean squares for the different models? Carefully examine and consider the expectations in the mixed-model column. One of the authors, when a student, looked at such a table and was convinced that it had an error in the mixed-model column; after all, if the row factor is the random one, shouldn't it be the *row* expected mean square that has the "extra" V_{int} term? Yet Table 6.13 says that it's the column expected mean square that has the extra V_{int} term, and it is correct. The logic derives from the definition of interaction. If the levels of the *row factor* are random, and the two factors interact, then the *column effects* will vary for each repeat of the experiment, because the column effects (essentially, the differences among the column means) depend on the levels of the row factor, and these levels will differ for each repeat of the experiment. Thus, the variability of the specific column effects we observe is larger, and so the mean square for columns on average is larger, because the column means vary for an additional reason besides the standard reasons of error and true column differences.

Table 6.13 **Expected Mean Square**

Mean Square	Fixed	Random	Mixed: Col. Fixed, Row Random
MSB_r	$\sigma^2 + V_r$	$\sigma^2 + V_r + V_{int}$	$\sigma^2 + V_r$
MSB_c	$\sigma^2 + V_c$	$\sigma^2 + V_c + V_{int}$	$\sigma^2 + V_c + V_{int}$
$MSI_{r,c}$	$\sigma^2 + V_{int}$	$\sigma^2 + V_{int}$	$\sigma^2 + V_{int}$
MSW	σ^2	σ^2	σ^2

To test for interaction, the proper ratio is the same for all three models. Table 6.14 summarizes the proper ratios. Again, for the mixed model we assume that the column factor is a fixed-level factor and the row factor is a random-level factor.

We offer no specific advice concerning whether a model is fixed, random, or mixed, except to consider the formal definitions of each model. As suggested earlier, sometimes it is clear that a factor is fixed (such as sex, or a specific set of mailing lists from which no generalizations can be or will be made, or two specific brands of battery); sometimes it is clear that a factor is random—the classic example is when the factor is "person." Imagine a two-factor experiment in which two brands are being tested (such as Coca-Cola versus Pepsi-Cola). A sample of each drink (unlabeled) is given to each person, who evaluates it on some measure, say, taste. In this case, a two-factor cross-classification might include "brand" and "person"[5] (although "person" would likely not be a very important factor in this experiment). We automatically consider "person" as a random factor—after all, the people in the experiment are likely to be a random sample of a large universe of people (for example, of all people, of men only, of heavy users of soft drinks), and the results are likely to be viewed as generalizable to that universe. Thus, in accordance with the mixed model (brand is a fixed-level factor, people is a random-level factor), we would compute all of the sum-of-squares terms the same way as we learned earlier in the chapter, but the F_{calc} for brands would be formed by taking the ratio of MSB_{brands} and $MSI_{brands \cdot people}$.

Depending on whether the model is fixed or random, there may be some difference in how the results are used, or at least thought about. When the random model is apropos, the experimenter is often more interested in the *universe* of μ values, as opposed to the specific μ values of the levels of the factor(s) that happened to be used in the analysis. The experimenter is also likely to be interested in the variance of these μ values, σ_μ^2—recall that in Chapter 3 we noted that in the random model the τ_j values are random variables; it follows that the $\mu_j = \mu + \tau_j$ values are also random variables. The experimenter can then go a step further and compute

$$\sigma_\mu^2/(\sigma_\mu^2 + \sigma^2)$$

Table 6.14 **Calculation of F_{calc}**

	Fixed	Random	Row Random
Row Factor	MSB_r/MSW	$MSB_r/MSI_{r,c}$	MSB_r/MSW
Column Factor	MSB_c/MSW	$MSB_c/MSI_{r,c}$	$MSB_c/MSI_{r,c}$
Interaction	$MSI_{r,c}/MSW$	$MSI_{r,c}/MSW$	$MSI_{r,c}/MSW$

which is a ratio between 0 and 1, and which can be considered a measure of the effect of that factor in proportion terms.

The situation is too context-dependent to provide further specific advice here. We simply note that if there is no interaction ($V_{\text{int}} = 0$), the three models converge to the same set of expected mean squares. Also, it is possible that all three models will yield the same ultimate "accept/reject" decision.

EXAMPLE 6.7 Brand Name Appeal for Men and Women

Now let's look at an interesting example of studying two cross-classified factors with replication. It is adapted from an article in the journal *Decision Sciences* (1978, vol. 9, p. 470). Two hundred college students participated in an experiment that was allegedly to test market a new cigarette, to determine what consumers thought of it and whether they would purchase it. However, the real reason for the experiment was to determine how, if at all, the proposed brand name of a new cigarette affected its attractiveness to potential customers. Of course, in 1978 smoking didn't carry the stigma it does today.

Two brand names were selected for testing: Frontiersman and April. It was suspected that the presumed "masculine" name Frontiersman would make the cigarette more appealing to men and less appealing to women, and that the presumed "feminine" name April would make the cigarette more attractive to women and less attractive to men. Four separate groups of 50 people were placed in a room and asked various questions about the cigarette under discussion; two of the groups were all male, two were all female. The cigarette was the same for each group and there was no brand indication on the cigarette paper. Fifty men and fifty women (two of the groups) were told that the brand name of the cigarette was to be "Frontiersman" and asked for various opinions about the cigarette. Another fifty men and fifty women (the other two groups) were told that the brand name was to be "April." Results for the two-factor experiment with $R = 2$ and $C = 2$, and $n = 50$ (replicates per cell), are in the following table. The entries in the cells represent average "intent to purchase" (that is, cell means) for each group on a seven-point scale with 7 representing nearly a certain purchase of the cigarette and 1 representing nearly zero chance of purchase of the cigarette. This question (intent to purchase) was a key question among several.

	Sex	
Brand Name	**Male**	**Female**
Frontiersman	4.44	2.04
April	3.50	4.52

Table 6.15 ANOVA Table

Source of Variability	SSQ	*df*	MS	F_{calc}
Sex	23.80	1	23.80	5.61
Brand name	29.64	1	29.64	6.99
Interaction	146.2	1	146.2	34.48
Error	831.0	196	4.24	
Total	1031	199		

The mean for Frontiersman, averaged over both sexes, is 3.24; for April, the average is 4.01. The means for male and female, respectively, each averaged over both brands, are 3.97 and 3.28. The ANOVA table is as shown in Table 6.15.

The model was assumed to be a fixed model. (For the sex factor, this is clear! For the brand name factor, the implication is that whatever results are found cannot be directly extrapolated to any other brand names.) For α = .05, and df = (1, 196), **c** = 3.84 and all three effects are significant. Clearly, the interaction effect is dominant. The main effect of brand name is −.77 (= 3.24 − 4.01). The clearest way to discuss the interaction might be: "Whereas the main effect of brand name is −.77, it's +.94 (= 4.44 − 3.50) for males, and −2.48 (= 2.04 − 4.52) for females. Clearly, the difference in the effects between the two sexes, 3.42 [= .94 − (−2.48)], jumps out." Of course, in the face of such strong interaction, the −.77 value of the brand effect is less useful than the separate effects by sex. An interaction plot is shown in Figure 6.7.

In Chapter 9 we further explore interaction effects, their interpretation, and the impact of a large interaction on the interpretation of the main effects.

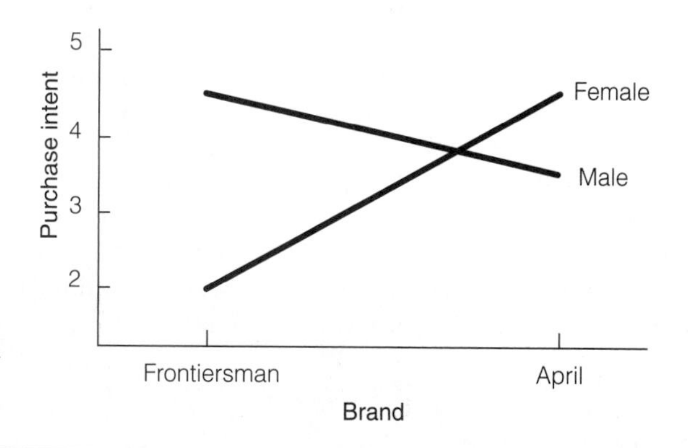

Figure 6.7 Interaction plot for brand name effect by sex

6.4 Two Factors with No Replication and No Interaction

As indicated earlier, without replication the error term is zero; there's no "pure" way to estimate error. Indeed, error is measured by considering more than one observation (replication) at the *same* treatment combination. Our general two-factor model from before, repeated here for convenience, is

$$Y_{ijk} = \mu + \rho_i + \tau_j + I_{ij} + \epsilon_{ijk},$$

which led to the sums-of-squares breakdown results of

$$\begin{aligned}\Sigma_i\Sigma_j\Sigma_k (Y_{ijk} - \bar{Y}_{...})^2 = {} & nC\,\Sigma_i (\bar{Y}_{i..} - \bar{Y}_{...})^2 + nR\,\Sigma_j (\bar{Y}_{.j.} - \bar{Y}_{...})^2 \\ & + n\,\Sigma_i\Sigma_j (\bar{Y}_{ij.} - \bar{Y}_{i..} - \bar{Y}_{.j.} + \bar{Y}_{...})^2 \\ & + \Sigma_i\Sigma_j\Sigma_k (Y_{ijk} - \bar{Y}_{ij.})^2\end{aligned}$$

We wrote the above equation symbolically as

$$\text{TSS} = \text{SSB}_\text{r} + \text{SSB}_\text{c} + \text{SSI}_\text{r,c} + \text{SSW}$$

However, as we mentioned then, without replication, $n = 1$, and the last term of both equations equals zero since, for every (i, j), $\bar{Y}_{ij.} = Y_{ijk}$; that is, with no replication, a cell mean equals the individual data value. Also, there is no longer a use for the third subscript, k. We then have

$$\begin{aligned}\Sigma_i\Sigma_j (Y_{ij} - \bar{Y}_{..})^2 = {} & C\,\Sigma_i (\bar{Y}_{i.} - \bar{Y}_{..})^2 + R\,\Sigma_j (\bar{Y}_{.j} - \bar{Y}_{..})^2 \\ & + \Sigma_i\Sigma_j (\bar{Y}_{ij} - \bar{Y}_{i.} - \bar{Y}_{.j} + \bar{Y}_{..})^2\end{aligned}$$

and

$$\text{TSS} = \text{SSB}_\text{r} + \text{SSB}_\text{c} + \text{SSI}_\text{r,c}$$

with allocation of degrees of freedom of

$$RC - 1 = (R - 1) + (C - 1) + (R - 1)(C - 1)$$

We know from our earlier work that with a fixed model,

$$\begin{aligned}E(\text{MSB}_\text{r}) &= \sigma^2 + V_r \\ E(\text{MSB}_\text{c}) &= \sigma^2 + V_c \\ E(\text{MSI}_\text{r,c}) &= \sigma^2 + V_\text{int}\end{aligned}$$

Considering our earlier discussion about the need to isolate an effect (such as V_r) in order to test it, we have a problem; there's no ratio available to test for row effects, column effects, or interaction effects. Unless some additional insight can be brought forth, we can't perform an ANOVA—which, in turn, suggests a major shortcoming of not replicating. Of course, there's one big advantage to not replicating—with fewer data values, it's cheaper! This suggests a useful question: Is there a way to gain the economic advantage of not replicating and yet avoid the shortcoming of not replicating? The answer is, potentially, yes.

Suppose we *assume,* based on the specific application at hand, that there is no interaction, and thus $V_\text{int} = 0$. Then $E(\text{MSI}_\text{r,c}) = \sigma^2$, and with

$$\begin{aligned}E(\text{MSB}_\text{r}) &= \sigma^2 + V_r \\ E(\text{MSB}_\text{c}) &= \sigma^2 + V_c \\ E(\text{MSI}_\text{r,c}) &= \sigma^2\end{aligned}$$

(now true for fixed and random models) we can use $SSI_{r,c}$ and $MSI_{r,c}$ to represent the role of SSW and MSW, respectively—the latter the denominator of F_{calc}. In other words, if there is really no interaction, the estimate of the "interaction" term is nonzero solely due to error, and we can use that value as an error estimate. It is traditional to change the notation to reflect the fact that there is no interaction, and that the term that is *calculated like an interaction term* is instead measuring *error*. With this substitution (and dropping the unnecessary subscript, k), our notation becomes

$$Y_{ij} = \mu + \rho_i + \tau_j + \epsilon_{ij}$$

and

$$TSS = SSB_r + SSB_c + SSW.$$

EXAMPLE 6.8 An Unreplicated Numerical Example

Let's illustrate the above model and its analysis with an example. Suppose we have an unreplicated two-factor experiment with the data array shown in Table 6.16. We presume that for good reason (usually the knowledge of the process expert) we can assume that there is no interaction between the two factors.

The sums-of-squares values are calculated to be

$$SSB_r = 28.67$$
$$SSB_c = 32$$
$$SSW \ (= \text{"}SSI_{r,c}\text{"}) = 1.33$$

and the ANOVA table is as shown in Table 6.17. If $\alpha = .01$, we find that the critical value, **c**, is equal to 9.78 with $df = (3, 6)$ and 10.93 for $df = (2, 6)$. Thus, the row and column factors are both highly significant.

As a comforting sidebar, consider the following: suppose we were wrong in assuming that there is no interaction. What happens to our conclusions? Are they all useless? Let's examine the row factor; the same reasoning holds for the column factor.

Table 6.16 Data Array for Numerical Example

Level of Factor B	Level of Factor A 1	2	3
1	7	3	4
2	10	6	8
3	6	2	5
4	9	5	7

Table 6.17 **ANOVA Table**

Source of Variability	SSQ	*df*	MS	F_{calc}
Row	28.67	3	9.55	43
Column	32.00	2	16.00	72
Error	1.33	6	.22	
Total	62.00	11		

We *thought* that our test statistic, F_{calc}, estimated the ratio with numerator and denominator expected mean squares of

$$(\sigma^2 + V_r)/\sigma^2$$

However, with the interaction not zero, we *actually* estimated

$$(\sigma^2 + V_r)/(\sigma^2 + V_{int})$$

This means that our calculated test statistic, F_{calc}, is, on average (that is, F_{calc} can be expected to be), smaller than it "deserves" to be. In other words, the quantity we want, which has the numerator and denominator expected values $(\sigma^2 + V_r)/\sigma^2$, is, on average, larger than the quantity we have calculated, which has the numerator and denominator $(\sigma^2 + V_r)/(\sigma^2 + V_{int})$. That is, without the V_{int} term, the denominator is less than or equal to what it was before, and the ratio is correspondingly greater than or equal to the value it was before. All of this, of course, concerns *expectations.* However, it can be shown that the basic idea probabilistically carries over to the actual results. That is, if with the assumption of no interaction we end up rejecting H_0 (as we did—for both rows and columns), there is even less chance of our rejection of H_0 occurring if, indeed, H_0 is true. If, on the other hand, we had accepted H_0 for a particular factor, the assumption works "against us." How nice! The consequences of having inappropriately assumed there was no interaction are, on average, potentially harmful only for a factor that is judged to have no effect.

EXAMPLE 6.9 Analysis Using Excel

We now analyze this numerical, no-replication example using Excel. First enter the input data, shown in Table 6.18 in a routine format for Excel. Because we clicked on "Two-Way Analysis without Replication," Excel knows that there

Table 6.18 **Excel Input Data**

7	3	4
10	6	8
6	2	5
9	5	7

Table 6.19 **Excel Output: ANOVA Table**

ANOVA: Two-Factor without Replication

Summary	*Count*	*Sum*	*Average*	*Variance*
Row 1	3	14	4.667	4.333
Row 2	3	24	8	4
Row 3	3	13	4.333	4.333
Row 4	3	21	7	4
Column 1	4	32	8	3.333
Column 2	4	16	4	3.333
Column 3	4	24	6	3.333

ANOVA

Source of Variation	*SS*	*df*	*MS*	*F*	*P-value*	*F crit*
Rows	28.667	3	9.556	43	0.000	4.757
Columns	32	2	16	72	0.000	5.143
Error	1.333	6	0.222			

is not only a column factor with three levels, but also a row factor with four levels.

Table 6.19 shows the output, which provides descriptive statistics for each row and column, and then the same ANOVA table information that our earlier numerical analysis yielded.

EXAMPLE 6.10 Analysis Using SPSS

We repeat the example above using SPSS. The input data in Table 6.20 is formatted typically for SPSS and, again, differently than it is for Excel. (Surprisingly, in the next SPSS case, Example 6.12, our illustration violates this usual SPSS format, and the required format for input data is similar to the Excel format.)

Table 6.21 shows the SPSS output, which is similar in format to Table 6.8, the SPSS output for the earlier example that did have replication.

Table 6.20 **SPSS Input Data**

Y	Row	Column
7.00	1.00	1.00
10.00	2.00	1.00
6.00	3.00	1.00
9.00	4.00	1.00
3.00	1.00	2.00
6.00	2.00	2.00
2.00	3.00	2.00
5.00	4.00	2.00
4.00	1.00	3.00
8.00	2.00	3.00
5.00	3.00	3.00
7.00	4.00	3.00

Table 6.21 **SPSS Output**

* * * ANALYSIS OF VARIANCE * * *

Y
by ROW
COLUMN

Source of Variation	Sum of Squares	DF	Mean Squares	F	Sig of F
Main Effects	60.667	5	12.133	54.600	.000
ROW	28.667	3	9.556	43.000	.000
COLUMN	32.000	2	16.000	72.000	.000
Explained	60.667	5	12.133	54.600	.000
Residual	1.333	6	.222		
Total	62.000	11	5.636		

12 cases were processed.
0 cases (.0 pct) were missing.

6.5 Blocking

One reason for having a second factor in an experiment, even if we are principally (or *only*) interested in one specific "prime factor," is the difficulty of studying the levels of the prime factor under homogeneous conditions. For example, suppose that we are studying worker absenteeism as a function of the age of the worker, and have different levels of ages—say, three of them: 25–30, 40–45, and 55–60. However, we may be concerned that a worker's gender may also affect his or her amount of absenteeism. Even though we are not particularly concerned with this potential impact of gender, we want to ensure that the gender factor does not pollute our conclusions about the effect of age. For example, if the absenteeism data of one age group happens to include a higher proportion of women than the absenteeism data of the other age groups, the absentee rate differences assigned to the factor "age group" could be including a gender effect along with it. What should we do?

One solution is to restrict our study to one gender. The downside of this strategy is that we would obtain results only on that gender. We could do two separate studies, one for each gender. The downside of this strategy could be that we don't get an overall measure of the impact of age, and the result for each gender would not have the reliability that a combined study might have. In addition, if there is any interaction effect, we have no opportunity to measure it.

What is often done in such a situation is to "block" on gender. To **block** we introduce a second factor (here, gender) and perform a two-factor study. Again, note that this would be true even if we have no interest at all in the effect of gender! Our goal in introducing gender as a factor—that is, blocking on gender—is to eliminate gender as a nuisance factor by accounting for the effects of a factor whose level may matter. We wish to ensure that, in the data, the levels of this nuisance factor are evenly distributed over the levels of the primary factor. The bottom line is that with respect to investigating the impact of age, it doesn't matter what the motivation is for introducing gender—because we want to study the effect of gender, or we don't want gender to pollute our conclusions about the impact of age. We can view the introducing of a blocking factor as something that is done reluctantly, and only when it is necessary to remove the blocking factor's nuisance value. Later in the text, notably in Chapter 10 on confounding, we dive more deeply into the issue of blocking.

6.6 Friedman Nonparametric Test

We now present a **nonparametric,** meaning assumption-free, test as an alternative to the two-factor ANOVA without replication. This test was developed by the Nobel Prize-winning economist Milton Friedman. As with the Kruskal-Wallis test discussed in Chapter 3, this test is less powerful than the ANOVA F test but doesn't require the normality assumption of the ANOVA F test. We illustrate the technique with an example.

EXAMPLE 6.11 Analysis of Angioplasty Equipment

The experiment was intended to see if there is a difference in burst pressure (that is, pressure at which the item bursts) as a function of type of angioplasty unit, for various types of balloon dilation catheters. The data for this unreplicated two-factor experiment are in Table 6.22. Note that the Friedman test considers differences in levels *only for the column factor*—in this case, type of angioplasty unit. By rewriting Table 6.22 as a 4×9 table, instead of its current 9×4 form, we could examine differences in the levels of type of balloon dilation catheter.

We wish to test the following hypotheses:

H_0: There are *no* differential effects among the different angioplasty-unit types with respect to burst pressure.

H_1: There *are* differential effects among the different angioplasty-unit types with respect to burst pressure.

As with many nonparametric tests, we first convert the data to ranks—here, ranks within each row. That is, we replace each data value by its rank

Table 6.22 Balloon Burst Pressure by Type of Angioplasty Unit

Balloon Dilation Catheter Type	Angioplasty Unit Type A	B	C	D
1	24	26	25	22
2	27	27	26	24
3	19	22	20	16
4	24	27	25	23
5	22	25	22	21
6	26	27	24	24
7	27	26	22	23
8	25	27	24	21
9	22	23	20	19

within its row (here, 1 through 4), as shown in Table 6.23. If there's a tie, we average the ranks. Finally, we sum the ranks for each column; these sums are designated $R_{\cdot j}$ and are noted in the bottom row of Table 6.23. For example, in the first row, the largest original value is 26; this is in column two, and receives the rank of 4; next highest is 25 in column three, which receives the rank of 3, and so on. For row two, there is a tie for the highest value, 27 in columns one and two; each gets a rank of 3.5, the average of 3 and 4.

The test statistic is as follows. In our problem, $C = 4$ and $R = 9$; the constants, 12 and 3, were fixed by Friedman and determined by scale considerations:

$$\begin{aligned} FR &= \{12/[RC(C+1)]\} \Sigma_{j=1\text{to}C} (R_{\cdot j}^2) - 3R(C+1) \\ &= [12/(9 \cdot 4 \cdot 5)](25^2 + 34.5^2 + 20^2 + 10.5^2) - 135 \\ &= 155.03 - 135 \\ &= 20.03 \end{aligned}$$

Under the null hypothesis, FR is well approximated by a χ^2 distribution with $(C - 1)$ degrees of freedom.[6] For our problem, $df = 3$, and with $\alpha = .05$, the

Table 6.23 Rank-Ordered Data for Friedman Test

	A	B	C	D
1	2	4	3	1
2	3.5	3.5	2	1
3	2	4	3	1
4	2	4	3	1
5	2.5	4	2.5	1
6	3	4	1.5	1.5
7	4	3	1	2
8	3	4	2	1
9	3	4	2	1
$R_{\cdot j}$	25	34.5	20	10.5

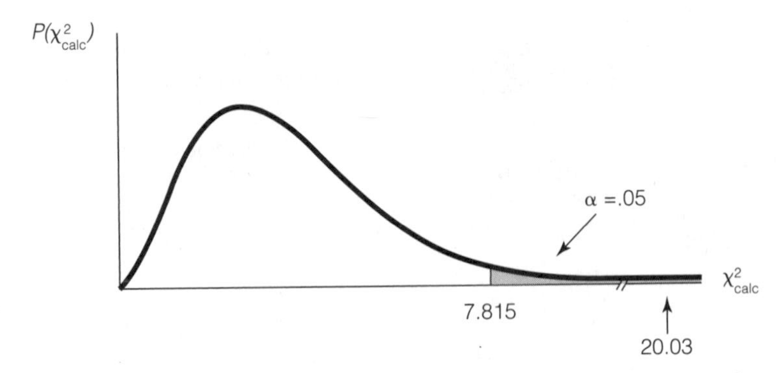

Figure 6.8 Chi-square distribution for Friedman test

critical value is 7.815. Since FR is well within the critical region (FR = 20.03 > 7.815), as shown in Figure 6.8, we reject H_0 and conclude that there are differences in angioplasty-unit type with respect to burst pressure.

EXAMPLE 6.12 Friedman Test Using SPSS

Excel, at least in its generic form, does not perform the Friedman test. As noted before, other software companies produce "add-on" software packages that work within Excel to allow it to do other analyses, including the Friedman test. We now illustrate the Friedman test using SPSS. Table 6.24 shows the in-

Table 6.24 SPSS Input Data for Friedman Test

24.00	26.00	25.00	22.00
27.00	27.00	26.00	24.00
19.00	22.00	20.00	16.00
24.00	27.00	25.00	23.00
22.00	25.00	22.00	21.00
26.00	27.00	24.00	24.00
27.00	26.00	22.00	23.00
25.00	27.00	24.00	21.00
22.00	23.00	20.00	19.00

Table 6.25 **SPSS Output for Friedman Test**

- - - Friedman Two-Way Anova

Mean Rank	Variable		
2.78	VAR00001		
3.83	VAR00002		
2.22	VAR00003		
1.17	VAR00004		
Cases	Chi-Square	D.F.	Significance
9	20.0333	3	.0002

put data. Note that it is in a form similar to what we have seen when using Excel. If it were in typical SPSS form it would be a column of the 36 data values, a second column of 1 through 9, repeated four times—to denote row membership of each data value—and finally a third column to designate column membership; this latter column would have nine 1s, and then nine 2s, nine 3s, and nine 4s.

Table 6.25 shows the output. It's relatively sparse but has the necessary information, including the p value of the test (called "significance"), which is well below .05, signifying that at $\alpha = .05$, we reject H_0.

Perspective on Friedman Test

If a regular two-factor ANOVA without replication were performed on this data, the same basic conclusions would result. This is often the case when a nonparametric test is used. One of the authors has frequently been an expert witness in the statistical area, often dealing with hypothesis-testing issues. Being able to show that the appropriate nonparametric test confirms the F test result is sometimes useful to prevent debate about the validity of assumptions from clouding the issue.

EXAMPLE 6.13 Planning Travel Packages at Joyful Voyages, Revisited

Let's return to the experimental design considered at the beginning of the chapter, with levels of the factors shown in Figure 6.1. The results of this study showed that the main effect of factor 1, "price," on "number of bookings generated" was highly significant. We found the same result in our discussion in Example 5.7 at the end of Chapter 5, when we examined an orthogonal contrast ("contrast 1") that essentially represented price. The main effect of "incentive" was also significant; this says that we cannot conclude that the four incentives (one of them being "no incentive") were equal with respect to number of bookings generated. This is consistent with the contrast 2 finding in Chapter 5; the analysis of this contrast indicated that priced-based incentives (financing and discount) differed from non-price-based incentives (flexibility, or no incentive).

What is added to the story by considering the eight offers in the two-factor cross-classification framework noted in Example 6.1 is the ability to formally statistically test for an interaction effect: Is the "price" effect different for different incentives? The answer is yes. There was significant interaction, and the direction indicated that the "price" effect was largest when there was no incentive, next largest when the incentive was the flexibility incentive, and not as prominent when the incentive was one of the priced-based incentives.

We could have addressed these issues to a lesser degree using the one-factor, orthogonal breakdown approach of Chapter 5. We could have inquired whether the price effect was different from one particular incentive to another incentive. For example, using the eight treatments in the order they appeared in Chapter 5 (Figure 5.1), a contrast of $(-1, 1, 1, -1, 0, 0, 0, 0)$ would have addressed whether the price effect when there is no incentive differs from the price effect when the incentive is the financing incentive; that is, the contrast represents

$$[(\text{mean } 2 - \text{mean } 1) - (\text{mean } 4 - \text{mean } 3)] = (-\text{ mean } 1 + \text{mean } 2 + \text{mean } 3 - \text{mean } 4)$$

However, this and similar contrasts do not address interaction questions in the aggregate (that is, "all together").

Exercises

1. Consider the data in Table 6EX.1 concerning the performance of an antenna aimed at a specific satellite. Management wishes to study the impact of two factors and their interaction: one factor is the type of mount of the antenna, and the other factor is the temperature at which it operates. The dependent variable is a measure of the distortion of the transmission. Three antennae are tested at each combination of mount type and temperature. Test for differences among temperatures, and mount types, and for the ex-

Table 6EX.1 Antenna Performance

Mount Type	Temperature 1	2	3
I	.80	1.10	.50
	1.00	1.15	.70
	1.05	1.20	.75
II	.60	.65	.55
	.80	.65	.90
	1.30	1.25	.95

Table 6EX.2 Two Sets of Data

Set 1

Rows	Column Factor a	b	c	d	e
I	35	39	37	43	41
II	43	45	33	57	47
III	39	51	35	59	51
IV	23	37	27	37	41

Set 2

Rows	Column Factor a	b	c	d	e
I	27	53	23	53	39
II	39	51	37	43	55
III	45	45	49	39	57
IV	21	23	31	45	45

istence of interaction between temperature and mount type. Use $\alpha = .05$. Assume a fixed model.

2. This exercise is designed partly to illustrate that you often can't tell what is going on *just* by looking at the data. Table 6EX.2 contains two sets of data—each has the same row means, column means, row SSQ, column SSQ, degrees of freedom, and about the same range of numbers.
   a. For each set of data, perform an ANOVA, with $\alpha = .05$. The model is

   $$y_{ij} = \mu + \rho_i + \tau_j + \epsilon_{ij}$$

   b. Why are the results so dramatically different from one set to the other?
3. For the exercise 2, set 1 data, perform Fisher's LSD analysis on the column factor. Use $\alpha = .05$.
4. For the exercise 2, set 1 data, perform Tukey's HSD analysis on the column factor. Use an experimentwise error rate, **a** = .05.
5. For the exercise 2, set 1 data, perform a Newman-Keuls analysis on the column factor. Use an experimentwise error rate, **a** = .05.
6. The data in Table 6EX.6 concerns time before failure of a special-purpose battery. It is important in battery testing to consider different temperatures and modes of use; a battery that is superior at one temperature and mode of use is not necessarily superior at other treatment combinations. Furthermore, other factors may be relevant in addition to these two factors, depending on circumstances. In a battery-production situation, even for a specific battery size (such as AA), management is acutely aware of the trade-offs that need to be made in choosing the properties of the battery to be produced. In this study, the special-purpose batteries were being tested at four different temperatures, for three different modes of use (intermittent [I], continuous [C], sporadic [S]). There were two replicates at each of the 12 temperature, mode-of-use combinations. Test for differences due to temperature, differences due to mode of use, and for the existence of interaction between temperature and mode of use. Assume a fixed model, and use $\alpha = .05$.

Table 6EX.6 Battery Time before Failure

Mode of Use	Temperature 1	2	3	4
I	12, 16	15, 19	31, 39	53, 55
C	15, 19	17, 17	30, 34	51, 49
S	11, 17	24, 22	33, 37	61, 67

7. For the data in exercise 6, conduct an orthogonal breakdown of the sum of squares associated with rows in order to test if there's a difference (1) between mode of uses I and C (the systematic use modes) and mode of use S (nonsystematic use mode), and (2) between mode of use I and mode of use C. Use an α value of .05.

8. Suppose that we have the data in Table 6EX.8 on mileage (miles per gallon), and management wishes to test whether mileage varies by gasoline type across various automobile makes. Perform an ANOVA to test whether gasoline type affects mileage. Test also whether auto make affects mileage. Use $\alpha = .01$.

Table 6EX.8 Miles per Gallon by Gas Type and Auto Make

Auto Make	Gas Type 1	2	3	4	5	6
F	26	32	26	20	28	18
B	38	37	27	30	33	33
Y	26	35	26	26	29	26
H	38	40	29	28	34	35

9. Perform a Friedman test on the data in exercise 8 to test for differences due to gasoline type. Use $\alpha = .01$. Do you get the same basic results as you did for exercise 8?

10. A major advertiser places the same ad in several different magazines; the ads are of the same size and inside the front cover in all cases. Consider $N = RC$ observations on the amount of responses received at each of R levels of the factor "magazine" (that is, R different magazines) and C different prices. Each of the RC published advertisements was sent to a random sample of 100,000 similar subscribers.

    Originally, because the same number of people received each advertisement, in the same location in the magazine, the differences in magazines were neglected, and a standard (one-way) ANOVA was performed to study the effect of price. An F_{calc} was routinely generated to test for price differences. Call this value F_1. Now a standard two-way ANOVA (without interaction and replication) has been performed, including the row factor, "magazine." Again, an F_{calc} is generated to test for price differences. Call this value F_2.

    In general, $F_1 \neq F_2$. Under what conditions on the sum of squares due to "magazines" will F_2 exceed F_1? Under what conditions on the sum of squares due to "magazines" will MSW_2 be less than MSW_1? These questions could form the basis of a consideration of when "blocking" (as described in section 6.5) is useful (or judicious).

11. In Example 6.6, we addressed an application using data based on teller training at the First United Federal Bank of Boston. The dependent variable was called the "error," and represented the larger of the sum of the overage errors and the sum of the underage errors. Another important performance measure, relating to productivity as opposed to accuracy, is the number of transactions per unit time. In retail banking, a transaction is any of a large variety of services provided to the customer; examples are check cashing, accepting deposits, and providing cashier's checks.

    Table 6EX.11 contains productivity data for the same 100 tellers listed in Table 6.10. The column headed Transactions indicates the average number of retail transactions per hour for each teller. The two factors, of course, are the same as in Example 6.6: the teller took a formal off-site training course (yes) or not (no), and received some number of weeks of one-on-one training (Weeks 1:1) by the supervisor. Does either (or both) of the training options factors affect produc-

Table 6EX.11 Number of Teller Transactions by Training Type

Teller	Course	Weeks 1:1	Transactions	Teller	Course	Weeks 1:1	Transactions
1	No	2	24.02	51	No	2	16.74
2	No	4	18.60	52	No	4	18.72
3	No	6	23.98	53	No	6	15.86
4	No	8	19.58	54	No	8	25.73
5	No	10	38.19	55	No	10	22.80
6	Yes	2	22.88	56	Yes	2	25.52
7	Yes	4	29.57	57	Yes	4	22.71
8	Yes	6	26.57	58	Yes	6	30.82
9	Yes	8	24.03	59	Yes	8	27.23
10	Yes	10	27.16	60	Yes	10	36.64
11	No	2	11.31	61	No	2	13.27
12	No	4	17.75	62	No	4	27.03
13	No	6	25.43	63	No	6	16.89
14	No	8	20.98	64	No	8	23.63
15	No	10	23.28	65	No	10	27.90
16	Yes	2	22.02	66	Yes	2	21.89
17	Yes	4	28.70	67	Yes	4	29.37
18	Yes	6	25.50	68	Yes	6	19.32
19	Yes	8	20.25	69	Yes	8	30.90
20	Yes	10	29.16	70	Yes	10	31.31
21	No	2	21.53	71	No	2	16.74
22	No	4	26.13	72	No	4	20.70
23	No	6	19.92	73	No	6	16.69
24	No	8	33.07	74	No	8	22.31
25	No	10	21.71	75	No	10	29.87
26	Yes	2	21.26	76	Yes	2	29.60
27	Yes	4	31.77	77	Yes	4	33.90
28	Yes	6	28.96	78	Yes	6	22.03
29	Yes	8	26.62	79	Yes	8	22.80
30	Yes	10	33.10	80	Yes	10	30.99
31	No	2	11.85	81	No	2	21.89
32	No	4	15.82	82	No	4	14.03
33	No	6	20.30	83	No	6	18.86
34	No	8	24.62	84	No	8	24.39
35	No	10	31.83	85	No	10	30.68
36	Yes	2	19.92	86	Yes	2	30.48
37	Yes	4	32.92	87	Yes	4	24.96
38	Yes	6	25.28	88	Yes	6	26.44
39	Yes	8	32.45	89	Yes	8	32.21
40	Yes	10	26.32	90	Yes	10	22.01
41	No	2	8.20	91	No	2	14.74
42	No	4	7.34	92	No	4	26.50
43	No	6	19.95	93	No	6	18.90
44	No	8	28.27	94	No	8	18.44
45	No	10	29.00	95	No	10	24.74
46	Yes	2	23.65	96	Yes	2	21.24
47	Yes	4	31.33	97	Yes	4	14.58
48	Yes	6	22.38	98	Yes	6	22.87
49	Yes	8	33.39	99	Yes	8	26.12
50	Yes	10	19.74	100	Yes	10	33.49

tivity? Is there interaction between the two? Use $\alpha = .05$. The data for the 100 tellers are also in the data sets, labeled Problem 6.11.

12. A survey was taken to see if a person's purchases based on infomercials on television differed by the level of several different factors. One study considered the two factors "household income" and "marital status." Household income was categorized into four categories: (1) under $30,000, (2) $30,000– $50,000, (3) $50,000–$100,000, and (4) over $100,000. Marital status was categorized into three levels: A, single (never married); B, married; and C, divorced/separated/widowed. For each of the 12 cells, ten people were surveyed and reported their estimated past purchases (in dollars) that were based on infomercials on television. The data are in Table 6EX.12; each cell presents the cell mean and the standard deviation of the ten replicates (rounded to nearest integer and labeled *SD* for standard deviation).

    Test for significance of the factor "household income."

13. For the exercise 12 data, test for significance of the factor "marital status."

14. For the exercise 12 data, test for significance of the interaction of "household income" and "marital status."

15. For the exercise 12 data, assume that there is no interaction and combine the sum of squares and degrees of freedom for the error with those of the (nominal) interaction effect. Do your results change with respect to conclusions about the main effects?

16. Three brands of batteries were tested with three low-power cassettes and three low-power flashlights. The 18 combinations are replicated eight times. The battery lifetimes are recorded in Table 6EX.16. The cassettes and flashlights were characterized as having similar power requirements, and test conditions were the same for each combination of battery brand and test device. Is there a difference in average lifetime among battery brands? Do the average lifetimes vary with test device? Is there an interaction between the test device and battery brand? (Data are in the data sets, labeled Problem 6.16.)

17. Table 6EX.17 contains data on four brands of batteries using four different types of similar test devices, each of which is considered to represent a moderate current load. Battery life is measured in minutes. Each combination of battery brand and test device is replicated 16 times (that is, with 16 batteries, one at a time). Test to see if brand or device affect battery life. Test for interaction as well. (Data are also in the data sets, labeled Problem 6.17.)

Table 6EX.12 Purchases Based on Infomercials

Marital Status	Household Income 1	2	3	4
A	Mean = 56 *SD* = 33	Mean = 73 *SD* = 39	Mean = 64 *SD* = 45	Mean = 62 *SD* = 44
B	Mean = 63 *SD* = 35	Mean = 80 *SD* = 23	Mean = 74 *SD* = 40	Mean = 65 *SD* = 20
C	Mean = 68 *SD* = 41	Mean = 81 *SD* = 32	Mean = 85 *SD* = 19	Mean = 73 *SD* = 27

Table 6EX.16 Battery Test with Low Load

Brand of Battery	Test Device	Useful Life (Hours)
ACME	Cassette A	5.94
Hi Power	Cassette A	6.25
Stancell	Cassette A	9.02
ACME	Cassette B	4.48
Hi Power	Cassette B	8.81
Stancell	Cassette B	6.72
ACME	Cassette C	7.12
Hi Power	Cassette C	8.76
Stancell	Cassette C	12.17
ACME	Flashlight A	6.48
Hi Power	Flashlight A	10.72
Stancell	Flashlight A	8.50
ACME	Flashlight B	2.82
Hi Power	Flashlight B	4.89
Stancell	Flashlight B	5.74
ACME	Flashlight C	4.59
Hi Power	Flashlight C	7.30
Stancell	Flashlight C	9.67
ACME	Cassette A	6.08
Hi Power	Cassette A	3.84
Stancell	Cassette A	6.31
ACME	Cassette B	6.98
Hi Power	Cassette B	6.29
Stancell	Cassette B	8.87
ACME	Cassette C	6.97
Hi Power	Cassette C	8.39
Stancell	Cassette C	9.89
ACME	Flashlight A	6.10
Hi Power	Flashlight A	2.22
Stancell	Flashlight A	8.64
ACME	Flashlight B	8.66
Hi Power	Flashlight B	5.51
Stancell	Flashlight B	7.15
ACME	Flashlight C	5.66
Hi Power	Flashlight C	8.69
Stancell	Flashlight C	10.32
ACME	Cassette A	10.10
Hi Power	Cassette A	8.37
Stancell	Cassette A	7.52

Brand of Battery	Test Device	Useful Life (Hours)
ACME	Cassette B	11.61
Hi Power	Cassette B	6.25
Stancell	Cassette B	4.84
ACME	Cassette C	7.35
Hi Power	Cassette C	4.71
Stancell	Cassette C	6.35
ACME	Flashlight A	2.08
Hi Power	Flashlight A	2.46
Stancell	Flashlight A	8.14
ACME	Flashlight B	4.89
Hi Power	Flashlight B	7.37
Stancell	Flashlight B	10.44
ACME	Flashlight C	6.78
Hi Power	Flashlight C	7.51
Stancell	Flashlight C	8.22
ACME	Cassette A	7.72
Hi Power	Cassette A	7.79
Stancell	Cassette A	6.74
ACME	Cassette B	7.57
Hi Power	Cassette B	8.37
Stancell	Cassette B	8.90
ACME	Cassette C	7.99
Hi Power	Cassette C	5.24
Stancell	Cassette C	6.33
ACME	Flashlight A	6.70
Hi Power	Flashlight A	10.20
Stancell	Flashlight A	5.80
ACME	Flashlight B	6.29
Hi Power	Flashlight B	10.72
Stancell	Flashlight B	5.46
ACME	Flashlight C	5.56
Hi Power	Flashlight C	6.76
Stancell	Flashlight C	6.20
ACME	Cassette A	9.30
Hi Power	Cassette A	7.41
Stancell	Cassette A	9.08
ACME	Cassette B	9.59
Hi Power	Cassette B	7.99
Stancell	Cassette B	5.67

(continued)

Table 6EX.16 **(continued)**

Brand of Battery	Test Device	Useful Life (Hours)	Brand of Battery	Test Device	Useful Life (Hours)
ACME	Cassette C	5.99	ACME	Cassette B	8.17
Hi Power	Cassette C	8.33	Hi Power	Cassette B	5.75
Stancell	Cassette C	12.32	Stancell	Cassette B	7.55
ACME	Flashlight A	7.36	ACME	Cassette C	5.87
Hi Power	Flashlight A	5.24	Hi Power	Cassette C	11.19
Stancell	Flashlight A	7.10	Stancell	Cassette C	8.73
ACME	Flashlight B	6.06	ACME	Flashlight A	5.82
Hi Power	Flashlight B	5.74	Hi Power	Flashlight A	8.58
Stancell	Flashlight B	7.50	Stancell	Flashlight A	7.81
ACME	Flashlight C	3.64	ACME	Flashlight B	3.87
Hi Power	Flashlight C	7.62	Hi Power	Flashlight B	5.63
Stancell	Flashlight C	8.05	Stancell	Flashlight B	9.52
ACME	Cassette A	5.62	ACME	Flashlight C	8.45
Hi Power	Cassette A	8.26	Hi Power	Flashlight C	6.15
Stancell	Cassette A	9.41	Stancell	Flashlight C	9.32
ACME	Cassette B	0.63	ACME	Cassette A	1.72
Hi Power	Cassette B	7.99	Hi Power	Cassette A	8.90
Stancell	Cassette B	9.34	Stancell	Cassette A	10.13
ACME	Cassette C	2.19	ACME	Cassette B	6.76
Hi Power	Cassette C	6.69	Hi Power	Cassette B	4.77
Stancell	Cassette C	8.34	Stancell	Cassette B	10.79
ACME	Flashlight A	6.97	ACME	Cassette C	7.77
Hi Power	Flashlight A	4.17	Hi Power	Cassette C	7.60
Stancell	Flashlight A	4.51	Stancell	Cassette C	7.60
ACME	Flashlight B	2.88	ACME	Flashlight A	4.33
Hi Power	Flashlight B	9.07	Hi Power	Flashlight A	6.23
Stancell	Flashlight B	8.86	Stancell	Flashlight A	8.95
ACME	Flashlight C	3.22	ACME	Flashlight B	7.51
Hi Power	Flashlight C	10.36	Hi Power	Flashlight B	8.62
Stancell	Flashlight C	7.44	Stancell	Flashlight B	5.91
ACME	Cassette A	9.09	ACME	Flashlight C	3.53
Hi Power	Cassette A	3.29	Hi Power	Flashlight C	4.83
Stancell	Cassette A	8.21	Stancell	Flashlight C	8.15

Table 6EX.17 Battery Life Test with Moderate Load

Battery Brand	Test Device	Runs 1	Runs 2	Runs 3	Runs 4	Runs 5	Runs 6	Runs 7	Runs 8
A	1	17	28	32	36	32	36	30	29
B	1	32	31	43	21	26	31	33	25
C	1	39	43	36	38	34	27	43	37
D	1	39	37	40	36	43	45	39	42
A	2	40	36	31	32	27	37	33	28
B	2	32	39	40	30	36	25	30	33
C	2	43	45	52	45	42	48	42	33
D	2	56	45	45	46	53	45	55	50
A	3	42	36	41	32	38	39	40	40
B	3	42	36	60	43	29	37	43	39
C	3	48	36	51	48	56	59	66	51
D	3	74	65	34	62	60	58	60	64
A	4	30	26	38	26	24	28	35	29
B	4	24	24	39	29	30	30	18	22
C	4	40	50	43	48	40	47	46	41
D	4	50	52	35	46	52	47	59	54
A	1	24	30	29	29	25	18	30	25
B	1	25	19	39	36	28	33	27	33
C	1	39	35	54	35	42	30	38	43
D	1	38	47	31	40	46	48	47	44
A	2	40	35	38	30	39	30	33	34
B	2	32	37	45	34	28	38	31	28
C	2	59	35	56	50	47	38	35	35
D	2	56	59	47	46	54	65	44	50
A	3	35	46	38	41	33	42	30	35
B	3	36	30	57	44	37	42	30	36
C	3	49	57	73	48	52	51	50	55
D	3	60	66	30	57	67	62	72	59
A	4	32	18	25	31	26	29	23	28
B	4	29	31	37	20	27	30	48	36
C	4	41	37	50	35	35	35	42	32
D	4	52	50	36	36	45	38	45	51

Notes

1. In fact, this is definitely the case. In general, a battery can be manufactured to cater to one type of usage, often at the expense of another kind of usage.

2. In this chapter we discuss experiments in which we actually run, with or without replication, all combinations of levels of the two factors. Later in the text, we consider experiments in which we could, but choose not to, run all combinations of levels of the two factors. The two factors are, however, still considered to be cross-classified. In addition, the definition of cross-classified extends to more than two factors.

3. Having an equal number of replicates in each cell is a sufficient, but not a necessary, condition for the design to be orthogonal.

4. An "add-in" is an additional piece of software that works seamlessly with the original piece of software (here, Excel) and enhances its capabilities, or at least makes the program more user-friendly. Many add-ins with experimental design capabilities are available for Excel, depending on the type of PC or Macintosh used.

5. This type of experiment, in which "people" is one of the factors, is often called a "repeated-measures" design, or a "within-subjects" design, to denote that the same person (say, as the row) is utilized for more than one level of the other factor(s).

6. This is true for sufficiently high values of R and C. For $R = 4$, it is considered true for $C > 4$ (our example has $R = 9$). Tables of exact values for lower values of R for different C's appear in various texts on nonparametric statistics, and in Friedman's original article in the *Journal of the American Statistical Association,* 1937, volume 32, pages 688–689 for the tables.

CHAPTER 7

Nested, or Hierarchical, Designs

EXAMPLE 7.1 Shaving Cream Evaluation at American Razor Corporation

A company that produces a large variety of shaving products wanted to test whether there were differences among three formulations of shaving cream they had produced in the lab, primarily with respect to closeness of the shave. Another measure they were interested in was the degree of irritation of the skin. An experiment that had 10 men randomly assigned to each shaving cream formulation was designed; that is, a total of 30 men were involved. Each man had been shaving routinely with a blade (as opposed to an electric shaver), was between 20 and 25 years old, and had the same type of beard hair and length of beard-hair growth; the beard-hair type categorization is a complex and technical classification scheme. Although it was acknowledged that using only one beard-hair type might limit generalization, this was thought to be outweighed by the reduction in variability achieved.

To reduce variability based on individual shaving habits (such as in what direction one strokes the razor and how often one strokes over the same area), six professional barbers were hired to do the shaving, each assigned to shave five men per morning, as described below.

Each man visited an independent research laboratory at precisely the same time each morning (six men at a time, at 15-minute intervals). The men received modest compensation for their inconvenience.

Each of the 30 men was shaved each morning by his "personal barber." All barbers used the same brand of razor and the same brand of blade (although both were debranded from a visibility point of view), used a new blade each morning, and used their assigned shaving cream formulation. There was a set protocol, beginning with washing the face with a designated soap, and then applying the shaving cream.

The protocol continued for seven consecutive days. On the morning of the eighth visit, the length of each man's beard was carefully measured, using a special device for this purpose. Measurements were taken separately on the cheek area and on the neck area, to hide the fact that the experiment was concentrating on the cheek area; the neck data were not utilized further. On each side of the cheek, 50 measurements of beard-hair length were taken (a random 50 hairs were chosen by the mechanism and individually measured), and the 100 measurements were averaged to arrive at a single number for each man. For these data, "smaller is better" in that a smaller length indicates less hair present, which is taken to indicate that the shave the previous morning was closer. Measurements of skin irritation were also taken.

One key issue in the analysis of this experiment is that we do not have a two-factor cross-classification design (that is, a barber by shaving cream formulation, with five replicates per cell), as we would if all the men used all the barbers (whatever that might actually mean). The company was concerned primarily about the "shaving cream formulation" effect, but it was also interested in the "barber" effect: Does how the shave is performed matter significantly to the beard-hair length issue?

Analysis of this experiment requires the methods discussed in this chapter. We return to this example at the end of the chapter.

7.1 Introduction to Nested Designs

EXAMPLE 7.2 Statistical Software and Professor Study

Suppose that we want to examine how four statistical software packages used in an MBA statistics course affect the statistical competence a student achieves. We also want to determine if the amount of statistical competence a student gains varies as a function of the professor teaching the course. The measure of competence will be the student's score on a comprehensive, standardized examination at the end of the course.

Let us assume that during the semester of interest, this required statistics course has 24 sections, and that 12 professors teach the course, each teaching two sections. It is a good idea to avoid conducting the experiment across different semesters; the authors' experience is that from semester to semester and year to year, students differ in aggregate. This may be due to self-selection based on when they take the course, different admission standards from year to year, or other reasons. This does not mean that we *can't* design an experiment across semesters; in Chapter 10 we focus on dealing with such issues by conducting an experiment across different "blocks."

Table 7.1 **Schematic for Two-Factor Hierarchical Study**

Software Package											
S_1			S_2			S_3			S_4		
Professor			Professor			Professor			Professor		
P_{11}	P_{12}	P_{13}	P_{21}	P_{22}	P_{23}	P_{31}	P_{32}	P_{33}	P_{41}	P_{42}	P_{43}
X_{111}	X_{121}	X_{131}	X_{211}	X_{221}	X_{231}	X_{311}	X_{321}	X_{331}	X_{411}	X_{421}	X_{431}
X_{112}	X_{122}	X_{132}	X_{212}	X_{222}	X_{232}	X_{312}	X_{322}	X_{332}	X_{412}	X_{422}	X_{432}

Note: *X*'s represent data points.

But for now, we decide that each professor will use the same statistical software package for his or her two sections of the course (in practice, the professors would lobby for this, simply as a matter of the effort required in terms of writing notes, holding computer labs, and so on), and that each of the four software packages will be randomly assigned to three of the twelve professors. Using X to stand for a data point (the average grade on the exam for the students in that section), S to stand for statistical software package, and P to stand for professor, Table 7.1 shows a schematic, or configuration, for the experiment. The first subscript of P indicates which software package is used, and the second subscript indicates which of the three professors is using that software package.

It is important to understand why Table 7.1 is not the same as the schematic in Table 7.2, which mirrors a two-factor cross-classification design with replication discussed in Chapter 6. The key is that the cross-classification configuration depicted in Table 7.2 implies that there are, essentially, only three distinct professors (or, instead, three levels of some other factor, such as "age of the professor"—in which case the factor is not "professor" but rather "age," with three professors at each age). However, in actuality there are *twelve* different professors. Also, there is no link from any professor using one software package to any professor using another software package. Saying this another way: in Table 7.1, there is no relationship among P_{11}, P_{21}, P_{31}, and P_{41}, even though they each have the same second subscript of 1. Indeed, for (or "within") each software package, the professors are arbitrarily ordered. P_{11}, P_{12}, and P_{13} (the three professors using software 1) could just as well be labeled Peter, Paul, and Mary, and P_{21}, P_{22}, and P_{23} (the three professors using software 2) could be labeled Larry, Moe, and Curly, and so forth. We notationally retain the letter P solely as a convenience for remembering that the issue is whether the Professor makes any difference to what we're measuring.

Table 7.2 Schematic for Replicated Two-Factor Cross-Classification Design

	Software Package			
Professor	S_1	S_2	S_3	S_4
P_1	X_{111}, X_{112}	X_{121}, X_{122}	X_{131}, X_{132}	X_{141}, X_{142}
P_2	X_{211}, X_{212}	X_{221}, X_{222}	X_{231}, X_{232}	X_{241}, X_{242}
P_3	X_{311}, X_{312}	X_{321}, X_{322}	X_{331}, X_{332}	X_{341}, X_{342}

Designs such as that depicted in Table 7.1 are called **nested designs,** or equivalently, **hierarchical designs.** The names derive from the view that the factors are in a hierarchy, and the levels of the so-called minor factor (here, professors) are nested under the levels of the so-called major factor. Not all texts use the labels *minor* and *major.* Also, nested designs can have more than two factors—or two stages of hierarchy. If, absurdly, each of the twelve professors used a different textbook for each of his or her two sections, so that 24 different textbooks were used in the course, "textbook" would be a third stage of the hierarchy (though, as described, it would not have replication). Perhaps "textbook" would be labeled the subminor factor? (Or perhaps this is why not all authors use the major/minor labels!)

The design in Table 7.2 is, of course, a perfectly fine design under many circumstances, and even here it could be a fine design—except that if the experiment were to be conducted all in the same semester, it would require a situation in which each of three professors taught eight sections of the course, two sections using each of the four software packages. We are not making any statement about which type of design is "superior," in general—that requires a context-dependent answer. Also, we are not trying to argue that the choice is primarily one of practical convenience—though sometimes that is the case, as here. Rather, we want to make clear the differences between the two types of designs (nested versus cross-classification), and under what set of circumstances each is appropriate.

Tables 7.1 and 7.2 are both schematics for two-factor designs. What we've now discovered is that the phrase *two-factor,* although perhaps necessary to describe the features of a design, is not sufficient to do so. For each type of design—the cross-classification and the nested—we have a somewhat different statistical model and corresponding breakdown of the total sum of squares and ANOVA. We have already explored the statistical model for cross-classification designs with two factors, and shall consider those with three or more factors in Chapter 8. Now we present the model and analysis for a replicated[1] nested design.

7.2 The Model

Our model, following the lead from previous chapters, is

$$Y_{ijk} = \mu + \rho_i + \tau_{(i)j} + \epsilon_{(ij)k} \tag{7.1}$$

where

i indexes the level of the **major factor** (here, the software package; for simplicity of notation later, let's also call it "factor A").

$(i)j$ indexes the level of the **minor factor,** j (here, the professor, to be called "factor B"), *nested within the level of the major factor, i.* The notation of putting i in parentheses (introduced here) is meant to be consistent with, and emphasize the concept of, the hierarchy as discussed previously.

$(ij)k$ indexes the replicate, k, nested within the (i, j) combination. Including the parentheses here in the error term is solely a matter of tradition. Obviously, replication in a nested design is not really different from replication in any design, in that the replication is *always* "within" the treatment combination.

We have

$i = 1, 2, 3, \ldots, M$ ($M =$ the number of levels of the major factor)
$j = 1, 2, 3, \ldots, m$ ($m =$ the number of levels of the minor factor for each level of the major factor)
$k = 1, 2, 3, \ldots, n$ ($n =$ the number of replicates per (i, j) combination)

This indexing description also follows the lead of earlier chapters by assuming a "balance," which, here, indicates that each level of the major factor has the same number of levels of the minor factor, and each combination of levels (i, j) has the same number of replicates. As before, we assume this balance for simplicity of exposition. We could write m_i to indicate that for each level of the major factor, j goes from 1 to m_i; we could also write n_{ij} to indicate the number of replicates at each (i, j) combination.

Note that when two factors are in a hierarchy, by definition the two factors do not interact, since the levels of the minor factor for one level of the major factor have no link to the levels of the minor factor for another level of the major factor. Hence, there is no explicit interaction term.

Our notation is

Y_{ijk} is the data value corresponding to the kth replicate at the ith level of A and the jth level of B within that ith level of A.
$\bar{Y}_{\cdots}$ is the grand mean of all data values.
$\bar{Y}_{i\cdot\cdot}$ is the mean for the ith level of A, averaged over all levels of B within that ith level of A, and all replicates therein.
$\bar{Y}_{ij\cdot}$ is the (cell) mean of the replicates of the jth level of B within the corresponding ith level of A.

The subscripts for the $\bar{Y}$ terms are consistent with our previous chapters, in that a . (dot) represents a dimension over which the data are averaged.

The parameter estimates (by common sense and by Gauss' least-squares principle) are

μ is estimated by $\bar{Y}_{\cdots}$
ρ_i is estimated by $(\bar{Y}_{i\cdot\cdot} - \bar{Y}_{\cdots})$
$\tau_{(i)j}$ is estimated by $(\bar{Y}_{ij\cdot} - \bar{Y}_{i\cdot\cdot})$
$\epsilon_{(ij)k}$ is estimated by $(Y_{ijk} - \bar{Y}_{ij\cdot})$

Note, however, that although parentheses are used to denote a nesting when writing the statistical model, as we did above, parentheses are usually *not* used in the subscript notation when referring to the data values and their means. Making that change and replacing the parameters of equation (7.1) with their estimates, we find

$$Y_{ijk} = \bar{Y}... + (\bar{Y}_{i}.. - \bar{Y}...) + (\bar{Y}_{ij}. - \bar{Y}_{i}..) + (Y_{ijk} - \bar{Y}_{ij}.) \tag{7.2}$$

As you can probably guess, after we subtract $\bar{Y}...$ from both sides of equation (7.2), we then square each side and sum both sides over all indices ($i = 1, \ldots, M$; $j = 1, \ldots, m$; $k = 1, \ldots, n$). The cross products cancel, yielding

$$\Sigma_i\Sigma_j\Sigma_k (Y_{ijk} - \bar{Y}...)^2 = \Sigma_i\Sigma_j\Sigma_k (\bar{Y}_{i}.. - \bar{Y}...)^2 + \Sigma_i\Sigma_j\Sigma_k (\bar{Y}_{ij}. - \bar{Y}_{i}..)^2 + \Sigma_i\Sigma_j\Sigma_k (Y_{ijk} - \bar{Y}_{ij}.)^2 \tag{7.3}$$

Since the first term on the right side of equation (7.3) does not depend on j and k, and the second term does not depend on k, equation (7.3) may be rewritten as

$$\Sigma_i\Sigma_j\Sigma_k (Y_{ijk} - \bar{Y}...)^2 = mn\,\Sigma_i (\bar{Y}_{i}.. - \bar{Y}...)^2 + n\,\Sigma_i\Sigma_j (\bar{Y}_{ij}. - \bar{Y}_{i}..)^2 + \Sigma_i\Sigma_j\Sigma_k (Y_{ijk} - \bar{Y}_{ij}.)^2 \tag{7.4}$$

The relationship involving these sums of squares may be summarized as follows:

$$\text{TSS} = \text{SSA} + \text{SSB/A} + \text{SSW}_{\text{error}}$$

where

SSA is the sum of squares associated with the major factor (A).
SSB/A is the sum of squares associated with the minor factor (B) within the major factor (A).
$\text{SSW}_{\text{error}}$ is the routine sum of squares due to error.

TSS has the corresponding degrees of freedom

$$Mmn - 1 = (M - 1) + M(m - 1) + Mm(n - 1)$$

Note that the total number of degrees of freedom is, yet again, one fewer than the number of data values. For M levels of the major factor, we have $(M - 1)$ degrees of freedom. Within each level of the major factor, we have m levels of the minor factor, producing $(m - 1)$ degrees of freedom within each level of the major factor, for a total of $M(m - 1)$ degrees of freedom. Each (i, j) combination includes n replicates, yielding $(n - 1)$ degrees of freedom each, for a total of $Mm(n - 1)$ degrees of freedom.

EXAMPLE 7.3 A Numerical Example: Statistical Software/Professor Study

Suppose, as in Example 7.2, that there are four software packages ($M = 4$ for the number of levels of A, the major factor), three professors assigned each software package ($m = 3$ for the number of levels of B, the minor factor, for each level of

A, for a total of 12 professors), and two replicates (sections of the statistics course) for each software-professor combination. The data are in Table 7.3. The quantities in the third row of data (in the double-boxed cells) are the means, $\bar{Y}_{ij\cdot}$, for each software-professor combination. As noted, the means for each software package are in the subsequent row, and the grand mean is in the last row.[2]

Now, we calculate the sums of squares for the ANOVA table:

$$\begin{aligned} SSA &= 3 \cdot 2 \cdot [(86.0 - 82.05)^2 + (77.2 - 82.05)^2 + (84.9 - 82.05)^2 \\ &\quad + (80.1 - 82.05)^2] \\ &= 6 \cdot [15.6025 + 23.5225 + 8.1225 + 3.8025] \\ &= 6 \cdot [51.05] \\ &= 306.3 \end{aligned}$$

$$\begin{aligned} SSB/A &= 2 \cdot [(82.1 - 86.0)^2 + (89.9 - 86.0)^2 + (86.0 - 86.0)^2 \\ &\quad + \cdots + (79.0 - 80.1)^2 + (82.3 - 80.1)^2 + (79.0 - 80.1)^2] \\ &= 2 \cdot [15.21 + 15.21 + 0 + \cdots + 1.21 + 4.84 + 1.21] \\ &= 2 \cdot [81.28] \\ &= 162.56 \end{aligned}$$

$$\begin{aligned} TSS &= (78.7 - 82.05)^2 + (85.5 - 82.05)^2 + (89.7 - 82.05)^2 \\ &\quad + \cdots + (79.1 - 82.05)^2 + (76.5 - 82.05)^2 + (81.5 - 82.05)^2 \\ &= 11.2225 + 11.9025 + 58.5225 + \cdots + 8.7025 + 30.8025 + .3025 \\ &= 646.22 \end{aligned}$$

$$\begin{aligned} SSW_{error} &= TSS - SSA - SSB/A \\ &= 646.22 - 306.3 - 162.56 \\ &= 177.36 \end{aligned}$$

Table 7.4 shows the ANOVA table, with the last column not filled in.

To know how to calculate F_{calc} (that is, what should be in ratio to what, in order to test the effects of interest), we need to know the expected value of each mean square. It can be shown that

$$\begin{aligned} E(MSW_{error}) &= \sigma^2 \\ E(MSB/A) &= \sigma^2 + V_{B/A} \end{aligned}$$

Table 7.3 Data for Software/Professor Study

S_1			S_4			S_3			S_4		
P_{11}	P_{12}	P_{13}	P_{21}	P_{22}	P_{23}	P_{31}	P_{32}	P_{33}	P_{41}	P_{42}	P_{43}
78.7	89.7	82.1	77.1	75.0	78.0	89.4	81.0	77.9	77.9	85.5	76.5
85.5	90.1	89.9	83.7	77.2	72.2	88.8	86.2	86.1	80.1	79.1	81.5
82.1	89.9	86.0	80.4	76.1	75.1	89.1	83.6	82.0	79.0	82.3	79.0
$\bar{Y}_{1\cdot\cdot} = 86.0$			$\bar{Y}_{2\cdot\cdot} = 77.2$			$\bar{Y}_{3\cdot\cdot} = 84.9$			$\bar{Y}_{4\cdot\cdot} = 80.1$		
$\bar{Y}_{\cdot\cdot\cdot} = 82.05$											

Table 7.4 ANOVA Table for Software/Professor Study

Source of Variability	SSQ	*df*	MSQ	F_{calc}
A (software)	306.3	3	102.1	
B/A (professor)	162.56	8	20.32	
Error	177.36	12	14.78	
Total	646.22	23		

and

$$E(\text{MSA}) = \sigma^2 + V_A \quad \text{if factor B is fixed}$$
$$E(\text{MSA}) = \sigma^2 + V_{B/A} + V_A \quad \text{if factor B is random}$$

where MSB/A is SSB/A divided by its degrees of freedom, and MSA is SSA divided by its degrees of freedom (see Table 7.4).

With B random, $V_{B/A} = n\,\sigma_\tau^2$; with B fixed, $V_{B/A} = n\,\Sigma_i\Sigma_j\,\tau_{(i)j}^2/[M(m-1)]$; with A random, $V_A = mn\,\sigma_\rho^2$; with A fixed, $V_A = mn\,\Sigma_i\,\rho_i^2/(M-1)$.

Note that the fixed-versus-random issue first encountered in Chapter 3, and first applied in Chapter 6, arises here also; however, only the expected mean square for factor A is affected (in terms of having a *V* present), and only by the type of levels of factor B.

The experiment in this example is using a random set of professors, in the sense that it could have been any professors and these simply happened to be scheduled for the course in the semester of the experiment, and also in the sense that our conclusions would pertain to whether, *in general*, the professor has an impact on competency. Therefore we view factor B (the professor) as a random-level factor. Thus the F_{calc} for factor A is the ratio of MSA (102.1) over SSB/A (20.32). The completed ANOVA table is shown in Table 7.5.

Table 7.5 ANOVA Table with F_{calc}

Source of Variability	SSQ	*df*	MSQ	F_{calc}
A (software)	306.3	3	102.1	5.02
B/A (professor)	162.56	8	20.32	1.37
Error	177.36	12	14.78	
Total	646.22	23		

At $\alpha = .05$, the *F*-table value with (3, 8) degrees of freedom is 4.07, whereas for (8, 12) degrees of freedom it is 2.85. Thus, at $\alpha = .05$, we conclude that there is a significant difference among the software programs ($p < .05$), but we cannot conclude that there are differences among the professors ($p > .25$). (Before the experiment, the authors believed that the results would come out just the opposite way.)

If we view the factor "software" as random also, then we have

$$E(\text{MSA}) = \sigma^2 + V_{\text{B/A}} + V_{\text{A}}$$

becomes

$$\sigma^2 + 2\sigma_\tau^2 + 6\sigma_\rho^2$$

and

$$E(\text{MSB/A}) = \sigma^2 + V_{\text{B/A}}$$

becomes

$$\sigma^2 + 2\sigma_\tau^2$$

We can now derive by subtraction estimates of σ_ρ^2 and σ_τ^2, to compare the effects of software and professor. We find that our estimates are $\sigma_\rho^2 = 13.63$ and $\sigma_\tau^2 = 2.77$. The former is about five times as large as the latter.

We believe that the impact of the software program was due to the varying degree of user-friendliness of the respective packages. The students indicated, on course evaluation sheets, that some of the software packages required much more time to learn than others, and that the total time spent on the course was (using our words, not theirs) a "zero-sum game"; the more time spent dealing with learning how to use the software, the less time spent gaining an understanding of the statistical concepts. With respect to the lack of impact of the professor, well—we suspect a Type II error!

We were not certain how to quantify the "degree of user-friendliness" of the four software packages; indeed, student opinion was not fully consistent. Hence, we did not attempt to break down the sum-of-squares associated with software packages (306.3) into "meaningful" subsets of user-friendliness. However, when we analyzed the four software means using Fisher's LSD approach with $\alpha = .05$, we found that packages one and three were not different from one another, but were each different from package two, and that the status of package four was not clear:

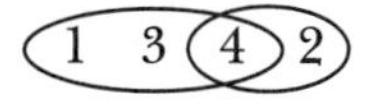

EXAMPLE 7.4 Using JMP for the Software/Professor Example

We now use the statistical software package JMP to analyze this same numerical example. Table 7.6 shows the output.

Table 7.6 **JMP Output: ANOVA Table for Software/Professor Study**

RSquare	0.725542
RSquare Adj	0.473956
Root Mean Square Error	3.844477
Mean of Response	82.05
Observations (or Sum Wgts)	24

Test wrt Random Effects

Analysis of Variance

Source	DF	Sum of Squares	Mean Square	F Ratio	Prob > F
Model	11	468.86000	42.6236	2.8839	0.0410
SW Pkg	3	306.3	102.1	5.0246	0.0302
Prof[w/SW Pkg]	8	162.56	20.32	1.3748	0.2986
Error	12	177.36000	14.7800		
C Total	23	646.22000			

EXAMPLE 7.5 A Larger Example: First United Federal Bank of Boston

We now revisit the venue of the First United Federal Bank of Boston, and again consider teller performance. The vice president of operations, M. L. Naughton, wants to investigate the performance of experienced tellers in each of four branch offices. Once again, the monthly gross overage and/or shortage for several tellers will be evaluated.

The bank staff has assembled the following data for four employees at each of four branches, all of whom have been with the respective branch for over ten years; there are 10 data points for each teller, each data point representing one week. A key question in this study is not only whether performance varies from branch to branch but also whether performance varies from teller to teller within branches. Table 7.7 shows the 160 data points (they are also in the data sets as Bank Example Chapter 7).

Table 7.7 **Teller Performance versus Branch**

Branch	Teller	Error	Branch	Teller	Error
Back Bay	1	$17.64	Charlestown	5	$26.30
Back Bay	2	$14.15	Charlestown	6	$21.98
Back Bay	3	$18.08	Charlestown	7	$18.14
Back Bay	4	$20.94	Charlestown	8	$14.77

Table 7.7 (continued)

Branch	Teller	Error	Branch	Teller	Error
State Street	9	$15.27	Charlestown	5	$20.11
State Street	10	$15.67	Charlestown	6	$27.27
State Street	11	$19.23	Charlestown	7	$31.51
State Street	12	$14.63	Charlestown	8	$17.34
Kenmore Square	13	$24.08	State Street	9	$22.17
Kenmore Square	14	$19.62	State Street	10	$20.14
Kenmore Square	15	$19.39	State Street	11	$17.48
Kenmore Square	16	$20.71	State Street	12	$26.45
Back Bay	1	$22.86	Kenmore Square	13	$12.96
Back Bay	2	$13.63	Kenmore Square	14	$16.56
Back Bay	3	$14.55	Kenmore Square	15	$15.16
Back Bay	4	$13.23	Kenmore Square	16	$15.71
Charlestown	5	$26.14	Back Bay	1	$20.26
Charlestown	6	$22.75	Back Bay	2	$24.36
Charlestown	7	$21.92	Back Bay	3	$12.22
Charlestown	8	$18.47	Back Bay	4	$22.26
State Street	9	$14.91	Charlestown	5	$25.75
State Street	10	$16.69	Charlestown	6	$37.35
State Street	11	$13.08	Charlestown	7	$27.23
State Street	12	$7.01	Charlestown	8	$17.98
Kenmore Square	13	$21.34	State Street	9	$17.01
Kenmore Square	14	$18.86	State Street	10	$9.11
Kenmore Square	15	$3.65	State Street	11	$23.07
Kenmore Square	16	$15.22	State Street	12	$12.08
Back Bay	1	$24.52	Kenmore Square	13	$8.56
Back Bay	2	$14.04	Kenmore Square	14	$23.85
Back Bay	3	$23.12	Kenmore Square	15	$5.85
Back Bay	4	$19.71	Kenmore Square	16	$13.85
Charlestown	5	$26.71	Back Bay	1	$22.93
Charlestown	6	$33.47	Back Bay	2	$13.37
Charlestown	7	$12.41	Back Bay	3	$14.13
Charlestown	8	$23.21	Back Bay	4	$22.05
State Street	9	$26.17	Charlestown	5	$20.75
State Street	10	$14.07	Charlestown	6	$23.82
State Street	11	$10.26	Charlestown	7	$26.10
State Street	12	$22.73	Charlestown	8	$23.99
Kenmore Square	13	$19.94	State Street	9	$21.89
Kenmore Square	14	$21.42	State Street	10	$21.82
Kenmore Square	15	$13.40	State Street	11	$14.65
Kenmore Square	16	$13.75	State Street	12	$4.96
Back Bay	1	$26.92	Kenmore Square	13	$18.03
Back Bay	2	$14.79	Kenmore Square	14	$18.56
Back Bay	3	$17.01	Kenmore Square	15	$15.20
Back Bay	4	$18.12	Kenmore Square	16	$10.13

(continued)

Table 7.7 (continued)

Branch	Teller	Error	Branch	Teller	Error
Back Bay	1	$17.00	Back Bay	1	$20.81
Back Bay	2	$12.30	Back Bay	2	$26.76
Back Bay	3	$22.38	Back Bay	3	$24.48
Back Bay	4	$12.60	Back Bay	4	$19.70
Charlestown	5	$18.27	Charlestown	5	$26.71
Charlestown	6	$25.19	Charlestown	6	$18.73
Charlestown	7	$22.30	Charlestown	7	$25.65
Charlestown	8	$21.15	Charlestown	8	$9.38
State Street	9	$23.74	State Street	9	$16.27
State Street	10	$13.84	State Street	10	$13.25
State Street	11	$22.27	State Street	11	$6.23
State Street	12	$17.07	State Street	12	$13.07
Kenmore Square	13	$26.35	Kenmore Square	13	$14.56
Kenmore Square	14	$6.78	Kenmore Square	14	$19.53
Kenmore Square	15	$14.06	Kenmore Square	15	$6.08
Kenmore Square	16	$7.39	Kenmore Square	16	$12.29
Back Bay	1	$17.68	Back Bay	1	$10.39
Back Bay	2	$14.18	Back Bay	2	$18.81
Back Bay	3	$17.36	Back Bay	3	$21.18
Back Bay	4	$22.38	Back Bay	4	$20.85
Charlestown	5	$31.30	Charlestown	5	$21.84
Charlestown	6	$24.83	Charlestown	6	$22.72
Charlestown	7	$22.06	Charlestown	7	$11.91
Charlestown	8	$22.64	Charlestown	8	$12.60
State Street	9	$14.61	State Street	9	$13.16
State Street	10	$15.78	State Street	10	$18.14
State Street	11	$8.41	State Street	11	$16.70
State Street	12	$19.30	State Street	12	$17.02
Kenmore Square	9	$20.98	Kenmore Square	13	$17.55
Kenmore Square	10	$24.68	Kenmore Square	14	$12.65
Kenmore Square	11	$10.84	Kenmore Square	15	$15.26
Kenmore Square	12	$11.12	Kenmore Square	16	$10.58

Table 7.8 shows the analysis of the data. The results indicate a significant difference between branches, and also a significant difference among tellers within branch. The output assumes that "tellers" is a random factor, which was indeed the case (the tellers were chosen at random from tellers at each branch that had at least ten years of service). The factor "branch" is actually a fixed factor; however, if we momentarily assume it is random, for the sake of comparison of variances, we calculate $\sigma^2_{\text{teller}} = 4.16$ and $\sigma^2_{\text{branch}} = 8.57$.

Table 7.8 **JMP Output: ANOVA Table for Bank Teller Study**

		RSquare Adj		0.301844	
		Root Mean Square Error		4.95981	
		Mean of Response		18.20285	
		Observations (or Sum Wgts)		160	
		Test wrt Random Effects			
		Analysis of Variance			
Source	DF	Sum of Squares	Mean Square	F Ratio	Prob > F
Model	15	2060.0467	137.336	5.5828	<.0001
Branch	3	1235.17	411.724	5.9896	.0098
Teller[w/Branch]	12	824.875	68.74	2.7943	.0019
Error	144	3542.3595	24.600		
Total	159	5602.4062			

7.3 Discussion

We noted earlier that nested experiments can have any number of stages of hierarchy. Especially with an equal number of levels of each lower-stage factor for each level of the higher-stage factor, and with an equal number of replicates for each hierarchical combination, the statistical model, formulas, computations, and ANOVA table logic and analysis follow somewhat directly from the discussion available in this chapter. Also, note that in a situation with three or more factors, we can have an experiment with both nesting and crossing. An example appears in exercise 6 at the end of the chapter.

EXAMPLE 7.6 Shaving Cream Evaluation at American Razor, Revisited

The experiment described at the beginning of the chapter was, indeed, conducted. One interesting issue which arose was that although an independent research laboratory was conducting the experiment, in terms of arranging for all of the protocols to be followed, the equipment used to measure the beard-hair length/growth was sufficiently complex that it had to be done by American Razor Corporation personnel. This raised the theoretical issue of independence of the testing. Hence, it was decided that the people doing the measuring would not know which men were assigned to which brand of shaving cream, so that they could not bias the results even were they so inclined. Of course, the men themselves did not know which shaving cream formulation they were using (and likely wouldn't have cared). This type of protocol is sometimes referred to as a "double-blind" experiment: both the subjects *and* the "beard-hair measurers" (or the subjects and the doctors in a medical experiment, and so on) do not know which data value pertains to which "treatment."

The results did indicate differences among the shaving cream formulations; using Tukey's HSD test, the two that were best in terms of enabling a closer shave (resulting in a shorter beard-hair length after a fixed period of time) were significantly better than the other shaving cream formulation. (By the way, it was assumed on technical grounds that a shorter beard-hair length was associated with a closer shave being delivered on the previous day, and not by some chemical or other action of the shaving cream that actually retards beard-hair growth. This is not a critical issue for the experiment at hand [a shorter beard-hair length is a shorter beard-hair length], but it would have implications for other uses—for example, a cream's potential role as a depilatory).

Of the two shaving cream formulations that were superior to the third, but not different from one another, one of them appeared to be superior in terms of engendering less skin irritation; this became the "winner" among the three competing shaving cream formulations.

Exercises

1. A study was performed to examine the use of emergency services. Of interest was variation among states and variation among counties within a state. Three bellwether states were selected, and four hospitals (that is, replicates) were chosen from each of two randomly selected counties from each state. Table 7EX.1 shows the data. Conduct an ANOVA to answer the study's questions. Use $\alpha = .05$.

Table 7EX.1 Use of Emergency Services by State and by County

State 1		State 2		State 3	
County		County		County	
A	B	C	D	E	F
31	43	31	19	35	15
55	49	33	39	45	49
55	59	37	35	37	45
31	53	31	31	47	31

2. Assuming for the sake of comparison that the two factors in exercise 1 are random factors, find the estimates of the variance components and compare them.

3. Perform a two-way ANOVA with replication for the exercise 1 data, now configured as shown in Table 7EX.3.

Table 7EX.3 County Emergency Services by State

County	State 1	State 2	State 3
A	31, 55, 55, 31	31, 33, 37, 31	35, 45, 37, 47
B	43, 49, 59, 53	19, 39, 35, 31	15, 49, 45, 31

4. Compare and discuss the results from exercises 1 and 3.

5. For a three-factor design in which a minor factor is nested within a major factor, and a

subminor factor is nested within the minor factor, the statistical model is

$$Y_{ijkl} = \mu + \rho_i + \tau_{(i)j} + \gamma_{(ij)k} + \epsilon_{(ijk)l}$$

Consider the data in Table 7EX.5, which correspond to the nested arrangement described above. There were two industries in the study, each was represented by two different companies, and each company provided data on manufacturing plants in two different geographical locations; there were eight distinct locations in the study. Two production managers from each plant were randomly selected (as replicates). The dependent variable was percent of income voluntarily contributed to a 401K pension plan. It was hoped that the analysis would shed light on whether there was variation between the industries, between companies within industries, and between (manufacturing plant) location within company.

Perform an ANOVA to examine the sources of variability noted above. Use $\alpha = .05$, and assume that company and location are random-level factors.

Table 7EX.5 Percent of Income Contributed to 401K Plan

Industry 1				Industry 2			
Company 1		Company 2		Company 1		Company 2	
Loc. 1	Loc. 2	Loc. 1	Loc. 2	Loc. 1	Loc. 2	Loc. 1	Loc. 2
2	4.5	5	6	9	9	7	5
4	5.5	4	5	7.5	8.5	7	7

6. We noted in section 7.3 that an experiment can have both nesting and crossing in it. Consider the following situation: We have a manufacturer with a three-shifts-per-day operation. During each shift there are four workers (in all, twelve different workers); we run an experiment over the same three-day period for each worker, and for each worker we collect four items (observations) per day. The dependent variable is the quality of the item observed. A layout of the experiment is in Table 7EX.6A; it is an experiment that combines nesting and crossing. (X stands for a data value.)

   Let "shift" be called factor A and indexed by i; "worker" be called factor B and indexed by j; and "day" be called factor C and indexed by k. The model for this experiment would be

$$Y_{ijkl} = \mu + \rho_i + \tau_{(i)j} + \gamma_k + I_{(i)jk} + I_{ik} + \epsilon_{(ijk)l}$$

Table 7EX.6A Experimental Layout Combining Nesting and Crossing

		Day		
Shift	Worker	1	2	3
1	1	XXXX	XXXX	XXXX
	2	XXXX	XXXX	XXXX
	3	XXXX	XXXX	XXXX
	4	XXXX	XXXX	XXXX
2	1	XXXX	XXXX	XXXX
	2	XXXX	XXXX	XXXX
	3	XXXX	XXXX	XXXX
	4	XXXX	XXXX	XXXX
3	1	XXXX	XXXX	XXXX
	2	XXXX	XXXX	XXXX
	3	XXXX	XXXX	XXXX
	4	XXXX	XXXX	XXXX

where

ρ_i = the effects of A
$\tau_{(i)j}$ = the effects of B, within A
γ_k = the effects of C
$I_{(i)jk}$ = interactions of B and C, within A
I_{ik} = interactions of A and C

Write out the sum-of-squares expression for each of the six sources of variability. Also note the degrees of freedom of each. That is, fill in Table 7EX.6B (which shows one example of a sum-of-squares expression).

Table 7EX.6B ANOVA Table

Source of Variability	SSQ	Degrees of Freedom
A	$48 \bullet \Sigma_i (\overline{Y}_{i...} - \overline{Y}_{....})^2$	
C		
B within A		
BC interaction within A		
AC interaction		
Error		

7. Table 7EX.7 contains productivity data for the same bank tellers and branches discussed in Example 5; the table is also in the data sets, labeled Problem 7.7. The column headed Transactions indicates the average number of retail transactions per hour for each teller. Does this measure of productivity vary from branch to branch? Does productivity vary for tellers within the same branch? Use $\alpha = .05$.

8. Suppose the twelve professors were using a total of only three software packages, instead of the four packages as noted in Table 7.3. That is, suppose that we have the same data as those in the software example in the chapter, but the data now represent the twelve professors using *three* software packages, four professors using each package, instead of the four packages, three professors using each package. The same but "reconfigured" data could then be shown as in Table 7EX.8. Is there a significant difference between professors "within" software package? Between software packages? The data are also in the data sets, labeled Problem 7.8. Use $\alpha = .05$.

9. The records of three physicians at each of four offices of the Merrimack Valley Pediatric Clinic were audited in an attempt to determine their use of generic versus brand-name drugs when generic drugs were available. Table 7EX.9 gives the percentage of times each physician insisted on brand-name drugs for each of four weeks during July 1998; assume that the different weeks are replicates. Is there a difference in percentage of brand-name drugs prescribed between office locations? Between physicians within office? Use $\alpha = .01$. The data are also in the data sets, labeled Problem 7.9.

10. The golf courses first discussed in Chapter 2, and then referred to in various exercises in Chapters 4 and 5, were designed by two retired professional golfers—Jim Fisher of Belmont, Massachusetts, and Edward "Fuzzy" Newbar of Medfield, Massachusetts. Fisher designed Near Corners and Meadow Brook, and Newbar designed Birch Briar and Fountainbleau. Fisher and Newbar have extensive design experience; these are only two of the many courses each has designed. With respect to the score of players "like those in the Eastern Electric golf league," based on these same golf score data, is there a difference between the two designers? Between courses "within" designer? Use $\alpha = .05$. The first couple of rows of the data set are in Table 7EX.10; the full data set is in the data sets, labeled Problem 7.10.

11. Suppose that in the exercise 10 example, Fisher and Newbar had each designed only the two courses ascribed to them above. How would your analysis change? Do your ultimate conclusions change?

Table 7EX.7 Teller Performance versus Bank Branch

Branch	Teller	Transactions
Back Bay	1	23.82
Back Bay	2	31.29
Back Bay	3	18.72
Back Bay	4	17.90
Charlestown	5	19.41
Charlestown	6	30.21
Charlestown	7	32.99
Charlestown	8	27.22
State Street	9	18.23
State Street	10	25.29
State Street	11	21.46
State Street	12	21.62
Kenmore Square	13	17.43
Kenmore Square	14	30.04
Kenmore Square	15	12.77
Kenmore Square	16	21.96
Back Bay	1	27.65
Back Bay	2	18.03
Back Bay	3	12.51
Back Bay	4	7.92
Charlestown	5	36.70
Charlestown	6	36.12
Charlestown	7	20.31
Charlestown	8	20.29
State Street	9	25.51
State Street	10	19.17
State Street	11	22.54
State Street	12	19.42
Kenmore Square	13	28.88
Kenmore Square	14	21.27
Kenmore Square	15	15.37
Kenmore Square	16	15.48
Back Bay	1	26.14
Back Bay	2	20.09
Back Bay	3	18.09
Back Bay	4	17.96
Charlestown	5	33.64
Charlestown	6	27.89
Charlestown	7	23.10
Charlestown	8	23.70
State Street	9	25.79
State Street	10	33.51
State Street	11	23.25
State Street	12	23.91

Branch	Teller	Transactions
Kenmore Square	13	18.70
Kenmore Square	14	19.31
Kenmore Square	15	15.31
Kenmore Square	16	12.97
Back Bay	1	28.27
Back Bay	2	21.71
Back Bay	3	23.30
Back Bay	4	24.88
Charlestown	5	23.29
Charlestown	6	37.54
Charlestown	7	27.17
Charlestown	8	26.65
State Street	9	30.33
State Street	10	32.35
State Street	11	21.04
State Street	12	14.02
Kenmore Square	13	21.31
Kenmore Square	14	15.96
Kenmore Square	15	12.52
Kenmore Square	16	21.60
Back Bay	1	22.12
Back Bay	2	24.78
Back Bay	3	29.89
Back Bay	4	21.51
Charlestown	5	40.23
Charlestown	6	45.13
Charlestown	7	23.81
Charlestown	8	23.21
State Street	9	23.15
State Street	10	30.19
State Street	11	18.30
State Street	12	19.90
Kenmore Square	13	22.39
Kenmore Square	14	32.78
Kenmore Square	15	23.27
Kenmore Square	16	18.58
Back Bay	1	26.33
Back Bay	2	21.01
Back Bay	3	18.83
Back Bay	4	24.05
Charlestown	5	32.96
Charlestown	6	30.46
Charlestown	7	30.87
Charlestown	8	27.54

(continued)

Table 7EX.7 (continued)

Branch	Teller	Transactions	Branch	Teller	Transactions
State Street	9	27.45	Kenmore Square	13	26.97
State Street	10	28.12	Kenmore Square	14	26.00
State Street	11	23.42	Kenmore Square	15	20.90
State Street	12	26.77	Kenmore Square	16	20.07
Kenmore Square	13	26.79	Back Bay	1	33.71
Kenmore Square	14	19.30	Back Bay	2	31.49
Kenmore Square	15	16.56	Back Bay	3	22.52
Kenmore Square	16	12.34	Back Bay	4	27.97
Back Bay	1	24.38	Charlestown	5	33.10
Back Bay	2	30.04	Charlestown	6	36.40
Back Bay	3	14.85	Charlestown	7	21.57
Back Bay	4	13.05	Charlestown	8	20.35
Charlestown	5	27.45	State Street	9	24.80
Charlestown	6	42.91	State Street	10	26.23
Charlestown	7	21.83	State Street	11	20.48
Charlestown	8	22.82	State Street	12	15.07
State Street	9	25.88	Kenmore Square	13	26.03
State Street	10	20.93	Kenmore Square	14	27.76
State Street	11	18.67	Kenmore Square	15	25.02
State Street	12	19.06	Kenmore Square	16	20.46
Kenmore Square	13	23.37	Back Bay	1	17.88
Kenmore Square	14	17.58	Back Bay	2	25.90
Kenmore Square	15	23.74	Back Bay	3	24.59
Kenmore Square	16	17.87	Back Bay	4	21.67
Back Bay	1	17.90	Charlestown	5	26.16
Back Bay	2	32.61	Charlestown	6	36.64
Back Bay	3	31.89	Charlestown	7	31.60
Back Bay	4	21.41	Charlestown	8	22.32
Charlestown	5	26.80	State Street	9	16.16
Charlestown	6	36.14	State Street	10	28.05
Charlestown	7	28.44	State Street	11	18.92
Charlestown	8	17.44	State Street	12	29.13
State Street	9	18.98	Kenmore Square	13	19.57
State Street	10	24.41	Kenmore Square	14	17.54
State Street	11	17.35	Kenmore Square	15	15.56
State Street	12	23.98	Kenmore Square	16	24.10

Table 7EX.8 Reconfigured Statistics Package Study

Software Package	Professor	Score
1	1	78.7
1	1	85.5
1	2	89.7
1	2	90.1
1	3	82.1
1	3	89.9
1	4	77.1
1	4	83.7
2	5	75.0
2	5	77.2
2	6	78.0
2	6	72.2
2	7	89.4
2	7	88.8
2	8	81.0
2	8	86.2
3	9	77.9
3	9	86.1
3	10	77.9
3	10	80.1
3	11	85.5
3	11	79.1
3	12	76.5
3	12	81.5

Table 7EX.9 Percentage Use of Generic Drugs

		Week			
Location	Physician	1	2	3	4
Amesbury	Dr. Barros	17	22	21	15
Amesbury	Dr. Russ	18	22	18	19
Amesbury	Dr. Rastogi	29	26	29	24
Andover	Dr. Maloney	12	14	15	15
Andover	Dr. Franks	21	20	21	18
Andover	Dr. Lawton	19	19	26	22
Methuen	Dr. Kohl	23	21	21	24
Methuen	Dr. Bruni-Masello	31	29	27	27
Methuen	Dr. Seth	32	28	28	26
Salem	Dr. Earle	19	15	14	15
Salem	Dr. Adams	16	23	21	20
Salem	Dr. Klonizchii	16	17	16	14

Table 7EX.10 **Golf Scores by Course and by Designer**

Designed by Fisher		Designed by Newbar	
Near Corners	Meadow Brook	Birch Briar	Fountainbleau
115	99	107	135
106	101	111	144
.	.	.	.
.	.	.	.
.	.	.	.

Notes

1. If in a nested design there is no replication, the lowest-stage factor in the hierarchy essentially disappears and takes the place of error. Hence, for an explicit set of factors to be studied in a nested design, with an ANOVA that includes them all, replication is necessary.

2. To precisely satisfy the "equal-variance" assumption, theoretically required for the *F* test to be appropriate, we would need to have the same number of students in each class, assuming that the σ^2 of each student's score (its propensity to vary from its truth) is identical. For each class to have the same number of students is somewhat unrealistic; however, for each to have *approximately* the same number of students is not. Given the robustness of the equal-variance assumption, minor differences in class size should not materially affect our results.

CHAPTER 8

Designs with Three or More Factors: Latin-Square and Related Designs

EXAMPLE 8.1 Maximizing Profit at Nature's Land Farms

Nature's Land Farms (NLF), a large farming cooperative, decided to conduct marketing research to determine the impact of certain factors on its sales and profits. The two products of principal interest were tomatoes and potatoes. For potatoes in particular, an experiment was designed in which three primary factors were studied, along with another factor: price.

NLF believed that price is a different kind of factor that might cloud the effect of other factors, and thus should be studied in a different way. This means that NLF suspected, rightfully so, that price could have significant interaction with some or all of the other three factors. Ultimately, the design chosen for studying the three main factors was repeated for each of three prices. Here, however, we focus on only the three main factors.

The three factors of interest to NLF were decided upon after extensive focus-group activity. First was *positioning*, the most prominently featured benefit. This involved highly visible material displayed wherever the potatoes are sold and had three levels:

POSITIONING

1. Environmentally friendly, with no sacrifice in taste
2. Superior taste, without sacrificing the environment
3. Biotech products, to provide extra nutrients and zero bacteria

A second factor was *cobranding* with another name well known to potato consumers; this has the potential of adding credibility and verisimilitude. Three other companies were considered as branding partners (for proprietary reasons we cannot list them; however, think of brands having the stature of Sunshine or Dole—neither of which were involved).

COBRANDING

1. Cobrand 1
2. Cobrand 2
3. Cobrand 3

The third factor was *packaging*—that is, the container in which the potatoes are sold. Three were of interest to NLF:

PACKAGING

1. Paper with no window
2. Paper with a window
3. Clear plastic

Of major concern to NLF was that three levels of three factors means 27 different possibilities. It would be prohibitively expensive to implement a study with all of them, even in a simulated test market. We return to this example at the end of the chapter.

8.1 Introduction to Multifactor Designs

When more than two factors are under study, the number of possible treatment combinations grows exponentially. For example, with only three factors, each at five levels, there are $5^3 = 125$ possible combinations. Although modeling such an experiment is straightforward, running it is another matter. It would be rare to actually carry out an experiment with 125 different treatment combinations, because the management needed and the money required would be great.

The cost issue for running the experiment is obvious. Usually, the cost of actually running the experiment is somewhat proportional to the number of data values obtained, although sometimes replicates of a given treatment combination are relatively cheap, so that cost is more in proportion to the number of treatment combinations.

But cost also includes setting up the experiment.[1] In practice, each experiment usually requires that a set of adjustments, process parameters, equipment settings, advertising agency arrangements, test market locations, and so on be established. Sometimes the equipment setup is far more expensive than the running of the experiment.

Managing the collection of data is also an issue. We recall an instance in which the gathering of sales data for a supermarket chain was partially compromised because an assistant manager at one store simply forgot to count how many bananas were sold that day! (This was before the era of price scanning.)

This isn't to say that, often, several factors should not be studied. But to be useful and affordable, a study must be designed carefully. Were that not the case, we would have no need for the discipline of experimental design; we could use the brute force approach of just running all combinations.

There are several ways to manage the size of experiments involving several factors. In subsequent chapters, we investigate powerful design techniques that restrict the number of levels of each factor as much as possible; fewer levels mean the experiment can be smaller and cheaper. Using factors with only two levels each allows the most powerful of the techniques to be applied with the most "delicacy,"[2] as we see in Chapter 11. In this chapter we consider procedures that allow more levels (actually, any number of levels), but are restrictive in other ways. Often, to design both effectively and efficiently an experiment that has many factors, we must make informed compromises.

8.2 Latin-Square Designs

Suppose we want to study three factors, each at three levels, as in the introductory NLF case. We can diagram the $3^3 = 27$ possible treatment combinations of factor levels as shown in Table 8.1. The notation is as follows: for any of the 27 cells, the row designates the level of A, the column designates the level of B, and the subscripts inside the cells designate the level of C. For example, the cell in the fourth row and third column has A at level 1, B at level 3, and C at level 2.

When we run an experiment such as this and wish to show the data, we follow the tradition of placing the data results inside the cell as well. Here, we can think of A as the row factor, B as the column factor, and C as the "inside factor." *Which is which makes no difference:* there is no advantage or disadvantage to being a row, column, or inside factor. Notice that the $3 \times 3 \times 3$ set of combinations is shown in Table 8.1 as a 9×3 matrix for reasons of simplicity. You can also think of this as a three-dimensional cube with 27 cells arranged $3 \times 3 \times 3$ and with C as the depth factor.

Suppose we run not all 27 combinations but only nine. Which nine? Consider Table 8.2; it shows only the nine treatment combinations that are printed in bold type in Table 8.1.

These nine treatment combinations have some important, desirable properties. They are not unique with respect to having these properties, but only 1/4000 of a percent of the possible groups of nine that can be formed by 27 treatment combinations have the desirable properties. What are these properties?

Notice that Table 8.2 is a square—after all, its design is called a Latin square, not a Latin rectangle! Several desirable properties of this design are possible only when we have a square—that is, when all three factors under

Table 8.1 **Three Factors, Three Levels**

	B_1	B_2	B_3
A_1	**C_1**	C_1	C_1
A_2	C_1	C_1	**C_1**
A_3	C_1	**C_1**	C_1
A_1	C_2	**C_2**	C_2
A_2	**C_2**	C_2	C_2
A_3	C_2	C_2	**C_2**
A_1	C_3	C_3	**C_3**
A_2	C_3	**C_3**	C_3
A_3	**C_3**	C_3	C_3

Note: Bold entries are those used in Table 8.2.

study have the same number of levels. As we shall see, there is no restriction on the number of levels of each factor except that each must be the same. A key property of the square in Table 8.2 is that it is balanced: each factor is at each level the same number of times (three); furthermore, each level of a factor is used in combination with each other level of a factor the same number of times—once.

This deserves a bit of elaboration, since it ensures unbiased estimates of the main effect of each of the factors. Note that A_1 is with B_1 once, B_2 once, and B_3 once. A_2 is also with B_1 once, B_2 once, and B_3 once; similarly for A_3. Furthermore, A_1 is with C_1 once, C_2 once, and C_3 once; similarly for A_2 and A_3. And the same can be said for the levels of factors B and C; each level of each of those factors is once and only once combined with each level of each other factor.

Table 8.2 **Nine of 27 Possibilities**

	B_1	B_2	B_3
A_1	C_1	C_2	C_3
A_2	C_2	C_3	C_1
A_3	C_3	C_1	C_2

This special balance in the set of nine treatment combinations of Table 8.2 is very important. As we mentioned above, it guarantees the unbiasedness of the main effects of the factors. For example, when we look at differences in the row means (that is, the mean for each level of A), we can note that row one includes exactly one data value at each level of B, and exactly one data value at each level of C.[3] The same holds for rows two and three. Thus, each row mean is on equal footing with respect to the levels of factors B and C. Consequently, differences among the row means can legitimately be examined in the traditional *F*-test, ANOVA way to determine whether the row differences are significant. For example, if the data values at A_1 all had factor C at C_1, and all data values at A_2 had factor C at C_2, and so on, row differences could not be attributed solely to the impact of factor A (along with the omnipresent error, of course). That is why the design in Table 8.3, although it consists of nine of the 27 treatment combinations of Table 8.1, is poor even though it does provide an unbiased evaluation of the effect of factor B.

Let's consider Table 8.3. In addition to error, differences in the row means can be due to the level of factor A or to the level of factor C; the two effects cannot be separated. When the impact of one factor can't be separated from that of a second, we say that the effects are **confounded.** We explore this concept more fully in Chapter 10.

If you know how chess pieces move, here's a way to understand how the levels of C are placed in the cells so as to balance them: if all the C_1's were rooks (castles) in chess, none of them could take each other off *in one move;* similarly for the C_2's and C_3's. This is true of the 3×3 square of Table 8.2, but not true of the 3×3 square of Table 8.3.

Actually, we're guilty of one sin of omission so far about Latin-square and related designs. In fact, for each level of each factor to be on equal footing in a design such as that depicted in Table 8.2, we must be willing to assume that there is no (or, in practice, negligible) interaction among the factors. After all, when we stated that in Table 8.2, A_1 is combined once with each level of B, and once with each level of C (and similarly for A_2 and A_3), we neglected to mention that A_1 includes, for example, one data point at (B_1, C_1), whereas neither A_2 nor A_3 have a data value at the (B_1, C_1) combination. So if B_1 and C_1 together induce a yield that is more (or less) than the sum of their separate

Table 8.3 A Poor Choice of Nine

	B_1	B_2	B_3
A_1	C_1	C_1	C_1
A_2	C_2	C_2	C_2
A_3	C_3	C_3	C_3

main effects (that is, factors B and C have a positive or negative interaction effect), the mean of row one (level A_1) includes the impact of that combination, but row means two and three do not include the impact of that combination. To repeat: for the design in Table 8.2 and throughout this chapter, to provide meaningful evaluation of the impact of each factor, we must assume that the factors do not interact. In subsequent chapters we present design procedures that also include only a portion of the total number of treatment combinations, yet do not require this restriction.

This point leads to the notion of a design trade-off. In Chapter 6 we saw that, if there were no interaction between two factors, we could, without negative consequences, design the experiment without replication. This can be viewed as a trade-off between making a limiting assumption and the cheaper cost of an unreplicated experiment with the same number of factors and levels. Likewise, the design depicted in Table 8.2 can be viewed as a trade-off, but one that is more dramatic: if the assumption of no interaction is incorrect, the consequences are far more severe than those mentioned in Chapter 6. But the benefits to be gained are also far greater!

If there really is interaction among the factors, the values of the main effects determined may not be valid. Consider the example in Table 8.2, in which A_1 included a data value that had levels (B_1, C_1), but A_2 and A_3 did not include such a value. If this (B_1, C_1) combination greatly increases the yield beyond the average effect of B_1 alone plus the average effect of C_1 alone, then the mean of row one may be much higher than the other row means solely due to including the (B_1, C_1) combination, *and not due at all to the level of A being A_1!* In the Chapter 6 situation, the effect and sum of squares of each factor were correctly estimated, even if there was interaction; only the F test had the possibility of being misleading. To repeat, for the design of Table 8.2, the assumption about interaction is more critical. However, zero interaction is not necessary as long as the interaction effects are very small.

Now consider the benefits. Notice that the design of Table 8.2 studies three factors at three levels each, with only nine combinations—the same number needed for just *two* factors at three levels each, as we saw in Chapter 6. For this reason we might say that the Table 8.2 design studies three factors for the cost of two factors. Sometimes this is called a "one-third replicate," because one runs (and pays for!) only nine of the 27 combinations.[4] This generally cuts the cost of running the experiment considerably; it is much less expensive than running all possible treatment combinations and simply not replicating them.

Designs of this sort, in which we study three factors, all at the same number of levels in this balanced manner and assuming that there is no interaction among factors, are called **Latin-square designs.** Latin-square designs may be replicated or unreplicated. An example of an unreplicated four-level Latin-square design is shown in Table 8.4. The dependent variable is the number of new car sales over a specified period (for car dealership franchises offering the same make and models of car, and that historically have sold about the same number of cars per year). The independent variables are A: service policy (for example, all scheduled services up to 30,000 miles are free); B: hours open for business (for example, open all day Sundays); and C: ancillary amenities (for example, a free car wash anytime, even if the car is not brought in for service that day). The numerical quantity in each cell of Table 8.4 is the yield—sales in units (that is, number of new vehicles sold).

Notice that Table 8.4 depicts only 16 of the 64 combinations possible. This unreplicated design (each of the 16 combinations is run only once) is a quarter replicate; it uses only a fourth of the 64 possible combinations. Also, it appears that the assumption of no interaction is reasonable in this application: the impact of each factor, whatever it is, would seem not to depend on the levels of the other factors.[5]

The Latin-Square Model and ANOVA

The model for the unreplicated Latin-square analysis, following the notation from earlier chapters, is as follows. Recall that, since this is a Latin-square design, interactions have been assumed to be zero or negligible. The model acknowledges this by omitting interaction terms. The data value for the *i*th level of the row factor, *j*th level of the column factor, and *k*th level of the inside factor is

$$Y_{ijk} = \mu + \rho_i + \tau_j + \gamma_k + \epsilon_{ijk}$$

where i, j, and k take on values 1, 2, 3, . . . , m. That is, we have three factors, each at m levels. The only new symbol is γ, analogous to ρ and τ and representing the third cross-classified factor.

As previously, we replace the parameters with their respective estimates:

$\bar{Y}...$ for μ
$(\bar{Y}_{i}.. - \bar{Y}...)$ for ρ_i
$(\bar{Y}_{.j.} - \bar{Y}...)$ for τ_j
$(\bar{Y}_{..k} - \bar{Y}...)$ for γ_k

Table 8.4 Data for New Car Sales Example

	B_1	B_2	B_3	B_4
A_1	C_4 855	C_3 877	C_2 890	C_1 997
A_2	C_1 962	C_2 817	C_3 845	C_4 776
A_3	C_3 848	C_4 841	C_1 784	C_2 776
A_4	C_2 831	C_1 952	C_4 806	C_3 871

This yields

$$Y_{ijk} = \bar{Y}_{...} + (\bar{Y}_{i..} - \bar{Y}_{...}) + (\bar{Y}_{.j.} - \bar{Y}_{...}) + (\bar{Y}_{..k} - \bar{Y}_{...}) + R$$

where R is a catchall term (the "remainder," to make the equation indeed an equality). We'll come back to R shortly. Rearranging the above equation slightly, we have

$$(Y_{ijk} - \bar{Y}_{...}) = (\bar{Y}_{i..} - \bar{Y}_{...}) + (\bar{Y}_{.j.} - \bar{Y}_{...}) + (\bar{Y}_{..k} - \bar{Y}_{...}) + R$$

where

$(Y_{ijk} - \bar{Y}_{...})$ is the total variability among yields
$(\bar{Y}_{i..} - \bar{Y}_{...})$ is the variability among yields associated with ρ, the row factor
$(\bar{Y}_{.j.} - \bar{Y}_{...})$ is the variability among yields associated with τ, the column factor
$(\bar{Y}_{..k} - \bar{Y}_{...})$ is the variability among yields associated with γ, the inside factor, and (by algebra):

$$\begin{aligned} R &= Y_{ijk} - \bar{Y}_{i..} - \bar{Y}_{.j.} - \bar{Y}_{..k} + 2\bar{Y}_{...} \\ &= (Y_{ijk} - \bar{Y}_{...}) - [(\bar{Y}_{i..} - \bar{Y}_{...}) + (\bar{Y}_{.j.} - \bar{Y}_{...}) + (\bar{Y}_{..k} - \bar{Y}_{...})] \end{aligned}$$

Recall the discussion in Chapter 6 of error and interaction when an unreplicated design is applied to a case where there is no interaction. Without replication, we have no pure way of estimating error. The model is "fully specified" without error. If we can assume that there is no interaction, then the interaction term must be measuring error. The second expression for R, above, shows that it is an interaction-like term; the first way of writing R, though less intuitive as a declaration of interaction, makes it easier to see that, algebraically, it is the correct expression to make the alleged equality into an actual equality.

By writing an equation for each i, j, and k, then squaring both sides of each equation, and finally adding all the left sides and all the right sides, we arrive at a breakdown of the sums-of-squares of

$$\text{TSS} = \text{SSB}_{\text{r(ows)}} + \text{SSB}_{\text{c(olumns)}} + \text{SSB}_{\text{inside·factor}} + \text{SSW}$$

The data array in general form for an m-level Latin square (three factors, each at m levels) would be as follows:

	τ_1	τ_2	. . .	τ_m
ρ_1	γ_1	γ_2	. . .	γ_m
ρ_2	γ_2	γ_3	. . .	γ_1
⋮	⋮	⋮	. . .	⋮
ρ_m	γ_m	γ_1	. . .	γ_{m-1}

The ingredients of the ANOVA table and the expected value of each mean square are summarized in Table 8.5.

The degrees of freedom for each factor are the same, $(m - 1)$, corresponding to the number of levels, m, minus one. The total number of degrees of freedom is equal to the total number of data points, m^2, minus one. We can, by subtraction, then calculate the number of degrees of freedom associated with the error term as follows:

$$\begin{aligned} \text{Error } df &= (m^2 - 1) - 3(m - 1) \\ &= m^2 - 3m + 2 \\ &= (m - 1)(m - 2) \end{aligned}$$

As indicated in Table 8.5, it is usually easier to calculate the sum of squares for error, SSW, by subtraction:

$$\text{SSW} = \text{TSS} - \text{SSB}_{\text{r(ows)}} - \text{SSB}_{\text{c(olumns)}} - \text{SSB}_{\text{inside·factor}}$$

The last column of Table 8.5 points the way for the ratios of mean squares used for calculation of F_{calc} for each factor. To test rows, use $\text{MSQ}_{\text{r}}/\text{MSW}$; to test columns, use $\text{MSQ}_{\text{c}}/\text{MSW}$; to test the inside factor, use $\text{MSQ}_{\text{inside·factor}}/\text{MSW}$. The detailed expressions for the V's are as we might expect them based on those of previous chapters; for example, $V_{\text{r(ows)}} = \sigma^2 + m\,\Sigma_i\,\rho_i^2/(m - 1)$ for a fixed factor, and for a random factor, $V_{\text{r(ows)}} = \sigma^2 + m\sigma_\rho^2$.

For the 4×4 Latin-square example of Table 8.4, the ANOVA results are shown in Table 8.6. None of the three factors are significant at $\alpha = .05$. This indicates that service policies, hours open, and amenity levels do not affect sales much. Note, however, that amenities, with a p value of .099, would be significant at $\alpha = .10$.

If the Latin-square design has replication, the model is similar, except that there is an additional, explicit term for error. Unlike the Chapter 6 discussion in which a two-factor design having replication allows separate capturing and testing of the interaction effect, a Latin-square design does not afford such a

Table 8.5 **ANOVA Table for *m*-Level Latin Square**

Source of Variability	Sums of Squares	*df*	Expected Value of MSQ
Rows	$m\,\Sigma_i\,(\bar{Y}_{i\cdot\cdot} - \bar{Y}_{\cdots})^2$	$m - 1$	$\sigma^2 + V_{\text{rows}}$
Columns	$m\,\Sigma_j\,(\bar{Y}_{\cdot j\cdot} - \bar{Y}_{\cdots})^2$	$m - 1$	$\sigma^2 + V_{\text{columns}}$
Inside factor	$m\,\Sigma_k\,(\bar{Y}_{\cdot\cdot k} - \bar{Y}_{\cdots})^2$	$m - 1$	$\sigma^2 + V_{\text{inside-ftr}}$
Error	By subtraction	$(m - 1)(m - 2)$	σ^2
Total	$\Sigma_i\Sigma_j\Sigma_k\,(Y_{ijk} - \bar{Y}_{\cdots})^2$	$m^2 - 1$	

Table 8.6 ANOVA Table for Car Dealership Example

Source of Variability	Sums of Squares	*df*	Mean Square	F_{calc}	*p* Value
Service policy	17566.5	3	5855.5	2.173	.192
Hours open	4678.5	3	1559.5	.579	.650
Amenities	26722.5	3	8907.5	3.306	.099
Error	16164.5	6	2694.1		
Total	65132.0	15			

luxury. It still requires us to assume that there is no interaction; recall that interaction would interfere with all levels of a factor being on equal footing.

Analyzing a replicated Latin-square design, therefore, is very similar to analyzing an unreplicated design. The sum of squares for each of the three factors is still determined by examining the mean at each factor level. Also, the expression for the sum of squares is the same as for an unreplicated design, except that there is an extra multiplicative value, n, representing the number of replicates per cell. For example, for the row factor,

$$SSB_{r(ows)} = nm\, \Sigma_i\, (\bar{Y}_{i\cdot\cdot} - \bar{Y}_{\cdot\cdot\cdot})^2$$

Algebra indicates that n belongs in the expression. From an intuitive point of view, it is present because the number of data values supporting each row mean (that is, the number of data values in the row) is no longer m but is now nm. Once we have SSB_r, SSB_c, and $SSB_{inside\text{-}factor}$, we simply determine the sum of squares associated with error, SSW, by subtraction, exactly as above. The degrees of freedom will still be $(m - 1)$ for each factor, but for error the degrees of freedom will change to

$$(nm^2 - 1) - 3(m - 1) = nm^2 - 3m + 2$$

which equals the sum of the previous degrees-of-freedom value of $[(m - 1)(m - 2)]$, and the additional degrees of freedom due to replication, $m^2(n - 1)$, which is the number of cells times the replication degrees of freedom per cell. The reason is that the total number of degrees of freedom changes from $(m^2 - 1)$ to $(nm^2 - 1)$. Still, the determination of the MSQ values proceeds in the usual way, and the same ratios form the respective F_{calc}'s. Following is another example of a nonreplicated case.

EXAMPLE 8.2 Latin-Square Analysis of Valet-Parking Use

We study the number of patients using valet parking at a large medical clinic near Boston. The factors, shown in Table 8.7, are the cost of the valet-parking

Table 8.7 Use of Valet Parking

		Number of Handicapped Spaces			
		4	3	2	1
Cost to Park	1	2 29	4 44	3 54	1 71
	2	3 22	1 22	2 59	4 100
	3	4 38	3 31	1 40	2 79
	4	1 29	2 27	4 83	3 100

Note: Top left number in cells is the level of the inside factor, number of valet-parking attendants on duty.

service (rows), the quantity of handicapped parking spaces in the parking lot closest to the clinic entrance (columns), and the number of valet-parking attendants on duty (inside factor, considered an indirect indicator of waiting time). The study's primary goal was to determine how best to accommodate the needs of patients efficiently.

Each factor is studied at four levels. For the number of spaces, levels 4, 3, 2, 1 are, respectively, ten, eight, six, and four spaces; for cost, levels 1, 2, 3, 4 are, respectively, $3, $4, $5, and $6. For number of attendants, the levels happen to be the actual values. Earlier studies have demonstrated that interaction effects are negligible. Table 8.7 shows the data in a Latin square, and Table 8.8 is the ANOVA table.

At $\alpha = .05$ and $df = (3, 6)$, **c** $= 4.76$. We conclude that the number of handicapped parking spaces nearby affects how many people use the valet-parking service, $p < .001$. The number of valet-parking attendants available, which is a surrogate for the amount of time a patient must wait to use the valet

Table 8.8 ANOVA Table for Valet-Parking Study

Source of Variability	Sums of Squares	*df*	Mean Square	F_{calc}
Cost	370.5	3	123.5	2.9
No. of spaces	9,025	3	3008.3	71.1
No. of attendants	1,389.5	3	463.2	10.9
Error	254	6	42.3	
Total	11,039	15		

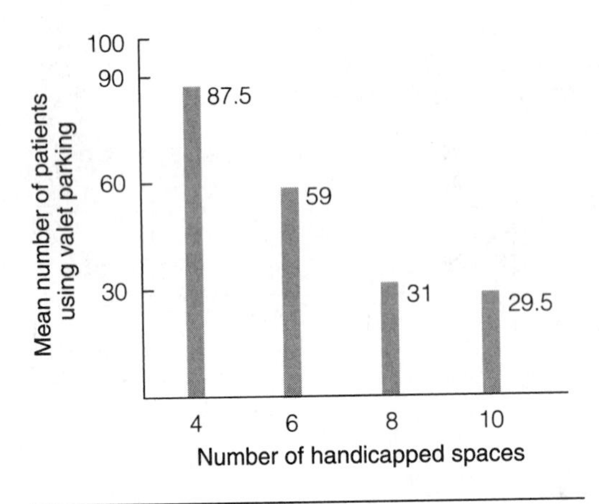

Figure 8.1

service, is also significant, $p < .01$. The cost of the valet service, in the range of costs studied, does not seem to affect its use: $p > .10$. The most significant factor is the number of handicapped spaces; as this number decreases from ten to four, the average demand increases, as shown in Figure 8.1. Also, as the number of attendants increases from one to four, the average demand increases from 40.5, to 48.5, to 51.75, to 66.25. The results indicate the importance of the number of handicapped spaces as well as of the number of valet-parking attendants.

Next we analyze the same problem using SPSS.

EXAMPLE 8.3 Using a Latin-Square Design with SPSS

Analysis of an experiment with a Latin-square design fits naturally within the framework of SPSS's ANOVA model analyses. For a two-factor design the input lists the data in one column and has two additional columns to indicate, for each data value, the level of the row factor and of the column factor. For a Latin-square design, one must simply add another column to indicate the level of the third factor. Table 8.9 shows this format for the input data from the valet-parking study of Example 8.2. The first column represents the dependent variable; the other three columns represent the levels, respectively, of the row, column, and inside factor.

When arranging for the SPSS output of a Latin-square analysis, enter the "Options" menu and be careful to specify no interaction effects. Table 8.10 shows the SPSS output.

Table 8.9 SPSS Format: Input Data for Valet-Parking Study

Demand	Cost	NumberSp	NumAtt
29	1.00	4.00	2.00
22	2.00	4.00	3.00
38	3.00	4.00	4.00
29	4.00	4.00	1.00
44	1.00	3.00	4.00
22	2.00	3.00	1.00
31	3.00	3.00	3.00
27	4.00	3.00	2.00
54	1.00	2.00	3.00
59	2.00	2.00	2.00
40	3.00	2.00	1.00
83	4.00	2.00	4.00
71	1.00	1.00	1.00
100	2.00	1.00	4.00
79	3.00	1.00	2.00
100	4.00	1.00	3.00

Table 8.10 SPSS Format: ANOVA Table for Parking Study

```
* * * ANALYSIS OF VARIANCE * * *
    DEMAND
  by  COST
      NUMBERSP
      NUMATT
```

Source of Variation	Sum of Squares	DF	Mean Squares	F	Sig of F
Main Effects	10785.000	9	1198.333	28.307	.000
COST	370.500	3	123.500	2.917	.123
NUMBERSP	9025.000	3	3008.333	71.063	.000
NUMATT	1389.500	3	463.167	10.941	.008
Explained	10785.000	9	1198.333	28.307	.000
Residual	254.000	6	42.333		
Total	1039.000	15	735.933		

EXAMPLE 8.4 Using a Latin-Square Design with JMP

Table 8.11 presents output for this same example using the JMP software.

Table 8.11 JMP Output: ANOVA Table

Analysis of Variance

Source	DF	Sum of Squares	Mean Squares	F Ratio	Prob > F
Model	9	10785.0000	1198.33	28.3071	0.0003
Cost	3	370.5000	123.50	2.9173	0.1225
Quantity	3	9025.000	3008.33	71.0630	<.0001
Number	3	1389.5000	463.17	10.9409	0.0076
Error	6	254.000	42.33		
Total	15	11039.000			

8.3 Graeco-Latin-Square Designs

Factors in a Latin-square design can have any number of levels, but strictly speaking a Latin square accommodates only three factors. Similar designs with four factors are called Graeco-Latin squares. Historically, similar designs with five factors were called Sino-Graeco-Latin squares. The names come from early forms of notation in which rows and columns were designated by numbers (1, 2, 3, . . .) and levels of the inside factor were designated by Latin-derived letters: a, b, c, ... When a fourth factor (second inside factor) was involved in a study, it was usually represented by Greek letters, such as α, β, γ, and the like. A fifth factor (third inside factor) might be represented by Chinese symbols. Wisdom ultimately prevailed, so that now any design involving more than three factors is called a Graeco-Latin square and, usually, all letters used are Latin letters (at least in the United States). An example of a Graeco-Latin square involving four factors, each at three levels, is shown in Table 8.12.

Notice that the balance requirement for a Latin square is satisfied here. Each level of A is paired once with each level of B; each level of A is paired once with each level of C; each level of A is paired once with each level of D. Hence, each level of A is on equal footing. The same holds for factors B, C, and D—all assuming that there are no interaction effects among the factors. We might use such a Graeco-Latin square if, for example, we had a variation on the valet-parking example: three levels per factor (cost, number of handicapped parking spaces, and number of valet-parking attendants), plus a fourth factor, also at three levels—perhaps the number of hours the clinic is open.

Table 8.12 **Example of a Graeco-Latin Square**

	B_1	B_2	B_3
A_1	C_1D_1	C_2D_2	C_3D_3
A_2	C_2D_3	C_3D_1	C_1D_2
A_3	C_3D_2	C_1D_3	C_2D_1

A necessary, but not sufficient, condition for the treatment combinations of factors A, B, C, and D (with C and D as inside factors) to form a Graeco-Latin square is that factors A, B, and C form a Latin square, and that factors A, B, and D form a Latin square. The reason these conditions are not sufficient is illustrated by the treatment combinations in Table 8.13. In that table, factors A, B, and C form a Latin square, and so do factors A, B, and D. However, factors C and D are confounded.

The Graeco-Latin square in Table 8.12 is a *complete* Graeco-Latin square. This means that the full capacity of the square (the maximum number of factors that can be included) is used. The relationship that defines this maximum involves the number of levels, m, and the resultant number of degrees of freedom. In an m-level Graeco-Latin square without replication, irrespective of the number of factors in the study, the number of data points is m^2 and the total number of degrees of freedom is $(m^2 - 1)$. Because the number of degrees of freedom associated with each factor is $(m - 1)$, the maximum number of factors that can be accommodated is

$$(m^2 - 1)/(m - 1) = m + 1$$

If $m = 3$, $(m + 1) = 4$. In the Table 8.12 example, we might say that we are analyzing four factors for the cost of two; the nine treatment combinations represent a one-ninth replicate of the possible $3^4 = 81$ treatment combinations.

Another example of a complete Graeco-Latin square appears in Table 8.14. It shows how, with five levels, six factors can be accommodated.

Table 8.13 **A Non-Graeco-Latin Square**

	B_1	B_2	B_3
A_1	C_1D_1	C_2D_2	C_3D_3
A_2	C_2D_2	C_3D_3	C_1D_1
A_3	C_3D_3	C_1D_1	C_2D_2

Table 8.14 **Graeco-Latin Square**

	B_1	B_2	B_3	B_4	B_5
A_1	$C_1D_1E_1F_1$	$C_2D_2E_2F_2$	$C_3D_3E_3F_3$	$C_4D_4E_4F_4$	$C_5D_5E_5F_5$
A_2	$C_2D_3E_4F_5$	$C_3D_4E_5F_1$	$C_4D_5E_1F_2$	$C_5D_1E_2F_3$	$C_1D_2E_3F_4$
A_3	$C_3D_5E_2F_4$	$C_4D_1E_3F_5$	$C_5D_2E_4F_1$	$C_1D_3E_5F_2$	$C_2D_4E_1F_3$
A_4	$C_4D_2E_5F_3$	$C_5D_3E_1F_4$	$C_1D_4E_2F_5$	$C_2D_5E_3F_1$	$C_3D_1E_4F_2$
A_5	$C_5D_4E_3F_2$	$C_1D_5E_4F_3$	$C_2D_1E_5F_4$	$C_3D_2E_1F_5$	$C_4D_3E_2F_1$

The creation of these arrays is not simple, especially when they get larger. Of the billions of subsets of 25 out of $5^6 = 15{,}625$ possible treatment combinations, less than a billionth of a percent qualify as Graeco-Latin squares. The issue is further complicated by the fact that some Graeco-Latin squares, even with only four factors (including two inside factors), are impossible to create. For example, it is impossible to find a 6×6 Graeco-Latin square (that is, one with four or more factors). It was thought for many years (even by Fisher, who invented the term "Latin square" in 1926) that it was impossible to create any Graeco-Latin square with dimension (that is, number of levels) $m = (4L + 2)$ for integer L; however, in 1959, Graeco-Latin squares of dimension 10 and 22 were published. Tables of Graeco-Latin-square designs appear in some of the references listed at the end of Chapter 15 (Peng, Table 6.4.2, p. 104; and others).

There is a certain elegance and efficiency inherent in complete Graeco-Latin-square designs, but they suffer a major flaw: all the degrees of freedom have been used to estimate the effects and none are left, in an unreplicated design, to allow the assessment of error. That is, if there are $(m + 1)$ factors, each "using up" $(m - 1)$ degrees of freedom, all $(m^2 - 1)$ degrees of freedom

Table 8.15 **Incomplete Graeco-Latin Square**

	B_1	B_2	B_3	B_4	B_5
A_1	C_1D_1	C_2D_2	C_3D_3	C_4D_4	C_5D_5
A_2	C_2D_3	C_3D_4	C_4D_5	C_5D_1	C_1D_2
A_3	C_3D_5	C_4D_1	C_5D_2	C_1D_3	C_2D_4
A_4	C_4D_2	C_5D_3	C_1D_4	C_2D_5	C_3D_1
A_5	C_5D_4	C_1D_5	C_2D_1	C_3D_2	C_4D_3

Table 8.16 ANOVA Table

Source of Variability	SSQ	df	MS	F_{calc}
A	SSB_A	4		
B	SSB_B	4		
C	SSB_C	4		
D	SSB_D	4		
Error	SSW	8		
Total	TSS	24		

By subtraction

are utilized; the SSW would come out zero, and the MSW would be zero divided by zero (its degrees of freedom), or, appropriately, an "indeterminate form." With an indeterminate MSW, it is not possible to perform significance testing (that is, the F test).

Accordingly, if we decide against replication, so-called *incomplete* Graeco-Latin squares may be used. These are Graeco-Latin squares, such as the one shown in Table 8.15, that have m levels (in this case, five) of each factor, but fewer than $(m + 1)$ factors (in this case, four).

The ANOVA table, Table 8.16, illustrates the testing of the four hypotheses related to the significance of each of factors A, B, C, and D. The value of SSW in Table 8.16 is determined by subtracting the other SSB terms from the TSS. In this case, the SSB for each factor has four degrees of freedom, corresponding to five levels. The number of degrees of freedom for SSW is also determined by subtraction. Given 25 data values, there are a total of 24 degrees of freedom; since 16 of them correspond to the four factors, that leaves 8 degrees of freedom for error. The mean squares and F_{calc} values for each factor follow arithmetically. Each F_{calc} would then be compared with an F-table value at the appropriate value of α and (4, 8) degrees of freedom. So we examine the impact of each of the four factors, in some sense studying four factors for the price of two: we are performing a 1/25th replicate (25 treatment combinations run, out of $5^4 = 625$ possible treatment combinations).

8.4 Other Designs with Three or More Factors

One need not use a Latin-square or Graeco-Latin-square design just because there are three or more factors. In many cases the situation isn't a square; factors under study simply do not have the same number of levels. Sometimes it may be possible to add or subtract a level to obtain a square, but not always.

If running a Latin square or Graeco-Latin square is not feasible or desirable, what do we do?

Possibly we could run every combination of factors and levels. This might work if the numbers of levels of the factors are relatively small. But if, for example, a study has four factors, with two having five levels and the other two having three levels (total combinations $5^2 \cdot 3^2 = 225$), the financial and management costs might be excessive. But if we can hone our study such that it considers two factors at three levels and two factors at two levels (total combinations $3^2 \cdot 2^2 = 36$), it may be cost-effective.

If all factors in a study have only two levels, running all treatment combinations of say, five factors ($2^5 = 32$) might not be a burden. Obviously, when all factors have two levels, the number of factors that can be studied for a fixed number of treatment combinations is maximized. This situation has been researched extensively, and we devote Chapters 9, 10, 11, and part of 13 directly to it, highlighting the running of fractional replicates (other than Latin squares or Graeco-Latin squares). These designs can also accommodate situations in which some factors have four levels and others have two levels. In all of these designs, we may encounter interaction effects among three or more factors at a time, not solely the interaction effects encountered thus far that are only two-factor interactions. We defer discussion of three-way and higher-order interactions until Chapter 9.

Some classes of designs in the incomplete-Latin-square family are beyond the scope of this text. For a design that is a Latin square, except that one factor has one level fewer than the other two factors, the analysis procedure is not too complicated (though far more so than for a routine Latin square), and discussion of that situation can be found in various texts on experimental design. If one row and one column are missing, the procedure is more complicated. However, certain patterns of incomplete Latin squares are systematic, in the sense that the designs and analyses can be generalized; for example, consider Table 8.17, an incomplete Latin square with three factors having numbers of levels three, seven, and seven for rows, columns, and inside factors respectively. Note that, within columns, each level of C occurs exactly once with each other level of C (for example, C_1 occurs in the same column with C_2 once [column B_7], with C_3 once [column B_1], and so on).

Table 8.17 Incomplete Latin Square

	B_1	B_2	B_3	B_4	B_5	B_6	B_7
A_1	C_7	C_6	C_5	C_4	C_3	C_2	C_1
A_2	C_1	C_7	C_6	C_5	C_4	C_3	C_2
A_3	C_3	C_2	C_1	C_7	C_6	C_5	C_4

Table 8.18 Incomplete Latin Square

	B_1	B_2	B_3	B_4
A_1	C_1	C_3	C_4	C_5
A_2	C_6	C_1	C_2	C_3
A_3	C_7	C_2	C_3	C_4
A_4	C_2	C_4	C_5	C_6
A_5	C_3	C_5	C_6	C_7
A_6	C_4	C_6	C_7	C_1
A_7	C_5	C_7	C_1	C_2

Now consider the incomplete Latin square in Table 8.18, with factors having numbers of levels seven, four, and seven, for rows, columns, and inside factors, respectively. In this incomplete Latin square, within rows each level of factor C occurs exactly *twice* with each other level of C (for example, in rows, C_4 occurs twice with C_1 [rows A_1 and A_6], twice with C_2 [rows A_3 and A_4], and so on). Designs exemplified by such *rectangles* are, curiously, called **Youden squares.** A Youden square is a special, systematic type of incomplete Latin square in which each level of the inside factor appears the same number of times in the incomplete dimension with each other level of the inside factor. In Table 8.17, each level of C appears once in a column with each other level; in Table 8.18, twice with each other level, in rows. The analysis of Youden squares involves systematic adjustments to the means of the levels of the factors not on equal footing: factors B and C in Table 8.17, and factors A and C in Table 8.18. The earlier-mentioned Peng reference in Chapter 15, among others, discusses Youden squares.

EXAMPLE 8.5 Maximizing Profit at Nature's Land Farms, Revisited

The experiment, as described at the beginning of the chapter, was run using a Latin square, so that of the 27 combinations of positioning, cobranding, and packaging, only nine combinations had to be formed. Remember, the use of a Latin square is helpful only if one has ascertained the appropriateness of the no-interaction assumption. Just what "interaction" meant in this situation was carefully explained to the management of the cooperative; people often confuse it with the concept of *correlation*. For example, we asked questions such as, "Whatever differences in results we get for different positioning options, if any, would you expect them to vary by cobrand?" The questions were repeated in various ways; for example, "Would you expect any differences due to the cobrand to vary with which positioning is chosen?"

A separate simulated test market was set up, with each treatment combination prominently displayed. A simulated test market is a storelike setting in which a "customer" (a participant in the study) is exposed to various possible products and brands and asked to spend a fixed amount of money or, in some cases, to buy a fixed number of items. Each of the nine treatment combinations were tested with the same set of competing products and brands. For each simulated test market/treatment combination, there were 100 "customers"/replicates. The primary dependent variable was the number of NLF packages of potatoes purchased; another measure was the attitude toward the NLF product, obtained by a questionnaire administered after the shopping experience was completed. The participants did not know during the shopping portion of the experiment which products and brands were being studied.

It turned out that the positioning factor and the packaging factor dominated the cobranding factor. In advance of the experiment, NLF managers agreed that the biotech issue would make a difference, although they were divided on the direction. "Biotech" potatoes decreased sales relative to other levels of the positioning factor, $p < .01$. "Superior taste" potatoes slightly outsold the "environmentally friendly" potatoes, but the difference was not statistically significant, with a p value of about .15.

The cobranding factor was not significant at all. It was decided afterward that since all three cobrands were well known and had a solid reputation, differences among them were too small to detect. In packaging, there was no difference between the clear plastic and the paper with a window. However, the paper packaging without a window was significantly inferior to both. Apparently, the ability to see the product is important to consumers. This did not seem to surprise NLF managers (although they did say that it was "too bad" because paper without a window is the cheapest packaging).

Exercises

1. The text example that studied the impact on demand for valet parking at the clinic included one day's data for each combination of parking cost, number of handicapped spaces available, and number of attendants. Suppose that there were, in fact, five random weekdays of data for each treatment combination of the three factors, as outlined in Table 8EX.1. Repeat the Latin-square analysis, now using the five replicates per cell. The data in the table are included in the data sets, labeled Problem 8.1.
2. Consider the Latin-square design in Table 8EX.2 in which the row-factor levels represent different magazines, the column-factor levels represent different sizes of print advertisements, and the inside-factor levels represent different times of the year. The dependent variable is "number of orders generated from the advertisement." Other fac-

Table 8EX.1 Valet-Parking Study

Cost to Park	No. of Spaces	No. of Attendants	Five Replicates (Days) 1	2	3	4	5
$3	10	2	29	18	28	27	21
$4	10	3	22	21	24	23	27
$5	10	4	38	42	39	46	29
$6	10	1	29	24	33	28	25
$3	8	4	44	43	49	50	48
$4	8	1	22	18	20	28	16
$5	8	3	31	32	36	36	24
$6	8	2	27	32	29	31	34
$3	6	3	54	56	52	54	58
$4	6	2	59	56	53	53	56
$5	6	1	40	49	45	47	34
$6	6	4	83	87	82	83	91
$3	4	1	71	73	60	68	72
$4	4	4	100	105	103	104	103
$5	4	2	79	81	81	78	76
$6	4	3	100	94	90	104	98

Table 8EX.2 Latin Square for Magazine Ad Study

	1	2	3	4
1	A 122	B 129	C 135	D 126
2	B 117	A 116	D 113	C 102
3	C 114	D 120	A 123	B 111
4	D 120	C 128	B 129	A 123

tors inherent in the ad—positioning, location in the magazine and so on—are held constant.

Analyze the experiment to determine if there are significant differences among magazines, among the different sizes of the advertisements, and among different times of the year. Use $\alpha = .05$. Note that we have used the "Latin-letters-on-the-inside" notation for the first time, to enhance the sense of the history of the field of experimental design. We also use old-style notation in Table 8EX.12.

3. For the exercise 2 data, perform Tukey's HSD analysis on the rows. Use **a** = .05.
4. For the exercise 2 data, perform a Newman-Keuls test on the rows. Use **a** = .05.
5. For the exercise 2 data, use an orthogonal breakdown of the sum-of-squares associated with the row factor to test whether (1) the average of magazines one and three differs from the average of magazines two and four; (2) the average of magazines one and two differs from the average of magazines three and four; (3) the average of magazines one and four differs from the average of magazines two and three. Use $\alpha = .01$.
6. The data in Table 8EX.6 expand upon exercise two and involve four factors—magazines

Table 8EX.6 Magazine Ad Study

Magazine	Print	Season	Colors	Issue 1	Issue 2	Issue 3	Issue 4
1	8 point	Fall	1	125	121	122	123
2	8 point	Winter	4	118	113	118	119
3	8 point	Spring	2	118	115	119	117
4	8 point	Summer	3	120	118	119	120
1	10 point	Winter	2	127	134	128	127
2	10 point	Spring	3	113	115	118	118
3	10 point	Summer	1	118	127	118	120
4	10 point	Fall	4	127	122	130	128
1	12 point	Spring	3	133	140	132	133
2	12 point	Summer	2	109	113	111	113
3	12 point	Fall	4	118	125	126	121
4	12 point	Winter	1	133	125	128	136
1	14 point	Summer	4	129	125	125	125
2	14 point	Fall	1	103	105	100	96
3	14 point	Winter	3	113	110	114	110
4	14 point	Spring	2	122	120	119	124

(row factor), size of print (column factor), season (first inside factor), and number of colors in the ad (second inside factor). For each treatment combination in the Graeco-Latin square arrangement, the table shows the results (number of orders generated) for four randomly chosen issues (replicates). The data are also in the data sets, labeled Problem 8.6.

Ignore the fourth factor, number of colors, and analyze as a Latin square. Find the *p* value when testing each factor.

7. Now analyze the exercise 6 data as a Graeco-Latin square as if there were no replication (that is, use only the first column of data).
8. Finally, analyze the exercise 6 data using all of the data in the table (that is, as a Graeco-Latin square with four replicates per cell). Find the *p* value when testing each factor.
9. The Merrimack Valley Pediatric Clinic has gathered data in order to determine what affects customer satisfaction. It is studying four factors: location (Andover, Methuen, Salem, and Amesbury), extended hours (none beyond the normal office hours, extra evening hours, extra Saturday hours, and extra hours on both evenings and Saturdays), follow-up calls by nurse-practitioners (levels: none, low, medium, and high), and clerical assistance for patients who need help with insurance forms (levels: none, minimum, medium, and maximum). Analyze the Table 8EX.9 data as a Graeco-Latin square; each data point represents a satisfaction score for one client, based on a lengthy discussion of the levels of the factors (higher is better). Use $\alpha = .05$. (Data are in the data sets as Problem 8.9.)
10. Prove that it is impossible to construct a Graeco-Latin square with four factors, each factor having two levels.
11. How many different 4×4 Latin squares are possible?
12. Suppose that in exercise 2 there had been a fourth factor (second inside factor), representing four different prices, thus forming a Graeco-Latin square, as shown in Table 8EX.12. Analyze the data as a Graeco-Latin square, using $\alpha = .05$.

Table 8EX.9 Analysis of MVPC Customer Satisfaction

Office	Hours	Follow-Up	Insurance Help	Score
Andover	None	None	None	47
Methuen	None	Low	Max	62
Salem	None	Med	Min	62
Amesbury	None	High	Med	68
Andover	Eve	Low	Min	68
Methuen	Eve	None	Med	67
Salem	Eve	High	None	85
Amesbury	Eve	Med	Max	62
Andover	Sat	Med	Med	72
Methuen	Sat	High	Min	68
Salem	Sat	None	Max	66
Amesbury	Sat	Low	None	73
Andover	Eve+Sat	High	Max	87
Methuen	Eve+Sat	Med	None	64
Salem	Eve+Sat	Low	Med	78
Amesbury	Eve+Sat	None	Min	82

Table 8EX.12 Old-Style Graeco-Latin Square

	1	2	3	4
1	Aα 122	Bβ 129	Cγ 135	Dδ 126
2	Bδ 117	Aγ 116	Dβ 113	Cα 102
3	Cβ 114	Dα 120	Aδ 123	Bγ 111
4	Dγ 120	Cδ 128	Bα 129	Aβ 123

13. Consider the analyses in exercises 2 and 12. Compare MSW values. Under what conditions, in general, will the MSW of the Graeco-Latin square be greater/less than the MSW of the Latin square?

14. Consider a $4 \times 4 \times 4$ Latin square without replication. Suppose that the upper-left cell has each of the three factors A, B, and C, at its respective level one; suppose further that of the 16 data values composing the Latin square, the upper-left value is missing—only the other 15 data values are available. What value would you substitute for this missing value, if the goal is to minimize the SSW of the experiment (when, with the substituted value, the experiment is analyzed as a "normal" Latin square)? In reality, this is exactly what would be done. However, the degrees of freedom for the error term would be reduced from 6, which it would routinely be, to 5, reflecting the fact that the total number of degrees of freedom is actually 14, since there were only 15 data values to begin with.

15. Suppose that you wanted to design an experiment in which there were four factors, A, B, C, and D, each having three levels. However, you cannot rule out a two-factor interaction effect between factors C and D. What set of treatment combinations would you suggest should be run?

16. Suppose that we revisit the data of exercise 1, but now we assume that the "replicates" are not true replicates but represent a random Monday, a random Tuesday, a random Wednesday, a random Thursday, and a

random Friday, in that order. Analyze the data on this basis. Hint: even if there were four days instead of five, this would not simply turn the Latin square into a Graeco-Latin square; after all, we have a Latin square for each "level" of day. For example, consider the upper-left cell, the (1, 4, 2) cell. Letting the fourth number represent the day, we have a data point for (1, 4, 2, 1), (1, 4, 2, 2), (1, 4, 2, 3), and (1, 4, 2, 4); a Graeco-Latin square would have a data point for only one day, X: (1, 4, 2, X).

17. Suppose in exercise 9 that four people were interviewed per treatment combination, and the data were as shown in Table 8EX.17. Analyze as a Graeco-Latin square with replication. Use $\alpha = .05$. (Data are also in the data sets as Problem 8.17.)

Table 8EX.17 Analysis of MVPC Customer Satisfaction

Office	Hours	Follow-Up	Insurance Help	Score
Andover	None	None	None	47, 58, 52, 45
Methuen	None	Low	Max	62, 54, 56, 50
Salem	None	Med	Min	62, 73, 73, 67
Amesbury	None	High	Med	68, 76, 69, 77
Andover	Eve	Low	Min	68, 57, 60, 63
Methuen	Eve	None	Med	67, 76, 63, 79
Salem	Eve	High	None	85, 75, 92, 73
Amesbury	Eve	Med	Max	62, 68, 58, 61
Andover	Sat	Med	Med	72, 76, 67, 73
Methuen	Sat	High	Min	68, 87, 78, 87
Salem	Sat	None	Max	66, 68, 75, 68
Amesbury	Sat	Low	None	73, 63, 72, 76
Andover	Eve+Sat	High	Max	87, 72, 75, 72
Methuen	Eve+Sat	Med	None	64, 66, 68, 81
Salem	Eve+Sat	Low	Med	78, 77, 69, 75
Amesbury	Eve+Sat	None	Min	82, 83, 80, 79

Notes

1. We do not include in our discussion the cost of analyzing the experiment. By and large, this cost doesn't materially vary with the size of the experiment. We do not view the data entry cost of 150 data points as materially more expensive than that of 32 data values, for example.

2. What we mean by *delicacy* is having a more finely grained resolution with respect to the ability to choose which subset of interaction effects is assumed to equal zero. The designs discussed in this chapter require the assumption that *all* interaction effects equal zero.

3. If there were replication, nothing material in our discussion would change. Instead of "exactly one data value" at each level of B, and of C, it would be "exactly n data values . . . ." The key is that it is the same number of data values at each level of each other factor.

4. A 3 × 3 Latin square can be said to be a "one-third replicate"; a 4 × 4 Latin square can be said to be a "one-fourth replicate"; a 5 × 5 a "one-fifth replicate"; and so forth.

5. One could argue the possibility that factors B and C interact; ancillary services might be a tad more attractive if hours open are more convenient. One could also argue that this interaction, if not zero, is negligible.

PART TWO

Primary Focus on the Number of Levels of a Factor

CHAPTER 9

Two-Level Factorial Designs

EXAMPLE 9.1 Pricing a Supplemental Medical/Health Benefit Offer

An insurance company, HealthMark, was interested in conducting a marketing research study to determine the "best" (that is, most profitable) offer of supplemental medical and health benefits. On the basis of previous studies, it already had determined the benefits that would be offered; there would be a core set of benefits priced as a package, along with three auxiliary benefits that had two levels each, a price and a specific degree of benefit. The key remaining step was the pricing of the core benefits and the price/specific degree of benefit of the three auxiliary benefits. The five core benefits were (in no particular order):

1. Pharmacy channel: savings of 25–50% for prescriptions at 25,000 pharmacies nationwide (HealthMark used the term "channel" to indicate a network of providers.)
2. Mail-order pharmacy service channel: a further 10% off the prices in benefit 1, for frequently ordered prescriptions
3. Vision channel: 25–50% off for eye glasses and contact lenses, major discounts on eye exams, especially at major chains, such as Pearle Vision, BJ's, and other warehouse stores offering optometry and ophthalmology services
4. Hearing aid channel: 50–75% off many hearing aid models, including most national brands, by mail-order, with a one-year full warranty
5. Dental channel: discount of 15–40% off a dentist's usual prices, with a national network, including periodontists and orthodontists

The three auxiliary benefits offered were

1. Chiropractic channel: discount of 25 or 50% (depending on the "level"—see Table 9.1) off chiropractic services, with a national network, and extensive referral services

Table 9.1 Prices and Benefits Offered

Factor	Levels (Low/High)
Price of core benefits	\$9.95/\$12.95 per month per adult
Price (and benefit) of chiropractic channel	\$0.50 (and 25% off)/\$100 (and 50% off) per month per adult
Price (and benefit) of dermatology channel	\$2 (and 20% off)/\$3 (and 40% off) per month per adult
Price (and benefit) of massage channel	\$1 (and 15% off)/\$2 (and 30% off) per month per adult

2. Dermatology/cosmetic surgery channel: discount of 20 or 40% off most dermatology and cosmetic surgery procedures, with a national network, and extensive referral service
3. Massage channel: a discount of 15 or 30% for massage treatments, with a national network and extensive referral network

The factors under study were the price to be attached to the core benefits, and the price and specific level of auxiliary benefits. In each case, previous research, competitive pressures, and marketing considerations narrowed each price to two options. The factors and the two levels of each are noted in Table 9.1.

There are 16 (2^4) treatment combinations. Five hundred people evaluated each of the 16 combinations by selecting a "purchase intent score" on a scale of 0–10, where 0 indicated "0 chances in 100 I would sign up," and 10 indicated "99 chances in 100 I would sign up." A score of 1 corresponded to 10 chances in 100, a score of 2 to 20 chances in 100, and so on, increasing by 10 chances at each step. We return to this example at the end of the chapter.

9.1 Introduction

We now change our focus from the number of factors in the experiment to the number of levels those factors have. Specifically, in this and the next several chapters, we consider designs in which all factors have two levels. Many experiments are of this type. This is because two is the minimum number of levels a factor can have and still be studied, and by having the *minimum number of levels (2)*, an experiment of a certain size can include the *maximum number of factors*. After all, an experiment with five factors at two levels each contains 32 combinations of levels of factors (2^5), whereas an experiment with these same five factors at just one more level, three levels, contains 243 combinations of levels of factors (3^5)—about eight times as many combinations! Indeed, studying five factors at three levels each ($3^5 = 243$ combinations) requires about the same number of combinations as are needed to study *eight* factors at two levels each ($2^8 = 256$). As we shall see in subsequent chapters, however, one does not always carry out (that is, "run") each possible combination; nevertheless, the principle that fewer levels per factor allows a larger number of factors to be studied still holds.

Two levels can also be chosen for continuous factors as a way to screen whether the level of the factor appears to affect the dependent variable. Designs that have many factors in order to narrow down the set of factors to use in a more detailed study are called, naturally enough, "screening designs."

9.2 Two-Factor Experiments

We now study a very simple, powerful, and efficient technique for the design and interpretation of experiments containing factors at two levels each; in Chapter 11 we detail the formal statistical analysis (hypothesis testing) of the results via ANOVA.

EXAMPLE 9.2 Direct-Mail Study

Suppose that we are conducting a direct-mail campaign in which the response rate is of prime interest. Management wishes to study the impact on response rate of three factors: size of the envelope, amount of the postage, and price of the product. Management wishes to assess not only the "main" effect of each factor but interaction among them as well. Each factor will be studied at only two levels, traditionally called "low" and "high." For relatively small experiments (two to five factors, perhaps), analysis requires no more than pencil and paper.

We start with an example of the simplest case in which only two of these three factors, envelope size and postage, are believed to potentially influence the response rate of the mailing. With two factors to study, each at two levels, we have four combinations: both low, the first low with the second high, the first high with the second low, and both high. Suppose that the two potential envelope sizes are #10 and 9×12, and the postage possibilities are third class and first class. The intuitively satisfying way to assign names would be to call third-class postage "low" and first-class postage "high." It need not be this way, however. We could just as well reverse the names. Furthermore, the levels being studied need not be quantitative—they could, for example, be the choice of envelope color (white versus pale gray), or the timing of the mailing (weekday versus weekend). The assignment of labels is arbitrary; we just need to know, for each factor, which level has been called high and which low, and we need to keep these consistent.

In our example, then, assume that we have the following assignment of labels:

FACTOR	LOW LEVEL	HIGH LEVEL
A: Envelope size	#10	9 × 12
B: Postage	Third class	First class

Each combination of factor levels is called a treatment, or a treatment combination; we designate them as follows:

a_0b_0 A at low level, B at low level
a_0b_1 A at low level, B at high level
a_1b_0 A at high level, B at low level
a_1b_1 A at high level, B at high level

For now, we assume that each of the four treatments is run only once (in a direct-mail situation it is typical to talk about the number of "1000s of names" mailed; hence, "once" in this example means one thousand names). Later, we address the use of several replications of each treatment, which would be more typical in this application. (Recall that a replication is another "data point" under the same combination of levels of factors.) For now, we just note that whether replication is present basically does not affect the details of the design and the calculation of the estimates of the impact of the factors in the experiment. However, as we know from earlier chapters, the *reliability* of the estimates, and the associated probabilities of Type I and Type II errors, are affected. We discuss this aspect further when we focus on the formal statistical analysis. What we now have is a **two-factor** (envelope size and postage), **two-level** (low and high), complete **factorial design without replication.** The term **factorial** signifies the inclusion of all combinations of levels of factors in the experiment; there is no connection between this term and the factorial function in mathematics.

One "trial" (here, the mailing to 1000 names) is "run" (conducted) at each of the four treatment combinations, and the response rate is determined. The more general term for the resulting value of the relevant performance measure, here the response rate, is "yield" or "response," or more generically, "dependent-variable value." The symbols used for the treatments, such as a_0b_1, are also used for the yields, generally without causing confusion. Thus, for example, a_0b_1 is used for the yield when treatment a_0b_1 is run.

Estimating effects in two-factor, two-level experiments Having run one trial at each of the four treatment combinations, we have four responses. We begin by estimating the effect of changing the envelope size from #10 to 9 × 12 (that is, going from envelope size "low" to envelope size "high"). We do so by averaging the change in response rate resulting from a changing envelope size, at the two levels of postage:

$$a_1b_1 - a_0b_1 = \text{estimate of effect of A at high B}$$
$$\underline{a_1b_0 - a_0b_0} = \text{estimate of effect of A at low B}$$
$$\text{Sum} \div 2 = \text{estimate of effect of A over all B}$$

Thus, the effect of A is estimated to be equal to

$$[(a_1b_1 - a_0b_1) + (a_1b_0 - a_0b_0)]/2 = [a_1b_1 - a_0b_1 + a_1b_0 - a_0b_0]/2$$

Similarly, the effect of B is estimated as follows:

$a_1b_1 - a_1b_0$ = estimate of effect of B at high A
$a_0b_1 - a_0b_0$ = estimate of effect of B at low A
Sum $\div$ 2 = estimate of effect of B over all A

Thus, the effect of B is estimated to be equal to

$$[(a_1b_1 - a_1b_0) + (a_0b_1 - a_0b_0)]/2 = [a_1b_1 - a_1b_0 + a_0b_1 - a_0b_0]/2$$

If we wish to estimate the overall average ("grand mean") of the four treatment combinations, so as to estimate the difference between this value and the mean of a particular treatment combination, we simply compute the arithmetic average of the four treatment combinations.

Estimating the interaction between A and B may be a bit less intuitive. Were there no interaction, we would expect the change in response rate resulting from a change in envelope size to be the same whether the postage were third class or first class (and vice versa). However, we could well have interaction, especially if there is a nonzero difference in the effect of envelope size at the two different levels of postage.[1] The estimate of the interaction of A and B is

$a_1b_1 - a_0b_1$ = estimate of effect of A at high B
$a_1b_0 - a_0b_0$ = estimate of effect of A at low B
Difference $\div$ 2 = estimate of the effect of B on the effect of A

This is called the interaction of A and B.

Alternately,

$a_1b_1 - a_1b_0$ = estimate of effect of B at high A
$a_0b_1 - a_0b_0$ = estimate of effect of B at low A
Difference $\div$ 2 = estimate of the effect of A on the effect of B

This is called the interaction of B and A.

The numerical values of these interactions are identical—they both equal $[(a_1b_1 + a_0b_0) - (a_0b_1 + a_1b_0)]$; the interaction of A and B is the same as the interaction of B and A. In practice, the preferred (or more sensible, or more intuitive) interpretation of the interaction effect depends on the situation. In the present example, the direct-mailer may prefer to think about the interaction as "the difference in response rate as a function of envelope size depends on the postage"; or equivalently, "the difference in response rate as a function of postage depends on the envelope size." Both statements are interpretable in a marketing sense, but in other cases one statement gives a more clear-cut interpretation. For example, it is understandable to hear that a fan's rating of a movie starring Michelle Pfeiffer and of one starring Tom Cruise differs depending on the fan's gender. But what if you hear that how a fan's rating of a movie differs by the fan's gender depends on whether the movie star is

Michelle Pfeiffer or Tom Cruise? The latter is not as clear a statement; yet an experiment would yield the same numerical result for each "interpretation."

9.3 Remarks on Effects and Interactions

Several important observations can be made about two-level factorial designs. All available data are used to estimate every effect. That is, we use all four yields (responses) to estimate the effect of A, we use all four yields to estimate the effect of B, and we use all four yields to estimate the effect of the interaction of A and B. Furthermore, every effect is estimated separately (and, as we shall discuss later, orthogonally), even though the experiment varied the levels of the factors simultaneously. Finally, we have obtained estimates of every possible aspect of the influence of the two factors on the response rate—the factors themselves and their interaction. Furthermore, the required computations and the logic behind them couldn't be simpler! These are among the merits of two-level factorial designs.

How else might we have designed this inquiry? We might have evaluated the effects on response rate by varying only one factor at a time. Then, however, we would not have been able to estimate interaction. We discuss this later; at this juncture, we merely draw attention to the following point: nonzero interaction is a possibility in any study with at least two factors that are cross-classified. If the effect of envelope size depends on the postage, it's a "fact of life"; it's not the creation of the statistician, the direct-mailer, or the U.S. Post Office. To consider its possibility is not to "make things more complicated." Indeed, to avoid addressing interaction in designing an experiment (for the misguided purpose of "keeping things simple"), as too often happens in practice, is counterproductive, possibly making the results less useful or even misleading; in fact, this avoidance is unnecessary. As we have seen from this example, explicit consideration of the interaction between two factors need not cause any appreciable increase in complexity. An important rule of thumb is that an interaction between two factors should be considered, and acknowledged in an experimental design, unless there is an explicit understanding of why it is acceptable to assume that it is zero. Admittedly, in practice, opinions may differ, and the "understanding" on occasion becomes somewhat subjective, or worse, murky. After all, it is rare to be 100% certain about the answer to this type of question in management decision making, although this certainty may not be quite as rare in the physical sciences.

9.4 Symbolism, Notation, and Language

A is called a main effect; our estimate of A is often simply written A. The same is true of B. The interaction effect between A and B is written AB; our estimate of *AB* is often simply written *AB*. Thus, we use the same letter(s) without italic to designate the factor and with italic to designate its effect and our estimate of the effect, generally without confusion. We should always remember that the quantities obtained from the yields (data) are (of course!) estimates. For economy in writing, we call a numerical result an "effect," but what we really mean is our *estimate* of that effect. Because we never know the true value of an effect (we need infinite data to determine the true value), why bother always writing $\hat{A}$ (A with a caret on top) or some other symbol for the estimate? It's simpler and more sensible to simply write A for the estimate.[2]

9.5 Table of Signs

There is a pattern to the calculations of *A*, *B*, and *AB* above. It is worth studying the nature of the previous calculations, before we move on, to gain insight into an evaluation technique that is very helpful in analyzing more complex designs. Note that, except for the multiplicative factor of 1/2, all of the effects are formed by sums and differences of the yields a_0b_0, a_0b_1, a_1b_0, and a_1b_1. We can illustrate these sums and differences in a table as follows:

	A	*B*	*AB*
a_0b_0	−1	−1	+1
a_0b_1	−1	+1	−1
a_1b_0	+1	−1	−1
a_1b_1	+1	+1	+1

We list the *effects* to be estimated as columns; we list the *yields* (treatment combinations) as rows. The order of listing the rows is not too important at this point, and we have chosen the order so that the subscripts are in numerical order (00, 01, 10, 11); later, we present a preferred order that offers certain conveniences. The table consists of the coefficients that multiply the yields to produce the effects (along with dividing by two). That is, for example,

$$2A = (-1)\,a_0b_0 + (-1)\,a_0b_1 + (+1)\,a_1b_0 + (+1)\,a_1b_1$$

A is estimated by subtracting the first and second terms from the third and fourth; the plus signs go with the high level of A. Similarly, B is estimated by subtracting the first and third terms from the second and fourth terms; here the plus signs go with the high level of B. In addition, note that the rightmost column, the column representing the interaction *AB*, has values that are the products of the values in the A and B columns. The interaction is obtained by subtracting the second and third terms from the first and fourth terms; a plus sign goes with the terms where factors A and B are both high or both low. A characteristic of this table is that the columns are **orthogonal;** that is, the dot product of any pair of columns is zero. (Remember, the dot product of two columns is formed by multiplying the columns, element by element—here, row by row—and summing the results. This is discussed in Chapter 5.) By way of example, the dot product of the first and second column is

$$(-1)(-1) + (-1)(+1) + (+1)(-1) + (+1)(+1) =$$
$$(+1) + (-1) + (-1) + (+1) = 0$$

and the two columns are orthogonal. When we defined orthogonality in Chapter 5, we also discussed the conditions that (1) the sum of the coefficients add to zero, which is satisfied here; and (2) the orthogonal matrix is made into an orthonormal one by scale factor. Since the sum of the squares of the coefficients in the second line of the equation above equals 4, to do the proper scaling, we should divide by the square root of 4, which is 2.

All of the coefficients in the sign table above have an absolute value of 1. Thus, it is traditional to omit the 1 from each coefficient, leaving only the sign (+ or −); the sign table would then be written as follows:

	A	*B*	*AB*
a_0b_0	−	−	+
a_0b_1	−	+	−
a_1b_0	+	−	−
a_1b_1	+	+	+

This omission of the 1's continues in sign tables throughout this chapter. This feature, that all of the coefficients are the same in absolute value, is true only for two-level experimentation, as we note in a subsequent chapter when we study three-level experiments.

EXAMPLE 9.3 Four Illustrations of Interaction

Suppose that the following four tables display the responses or yields of a two-by-two factorial design:

EXAMPLE 1

A \ B	Low	High
Low	10	12
High	13	15

EXAMPLE 2

A \ B	Low	High
Low	10	15
High	15	15

EXAMPLE 3

A \ B	Low	High
Low	10	13
High	13	10

EXAMPLE 4

A \ B	Low	High
Low	15	15
High	15	15

In the first example, the change in response for a change in the level of A (the word "change" always refers to going from the low level to the high level) is +3, independent of the value of B. Similarly, the change in response for a change in the level of B is +2, independent of the level of A. The interaction is calculated to be zero, and logically so, since the change in yield associated with the change in level of each factor is a constant, regardless of the level of the other factor. We can represent these results graphically, as shown in Figure 9.1.

Such is not the case for the interaction in the next two examples. In example 2, a change in yield of +5 results from a change in the level of A when the level of B is low, but no change in yield results from changing the level of A when B is at its high level. Thus, the main effect of A (that is, the average change) is +2.5. Similarly, a change of +5 in the yield is associated with a change in the level of B when the level of A is low, but no change in yield results from changing the level of B when the level of A is high, and again the main effect (of B) is 2.5. However, here we have substantial interaction. *AB* is estimated to be $(10 - 15 - 15 + 15)/2 = -2.5$. After all, the change in yield associated with changing the level of A *decreases* as we go from low B to high B (from +5 to 0). Another

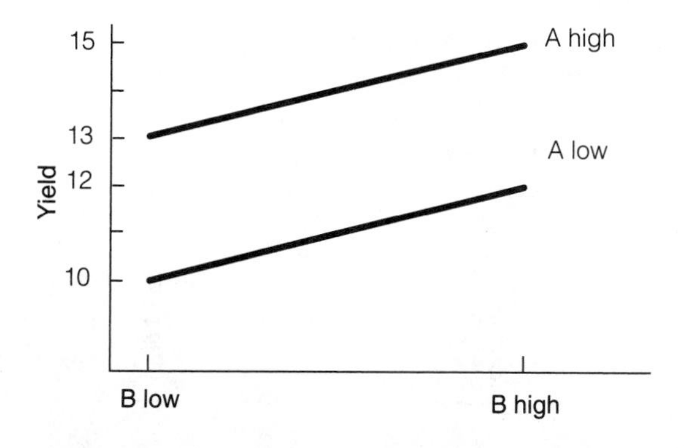

Figure 9.1 Example 1: zero interaction.

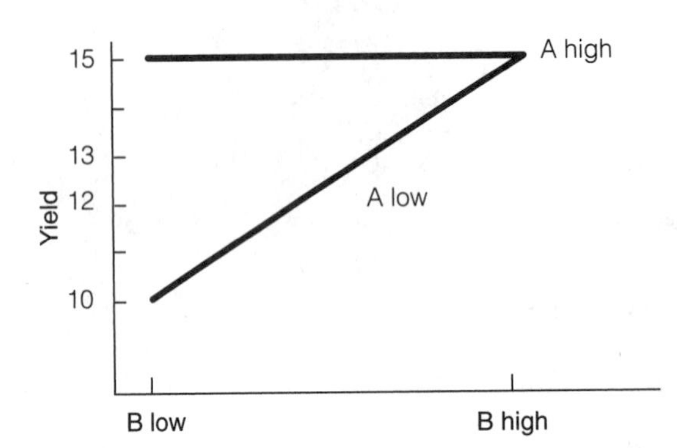

Figure 9.2 Example 2: interaction.

way to possibly intuit this value is to suppose that someone had tested the effects of the two factors separately, by starting in the upper left corner (the "low, low" cell), and changing the level of each factor. They might have concluded that response could be increased by 5 by changing the level of A to high, and another 5 by changing the level of B to high (and might have spent substantial capital to implement the high levels), only to be disappointed that the *total increase* is not 10 but only 5. Figure 9.2 gives a view of the interaction effect in example 2.

In example 3, starting in the upper left corner and changing the level of either A or B separately gives a change in yield of 3, but simultaneous change of both (that is, changing the level of both factors) gives a net result of no change in yield! More careful inspection leads to some interesting observations. When B is *low,* changing the level of A gives a change in yield of +3; however, when B is *high,* changing the level of A gives a change in yield of −3; on the average, changing the level of A doesn't change yield at all! This reminds us of the guy who had his head in the refrigerator, and his feet in the stove, and on the average he felt fine. The main effects of A and B are both zero. All the "action" is in the interaction! Indeed, $AB = -3$; as we go from low B to high B, the main effect of A goes from +3 to −3, decreasing by 6; dividing this by 2 (so it retains the same scale as main effects), the result is −3. This example is graphically represented in Figure 9.3.

In the fourth example, the yield doesn't change as the levels of A and B vary. Apparently, the process is not influenced by these factors, directly or through interaction. The purpose of this example 4 is to emphasize that effects are measuring *incremental* changes; effects of zero do not mean the yields are zero! The graphical representation would be simply two coincident horizontal lines at yield = 15.

We summarize the results of the analysis of these examples below:

Example	*A*	*B*	*AB*
1	3	2	0
2	2.5	2.5	−2.5
3	0	0	−3
4	0	0	0

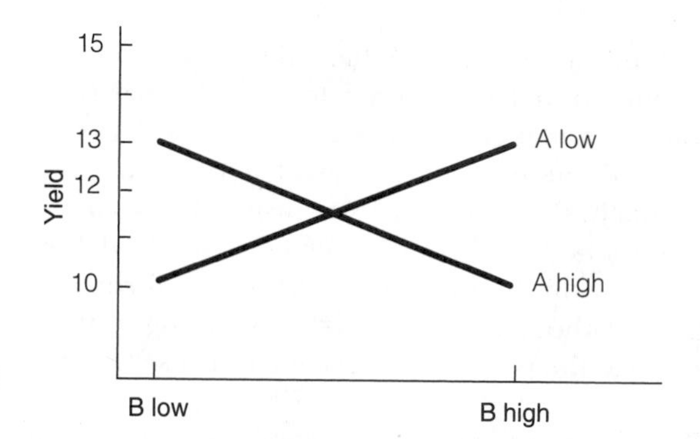

Figure 9.3 **Example 3: interaction.**

A final note relates to the scaling of the numerical quantities. Suppose a constant were added to each response, for example, as a result of weighing each of several products, packaging and all, that have the same packaging weight. The packaging weight would be netted out in the calculation of the effects. Now suppose instead that all responses were multiplied by some constant, as would happen in a change in scale from kilograms to pounds. The change in units would carry through the calculation of the effects, yielding equivalent results (for example, an effect is 5 kilograms, or 11 pounds). In mathematical terms, the analysis is "tolerant" of linear transformations. What matters is not the numerical values of the effects themselves but rather their relative magnitudes.

9.6 Modern Notation and Yates' Order

We can, with no loss of generality, rename or "relabel" our responses; the modern notation, used in virtually all textbooks, is as follows:

Former Notation	Modern Notation
a_0b_0	1
a_0b_1	b
a_1b_0	a
a_1b_1	ab

We are not giving these responses new *values*. That is, the quantity represented by a_0b_0 did not just get set equal to 1. It merely got renamed. This notation is more compact and makes it a lot easier to extend our analysis. The explicit inclusion of the letter in modern notation indicates that the factor is at its high level; thus, a represents A high and (because of the absence of the b) B low. Then, b represents B at high level and A at low level; ab represents both factors at high level. When all factors are at low level, we have no letters. Yet, some symbol is required to indicate that situation. The symbol 1 was adopted as a matter of convention; it could have been ϕ, the symbol for the "empty set" (here, the lack of any letters), or any other symbol.

Actually, the choice of 1 was inspired by its making easier some algebraic steps explained in a later chapter. From the mathematical point of view, it's interesting to think of the subscripts in the former notation as exponents, so that a_0b_1 is thought of as $a^0b^1 = b$. Anything to the power of 0 equals 1, and anything to the power of 1 is the quantity itself. As other examples, a_1b_1 would be $a^1b^1 = ab$, and a_0b_0 would be $a^0b^0 = 1$.

We also introduce Yates' (standard) order of listing treatments and responses in two-level experimentation. Each letter is followed by all combinations of that letter and letters previously introduced, before any additional letter is introduced; and new letters are introduced in alphabetical order. This is the order preferred for analysis (we discuss later in this chapter why this is true) and need not be related to the actual order in which the treatment combinations are performed (which should be randomly chosen).

For a two-level, two-factor design, Yates' order is

$$1 \quad a \quad b \quad ab$$

Note that we start with the 1. Then a is introduced. Next, a appears in combination with all previous letters (but there are none because a is the first letter of the alphabet). Then b is added. Next, b is listed in combination with all other previously introduced letters (of which there is only one, a). Using the modern notation and Yates' order, the estimates of the effects are as follows:

$$A = (-1 + a - b + ab)/2$$
$$B = (-1 - a + b + ab)/2$$
$$AB = (1 - a - b + ab)/2$$

The results are, of course, exactly as before.

9.7 Three Factors, Each at Two Levels

To examine how a third factor is studied, let's return to the direct-mail example and extend it.

EXAMPLE 9.4 Direct-Mail Study with Third Factor

As before, we are trying to sell a product by mail (say, a new book on gardening), and the yield or response variable is the rate of response to the offer.

The factors under study are the postage on the envelope, the price of the product being sold, and the size of the envelope. Here are the specifics of the factors and their levels:

Factor	Low	High
A: Postage	Third-class	First-class
B: Product price	$9.95	$12.95
C: Envelope size	#10	9 ×12

The former style of notation (in numerical order of subscripts) and the corresponding modern notation are

Former:	$a_0b_0c_0$	$a_0b_0c_1$	$a_0b_1c_0$	$a_0b_1c_1$	$a_1b_0c_0$	$a_1b_0c_1$	$a_1b_1c_0$	$a_1b_1c_1$
Modern:	1	c	b	bc	a	ac	ab	abc

A reordering of the modern row above, in Yates' standard order, is

Yates' order: 1 a b ab c ac bc abc

Observe again that the notation shows the letter only when it corresponds to a factor at high level.

Note that in Yates' order, after c is introduced, it is combined with previous treatment combinations in the order in which those were introduced (ac before bc, and so forth).

Estimating Effects in Three-Factor, Two-Level Designs

We begin with some sample calculations. We then reveal the underlying pattern that is facilitated by Yates' order.

Estimate of *A*

1. $a - 1$ = estimate of A, with B low and C low
2. $ab - b$ = estimate of A, with B high and C low
3. $ac - c$ = estimate of A, with B low and C high
4. $abc - bc$ = estimate of A, with B high and C high

Sum ÷ 4 = overall estimate of A

Thus,

$$\begin{aligned}\text{Estimate of } A &= (a + ab + ac + abc - 1 - b - c - bc)/4 \\ &= (-1 + a - b + ab - c + ac - bc + abc)/4 \\ &\quad \text{in Yates' order}\end{aligned}$$

Notice that all terms with A high have plus signs, and all with A low have minus signs. The estimate of *A* is determined by averaging over the four combinations of B and C. The estimates of *B* and *C* are determined similarly.

Estimate of *AB*

Effect of A with B high, minus effect of A with B low, *all with C high*

plus

Effect of A with B high, minus effect of A with B low, *all with C low*

Note that this interaction is an average (as are all interactions). Just as our estimate of *A* is an average of response to A *over all B and all C*, so our estimate of *AB* is an average of response to AB *over all C*.

Referring to the four numbered terms in the estimate of *A* above,

$$AB = \{[(4) - (3)] + [(2) - (1)]\}/4$$
$$= (1 - a - b + ab + c - ac - bc + abc)/4 \qquad \text{in Yates' order}$$

Estimate of *ABC* The *ABC* effect is a "three-way" or "three-factor" interaction. Basically, it represents a three-way joint effect; one interpretation of *ABC* is, for example, how the level of factor C affects the *AB* interaction. *ABC* is discussed further, after we present its estimate (based on the interpretation of the previous sentence):

Effect of A with B high, minus effect of A with B low, *all with C high*

minus

Effect of A with B high, minus effect of A with B low, *all with C low*

Again referring to the numbers assigned in the estimate of *A* calculation above,

$$ABC = \{[(4) - (3)] - [(2) - (1)]\}/4$$
$$= (-1 + a + b - ab + c - ac - ab + abc)/4 \qquad \text{in Yates' order}$$

This is our first encounter with a three-factor interaction. Extending our previous discussion of two-factor interactions: although the numerical value of the estimate is unique, there are different possible interpretations, all of which are technically correct, but one of which is likely more intuitive for the context. For a three-factor interaction we have three ways of "interpreting" the numerical result. In the direct-mail example, it measures the impact on the response rate of interaction *AB* when factor C (envelope size) goes from #10 to 9 × 12; or it measures the impact on the response rate of interaction *AC* when factor B (price) goes from \$9.95 to \$12.95; or, finally, it measures the impact on the response rate of interaction *BC* when factor A (postage) goes from third class to first class. (We'll not try to extend here the earlier example with Michelle Pfeiffer and Tom Cruise.) As mentioned before, the derivation of the signs of the treatment combinations in the above equation are based on the first of these interpretations: the impact of the level of C on the *AB* interaction.

Table 9.2 **Signs for Effects of Three Factors at Two Levels Each**

	A	*B*	*AB*	*C*	*AC*	*BC*	*ABC*
1	−	−	+	−	+	+	−
a	+	−	−	−	−	+	+
b	−	+	−	−	+	−	+
ab	+	+	+	−	−	−	−
c	−	−	+	+	−	−	+
ac	+	−	−	+	+	−	−
bc	−	+	−	+	−	+	−
abc	+	+	+	+	+	+	+

As with two-factor, two-level factorial designs, the formation of estimates in three-factor, two-level factorial designs can be summarized in a table of signs like that in Table 9.2. Note that both columns and rows follow Yates' order. The array is not square—there are seven columns, one for each of the seven effects, and there are eight rows, one for each of the eight treatments. Notice the pattern of the signs. Column *A* alternates signs. Column *B* alternates two at a time. Column *C* alternates four at a time. (Were there a column *D*, as would be present in a four-factor design, its signs would alternate eight at a time.) Next, observe that the signs for the interactions are products of the signs for the factors interacting. For example, the signs in column *AB* are, respectively, the product of the sign in column *A* and the sign in column *B*; the signs in column *ABC* are, respectively, the product of the sign in column *A*, the sign in column *B*, and the sign in column *C*. Finally, all columns are mutually orthogonal (that is, the dot product of every two columns is zero, using −1 and +1 as the column entries).

EXAMPLE 9.5 Response Rate in Direct-Mail Study

We continue with our earlier example. Suppose the response rates were as follows (for each treatment combination, the proportion of people who responded):

1	*a*	*b*	*ab*	*c*	*ac*	*bc*	*abc*
.062	.074	.010	.020	.057	.082	.024	.027

Using the table of signs (Table 9.2) and dividing by 4, we calculate the effects as follows:

A = main effect of postage = .0125
B = main effect of price = −.0485
AB = interaction of postage and price = −.0060

C = main effect of envelope size = .0060
AC = interaction of postage and envelope size = .0015
BC = interaction of price and envelope size = .0045
ABC = interaction of postage, price, and envelope size = −.0050

Note that the largest response rate was generated at (treatment) *ac*, while (effect) *AC* is the smallest effect (in absolute value). This highlights that the lowercase letters (treatments) and the capital letters (effects) represent very different quantities, so it's important to remember which is which. Note also that it is yet to be determined if any of these effects are actually statistically significant (that is, nonzero beyond a reasonable doubt, as opposed to simply an aberration due to the random variation of uncontrolled factors).[3] We return to this in a later chapter.

The largest estimate is −.0485, for the main effect *B*. (Remember: when we speak of the size of an effect, we mean its magnitude, independent of sign.) An increase in price from $9.95 to $12.95 is, not surprisingly, associated with a decline in response rate (although this is only one consideration in determining price; the direct-mailer also has to consider the margin and other issues). The interaction $AB = -.006$; an increase in price (factor B) reduces the effect of A (which, on average, equals .0125) on response rate. Or, equivalently, *AB* may be viewed this way: a change in postage (factor A) from third class to first class reduces (makes more negative) the already negative effect of price (−.0485) on response rate. Do not assume that a change in response rate of "only" .006, or even less, is not important from a practical point of view; in many direct-mail campaigns the mailer would be delighted to receive a total response rate of .006, not to mention an increase of .006. After all, if it costs, say, $0.50 to send out a mail piece, and the margin on a sale (perhaps including potential resales) is, say, $200, a mailing to a million names, with a response rate of .006, will yield a profit of $700,000!

Finally, the interaction of all three factors is $ABC = -.0050$. This may be viewed in three ways. Going from a #10 envelope to a 9 × 12 envelope (factor C), the negative interaction effect between postage and price (*AB*) becomes even more negative. Or, increasing price (B) reduces the positive interaction between postage and envelope size (*AC*). Or, changing postage from third class to first class (A) reduces the positive interaction between price and envelope size (*BC*). All three descriptions of the interaction *ABC* have the same numerical value, but the direct-mailer would select the one that made the most intuitive sense in the direct-mail industry.

Using the results Now that we have analyzed the results of the experiment, what do we do with them? (We assume for the moment that we have determined that all our estimates of effects, both main effects and interaction effects, are statistically significant; if they were not, we would replace them with zero.) The point of an experiment *can* be merely to point the way to the next experiment, perhaps to narrow down how many factors need to be studied so that additional levels can be added for more detail. However, in the world of management, the *raison d'être* of the experiment is more likely to directly guide the selection of the levels of the factors that optimize the yield. In our example, the yield—often called the "performance measure" in management studies—is the response rate.

Assume for the moment that we need to limit our choices to the levels of the factors in the experiment. For postage and envelope size, those are probably sensible choices. For price, the choice may not be so clear; even if we insist for psychological reasons that price end in .95, and even if market forces would make unworkable a price outside the range of our low ($9.95) and high ($12.95) levels, why not $10.95 or $11.95? In theory, by having only two levels of the price factor, we cannot say anything about what the response rates would be for the intermediate values, except what those values would be if we assumed that the relationship were linear. (If we assumed the relationship to be quadratic or higher order, we would not have sufficient information to determine the intermediate values. A quadratic function, for example, has three constants, the c_1, c_2, and c_3 of $c_1x^2 + c_2x + c_3$, requiring us to have data for three points; we have data for only two points.) Again, we shall assume that we are limiting choice to the levels of the factors in the experiment.

Suppose $a_0b_0c_0$ (now called 1) corresponds to the previous mailing conditions. Were we to change postage to first class, price to $12.95, and envelope size to 9×12, we would change response rate by $-.035$ $(= .027 - .062)$. If a higher response rate were the only consideration, we would prefer not to have this decrease; however, as mentioned earlier, a lower response rate with a higher price can sometimes be preferable. In addition, there are differences in cost between different sizes of envelopes, and an obvious cost difference in the choice of postage. But suppose, just for illustration, that we wish to maximize response rate. What combination of levels of factors would we choose? Actually, in this case, the choice is obvious. We would examine the results from the eight treatment combinations and choose the largest. Hence, we would choose ac, which is A-high, B-low, C-high: first-class postage, price of $9.95, and 9×12 envelope; response rate would be predicted to be about .082. Of course, we would not be naive enough to expect to get exactly .082, but we could find, for example, a 95% confidence interval for the response rate, to obtain limits that have a .95 probability of containing the true response rate.

In Chapter 11, when we choose to run only a fraction of the total treatment combinations possible (reasonably enough called a fractional-factorial design), the process of determining the optimal levels of the factors is not quite as simple. Because we will not have available all treatment combinations, we cannot simply observe which has the highest (or lowest) value. At that point, we introduce a general way to do the optimization. The ultimate result would be equivalent to maximizing (or minimizing, in some contexts) the equation

$$\begin{aligned}\text{Yield} = \text{response rate} = {} & .0445 + (.0125/2)A + (-.0485/2)B \\ & + (-.0060/2)AB + (.0060/2)C + (.0015/2)AC \\ & + (.0045/2)BC - (.0050/2)ABC\end{aligned}$$

where .0445 is the estimate of the grand mean, the mean of all eight treatment combinations; and A, B, and C are each 1 if the factor is at high level,

−1 if at low level. For example, if all three factors are at their low level, the equation predicts a response rate of .062.

In practice, if we were predicting the response for a treatment combination not actually in the experiment, we would want to try out that treatment combination to be certain that the response is close to what we predict. This is called "running a confirmation test," and is discussed in Chapter 13.

9.8 Number and Kinds of Effects

In a two-factor, two-level design, each factor is studied at two levels, yielding four treatment combinations. Three factors raises this to eight treatment combinations. Four makes it 16 treatment combinations. In general, with k factors at two levels, there are 2^k treatment combinations. For this reason, a design with each of k factors at two levels is called a **2^k design.** We can present the number of each kind of effect that exists, and can be studied, in an array known as Pascal's triangle; Table 9.3 displays these numbers for designs up to and including 2^7.

In a 2^k design, the number of r-factor main/interaction effects is

$${}_kC_r = k!/[r!(k-r)!]$$

where ${}_kC_r$ stands for the number of combinations of k objects taken r at a time.

Notice that the total number of effects estimated in any of these designs is always one fewer than the number of treatment combinations:

In a 2^2 design, we estimate $2^2 - 1 = 3$ effects.

In a 2^3 design, we estimate $2^3 - 1 = 7$ effects, and so on.

This is to be expected because the "one fewer than the number of treatment combinations" is a special case of the $(n-1)$ rule for degrees of freedom that

Table 9.3 Number of Effects Estimated

	Type of Design					
Type of Effect	2^2	2^3	2^4	2^5	2^6	2^7
Main	2	3	4	5	6	7
Two-factor interaction	1	3	6	10	15	21
Three-factor interaction	—	1	4	10	20	35
Four-factor interaction	—	—	1	5	15	35
Five-factor interaction	—	—	—	1	6	21
Six-factor interaction	—	—	—	—	1	7
Seven-factor interaction	—	—	—	—	—	1
Total	3	7	15	31	63	127

Table 9.4 Sign Table

	Effect																													
			2^2				2^3								2^4															2^5
Treatment	***A***	***B***	***AB***	***C***	***AC***	***BC***	***ABC***	***D***	***AD***	***BD***	***ABD***	***CD***	***ACD***	***BCD***	***ABCD***	***E***	***AE***	***BE***	***ABE***	***CE***	***ACE***	***BCE***	***ABCE***	***DE***	***ADE***	***BDE***	***ABDE***	***CDE***	***ACDE***	***BCDE***	***ABCDE***
1	−	−	+	−	+	+	−	−	+	+	−	+	−	−	+	−	+	+	−	+	−	−	+	+	−	−	+	−	+	+	−
a	+	−	−	−	−	+	+	−	−	+	+	+	+	−	−	−	−	+	+	+	+	−	−	+	+	−	−	−	−	+	+
b	−	+	−	−	+	−	+	−	+	−	+	+	−	+	−	−	+	−	+	+	−	+	−	+	−	+	−	−	+	−	+
ab	+	+	+	−	−	−	−	−	−	−	−	+	+	+	+	−	−	−	−	+	+	+	+	+	+	+	+	−	−	−	−
c	−	−	+	+	−	−	+	−	+	+	−	−	+	+	−	−	+	+	−	−	+	+	−	+	−	−	+	+	−	−	+
ac	+	−	−	+	+	−	−	−	−	+	+	−	−	+	+	−	−	+	+	−	−	+	+	+	+	−	−	+	+	−	−
bc	−	+	−	+	−	+	−	−	+	−	+	−	+	−	+	−	+	−	+	−	+	−	+	+	−	+	−	+	−	+	−
abc	+	+	+	+	+	+	+	−	−	−	−	−	−	−	−	−	−	−	−	−	−	−	−	+	+	+	+	+	+	+	+
d	−	−	+	−	+	+	−	+	−	−	+	−	+	+	−	−	+	+	−	+	−	−	+	−	+	+	−	+	−	−	+
ad	+	−	−	−	−	+	+	+	+	−	−	−	−	+	+	−	−	+	+	+	+	−	−	−	−	+	+	+	+	−	−
bd	−	+	−	−	+	−	+	+	−	+	−	−	+	−	+	−	+	−	+	+	−	+	−	−	+	−	+	+	−	+	−
abd	+	+	+	−	−	−	−	+	+	+	+	−	−	−	−	−	−	−	−	+	+	+	+	−	−	−	−	+	+	+	+
cd	−	−	+	+	−	−	+	+	−	−	+	+	−	−	+	−	+	+	−	−	+	+	−	−	+	+	−	−	+	+	−
acd	+	−	−	+	+	−	−	+	+	−	−	+	+	−	−	−	−	+	+	−	−	+	+	−	−	+	+	−	−	+	+
bcd	−	+	−	+	−	+	−	+	−	+	−	+	−	+	−	−	+	−	+	−	+	−	+	−	+	−	+	−	+	−	+
abcd	+	+	+	+	+	+	+	+	+	+	+	+	+	+	+	−	−	−	−	−	−	−	−	−	−	−	−	−	−	−	−
e	−	−	+	−	+	+	−	−	+	+	−	+	−	−	+	+	−	−	+	−	+	+	−	−	+	+	−	+	−	−	+
ae	+	−	−	−	−	+	+	−	−	+	+	+	+	−	−	+	+	−	−	−	−	+	+	−	−	+	+	+	+	−	−
be	−	+	−	−	+	−	+	−	+	−	+	+	−	+	−	+	−	+	−	−	+	−	+	−	+	−	+	+	−	+	−
abe	+	+	+	−	−	−	−	−	−	−	−	+	+	+	+	+	+	+	+	−	−	−	−	−	−	−	−	+	+	+	+
ce	−	−	+	+	−	−	+	−	+	+	−	−	+	+	−	+	−	−	+	+	−	−	+	−	+	+	−	−	+	+	−
ace	+	−	−	+	+	−	−	−	−	+	+	−	−	+	+	+	+	−	−	+	+	−	−	−	−	+	+	−	−	+	+
bce	−	+	−	+	−	+	−	−	+	−	+	−	+	−	+	+	−	+	−	+	−	+	−	−	+	−	+	−	+	−	+
abce	+	+	+	+	+	+	+	−	−	−	−	−	−	−	−	+	+	+	+	+	+	+	+	−	−	−	−	−	−	−	−
de	−	−	+	−	+	+	−	+	−	−	+	−	+	+	−	+	−	−	+	−	+	+	−	+	−	−	+	−	+	+	−
ade	+	−	−	−	−	+	+	+	+	−	−	−	−	+	+	+	+	−	−	−	−	+	+	+	+	−	−	−	−	+	+
bde	−	+	−	−	+	−	+	+	−	+	−	−	+	−	+	+	−	+	−	−	+	−	+	+	−	+	−	−	+	−	+
abde	+	+	+	−	−	−	−	+	+	+	+	−	−	−	−	+	+	+	+	−	−	−	−	+	+	+	+	−	−	−	−
cde	−	−	+	+	−	−	+	+	−	−	+	+	−	−	+	+	−	−	+	+	−	−	+	+	−	−	+	+	−	−	+
acde	+	−	−	+	+	−	−	+	+	−	−	+	+	−	−	+	+	−	−	+	+	−	−	+	+	−	−	+	+	−	−
bcde	−	+	−	+	−	+	−	+	−	+	−	+	−	+	−	+	−	+	−	+	−	+	−	+	−	+	−	+	−	+	−
abcde	+	+	+	+	+	+	+	+	+	+	+	+	+	+	+	+	+	+	+	+	+	+	+	+	+	+	+	+	+	+	+

was introduced in Chapter 2. In essence, an unreplicated 2^3 can be examined from the perspective of degrees of freedom as an eight-column, one-factor analysis, with the eight columns yielding seven degrees of freedom.

One need not repeat the logic for deriving the signs of the treatment combinations in the estimates for 2^k designs for higher values of k. Table 9.4 is a table of signs going up to 2^5.

9.9 Yates' Forward Algorithm

A systematic method of calculating the estimates of effects in 2^k designs was developed by Yates in 1937; using his algorithm requires dramatically fewer calculations compared with calculating each effect using the sign tables. For 2^k complete factorial designs, the subject of this chapter, first arrange the responses in Yates' (standard) order in a column. Now create a second column from the first column by taking sums of adjacent elements and then differences of adjacent elements of the first column. That is, add the first two elements (place the result as the first element of the second column), then add the next two (place the result as the second element of the second column), then the next two, and so on, until you have gone through the entire first column. Then go through the first column again, *subtracting* the first from the second, then *subtracting* the third from the fourth, and so on, placing the differences in the second column. At this point you have formed the entire second column. (We find it useful to refer to what was just done as having "Yates'd the data once," or in present tense, as "Yatesing the data." No disrespect of Yates is intended.) Repeat this process to form additional columns (that is, use the values of the second column to determine a third column, and so forth). For a 2^k design, k columns will be calculated; that is, the data must be Yates'd k times. Table 9.5 illustrates this for a 2^3 design.

Note the pattern in Table 9.5—the elements in the top half of the columns are formed by adding pairwise the elements from the preceding col-

Table 9.5 Yates' Forward Algorithm

Response	First Column	Second Column	Third Column	Estimate
1	$a + 1$	$ab + b + a + 1$	$abc + bc + ac + c + ab + b + a + 1$	8μ
a	$ab + b$	$abc + bc + ac + c$	$abc - bc + ac - c + ab - b + a - 1$	$4A$
b	$ac + c$	$ab - b + a - 1$	$abc + bc - ac - c + ab + b - a - 1$	$4B$
ab	$abc + bc$	$abc - bc + ac - c$	$abc - bc - ac + c + ab - b - a + 1$	$4AB$
c	$a - 1$	$ab + b - a - 1$	$abc + bc + ac + c - ab - b - a - 1$	$4C$
ac	$ab - b$	$abc + bc - ac - c$	$abc - bc + ac - c - ab + b - a + 1$	$4AC$
bc	$ac - c$	$ab - b - a + 1$	$abc + bc - ac - c - ab - b + a + 1$	$4BC$
abc	$abc - bc$	$abc - bc - ac + c$	$abc - bc - ac + c - ab + b + a - 1$	$4ABC$

umn, and the elements in the bottom half of the columns are formed by taking the differences, again pairwise, from the elements in the preceding column. Note also that the differences are the second minus the first, the fourth minus the third, and so on; for adding, the order does not matter, but for subtraction it is important.

In Table 9.5, note the line-by-line correspondence between yields (lowercase letters in the left column of the table, which are also treatment combinations) and effects estimated (capital letters in the last column). Treatment combinations and estimates of effects are both in Yates' order. We made a point earlier in the chapter that there is no direct connection between the "lowercase letters" (treatment combinations) and the "capital letters" (effects); this is still true, but there must be some way to link together what is estimated by what. How useful would an algorithm be if it gave us all the estimates but never informed us which estimate is which? We could check the 2^3 section within the table of signs (Table 9.4) and observe that each entry in what is labeled "Third Column" in Table 9.5 does indeed estimate the listed effect. Of course, whether using Yates' algorithm or the sign table, we must divide the estimates obtained by the adding and subtracting, in this case by 4 (or, in general, by 2^{k-1}).

Applying Yates' forward algorithm to the response rate data, we get the results shown in Table 9.6. Again, note the line-by-line correspondence between treatment combinations and estimates; both are in Yates' order. Although there are only seven effects to be determined, we have eight values in the last column of numbers. The top value, .356, is the sum of all eight responses. Dividing this sum by 8, we have $.356/8 = .0445$; this is the average response rate (estimate of μ) across all eight treatment combinations.

Although the wide availability and usage of statistical software might seem to belie the necessity for learning Yates' algorithm, an appreciation of the workings of Yates' algorithm offers insights into how one goes from the data to the results (that is, what's going on inside the software). Indeed, the use of Yates' standard order and Yates' algorithm helps clarify some topics in Chapters 10 and 11.

Table 9.6 Yates' Forward Algorithm for Direct-Mail Study

Treatment	Yield	First Column	Second Column	Third Column	Estimate
1	.062	.136	.166	.356	8μ
a	.074	.030	.190	.050	$4A$
b	.010	.139	.022	−.194	$4B$
ab	.020	.051	.028	−.024	$4AB$
c	.057	.012	−.106	.024	$4C$
ac	.082	.010	−.088	.006	$4AC$
bc	.024	.025	−.002	.018	$4BC$
abc	.027	.003	−.022	−.020	$4ABC$

9.10 A Note on *Replicated* 2^k Experiments

If a 2^k experiment is replicated (with equal replicates for each treatment combination), the way one determines the effects is substantially the same as for nonreplicated experiments. This is true whether we determine the effects using Yates' algorithm, using simple successive adding and subtracting as indicated by the sign table, or doing the problem using software. In essence, using Yates' algorithm or the sign tables, we have two equivalent methods when replication is present. One method is to compute the cell mean for each treatment combination, and then to proceed as if each cell mean was the single data value of the cell. The other method is to compute the effects separately for each replicate, and then take the average of each effect over the replicates, to arrive at the "final" estimate of each effect. Both methods will produce the exact same results. When using software, the data for all replicates are simply included as part of the input data.

Although the fact that the data are replicated does not change the effects, it does impact the *F* test as to whether the effect in question can reasonably be said to be nonzero. We will see the impact of replication on the *F* test in Chapter 11.

EXAMPLE 9.6 Using SPSS for the Direct-Mail Study

To analyze our direct-mail example using SPSS, we would input the data as indicated in Table 9.7. The first column contains the dependent variable values, and the second, third, and fourth columns represent, respectively, the level of

Table 9.7 SPSS Input Data

.062	1.00	1.00	1.00
.074	2.00	1.00	1.00
.010	1.00	2.00	1.00
.020	2.00	2.00	1.00
.057	1.00	1.00	2.00
.082	2.00	1.00	2.00
.024	1.00	2.00	2.00
.027	2.00	2.00	2.00

Table 9.8 SPSS Output for Three-Factor Study

- - Description of Subpopulations - -

Summaries of DEP VAR

By levels of VAR A

Variable	Value Label	Mean	Std Dev	Cases
For Entire Population		.0445	.0274	8
VAR A	1.00	.0383	.0253	4
VAR A	2.00	.0507	.0318	4

Total Cases = 8

- - Description of Subpopulations - -

Summaries of DEP VAR

By levels of VAR B

Variable	Value Label	Mean	Std Dev	Cases
For Entire Population		.0445	.0274	8
VAR B	1.00	.0688	.0114	4
VAR B	2.00	.0203	.0074	4

Total Cases = 8

- - Description of Subpopulations - -

Summaries of DEP VAR

By levels of VAR C

Variable	Value Label	Mean	Std Dev	Cases
For Entire Population		.0445	.0274	8
VAR C	1.00	.0415	.0313	4
VAR C	2.00	.0475	.0274	4

Total Cases = 8

factors A, B, and C. We use 1 for the low level and 2 for the high level; we could just as well use 0 and 1, or 7 and 8—any two consecutive integers.

The output appears in Table 9.8. SPSS is not oriented toward providing output in a form traditionally associated with two-level experimentation. In fact, the output does not actually provide the effects. For example, for factor A (VAR A in Table 9.8), the output tells us that the mean is .0507 for high A, .0383 for low A. The difference between the two values, .0124, is the effect of A. The value resulting from Yates' algorithm in the previous section was .0125 (that is, $4A = .05$, $A = .0125$); the difference is rounding error, as the value .0383 in Table 9.8 is actually .03825, and .0507 is actually .05075.

EXAMPLE 9.7 Using JMP for the Direct-Mail Study

The statistical software JMP is more oriented toward two-level experimentation than SPSS is. Using JMP to design and analyze the same study, begin by selecting the "Design Experiment" command from the Tables menu. Next, choose "2-level Design"; this causes the window titled "Two Level Design Selection" to appear. Define the "Number of Factors" to be 3, then click on "Search for Designs." Six design choices are available; we want "8, Full Factorial, Full Resolution." Highlight this selection and then click on "Generate Selected Design." The design template in Table 9.9, minus column headings and experimental results, appears. In a real experimental-design situation, we would use this as our starting point for running the experiment to obtain the data for subsequent analysis. Having already worked through this example, we have the data to enter in the last column, as shown in Table 9.9. The input is completed by adding the factor labels as column headings.

To analyze the experimental results, we click on "Analyze," then select "Fit Model." We highlight Y and input it as Y (thereby identifying the results column). We highlight Postage, Price, and Size and use "ADD" to indicate that these are the sought-after main effects. Next we highlight "Postage and Price" and use "CROSS" to indicate this as a desired interaction effect. We do the same for "Postage and Size," "Price and Size," and "Postage, Price and Size." This procedure tells the software which effects to include in the model and which interactions to assume are zero (these will be put into the error term).

Table 9.9 JMP Template for Direct-Mail Study

	Pattern	Postage	Price	Size	Y
1	---	-1	-1	-1	0.062
2	+--	1	-1	-1	0.074
3	-+-	-1	1	-1	0.01
4	++-	1	1	-1	0.02
5	--+	-1	-1	1	0.057
6	+-+	1	-1	1	0.082
7	-++	-1	1	1	0.024
8	+++	1	1	1	0.027

Table 9.10 **JMP Output for Direct-Mail Study**

Response: Y

Summary of Fit

RSquare	1
RSquare Adj	?
Root Mean Square Error	?
Mean of Response	0.0445
Observations (or Sum Wgts)	8

Parameter Estimates

Term	Estimate	Std Error	t Ratio	Prob>ltl
Intercept	0.0445	?	?	?
Postage	.00625	?	?	?
Price	-0.02425	?	?	?
Size	0.003	?	?	?
Postage*Price	-.003	?	?	?
Price*Size	.0045	?	?	?
Postage*Size	0.00075	?	?	?
Postage*Price*Size	-0.0025	?	?	?

Effect Test

Source	Nparm	DF	Sum of Squares	F Ratio	Prob>F
Postage	1	1	0.00031250	?	?
Price	1	1	0.00470450	?	?
Size	1	1	0.00007200	?	?
Postage*Price	1	1	0.00007200	?	?
Price*Size	1	1	0.00004050	?	?
Postage*Size	1	1	0.00000450	?	?
Postage*Price*Size	1	1	0.00005000	?	?

Finally, we then click on "Run Model" to get our output, which appears in Table 9.10.

In Table 9.10, note that the effects (such as Postage = .00625) are half of the values obtained earlier (Main effect of postage = .0125). This is because JMP provides the change in *Y* per unit change in the factor level, whereas the traditional approach provides the change in *Y* per two-unit change in the factor level (from -1 to 1). Note also that a question mark appears in many places. This is because no error estimate is available, so all ratios requiring such a value cannot be determined. Various plots are available to show the output, but for this study most are simply two points on a grid.

9.11 Main Effects in the Face of Large Interactions

One must be cautious in interpreting a main effect when an interaction involving that factor is large. Consider the 2^2 study design in Example 6.7, which we explore further in Example 9.8.

EXAMPLE 9.8 Cigarette Brand Names and Gender

We encountered a large interaction effect in the cigarette marketing study in which one factor was the sex of the person rating a proposed new cigarette (Male, Female), and the other factor was the proposed brand name of the cigarette (Frontiersman, April). Recall that Frontiersman was chosen to represent a masculine name, and April a feminine name. There were 50 data values (that is, people) for each treatment combination, though only the means are relevant to make our point here. The treatment combinations and means appear in the following table. Each person was told the proposed brand name of the cigarette, smoked it, and rated the cigarette on a variety of attributes, though only the overall "purchase intent" values are reported (rated on a scale from 1 to 7, higher being more likely to purchase).

	Brand Name (B)	
Sex (S)	**Frontiersman (Low)**	**April (High)**
Male (low)	4.44	3.50
Female (high)	2.04	4.52

The estimates of the effects are

$$S = -.69 \qquad B = +.77 \qquad \text{and} \qquad BS = +1.71$$

Remember that interactions, like main effects, are averages. Furthermore, in this example, the interaction is an average of widely disparate numbers.

Effect of B at high S = $bs - s = 4.52 - 2.04 = +2.48$
Effect of B at low S = $b - 1 = 3.50 - 4.44 = -.94$
$B = [2.48 + (-.94)]/2 = +1.54/2 = +.77$
$BS = [2.48 - (-.94)]/2 = +3.42/2 = +1.71$

Although the average effect of B is + .77 (that is, averaging over both sexes, the cigarette is rated more highly when the purported brand name is April than when it's Frontiersman), that knowledge is not nearly as helpful as knowing that the effect of B at high S is +2.48 whereas at low S it's −.94. In other words, for females the cigarette is rated a lot more highly when the brand name is to be April than when it is to be Frontiersman, and for males the cigarette is rated somewhat more highly when the brand name is to be Frontiersman than when it is to be April.

(In terms of the movie example, it might be more insightful to say that males have a strong preference for Michelle Pfeiffer over Tom Cruise, and females have a strong preference for Tom Cruise over Michelle Pfeiffer, rather than to observe that, averaging over the sexes, Michelle and Tom are about equally popular.)

Thus, in the face of large interactions, it seems more useful to "specialize" the main effect of each factor to particular values of the other factors. Of course, in doing so, we have two (instead of four) treatment combinations contributing to each specialized estimate of *S* at specific levels of B and of *B* at particular levels of S.

In general, we accept high interactions where we find them, and seek to explain them. In the process of explaining them, we may find it beneficial to replace main effects by more meaningful specialized effects (and, possibly, lower-order interactions by more specialized higher-order interactions).

9.12 Levels of Factors

Sometimes in an experiment, a factor may have only two (or some other specific number of) levels at which it is practical to test. In other cases, especially when the factor is measured along a continuum, one must choose the particular (say, two) levels at which to set the factor for purposes of the experiment. Does it matter which levels are chosen? Let's attempt to answer this question by considering an example. Suppose that responses follow the curves in Figure 9.4 and yield represents the suppleness of leather as a function of pressure (P, at two levels, p_1 and p_2) and temperature (T, at two of the four levels t_1, t_2, t_3, and t_4). Yield is graphed as a function of T at each level of P. First consider the effect *P* at various levels of *T*:

Temperature Level	*P* Effect ($p_2 - p_1$)
t_1	22 − 30 = −8
t_2	29 − 58 = −29
t_3	47 − 20 = +27
t_4	18 − 15 = +3

Then, at (t_1, t_2):

$$P = (-8 - 29)/2 = -18.5$$
$$T = [(29 - 22) + (58 - 30)]/2 = 17.5$$

and

$$PT = [-29 - (-8)]/2 = (7 - 28)/2 = -10.5$$

At (t_3, t_4):

$$P = (27 + 3)/2 = 15$$
$$T = [(18 - 47) + (15 - 20)]/2 = -17$$

and

$$PT = (3 - 27)/2 = [-29 - (-5)]/2 = -12$$

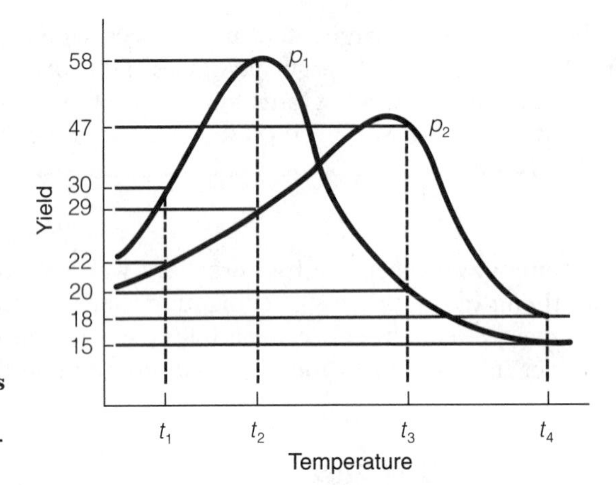

Figure 9.4 **Suppleness of leather as a function of pressure and temperature.**

If we considered (t_1, t_3), or (t_2, t_4), or (t_1, t_4), we would obtain yet other answers for P, T, and PT. It is only when the conjectured responses in a diagram like Figure 9.4 are linear and parallel that the choice of levels is unimportant. We must acknowledge the essentially circular nature of this observation. One needs to understand the nature of the response curves to set up the experiment (that is, to choose the levels of the factors to implement in the experiment)—for an experiment designed to reveal that nature! This kind of circularity characterizes all experimental science.

We need not, however, throw up our hands in despair. There are two approaches to resolving this dilemma. One is to use the knowledge of people who are experts in the process under study. Typically, experts in the process or product to be optimized are not experts in statistics or experimental design; that's why we, the experimental-design consultants, are involved; however, the relationship can and should be nicely symbiotic. The second approach is to do a series of experiments. Usually, we begin by getting the best starting point we can, often through careful, probing questions. Then we do a few simple experiments, in sequence, often with each factor at two levels, to get a sense of how to change the levels of the factors to levels that perform better. We continue to run sequences of experiments until we are confident that we are in the neighborhood of the best choices of levels. Finally, we run a more detailed (that is, larger) experiment to home in on the best levels. We revisit this concept in Chapter 14 on response-surface methodology.

9.13 Factorial Designs versus Designs Varying Factors One at a Time

We have spent many pages discussing the logic of two-level factorial designs, for two good reasons. One is that factorial designs (whether two levels or more) are very efficient, as we mentioned earlier. In this section we make more explicit the second reason, the enormous advantages that the factorial design has over the other type of design frequently used: designs in which factors are varied one at a time. These are designs that vary the first factor while holding all others constant (thus measuring the impact of the first factor on yield), then varying the second factor while holding all other factors, including the first factor, constant (thus measuring the impact of the second factor), and so on, studying each factor individually. This seems, on the surface, to be a safe way to experiment, but it is very inefficient, and the safety is an illusion.

Let us consider a study inquiring whether future profitability of a company is related to (that is, can be predicted from) two factors that are characteristics of a company's annual report: principal time frame discussed in the company president's message, and principal presentation mode of the company president's message. (A study of the impact of these factors, among others, was actually conducted.) We study both factors at two levels:

Factor	Low Level	High Level
Time frame	Past year	Future
Mode	Numerical	Nonnumerical

First consider a design for studying the impact of the factors on profitability by varying one factor at a time. Hold Time Frame constant at "past year" and examine two (randomly chosen) companies at each level of Mode of presentation. Then hold Mode constant at "numerical" and add two randomly chosen companies with Time Frame at the level "future." (We take two observations at each condition tested in order to facilitate comparison with the factorial design.) Thus, we have the following data points, as represented by the dots:

Mode	Time Frame: Past Year	Time Frame: Future
Numerical	••	••
Nonnumerical	••	

Now consider an unreplicated 2^2 factorial design. We have one response at each of the four treatment combinations, one of which is both factors at high level (that is, future, nonnumerical), not included in the previous study. The data points are as follows:

Mode	Time Frame: Past Year	Time Frame: Future
Numerical	•	•
Nonnumerical	•	•

In a comparison of designs, we can note the following:

1. In spite of having (and paying for!) six data values in the one-at-a-time design, and only four in the factorial design, the one-at-a-time design has the same number of data points contributing to the estimate of the main effects as does the factorial design—four. (In the one-at-a-time design, only the four "vertical" data points can contribute to estimating the impact of Mode; only the four "horizontal" data points can contribute to estimating the impact of Time Frame.) Thus, the estimate of each main effect has the same amount of supporting data (and in essence the same reliability) in both designs. (Section 9.15 examines more closely the direct linkage between the reliability of an estimate—its variance [σ^2]—and how many data points support the estimate.)

2. Interactions, routinely estimable in the factorial design, cannot be estimated in the one-at-a-time design; this is a serious defect. Even if we were to say that there may not be interaction, there is no decline in the merit of the factorial design—see item 1 above. Indeed, in the one-at-a-time design, there would be no way to determine that there is no interaction.

3. In the factorial design, each main effect is estimated over *both* levels of the other factor, not at one level as in the case of the one-factor-at-a-time design; this increased generality is usually, though not always, attractive. If interaction is high, we may want to determine the effect of each factor at *each* level of the other factor. The one-factor-at-a-time design provides more of this information for two situations (the Mode effect at the Time Frame level "past year," and the Time Frame effect at the Mode level "numerical") than does the factorial design (four data points of support, instead of two). However, as noted earlier, in an instance of high interaction, the one-factor-at-a-time design will not reveal the magnitude of the interaction in the first place.

4. An estimate of the effect of factors other than the two factors studied is possible in the experiment with six data points. The differences in yields at a given treatment combination cannot be due to Mode, or Time Frame, or their interaction, since Mode and Time Frame were fixed throughout each set of replicates. These differences must be due to the random variation of other, uncontrolled, factors. A replicated factorial design experiment can, of course, provide such an estimate, but the cost increases because replicating this factorial design with equal replication per treatment combination requires a minimum of eight data points instead of six; hence its advantage over the competing one-at-a-time design is reduced.

5. One-factor-at-a-time designs are less vulnerable to missing yields.

In summary, the general judgment, particularly in recent years, is that factorial designs are definitely superior to one-factor-at-a-time experimentation.

EXAMPLE 9.9 Factorial versus One-at-a-Time Designs

We end this section by considering an example of a 2^5 design. Using five factors demonstrates more dramatically the advantage of a factorial design over a one-factor-at-a-time design. In a 2^5 factorial design, we have 32 treatment combinations and, without replication, 32 data values. Each data value contributes to the estimate of each effect. Thus, each effect has a "reliability" corresponding to 32 data values. To achieve the same reliability doing one-factor-at-a-time experimentation, we would need 96 data values!

Consider the following six treatment combinations:

1. A-low B-low C-low D-low E-low
2. A-**high** B-low C-low D-low E-low
3. A-low B-**high** C-low D-low E-low
4. A-low B-low C-**high** D-low E-low
5. A-low B-low C-low D-**high** E-low
6. A-low B-low C-low D-low E-**high**

In a one-factor-at-a-time design, we would estimate the effect of A by comparing treatment combinations 1 and 2 (that is, subtracting the yield at treatment combination 1 from the yield at treatment combination 2); the yields of the other four treatment combinations would shed no light on the effect of A. To achieve the "reliability" of 32 data points of support, we would need 16 replicates of each of the two treatment combinations. Similarly, we would compare the yields of treatment combinations 3 and 1 for the effect of B, 4 and 1 for the effect of C, 5 and 1 for the effect of D, and 6 and 1 for the effect of E. Needing 16 of each of the six treatment combinations above would require a total of 96 data values simply to achieve the same reliability of the factorial design, which requires only 32 data points in total. Furthermore, the one-at-a-time design won't inform us about the presence of interaction effects![4]

9.14 Factors Not Studied

We mentioned that there may be variation among the yields due to factors other than the main effects and interactions of the factors we are studying. This is, of course, what in the previous chapters we called "error." In virtually any experiment, factors other than those studied may be influential. These other factors may be neglected, and their impact ignored, but the cost of neglect can be high. These other factors often have uneven impact, affecting some treatments more than others, and thereby seriously confounding inferences about the studied factors. It is important to deal explicitly with them;

even more, it is important to measure their impact. The primary ways of addressing these uncontrolled factors are as follows:

1. *Hold them constant.* Suppose that one wished to study how bakery sales were affected by price and promotion. It is well known that bakery sales differ also by day of the week, even though this effect may not be of prime interest to the manager (in part because the effect is known). One option for "neutralizing the nuisance value" of day of the week is to hold it constant. For example, vary price and promotion from day to day, using only Mondays. However, the downside of this approach is that generalizability may be compromised; perhaps the results found are valid only for Mondays.

2. *Randomize their effects.* In the bakery example above, one could randomly assign different prices and promotional expenditures to different days; this approach is taken quite often. At least on an a priori basis, estimates of effects are unbiased. There is not complete agreement in the statistical community exactly what to do if, for example, by chance all low prices were randomly assigned to weekdays and all high prices to weekends.

3. *Estimate their magnitude by replicating the experiment.* If one replicates each treatment combination enough, variation due to these uncontrolled factors "averages out," or "evens out," or "balances out." The degree to which the variations even out depends on how much variation is present and how much one replicates. For example, if a low price were randomly assigned to 180 days of the year, and a high price to the other 180 days of the year, it is likely (due to considerations of immutable [and helpful] statistical laws) that the distribution of days using each price will not differ materially.

4. *Estimate their magnitude via side or previous experiments.* In theory, especially if the assumption of equal variance for each treatment combination is tenable, one could estimate the impact of the uncontrolled factors by replicating one treatment combination in a side experiment. We say "in theory" because it is infrequent that one sees this approach implemented.

5. *Argue (convincingly) that the effects of some of these unstudied factors and/or some of their interactions with studied factors are zero, either in advance of the experiment or in light of the yields.* This is done all the time, at least for interaction effects between unstudied factors and the factors under study. (This would be analogous to having "covariates" in a regression model and not including their interactions with the primary factors as separate "independent variables.") Input from process experts is usually necessary here.

6. *Confound certain nonstudied factors.* This means to design the experiment so that the "nuisance impact" of some specific nonstudied factors is "neutralized." This is the subject of the next chapter.

9.15 Errors of Estimates in 2^k Designs

Assume that the probability distribution of data points has a variance, σ^2. Further assume that the σ^2 value is constant for each treatment combination.

Of course, if we ran several replications of this treatment combination, they would not all be equal, and would form a probability distribution (which we often assume is a normal [bell-shaped, Gaussian] distribution, though that assumption is not necessary for this section). Because of this variability, which characterizes all of the treatment combinations, our estimates (which, as we know, combine treatment combinations by addition and subtraction, and then division by a constant) are subject to error, and have their own probability distribution and variance. This variance (and its square root, the standard deviation) speaks directly to the reliability of the estimate. Sometimes it is relatively small; other times it is substantial; everything else equal, of course, smaller variance is better! The smaller the variance, the narrower (more precise) the confidence interval is, and the smaller the Type II error (or, equivalently, the larger the power) is for a fixed significance level.

What is the variance of an estimate? In a 2^k design without replication, 2^k treatment combinations go into each estimate. The signs of the treatment combinations are half "+" and half "−", the assignment of which depends on the effect being estimated. So:

$$\text{Any estimate} = (1/2^{k-1}) \cdot [\text{generalized } (+ \text{ or } -) \text{ sum of } 2^k \text{ treatments}] \tag{9.1}$$

It can be proved that, in general, if K is a constant,

$$\sigma^2(KX) = K^2\sigma^2(X)$$

Also, for Xs that are independent (as we assume that the treatment combinations are) and have a common variance, σ^2 (which, as noted in earlier chapters, is one of the "standard assumptions" usually made),

$$\sigma^2(X_1 \pm X_2 \pm \cdots \pm X_n) = n\sigma^2(X)$$

Thus we can show that

$$\sigma^2(\text{any estimate}) = (1/2^{2k-2})(2^k\sigma^2) = \sigma^2/2^{k-2} \tag{9.2}$$

Note that the larger the number of factors, k, the smaller the variance of any estimate.

What if we have replication? That is, suppose that instead of one value at each treatment combination, we have r replicates at each treatment combination. Then a similar analysis indicates that the result is the same as that in equation (9.2), except that the σ^2 term becomes σ^2/r, since the variance of the average is the variance of the individual value, divided by the number of data points composing the average (this "sample size" is, here, the number of replicates). Thus, we have

$$\sigma^2(\text{any estimate}) = (1/2^{2k-2})[2^k(\sigma^2/r)] = \sigma^2/(r2^{k-2}) \tag{9.3}$$

and the larger the replication per treatment combination, the smaller the variance of each estimate.

So, the variance of an estimate depends on k (the number of factors studied) and r (the replications per treatment combination). It also obviously depends on the variance, σ^2. The variance can be reduced by holding some of the nonstudied factors constant; however, as mentioned before, this gain may be offset by the reduced generality of the conclusions.

9.16 Comment on Testing the Effects in 2^k Designs

As noted earlier, we consider the analysis of the 2^k design via an F test (or, equivalently, a t test) in Chapter 11. We have chosen this separate presentation because the methodology used for testing whether effects are nonzero is a key component of the *analysis*, as opposed to the *design*, of the experiment. The methodology is essentially the same for the topic of this chapter and those of Chapters 10 and 11. Yet, the design issues of this and the subsequent two chapters, though related, are distinct enough to warrant separate chapters. We believe that our approach is efficient and helps point out the overall umbrella under which the hypothesis testing process operates in these experimental-design settings.

EXAMPLE 9.10 Pricing a Supplemental Medical/Health Benefit Offer, Revisited

As mentioned in Chapter 1, self-reported purchase intent scores are nearly always overstated, and need to be lowered in a nonlinear way that usually differs by product category; here the product category is health care products and insurance. Most algorithms for lowering these scores are proprietary, based on experience. Details of this process cannot be discussed here.

Perhaps to nobody's surprise, all main effects were significant. This finding in itself was of little value (although had it *not* been true, that would be of major interest). The real question was whether the additional demand generated by having the lower price resulted in a higher versus lower profit. That is, the issue's resolution is primarily one of elasticity (along with the issue that different levels of benefits of the auxiliary factors *cost* different amounts). Adding to the complexity, all two-factor interactions were also significant, in the anticipated direction—when one component (a core or an auxiliary benefit) was at a higher price, using the higher price of another component had a larger negative effect on demand.

There were significant, but not really material, three-way interactions in the same direction. This is a case when some differences (here, three-factor interactions) observed are *statistically* significant, which strongly indicates that these interactions aren't literally zero, but they are so small relative to the two-factor interactions and main effects that they can be ignored in the decision-making process and viewed as having no *practical* significance.

The results clearly indicated that of the auxiliary benefits, the massage network had the least diminution of demand as price went from low to high

($1 to $2, per month), along with raising the level of benefit. The chiropractic network was next in terms of diminution of demand (even though its going from low to high was only a $0.50 increase per month, as opposed to a $1 difference for the other auxiliary networks), and the dermatology network was last. The diminution of demand for the core benefits resulting from the $3 increase per month in going from low price to high price indicated that the low price would be the wisest choice (there was discussion that this effect could be partly due to the consumer appeal of the promotion phrase "for under $10 per month . . .").

The overall analysis, taking into account all the main effects and two-factor interactions, indicated an optimal strategy of core benefits at low price, the optional dermatology and massage networks at high price, and the chiropractic network at the low price. For the chiropractic network decision, the difference of only $0.50 per month between the low and high price played a key role in the result (that is, going to the high price would gain less additional profit).

This HealthMark benefit offer has not yet been implemented. Reasonably enough, the company first undertook a "confirmation test" to inquire whether the results would be verified (in Chapter 13, we look more deeply into the issue of confirmation testing). The results were, indeed, verified. However, by the time the confirmation test was completed, HealthMark had decided to add another auxiliary benefit, and was so pleased with the experimental design/marketing research process that it decided to undertake a next (two-level factorial) experiment, now with five factors (the pricing of the core benefits and of the *four* auxiliary benefits). This next experiment, and some additional issues that arose with it, are discussed in Chapter 10.

Exercises

1. In Example 9.8, involving the brand name of the cigarette and the sex of the person rating the cigarette, the treatment combinations can be assigned letters *A–D* as follows:

	Brand Name (B)	
Sex (S)	**Frontiersman**	**April**
Male	*A*	*B*
Female	*C*	*D*

The responses presented in Example 9.8 were the purchase intent averages per treatment combination. Participants also rated several other dependent variables, all on a seven-point scale. Among these were (1) the richness (versus blandness) of the flavor, (2) the strength (versus weakness) of the taste, and (3) the "masculinity" (versus "femininity") of

Table 9EX.1 Cigarette Ratings

Variable	A	B	C	D
Richness	4.42	3.84	3.30	4.26
Strength	4.40	4.60	6.06	4.32
Masculinity	3.94	1.60	4.68	0.88

the cigarette. Using the treatment combinations above, the results for the three other dependent variables (that is, cell means) were as shown in Table 9EX.1.

For each of the three additional dependent variables, find the sex effect, the brand name effect, and the sex/brand name interaction effect.

2. In exercise 1, which dependent variable has the largest sex effect? Does this make sense? Explain.
3. In exercise 1, which dependent variable has the largest interaction effect? Does this make sense? Explain.
4. In an experiment conducted by Laurence Baker at the University of Indiana as part of his doctoral dissertation, a large number of (randomly chosen) people were called on the telephone, and were requested to give help, as described below. The yield was the proportion of persons who agreed to help.

   The experiment was conducted as follows: a male made a phone call to a person in the general Bloomington, Indiana, region and said that his name was Larry Baker, and that he was trying to call a taxi to visit a certain location but had apparently misdialed, using up his last coins. Could the person who answered the phone please call the location and inform the person there that Larry was stranded at a specified street corner? The three factors under study and their levels (L = low, H = high) were:

A: Sex of person called	Male (L) Female (H)
B: Location to be visited (a matter of sordid/respectable)	Clara's Rooming House (L) Well-Known Hotel (H)
C: Whether the caller seeking help said that he was blind	No (L) Yes (H)

   The results were:

1	*a*	*b*	*ab*	*c*	*ac*	*bc*	*abc*
.64	.45	.72	.65	.59	.55	.73	.78

   Use Yates' algorithm to determine the effects of the three factors, and all interaction effects. Incidentally, Larry Baker is blind.

5. In an analysis of return on investment (ROI) of certain investment options, researchers studied three factors:

A: Type of option	Put (L)	Call (H)
B: Length of option	60 days (L)	6 months (H)
C: Market trend during which option was purchased	Bull (L)	Bear (H)

   The bull market period studied was March 1968 to August 1968; the bear market period studied was April 1966 to September 1966. (A neutral market period was also studied, but results for this period are not included here. For more information, see "Profitability in Buying Puts and Calls," by Hawkins and Halonin, *Decision Sciences*, 1973, 4:109–118.) The yields recorded were the yearly percent returns. The results were:

1	*a*	*b*	*ab*	*c*	*ac*	*bc*	*abc*
−363	1056	−85	194	410	81	321	−144

   Find the effect of each factor and their interactions.

6. In exercise 5, the statistical analysis of the data indicated that the main effects, *A* and *C*, were not "significant." Can you devise a way to determine what the yields would be if the *A* effect and *C* effect were zero while the other effects stay the same as they are based on the above yields? Hint: what you need is a "reverse" Yates' algorithm.
7. For a 2^4 design, with four factors, A, B, C, and D, suppose that we have the following data in Yates' standard order:

   7.2 6.1 7.5 5.9 6.9 5.8 7.6 6.2 7.0 6.0 8.0 5.5 7.0 5.5 8.0 5.3

Use Yates' algorithm to find all the main effects and interaction effects.

8. Consider a 2^3 design with no replication. Give an example of values for the eight treatment combinations so that all two-factor interactions equal zero, but the three-way interaction is (strictly) positive:

   $1 =$ ______ $c =$ ______

   $a =$ ______ $ac =$ ______

   $b =$ ______ $bc =$ ______

   $ab =$ ______ $abc =$ ______

   After accomplishing your goal, consider the issue of more versus less efficient ways of achieving this goal.

9. Whether a person redeems a coupon or not is suspected to be, in part, affected by four factors: Face Value, Product, Customer Willingness To Use (CWTU), and Ease of Use. Suppose that the 2^4 factorial design in Table 9EX.9 is run to test this suspicion. The numbers in the table cells represent the number of coupons redeemed out of 1000 coupons. Find the effects and interpret them.

10. In exercise 7, what is the variance of an effect, as a function of the (unknown) σ^2?

11. We now revisit a problem first encountered in the exercises of Chapter 2, in which a series of one-factor-analyses were called for. Now we have the ammunition to consider the same problem in a more sophisticated way, as a factorial design. We repeat the exercise's statement:

    One of the authors recently taught two sections of the same course, called Quantitative Methods, in the same semester. This course was a core MBA course covering the basics of introductory statistics, ranging from probability, through discrete and continuous distributions, confidence intervals, hypothesis testing, and extensive model-building techniques, including multiple regression and stepwise regression. One class was taught on Tuesday evenings, the other Wednesday evenings (each class of three hours was held once per week for 14 weeks, plus a final exam week).

    The distribution (in alphabetical order) of the final numerical grades (prior to translating them into letter grades) was tabulated by evening, status (part-time/full-time), and gender. The results for the 55 students are in Table 9EX.11 (and, for convenience, also repeated in the data sets as Problem 9.11).

    Note that the data, not having come from a *designed* experiment, do not have equal frequencies in the eight cells that the treatment combinations would form. It isn't viable to force the same frequency, and even if it were possible and desirable, students sometimes drop a course during the semester. Of course, the analysis can still be done. We assume the use of a software package,

Table 9EX.9 Coupons Redeemed: Four-Factor Design

		Low Value &		High Value &	
CWTU	Ease of Use	Food Product	Paper Product	Food Product	Paper Product
Low	Low	4	2	8	6
Low	High	4	4	8	8
High	Low	4	5	9	9
High	High	7	6	8	8

Table 9EX.11 **Course Grade, Fall Semester**

Student	Evening	Status	Gender	Grade	Student	Evening	Status	Gender	Grade
1	Tuesday	Part-time	Male	61.39	31	Wednesday	Part-time	Female	54.80
2	Tuesday	Part-time	Male	81.47	32	Wednesday	Part-time	Male	54.30
3	Tuesday	Part-time	Male	65.59	33	Wednesday	Full-time	Female	77.80
4	Tuesday	Full-time	Female	70.30	34	Wednesday	Full-time	Female	67.30
5	Tuesday	Part-time	Male	85.20	35	Wednesday	Full-time	Female	72.20
6	Tuesday	Part-time	Female	63.70	36	Wednesday	Part-time	Male	62.51
7	Tuesday	Part-time	Female	51.94	37	Wednesday	Full-time	Female	81.90
8	Tuesday	Part-time	Male	68.74	38	Wednesday	Part-time	Female	80.90
9	Tuesday	Part-time	Female	69.28	39	Wednesday	Part-time	Female	39.00
10	Tuesday	Full-time	Male	66.90	40	Wednesday	Part-time	Male	65.85
11	Tuesday	Full-time	Male	64.30	41	Wednesday	Full-time	Female	52.58
12	Tuesday	Full-time	Male	58.88	42	Wednesday	Full-time	Male	86.30
13	Tuesday	Part-time	Male	86.09	43	Wednesday	Part-time	Female	30.56
14	Tuesday	Full-time	Female	74.40	44	Wednesday	Part-time	Male	75.29
15	Tuesday	Part-time	Female	56.70	45	Wednesday	Part-time	Male	71.11
16	Wednesday	Full-time	Female	61.80	46	Wednesday	Part-time	Female	56.04
17	Wednesday	Part-time	Male	62.08	47	Wednesday	Part-time	Male	71.10
18	Wednesday	Part-time	Male	83.80	48	Wednesday	Part-time	Male	63.60
19	Wednesday	Full-time	Female	76.10	49	Wednesday	Part-time	Male	47.90
20	Wednesday	Part-time	Male	60.59	50	Wednesday	Full-time	Male	83.30
21	Wednesday	Full-time	Female	71.70	51	Wednesday	Part-time	Male	74.60
22	Wednesday	Full-time	Female	83.50	52	Wednesday	Part-time	Male	59.68
23	Wednesday	Part-time	Male	50.15	53	Wednesday	Part-time	Female	50.51
24	Wednesday	Part-time	Male	88.90	54	Wednesday	Full-time	Male	87.30
25	Wednesday	Part-time	Female	53.08	55	Wednesday	Part-time	Male	70.59
26	Wednesday	Part-time	Male	66.05					
27	Wednesday	Full-time	Female	62.49					
28	Wednesday	Part-time	Male	66.24					
29	Wednesday	Part-time	Female	61.72					
30	Wednesday	Part-time	Male	72.49					

which eliminates the arithmetic burden resulting from the uneven cell sizes.

Perform a 2^3 analysis, with replication, to find all effects. On the basis of either the software output or your intuition after examining the magnitudes of the effects (we haven't formally covered the F test in two-level factorial designs, but we have shown some output of said tests), which effects appear to be statistically significant?

12. Now consider a subset of the data of exercise 11. These data represent the first two students on the list having each of the eight treatment combinations of the three factor levels; it is viewed as a random sample of the data values in each cell. These data, shown in Table 9EX.12, represent the students numbered 1, 2, 4, 6, 7, 11, 12, 14, 16, 17, 18, 19, 25, 29, 42, 50.

    The experimental data now represent a balanced three-factor, two-level design with two replicates per treatment combination. Find all the effects. Which are, or appear to be, statistically significant?

Table 9EX.12 **Two-Replicate Sample of Grade Study**

Evening	Status	Gender	Grade
Tuesday	Part-time	Male	61.39
Tuesday	Part-time	Male	81.47
Tuesday	Full-time	Female	70.30
Tuesday	Part-time	Female	63.70
Tuesday	Part-time	Female	51.94
Tuesday	Full-time	Male	64.30
Tuesday	Full-time	Male	58.88
Tuesday	Full-time	Female	74.40
Wednesday	Full-time	Female	61.80
Wednesday	Part-time	Male	62.08
Wednesday	Part-time	Male	83.80
Wednesday	Full-time	Female	76.10
Wednesday	Part-time	Female	53.08
Wednesday	Part-time	Female	61.72
Wednesday	Full-time	Male	86.30
Wednesday	Full-time	Male	83.30

13. How do the results from exercise 11 compare with those of exercise 12?
14. We now extend the data from exercise 11 to include the individual components of the students' grades. The components were two term papers (P1 and P2), graded for a data analysis and write-up of the results, two quizzes (Q1 and Q2) during the semester, homework assignments (HW) that were required to be handed in each week, class participation (CP), and the final exam (F). The first four columns of the data in Table 9EX.14 are the same as the data of exercise 11. (The data are also in the data sets as Problem 9.14.)

    Repeating exercise 11, but now for the grade on *P1 in particular,* perform a 2^3 analysis, with replication, to find all effects.
15. Repeating exercise 11, but now for the grade on *Q1 in particular,* perform a 2^3 analysis, with replication, to find all effects.
16. Repeating exercise 11, but now for the grade on *P2 in particular,* perform a 2^3 analysis, with replication, to find all effects.
17. Repeating exercise 11, but now for the grade on *Q2 in particular,* perform a 2^3 analysis, with replication, to find all effects.
18. Repeating exercise 11, but now for the

Table 9EX.14 **Course Grades and Components**

Student	Day	Status	Sex	P1	Q1	P2	Q2	HW	CP	F	Grade
1	Tues	Part	M	8.0	30	7.0	15	8.9	3	40	61.39
2	Tues	Part	M	8.5	25	9.0	25	8.7	9	71	81.47
3	Tues	Part	M	9.0	25	8.0	20	8.9	1	54	65.59
4	Tues	Full	F	10.0	20	9.0	15	10.0	10	46	70.30
5	Tues	Part	M	8.5	25	10.0	28	10.0	5	84	85.20
6	Tues	Part	F	9.0	19	8.0	19	10.0	0	59	63.70
7	Tues	Part	F	9.0	16	7.0	16	6.4	3	35	51.94
8	Tues	Part	M	7.5	24	9.5	20	8.4	9	41	68.74
9	Tues	Part	F	9.0	28	7.0	23	9.8	3	50	69.28
10	Tues	Full	M	8.5	28	10.0	17	10.0	3	43	66.90
11	Tues	Full	M	8.5	24	10.0	13	10.0	5	41	64.30
12	Tues	Full	M	8.5	22	9.0	20	9.8	1	32	58.88
13	Tues	Part	M	9.0	27	9.0	26	8.9	9	79	86.09
14	Tues	Full	F	9.0	27	8.0	25	10.0	1	68	74.40
15	Tues	Part	F	8.0	15	7.0	22	10.0	3	34	56.70

(continued)

Table 9EX.14 (continued)

Student	Day	Status	Sex	P1	Q1	P2	Q2	HW	CP	F	Grade
16	Wed	Full	F	9.5	30	10.0	12	10.0	2	31	61.80
17	Wed	Part	M	9.0	29	9.0	14	8.2	3	38	62.08
18	Wed	Part	M	10.0	26	10.0	25	10.0	10	61	83.80
19	Wed	Full	F	9.5	23	10.0	21	10.0	9	52	76.10
20	Wed	Part	M	8.0	26	8.0	19	9.1	1	40	60.59
21	Wed	Full	F	9.5	30	7.0	23	10.0	1	59	71.70
22	Wed	Full	F	9.5	26	10.0	30	10.0	2	80	83.50
23	Wed	Part	M	8.0	16	7.0	12	4.5	1	52	50.15
24	Wed	Part	M	8.0	27	10.0	27	10.0	9	83	88.90
25	Wed	Part	F	8.5	13	10.0	13	8.2	2	38	53.08
26	Wed	Part	M	8.0	24	10.0	28	4.5	1	55	66.05
27	Wed	Full	F	9.0	24	10.0	19	9.1	0	43	62.49
28	Wed	Part	M	8.5	11	8.0	23	9.6	6	57	66.24
29	Wed	Part	F	9.5	16	9.0	22	9.8	0	48	61.72
30	Wed	Part	M	9.0	26	9.0	23	9.1	2	63	72.49
31	Wed	Part	F	9.0	12	10.0	18	10.0	0	36	54.80
32	Wed	Part	M	9.0	19	10.0	11	10.0	1	31	54.30
33	Wed	Full	F	9.5	22	10.0	14	10.0	9	71	77.80
34	Wed	Full	F	9.5	18	7.0	27	10.0	0	61	67.30
35	Wed	Full	F	9.5	29	9.5	20	10.0	1	59	72.20
36	Wed	Part	M	10.0	20	7.0	16	8.9	3	52	62.51
37	Wed	Full	F	10.0	24	10.0	21	10.0	9	68	81.90
38	Wed	Part	F	9.0	30	10.0	28	10.0	1	73	80.90
39	Wed	Part	F	9.0	14	7.0	2	8.0	1	20	39.00
40	Wed	Part	M	10.0	16	10.0	16	5.5	7	58	65.85
41	Wed	Full	F	8.5	26	10.0	9	8.2	0	28	52.58
42	Wed	Full	M	10.0	25	10.0	27	10.0	9	71	86.30
43	Wed	Part	F	8.5	3	0.0	7	6.4	2	29	30.56
44	Wed	Part	M	9.0	30	9.0	22	9.1	6	54	75.29
45	Wed	Part	M	9.0	12	10.0	27	8.9	9	49	71.11
45	Wed	Part	F	10.0	20	7.0	15	3.6	2	53	56.04
47	Wed	Part	M	8.5	30	8.0	16	10.0	3	62	71.10
48	Wed	Part	M	9.5	25	9.0	14	10.0	0	52	63.60
49	Wed	Part	M	10.0	13	7.0	15	10.0	0	23	47.90
50	Wed	Full	M	10.0	30	10.0	22	10.0	9	61	83.30
51	Wed	Part	M	9.0	26	10.0	23	10.0	1	67	74.60
52	Wed	Part	M	9.5	22	7.0	27	8.2	0	35	59.68
53	Wed	Part	F	8.0	20	9.0	4	8.9	0	42	50.51
54	Wed	Full	M	9.5	30	10.0	28	10.0	9	66	87.30
55	Wed	Part	M	9.5	30	7.0	23	9.1	2	55	70.59

grade on *HW in particular,* perform a 2^3 analysis, with replication, to find all effects.

19. Repeating exercise 11, but now for the grade on *CP in particular,* perform a 2^3 analysis, with replication, to find all effects.

20. Repeating exercise 11, but now for the grade on *F in particular,* perform a 2^3 analysis, with replication, to find all effects.

21. The analyses of exercises 14–20 allow discussion of which portions of the overall grade differ by evening, status, and gender. Which components of the overall grade differ the most by evening? Which differ the most by status? Which differ the most by gender? Is there any support for the notion that part-time MBA students, who generally have more work experience and are older, are able to perform better on quizzes and exams, whereas full-timers, who surely have more time available to study, are able to perform better on handed-in assignments (papers and homework)?

22. Edward "Fuzzy" Newbar, a former professional golfer and a golf course designer, funded a series of studies to determine if various club offerings affected golf scores for the players at several New England industrial golf leagues like that at Eastern Electric. Four factors were examined: the provision of steeply discounted golf lessons (Teach), the provision of steeply discounted practice sessions (Prac), the provision of generous prizes for improved performance (Prize), and required participation in a winners' pool where each player had to pay an extra $100 entrance fee above all other fees (Risk). In each case, the "low" factor level, represented by -1, was the presence of the offering, and the "high" level of each factor, represented by $+1$, represented the absence of the offering. (The reasoning for this as-

Table 9EX.22 Edward "Fuzzy" Newbar Golf Study

	Factors				Scores							
Comb	Teach	Prac	Prize	Risk	1	2	3	4	5	6	7	8
1	−1	−1	−1	−1	82	78	84	76	84	77	80	76
2	−1	−1	−1	1	89	82	83	93	90	79	80	84
3	−1	−1	1	−1	83	85	87	90	87	87	84	78
4	−1	−1	1	1	88	88	81	91	89	93	88	85
5	−1	1	−1	−1	91	95	95	95	97	96	99	93
6	−1	1	−1	1	105	97	97	102	101	95	104	103
7	−1	1	1	−1	101	100	99	102	101	100	106	101
8	−1	1	1	1	101	105	110	108	101	101	110	109
9	1	−1	−1	−1	94	97	93	94	99	95	95	93
10	1	−1	−1	1	98	98	99	101	95	101	99	102
11	1	−1	1	−1	102	99	105	97	103	98	102	97
12	1	−1	1	1	109	109	103	97	101	106	109	106
13	1	1	−1	−1	108	105	113	98	103	102	107	109
14	1	1	−1	1	108	108	108	107	112	109	109	111
15	1	1	1	−1	112	107	108	109	113	111	111	114
16	1	1	1	1	110	118	118	112	113	119	114	113

signment of levels was that lower scores correspond with more improvement.) The study involved 128 players, a random selection of eight players who had experienced each of the 16 possible treatment combinations. The results are in Table 9EX.22 and in the data sets as Problem 9.22.

Analyzing the results as a 2^4 factorial design, find all the effects. Ignoring all interactions, which main effect has the highest value (and hence the highest reduction in average score)?

23. Would consideration of interaction effects change the result found in exercise 22?

Notes

1. Of course, the observed difference in response rate between any of the two treatment combinations is not necessarily the true difference, because the data are from a sample that *estimates* the true values.

2. One reviewer of this book was adamant that using the same notation for the factor, the true effect, and the estimate of the effect was "very sloppy and will be confusing to the student." We respectfully disagree. Virtually all treatises on this subject use this notation; what would be confusing would be to change a notation that is so universal. But we thank the reviewer for reinforcing our interest in alerting the student to what we and other experimental designers are doing.

3. Because the dependent variable is a proportion, the normality assumption traditionally made when performing and interpreting an ANOVA may not be strictly true. However, this is not the primary issue here, and the robustness of the normality assumption is likely to keep this consideration from having a material effect on the results of an ANOVA. If necessary, we can transform the data using the "standard" transformation of $Y_{\text{transformed}} = \arcsin \sqrt{Y}$.

4. In situations where replication is relatively inexpensive, and most of the cost of a data point is driven by changes in setup, the cost of 32 different treatment combinations may *not* be less than the cost of 16 replicates of six treatment combinations.

CHAPTER 10

Confounding/Blocking in 2^k Designs

EXAMPLE 10.1 Pricing a Supplemental Medical/Health Benefit Offer—Phase II

As noted in the previous chapter, the insurance company HealthMark was interested in conducting a marketing research study to determine the best (that is, most profitable) offer of supplemental medical and health benefits. Example 9.10 reported the results of an experiment to study pricing for a set of core benefits and three auxiliary benefits. Those four factors, and their levels, are the first four factors listed in Table 10.1. After confirming the results of the experiment, HealthMark management decided to add a fourth auxiliary benefit, which is the fifth factor listed in Table 10.1. Descriptions of the first four factors are in Example 9.1. The emergency care channel provides a discount of 25% or 50% at a national chain of emergency care centers, for emergency care, with an extensive referral service.

There are now 32 treatment combinations, 2^5. The previous experiment found statistically significant two-factor and three-factor interaction effects, as we noted, so an experiment that does not allow each of these interaction effects to be separately and routinely identified would not be appropriate. In essence, all 32 treatment combinations need to be examined. (In Chapter 11 we examine designs in which not all treatment combinations are run, an expansion of some ideas we considered with Latin-square and Graeco-Latin-square designs.)

However, it is unwise to show a respondent 32 scenarios to solicit his or her purchase intent. It is generally accepted that after 16 scenarios (or fewer, if a large amount of demographic and other information is also solicited), respondents get tired and the quality of their responses decreases dramatically. Hence, the 500 available respondents were split into two groups, each of which was shown only 16 of the 32 scenarios. That led to another issue: the potential bias induced when

Table 10.1 Prices and Benefits Offered

Factor	Levels (Low/High)
Price of core benefits	\$9.95/\$12.95 per month per adult
Price (and benefit) of chiropractic channel	\$0.50 (and 25% off)/\$1.00 (and 50% off) per month per adult
Price (and benefit) of dermatology channel	\$2 (and 20% off)/\$3 (and 40% off) per month per adult
Price (and benefit) of massage channel	\$1 (and 15% off)/\$2 (and 30% off) per month per adult
Price (and benefit) of emergency care channel	\$2.50 (and 25% off)/\$3.50 (and 50% off) per month per adult

one set of respondents evaluates 16 scenarios and a different set of respondents evaluates the other 16 scenarios. Try as you might to make the respondents in each group alike, experience indicates that there will likely still be bias.

What can be done to eliminate this potential bias? Are there ways to systematically split respondents into two groups (or, more generically, blocks) such that there is no bias? We return to this example at the end of the chapter.

10.1 Introduction

The topic of this chapter is useful in its own right and absolutely essential to understanding the subject of fractional-factorial designs, discussed in Chapter 11. Imagine coming to a point in designing our experiment where we have settled on the factors and levels of each factor to be studied. Usually this will not be an exhaustive list of all the factors that might possibly influence the experimental yield, but a bigger list would likely be prohibitive and, even then, not truly exhaustive. There are always factors that affect the yield but that cannot be fully identified. Of course, if we are fortunate, these unidentified factors are not among the most influential (often the intuition of good process experts contributes to such "luck"). Ideally, we would like to have all of these other factors held constant during the performance of our experiment. However, this cannot always be done, and in this chapter we discuss a potentially powerful way to partially mitigate the consequences if we can't. We focus on 2^k factorial designs; however, the concepts and reasoning involved apply to all experimental designs.

10.2 Simple Confounding

An unreplicated 2^k experiment has 2^k treatment combinations. If we study three factors (that is, $k = 3$), each at two levels, we have eight treatment combinations; if each takes two hours to run, 16 hours will be required to complete the experiment. Over such a long period many influences could occur that are not of interest to us in this experiment, but that might make the interpretation of our results unclear. For example, personnel could change (the day-shift radiologist might be replaced by the night-shift radiologist in a hospital radiology experiment), the humidity in the photo lab might vary, line voltages might fluctuate due to the evening air-conditioning demand, water pressure might drop, traffic might increase, and the like. These "nuisance" factors can pollute our data, rendering the interpretation problematic.

Suppose we have only eight hours available in a day, and so are forced to run our 16-hour experiment over two days: a block of four treatment combinations on Monday and another block of four on Tuesday. (This is just one example; instead of two time periods, with "time" or "day" as the nuisance factor, we could be dealing with two radiologists sharing eight sets of X-rays on the same shift, or two auto-parts stores, or two groups of responders evaluating subsets of alternative product configurations in a marketing-research study, and so on.)

Whatever the situation, imagine that we have not been able to run all eight treatment combinations as one large block under homogeneous conditions; instead, we have had to split the one large block of eight treatment combinations into two smaller blocks of four.

The eight treatment combinations, in Yates' standard order, are 1, *a*, *b*, *ab*, *c*, *ac*, *bc*, *abc*; there are 70 ways in which we can assign them to the two days (assuming an even split of four and four). Does it matter which of these we choose? Should they be assigned randomly? Table 10.2 shows 3 of the 70 possible arrangements. Before continuing, consider the three arrangements (designs) and form your own intuitive answers to the above questions.

Is one of the three arrangements in Table 10.2 preferable to the others? Assume that, if there is a "block effect" (here, the systematic difference, if any, between a measurement made on Monday and the same one made on Tuesday), it causes all Monday yields to be higher, by some unknown but approximately constant amount, *X*, than if they had been obtained on Tuesday; *X* can be positive or negative.

It's useful to identify the block that includes the "1" observation (that is, all factors at low level) by a name; it is usually called the **principal block.** This

Table 10.2 **Three Possible Designs**

	Design 1		Design 2		Design 3	
	Monday	**Tuesday**	**Monday**	**Tuesday**	**Monday**	**Tuesday**
Treatments	1 *a* *b* *ab*	*c* *ac* *bc* *abc*	1 *ab* *c* *abc*	*a* *b* *ac* *bc*	1 *ab* *ac* *bc*	*a* *b* *c* *abc*

is simply a way to identify this block; in no way is the principal block different from (except that it has the "1" observation in it), or of greater or lesser stature than, any other block.

A cursory examination of Design 1 in Table 10.2 reveals that all treatments with C high occur on Tuesday; all with C low are on Monday. Recall from the previous chapter that the C effect is estimated by taking the difference between the average of the yields with C high and the average with C low. However, with this "blocking design", we can't tell how much of this difference is associated with the main effect C and how much is associated with the block (here, day) effect (and, of course, "noise" is present also). Our estimate of C has thus been *confounded* with the block effect, and other effects could be confounded as well (in this case, they aren't). Usually, a design that confounds a main effect is not desirable. **Confounded** is the word used to describe a situation in which two or more effects (here, the main effect, *C*, and the block effect, *X*) are estimated together, in sum or difference, and are not separable. In many texts, the word "confounded" is used narrowly to indicate the nonseparability of a factor (main or interaction) effect and a block effect.

It's more difficult to spot the flaw in Design 2. The main effects seem to be (and in this case are) unaffected by the block effect—the plus and minus signs or, equivalently, the number of high and low levels of the factor are equally balanced over the two days. But notice that all terms with plus signs in the calculation of *AB* are on Monday, and all those with negative signs for *AB* are on Tuesday. Hence, the *AB* interaction term is confounded with the block effect.

The third arrangement is even more difficult to analyze. The main and two-factor interactions appear to be unaffected by the block effect (that is, the signs of the effects are "balanced"). However, a more rigorous analysis of the block effect could proceed as follows: calculate the three main effects, the three two-factor interaction effects, and the three-factor interaction effect as in Chapter 9, with each Monday yield shown with the constant *X* added explicitly.

For example, the usual estimate of main effect *A* is given by

$$A = (1/4)[-1 + a - b + ab - c + ac - bc + abc]$$

For Design 3 in Table 10.2, the usual estimate of A would become

$$\begin{aligned} A = {} & (1/4) \\ & [-(1 + X) + a - b + (ab + X) - c + (ac + X) - (bc + X) + abc] \\ = {} & \text{usual estimate of } A \end{aligned}$$

because the *X*'s cancel. As the *X*'s cancel out, no "pollution" is introduced into our estimate of *A* due to a possible block effect.

Continuing with Design 3, the usual estimate of *ABC* is

$$ABC = (1/4)[-1 + a + b - ab + c - ac - bc + abc]$$

In Design 3 this becomes

$$\begin{aligned} ABC = {} & (1/4) \\ & [-(1 + X) + a + b - (ab + X) + c - (ac + X) - (bc + X) + abc] \\ = {} & \text{usual estimate} - X \end{aligned}$$

That is, our estimate of the three-factor interaction effect, *ABC*, is polluted by (confounded with) the block effect; it is equal to the difference of two quantities—the "pure" estimate of *ABC* (the word *pure* is in quotes because unfortunately the "noise," due to factors other than day, would be present even if the *X*'s cancel) and the block effect, *X*. Since we don't know *X*, we don't know *ABC*. Were we to calculate all seven effects for Design 3, we would find that only the estimate of *ABC* has been confounded; all main and two-factor interactions are unaffected by the separation of the eight treatment combinations into two smaller blocks of four. Estimates that are polluted by the block effect are simply said to be confounded. We would say that estimates *A*, *B*, *C*, *AB*, *AC*, and *BC* are "clean," whereas *ABC* is "confounded."

What happens if we reverse the order of the two blocks, such that the principal block is run on Tuesday and the other on Monday? All main and two-factor interaction effects would still be clean, and *ABC* = usual estimate + *X*. Since the sign of *X* is generally unknown (and doesn't really matter), in the final analysis it makes no difference which block is on Monday and which is on Tuesday; it's the way they're grouped (blocked) that matters!

Similar analysis of the other two designs yields the results shown in Table 10.3. Normally, main effects are most important, two-factor interactions are of next importance, and three-factor interactions are of least interest, in an experiment with three factors. Thus, these three designs are likely not equivalent in the eyes of the experimenter. All else being equal, Design 3 is probably superior to the others.

Replacement of one block by two smaller blocks requires the "sacrifice" (confounding) of at least one effect.

Consider another design—the one shown in Table 10.4. The confounded effects are *B*, *C*, *AB*, and *AC*; only three of the seven effects, *A*, *BC*, and *ABC*, are clean. Of the 70 arrangements possible, only 14 have just one effect confounded (the seven designs corresponding to each effect, respectively, and their mirror images—switching Monday and Tuesday). The rest of the designs

Table 10.3 Confounding in the Three Possible Designs

	Design 1		Design 2		Design 3	
	Monday	**Tuesday**	**Monday**	**Tuesday**	**Monday**	**Tuesday**
Treatments	1 *a* *b* *ab*	*c* *ac* *bc* *abc*	1 *ab* *c* *abc*	*a* *b* *ac* *bc*	1 *ab* *ac* *bc*	*a* *b* *c* *abc*
Confounded Effects	Only *C*		Only *AB*		Only *ABC*	

Table 10.4 **A Fourth Design**

	Monday	Tuesday
Treatments	1 *a* *b* *ac*	*ab* *c* *bc* *abc*
Confounded Effects	*B*, *C*, *AB*, and *AC*	

have more than one effect confounded, in most cases four of them. Of course, we would never prefer to confound more effects than is minimally necessary. Fortunately, *we can choose which effect we wish to confound* (given that there must be one!). Thus, randomly assigning treatments to blocks is silly because it abrogates an opportunity to design a superior experiment.

To summarize thus far, it's best to run all treatment combinations at the same time, under the same conditions. When we can't, we partition the original block into two (or, as we discuss later, more than two) equal-sized smaller blocks. By doing so, we reduce experimental error (that portion due to the block effect) for the effects we care about most at the expense of confounding one effect (or more than one, when more than two blocks are necessary). All "clean" estimates (those that are not confounded) can be judged against reduced variability, with the corresponding narrower confidence limits, increased power, and the like; for these more important effects, it's as though there were no block effect. In the appendix to this chapter we present a more explicit and detailed discussion and example of the mechanics of this error reduction.

We choose a design that confounds only one effect—the most expendable one. Generally, the most expendable effect is the highest-order interaction effect. To be prudent, we often allow for the existence of a block effect unless we are confident that there is none (or that it's negligible). The cost of designing an experiment to deal as efficiently as possible with a block effect that doesn't exist is less than the cost of ignoring a block effect when one is present.

Underlying the approach of this chapter is the assumption that X is approximately constant. If there is reason to suspect that the block effect is not independent of the treatment combinations, but rather varies for different treatment combinations (that is, the blocking factor interacts with primary factors and/or their interactions), then the blocking factor (such as day or time) must be considered explicitly as another factor under study.

As the size of an experiment grows (k becomes larger in a 2^k design), confounding becomes more popular for two reasons: it is difficult to create large homogeneous blocks (hence one may have no choice but to confound), and the loss of one effect may not be of great consequence—for example, in a 2^7 design, giving up one out of 127 effects, perhaps $ABCDEFG$.

10.3 Partial Confounding

We began our discussion of confounding with an unreplicated 2^3 design in which we had to run the experiment over two days; we assumed that we could run only four treatments per day. We now extend this concept of confounding to a 2^3 design with replication. Our first illustration has four replications. Clearly, more than two days are involved. Assume that we decide to run the experiment over a four-week period, and that each replicate consists of one block of four treatment combinations run on Monday and the other block on Tuesday. We could, of course, repeat the same design each week, thereby confounding the same effect, likely *ABC*. If we did so, each main effect and each two-factor interaction effect could be estimated cleanly, based on all four replications; that is, each effect would be based on all 32 data values. We might loosely say that "32 units of reliability" contribute to each estimate.[1] We could arrive at each effect by averaging the four estimates of each effect, one per replicate, or equivalently, averaging each of the four yields for the eight treatment combinations before calculating the effects. In either case, we would not get any useful information about *ABC*.

Rather than confounding the same effect for each replicate, we might instead run a different "confounding scheme" or "design" for each replicate, as shown in Table 10.5. This design "partially" confounds each of the interaction effects. We can still base our estimate of each of the three main effects on all 32 data values. The interaction effects can all be estimated cleanly, albeit with reduced reliability. We use all but the second replicate for *AB*, all but the third for *AC*, all but the fourth for *BC*, and all but the first for *ABC*. The interaction effects are each based on 24 data values. No effect is completely confounded (or "lost"); instead, each of the interaction effects is partially confounded. For the loss of *some* reliability for *some* effects, we don't have to *completely* sacrifice any one effect. And, once again, we have the latitude to select which effects are clean and which are partially confounded.

Table 10.5 Four Replicates with Different Designs

	Week 1		Week 2		Week 3		Week 4	
	Monday	Tuesday	Monday	Tuesday	Monday	Tuesday	Monday	Tuesday
Treatment	1 ab ac bc	a b c abc	1 ab c abc	a b ac bc	1 b ac abc	a ab c bc	1 a bc abc	b ab c ac
Confounded Effect	*ABC*		*AB*		*AC*		*BC*	

In Table 10.5, we have a rather specific scenario of the timing of the blocks. Is it important that we use Monday and Tuesday of successive weeks (or some other comparable pattern)? What would happen if we ran all replications within a two-week period? Suppose we ran the principal block of each replicate one week and the nonprincipal block of each replicate the next week? Haven't we introduced another block effect—the "week" effect? The answer to this question is "Yes, but it doesn't matter." Recall that block effects relate to an (approximately) constant difference between the same treatment combination run in two different blocks. If the first replication in Table 10.5 is conducted on two successive Mondays, the constant difference might be different from the constant difference between Monday and Tuesday of the first week, but so what? Only *ABC* will be confounded in either case, and we agreed to sacrifice *ABC* even before we ran the first replication. That is, the "block effect" might be week-to-week, instead of day-to-day, but when it cancels for an effect (in the sense of the illustrations with *X* earlier in the chapter), it cancels. And when it doesn't, it doesn't. In fact, even if the principal blocks were run on Monday through Thursday of the first week, and the nonprincipal blocks were run on Monday through Thursday of the next week—*in reverse order* (perhaps a silly plan, but just to prove a point), it wouldn't matter. Again—when the block effect cancels, it cancels, and when the block effect doesn't cancel, the estimate is confounded.

To emphasize this subtle but important (and often confusing) point, assume that the yield being measured for each treatment combination is a weight, in pounds, to be measured on one of eight different scales. Each scale has some (fixed, but unknown) calibration error. Any of the eight columns in Table 10.5, currently designated by day and week, could correspond to any one of the eight scales. The column's mate (the other block of that replicate) corresponds to some other scale. Each scale has some fixed calibration error that is applied to every yield measured on that scale, but, from the previous section, we know that if the experiment is designed thoughtfully, for each replicate the calibration errors cancel for every effect but one—the chosen confounded (sacrificed) effect.

Finally, remember that block effects are effects due to the partitioning (or replication) of an experiment over two or more sets of experimental conditions (blocks) that potentially have some nearly constant difference in yield. The requirement to partition is often due to some constraint of resources. For example, eight pieces of ceramic tile must each be weighed within a specific time after removal from the kiln to measure shrinkage on a consistent basis, but one set of scales cannot do the weighings quickly enough; two sets of scales are needed. The constraint is often time, but it might be anything.

EXAMPLE 10.2 Grading X-Ray Film

By way of further example, consider an experiment involving a new X-ray film. The dependent variable is "film readability," a subjective evaluation by an ex-

Table 10.6 **Two Partially Confounded Effects in 2^3 Design**

	First Replication		Second Replication	
	Dr. McGwire	**Dr. Sosa**	**Dr. McGwire**	**Dr. Sosa**
Treatments	1 *ab* *ac* *bc*	*a* *b* *c* *abc*	1 *ab* *c* *abc*	*a* *b* *ac* *bc*
Confounded Effect	*ABC*		*AB*	

perienced radiologist using a reference tumor pattern on a simulated torso. The radiologist grades each image on a 100-point scale; the higher the score, the greater the readability. Factors are photo-resist thickness, exposure time, and development time. We label these as follows:

Designation	Factor
A	Photo-resist thickness
B	Exposure time
C	Development time

Two radiologists are available to evaluate the films; each will examine four treatment combinations for each of two replications. (The two replications are available at different times, which eliminates the option of letting each radiologist evaluate a complete replicate; as indicated earlier, a different time for a different complete replicate is not a concern.) We want to allow for the possibility that the radiologists may differ in their assessments—one might be a bit more demanding than the other. We assume that this will introduce an approximately constant difference in evaluation; the block effect is associated with the use of two different radiologists.

The design of this 2^3 experiment, with two replicates, is shown in Table 10.6. Five of the seven effects, *A*, *B*, *C*, *AC*, and *BC*, are clean in both replications; they will be estimated using all 16 yields. *ABC* and *AB* are partially confounded. *ABC* is estimated using data only from the second replication, and *AB* is estimated with data only from the first replication; estimates of the partially confounded effects are based on eight data points.

By the way, as noted earlier in the chapter, there are two equivalent arithmetic ways to estimate the five effects that are based on both replicates. One way to estimate, for example, the *A* effect is to calculate it separately for each of the two replicates in the routine way (using, say, Yates' algorithm); then take the average of the two estimates. An equivalent way is to first average the two 1's, the two *a*'s, the two *b*'s, and so on, and apply Yates' algorithm once to these "cell means." Both arithmetic procedures yield the same result. Software packages don't usually allow for explicit partial confounding. Hence, it may be necessary to use the first approach mentioned, with each replicate "run" separately.

10.4 Multiple Confounding

As experiments grow larger, it is sometimes necessary to have more than two blocks. Consider an unreplicated 2^4 design in which we must partition the experiment into four blocks of four treatment combinations each. Table 10.7 shows one possible design.

Imagine that yields in the four blocks of Table 10.7 differ by constants in terms of the variable being measured—yields in the first block are too high (or low) by *R*, those in the second block by *S*, those in the third by *T*, and those in the fourth by *U*. (These letters play the role of *X* in the discussion of confounding into two blocks.) The exact values of *R*, *S*, *T*, and *U* are irrelevant; however, we assume, with no loss of generality, that $R + S + T + U = 0$. Tedious examination of the expressions for all 15 effects ($15 = 2^4 - 1$) reveals that the (unknown, but constant and systematic) block effects confound three estimates—*AB*, *BCD*, and *ACD*. (As we will see in more detail later, the minimum number of confounded effects is one fewer than the number of blocks.) The remaining 12 effects in this 2^4 design are clean.

This result is illustrated for *ACD* (a confounded effect) and *D* (a "clean" effect) in Table 10.8. The treatments are listed in Yates' order, and the columns of signs are from our previous work. The result is to yield the block effects, *R*, *S*, *T*, and *U*, with the signs as shown. We observe that all block effects cancel in our calculation of *D*, but they do not cancel for *ACD*. Indeed, our estimate of *ACD* is

$$(1/8)\,[8ACD - 4R + 4S - 4T + 4U] = ACD - R/2 + S/2 - T/2 + U/2$$

Clearly, the *ACD* estimate is hopelessly confounded with block effects.

For illustrative purposes, we began our discussion of multiple confound-

Table 10.7 Four Blocks in a 2^4 Design

Block 1	Block 2	Block 3	Block 4
1	a	b	c
cd	acd	bcd	d
abd	bd	ad	abcd
abc	bc	ac	ab

Table 10.8 **Signs for a Confounded and a Clean Effect**

	ACD		*D*	
Treatment	Sign of Treatment	Block Effect	Sign of Treatment	Block Effect
1	−	$-R$	−	$-R$
a	+	$+S$	−	$-S$
b	−	$-T$	−	$-T$
ab	+	$+U$	−	$-U$
c	+	$+U$	−	$-U$
ac	−	$-T$	−	$-T$
bc	+	$+S$	−	$-S$
abc	−	$-R$	−	$-R$
d	+	$+U$	+	$+U$
ad	−	$-T$	+	$+T$
bd	+	$+S$	+	$+S$
abd	−	$-R$	+	$+R$
cd	−	$-R$	+	$+R$
acd	+	$+S$	+	$+S$
bcd	−	$-T$	+	$+T$
abcd	+	$+U$	+	$+U$

ing with four treatment combinations allocated to each of four blocks. We then determined which effects would, as a result of that allocation scheme (design), be confounded. In practice, the order of these steps is reversed; the experimenters begin by deciding which effects they would be willing to sacrifice (confound), and select the design to achieve that result. Indeed, this order embodies the objective of experimental design. Presumably the 2^4 experiment above was designed to confound one two-factor interaction, *AB*, and two three-factor interactions, *BCD* and *ACD*.

Knowing the relatively decreasing importance of multifactor interaction effects in most typical situations, we might wonder why, in the 2^4 experiment, the designers didn't elect to sacrifice, for example, *ABCD* rather than *AB*. The answer, due to a theorem by Barnard, is that they can't—at least not without other undesirable ramifications. Barnard proved that in a four-block design, where three effects (at minimum) must be confounded, only two of the three confounded effects can be freely chosen by the designer; the third is defined (that is, mandated) by the choice of the first two. As we shall see, insisting on confounding *ABCD* will result in the loss of either two two-factor interactions, or one main effect; both of these alternatives are likely inferior to the choice made in the above example.

10.5 Mod-2 Multiplication

We now describe a technique whose roots are in the area known as polynomial rings and extension fields in the discipline of modern abstract algebra. We use only the fruits of this field of study, in a down-to-earth manner, and do not concern ourselves with "proofs"; however, everything stated is rigorously supported mathematically.

First, we develop some necessary notation. We need to define an operation of multiplying two effects by one another, but not exactly in the usual sense of arithmetically multiplying two numbers. We call this operation **mod-2 multiplication;** notice the mod-2 or base-2 element of the operation. Some people refer to the operation as the "symbolic product," and others as the "exclusive union." The mod-2 multiplication of A and B is $A \cdot B = AB$. The mod-2 multiplication of AB and CD is then $AB \cdot CD = ABCD$. So far, it's like "regular" multiplication. The mod-2 multiplication of A and A is $A \cdot A = A^2 = A^0 = 1$. That is, exponents are from the binary field; the only exponents allowed are 0 and 1. Any odd-integer exponent is equivalent to the exponent 1 and any even-integer exponent is equivalent to the exponent 0. Finally, we observe that anything to the 0 power is, of course, equal to 1.

Barnard's theorem says that, relative to the example above of a 2^4 in four blocks, whenever two effects are specified to be confounded, the mod-2 multiplication of these two effects is automatically (that is, must be!) the third effect confounded. For a threesome of effects, it doesn't matter which two are specified; the three form a "closed group" of sorts. Table 10.9 illustrates this for our example. Note that whichever two effects we start with, the third comes out as the remaining member of the group. With a little practice, one can easily learn to do mod-2 multiplication rapidly in one's head.

We need to choose the selected effects to be confounded with care; otherwise the resulting third confounded effect may not be a desirable one, rendering the set of three confounded effects an unwise choice. Suppose in a 2^5 design we had first specified the confounding of $ABCDE$ (perhaps thinking "who cares about a five-factor interaction"), and then $ABCD$—"who cares about that one either." Oops! We would then lose $ABCDE \cdot ABCD = E$, a main effect. It would be better to confound, say, ABD, ACE, and $BCDE$, so as not to lose any main effects or any two-factor interactions. One approach often taken

Table 10.9 Closed Group of Confounded Effects

First Specified Confounded Effect	Second Specified Confounded Effect	Resultant (Third) Confounded Effect
AB	BCD	$AB \bullet BCD = AB^2CD = AB^0CD = ACD$
AB	ACD	$AB \bullet ACD = A^2BCD = A^0BCD = BCD$
BCD	ACD	$BCD \bullet ACD = ABC^2D^2 = ABC^0D^0 = AB$

for selecting confounded effects might be called a minimax strategy—pick a design that minimizes the importance of the most important effect confounded.

10.6 Determining the Blocks

Once effects to be confounded are chosen, treatment combinations that go into each block are determined as follows:

> Treatment combinations that have an even number of letters in common with each of the confounded effects go into one block—the principal block.
>
> Treatment combinations to fill the other blocks are determined by mod-2 multiplication of any treatment combination not in the principal block times each of the treatment combinations in the principal block.

This will become clear with an example. Consider a 2^5 experiment with four blocks. Three effects must be confounded; we can specify two of the three. We specify *ABD* and *ACE*, and thereby also confound *BCDE*, as in the example above. In this design, all main effects and two-factor interactions are clean.

Having determined which effects are to be confounded, we now develop the design to achieve that end. The principal block contains all treatment combinations that have an even (remember—zero is even) number of letters in common with all confounded effects. Actually, we need be concerned only with the two effects that are "independent generators," as they're sometimes called. If a treatment combination has an even number of letters in common with these, it will automatically have an even number of letters in common with any effect that is the mod-2 multiplication of these.

We might first list all 32 treatment combinations in Yates' order: 1, *a*, *b*, *ab*, . . . , *abcde*. We can show that the principal block is

Principal block 1 ***abc*** ***bd*** *acd* ***abe*** *ce* *ade* *bcde*

Follow carefully the steps used to arrive at this principal block:

1. Clearly, the treatment combination 1 has an even number of letters in common with *ABD*, *ACE*, and *BCDE*—zero. Indeed, 1 has zero letters in common with all effects; thus, we always begin with 1 in forming the principal block—it's definitional. We ignore, in terms of describing "letters in common," that the treatment combinations are lowercase and the effects are capitalized.

2. We next identified ***abc***, which has two letters in common with *ABD*, *A* and *B*; it also has two in common with *ACE*, and consequently an even num-

ber, in this case two, in common with *BCDE*. How did we identify ***abc***? By "educated" trial and error! In this specific case, we looked at the first confounded effect listed, *ABD*, and arbitrarily picked the *A* and *B* as two (an even number) in common; we then hoped that the treatment combination ***ab*** had an even number in common also with the next confounded effect listed, *ACE*. Alas, it didn't! So, we added *c*, yielding ***abc***, which did have an even number of letters in common with *ACE*, as well as being "even" with *ABD*. This may seem like an ordeal; it's not. With a minimal amount of practice, this kind of reasoning becomes easy and quick. Nevertheless, we boldfaced the ***abc*** term in the principal block list because it did need to be "innovated" (our term for a treatment combination that we had to figure out with the "educated trial and error process," instead of determining it mechanically, as described below). As we'll see, relatively little innovating is needed.

3. Then, again somewhat arbitrarily, and again using educated trial and error, we identified ***bd***, which has two letters in common with *ABD*, zero with *ACE* (and two with *BCDE*). ***bd*** is also boldfaced above.

4. Next we take advantage of a very helpful property of the principal block. Any two members of the principal block, when mod-2 multiplied, always yield a member of the principal block. (For the algebraically inclined, the principal block forms what is called a "closed group under the operation of mod-2 multiplication.") This yields ***abc*** · ***bd*** = *acd* as the next listed member of the principal block.

5. Then we innovated one more treatment combination: ***abe***.

6. Finally, we used the "closed group" property to multiply ***abe*** times the previous three treatment combinations listed, to generate, respectively, *ce*, *ade*, and *bcde*.

The elements ***abc***, ***bd***, and ***abe***, the treatment combinations that were innovated, are said to be "generator" elements. The generators are not unique; we could have ended up with other sets of three, depending on which ones we happened to think of first. We now have all eight treatment combinations that have an even number of letters in common with the three confounded effects, and the set of eight is unique!

A second block is formed from the first by picking *any* yet-unused treatment combination and mod-2 multiplying it by each element of the principal block, respectively. If we pick ***a*** (why not pick a treatment combination that has as few letters as possible?), we get the following:

Principal block (block 1)	1	*abc*	*bd*	*acd*	*abe*	*ce*	*ade*	*bcde*
Multiply by ***a*** (block 2)	*a*	*bc*	*abd*	*cd*	*be*	*ace*	*de*	*abcde*

Note that the mod-2 product of any two elements in block 2 does *not* yield a member of block 2. Indeed, it yields an element of the principal block. (Can you see why?)

We continue the process by selecting some still-unused treatment combination, say ***b***, and mod-2 multiplying it by the elements of the principal block:

Principal block (block 1)	1	*abc*	*bd*	*acd*	*abe*	*ce*	*ade*	*bcde*
Multiply by ***a*** (block 2)	*a*	*bc*	*abd*	*cd*	*be*	*ace*	*de*	*abcde*
Multiply by ***b*** (block 3)	*b*	*ac*	*d*	*abcd*	*ae*	*bce*	*abde*	*cde*

The final block is made up of those treatment combinations still unassigned. It may be easier to simply generate the last block by picking one more still-unused treatment combination, say ***e***, and mod-2 multiplying it by the elements of the principal block. The result is as follows:

Principal block (block 1)	1	*abc*	*bd*	*acd*	*abe*	*ce*	*ade*	*bcde*
Multiply by ***a*** (block 2)	*a*	*bc*	*abd*	*cd*	*be*	*ace*	*de*	*abcde*
Multiply by ***b*** (block 3)	*b*	*ac*	*d*	*abcd*	*ae*	*bce*	*abde*	*cde*
Multiply by ***e*** (block 4)	*e*	*abce*	*bde*	*acde*	*ab*	*c*	*ad*	*bcd*

We compute all effects as before—by using either Yates' algorithm, the sign table, or, of course, software. When we get the results, we basically drop the confounded effects, here, *ABD*, *ACE*, and *BCDE*, from further consideration. We illustrate the confounded nature of *ABD*, *ACE*, and *BCDE* and the unconfounded nature of two arbitrarily chosen clean effects, *AB* and *D*, in Table 10.10. Note that for the clean effects, block effects cancel out within each block. That is, for both *AB* and *D* (and any other clean effect we might have chosen to examine) the table of signs has four plus signs and four minus signs *within each block*. This equality is required to cancel out the block effects. Note, as well, that this is not true in the case of the confounded effects. Within each block, for any given confounded effect, the signs are all the same; rather than cancel, the block effects accumulate.

Table 10.10 **Sign Table for Confounded and Clean Effects**

Block (and constant)	Treatment	Confounded Effects			Clean Effects	
		ABD	*ACE*	*BCDE*	*AB*	*D*
Block 1	1	−	−	+	+	−
	abc	−	−	+	+	−
(Too high or too low by *R*)	*bd*	−	−	+	−	+
	acd	−	−	+	−	+
	abe	−	−	+	+	−
	ce	−	−	+	+	−
	ade	−	−	+	−	+
	bcde	−	−	+	−	+
Block 2	*a*	+	+	+	−	−
	bc	+	+	+	−	−
(Too high or too low by *S*)	*abd*	+	+	+	+	+
	cd	+	+	+	+	+
	be	+	+	+	−	−
	ace	+	+	+	−	−
	de	+	+	+	+	+
	abcde	+	+	+	+	+
Block 3	*b*	+	−	−	−	−
	ac	+	−	−	−	−
(Too high or too low by *T*)	*d*	+	−	−	+	+
	abcd	+	−	−	+	+
	ae	+	−	−	−	−
	bce	+	−	−	−	−
	abde	+	−	−	+	+
	cde	+	−	−	+	+
Block 4	*e*	−	+	−	+	−
	abce	−	+	−	+	−
(Too high or too low by *U*)	*bde*	−	+	−	−	+
	acde	−	+	−	−	+
	ab	−	+	−	+	−
	c	−	+	−	+	−
	ad	−	+	−	−	+
	bcd	−	+	−	−	+

10.7 Number of Blocks and Confounded Effects

We now summarize the consequences of partitioning a 2^k experiment into 2^r blocks. As a function of the size of the experiment and the number of blocks, we list in Table 10.11 the minimum (which should be the actual) number of effects that will be confounded, how many of these the designer can specify, and how many are an automatic consequence of the designer's choice.

Table 10.11 **Number of Blocks and Confounded Effects**

Number of Smaller Blocks 2^r	Number of Confounded Effects 2^r-1	Number Designer May Choose r	Number Defined by Consequence 2^r-1-r
2	1	1	0
4	3	2	1
8	7	3	4
16	15	4	11

We have already explicitly discussed the first two rows of Table 10.11, for two blocks and four blocks. For the case of eight blocks, the experimenter may initially choose three of the necessary seven effects to be confounded. Call these effects X, Y, and Z. Then, the four consequentially confounded effects will be (where the symbolic operation is "mod-2 multiplication")

$$X \cdot Y, \quad X \cdot Z, \quad Y \cdot Z, \quad X \cdot Y \cdot Z$$

For the case of 16 blocks, the experimenter may (initially) choose four of the fifteen necessary effects to be confounded. Call these effects X, Y, Z, and V. Then the eleven consequentially confounded effects will be these four effects multiplied two at a time, three at a time, and four at a time:

$$X \cdot Y, \quad X \cdot Z, \quad X \cdot V, \quad Y \cdot Z, \quad Y \cdot V, \quad Z \cdot V, \quad X \cdot Y \cdot Z,$$
$$X \cdot Y \cdot V, \quad X \cdot Z \cdot V, \quad Y \cdot Z \cdot V, \quad \text{and} \quad X \cdot Y \cdot Z \cdot V$$

You might think there would be little interest in designs that confound as many as, say, seven effects. In practice, such is not the case. For example a 2^6 design has $2^6 - 1 = 63$ effects; confounding seven of 63 effects might well be tolerable.

When the size of an experiment is quite large, it can be somewhat cumbersome (though not really difficult) to explicitly determine and lay out the results of a confounding scheme. Various publications provide lists of designs with confounding, as a function of various parameters. These sources combine full-factorial designs and fractional-factorial designs (the latter being the subject of Chapter 11). Most notable of these are a series of books, one for two-level designs (Statistical Engineering Laboratory, *Fractional Factorial Experiment Designs for Factors at Two Levels,* National Bureau of Standards, Applied Mathematics Series, 48, April 1957), one for three-level designs—covered in Chapter 12 (W. Connor and M. Zelen, *Fractional Factorial Experiment Designs for Factors at Three Levels,* National Bureau of Standards, Applied Mathematics Series, 54, May 1957), and one for experiments that mix two-level and three-level factors (W. Connor and S. Young, *Fractional Factorial Designs for Experiments with Factors at Two and Three Levels,* National Bureau of Standards, Applied Mathematics Series, 56, September 1961). The designs are indexed by what we would expect: the number of factors, number of levels, number of blocks, the fractional replicate (when applicable), and the number of treatment combinations run.

In some cases, it is possible to use software to provide the design of a confounding scheme. The logic and process used are generally similar to those for using software to provide a fractional-factorial design. Hence, we defer such discussion to the next chapter.

10.8 Comment on Calculating Effects

This chapter has dealt with the topic of confounding. We did not include any explicit numerical examples. This is because the numerical calculations that would be utilized in this chapter are identical to those of the previous chapter. That is an important point that bears repeating: effects are calculated in the routine 2^k way; the values of the effects that are confounded are simply not accorded the status of an unbiased estimate.

EXAMPLE 10.3 Pricing a Supplemental Medical/Health Benefit Offer, Revisited

Based on its earlier experience, HealthMark was not willing to assume three-factor interactions were zero; however, it *was* willing to assume that all four-factor interactions and the five-factor interaction were zero. Thus, an experiment was designed in which the 32 treatment combinations were split into two blocks of 16 treatment combinations each, using *ABCDE* as the confounded effect. This meant that each of the treatment combinations was evaluated by only 250 respondents, instead of all 500 respondents seeing each treatment combination. However, the reliability of the results was similar to that of the experiment in Chapter 9, in which 500 respondents evaluated 16 scenarios: $500 \cdot 16 = 250 \cdot 32$ (see the discussion of reliability in section 9.15).

The results mirrored what was found from the previous experiment, which included the core benefits and the first three of the four optional factors of this experiment: the optimal prices were \$9.95 for the core benefits, \$.50 (and 25% off) for the chiropractic channel, \$3 (and 40% off) for the dermatology channel, and \$2 (and 30% off) for the massage channel. For the factor that was new to this experiment, the emergency care channel, the optimal price was the high price, \$3.50 (and 50% off) per adult per month.

Alas, HealthMark has decided to experiment further before arriving at a final configuration for its offering. Also, it wishes to consider other possible types of product designs, such as offering more products (at HealthMark's optimal price) but insisting, for example, that a purchaser choose at least two of the optional channels from the, say, seven offered. It remains to be seen exactly how HealthMark finalizes its offering.

Appendix Detailed Example of Error Reduction through Confounding

The following example illustrates in more detail the reduction of error that designs with confounding provide for estimates of the clean effects. Reconsider the example of a 2^3 design with no replication (to simplify the example) and the necessity to run the eight treatment combinations in two blocks of four. Again suppose, without loss of generality, that we run four treatment combinations on Monday (M) and four on Tuesday (T).

Consider an effect, say A. We know that the estimate of A is

$$(1/4)(-1 + a - b + ab - c + ac - bc + abc)$$

Suppose that $\sigma^2 = 4$ for any and every data value (this discussion loses no generality by assuming that σ^2 is known). Then, we know from Chapter 9 that if the experiment is routinely carried out in *one* complete block, the variance of A, $V_1(A)$, is

$$(1/16)8\sigma^2 = \sigma^2/2$$

However, in our example we must, as noted above, run the experiment with four treatment combinations on M and four treatment combinations on T. As before, X represents the difference in the same yield on M from that on T. Further suppose that we allocate the eight treatment combinations into two sets of four *randomly* (reflecting, perhaps, the lack of knowledge to use a nonrandom design). Of the 70 ways to allocate eight treatment combinations into two sets of four treatment combinations $[70 = 8!/(4! \cdot 4!)]$, there is one way that results in an estimate of A of $(A + X)$:

M: *a, ab, ac, abc*

T: 1, *b, c, bc*

There is also one way that results in an estimate of A of $(A - X)$, the reverse of the allocation above.

There are 36 ways $[4!/(2! \cdot 2!) \cdot 4!/(2! \cdot 2!) = 6 \cdot 6]$ of allocating eight treatment combinations into two sets of four treatment combinations that result in an estimate of A of A (that is, A is clean); the + terms of A and the − terms of A are each two on Monday and two on Tuesday. There are 16 ways $[4!/(3! \cdot 1!) \cdot 4!/(1! \cdot 3!) = 4 \cdot 4]$ of allocating eight treatment combinations into two sets of four treatment combinations that result in an estimate of A of $(A + X/2)$; the + terms of A are three on Monday and one on Tuesday, and the − terms are one on Monday, three on Tuesday. One example would be

M: 1, *a, ab, ac*

T: *b, c, bc, abc*

Similarly, the mirror images of each of these 16 allocations result in an estimate of A of $(A - X/2)$.

Hence, if we randomly allocate four treatment combinations to Monday and four to Tuesday, we have the following probability distribution of estimates of A (ignoring, for the moment, the $\sigma^2 = 4$ alluded to earlier):

Estimate of A	Probability
$A - X$	1/70
$A - X/2$	16/70
A	36/70
$A + X/2$	16/70
$A + X$	1/70
	1

This distribution has a variance, associated with "day of week," $V_{day}(A)$, of

$$V_{day}(A) = (1/70)(-X)^2 + (16/70)(-X/2)^2 + 0 + (16/70)(X/2)^2 + (1/70)X^2$$
$$= (X^2/70)(1 + 16/4 + 0 + 16/4 + 1)$$
$$= X^2/7$$

Now suppose that X, the Monday/Tuesday difference, also equals 4. Just for the sake of the example, we selected X to equal 2σ (if $\sigma^2 = 4$, $\sigma = 2$, and $2\sigma = 4$)—we could have used any value. Then,

$$V_{day}(A) = X^2/7 = 16/7 = 2.29$$

Assuming that the variability associated with day of the week and the variability associated with the other components of error (that exist even if the entire experiment is run on one day) are independent of one another, we have, with the random allocation,

$$V_{total}(A) = V_1(A) + V_{day}(A)$$
$$= 4 + 2.29$$
$$= 6.29$$

Of course, with the proper confounding design in which A is clean, $V_{total}(A)$ would revert to 4. This would represent about a 20% reduction in standard deviation (from 2.51, the square root of 6.29, to 2, the square root of 4). In turn, this would result in about a 20% reduction in the width of (a 20% increase in the precision of) a confidence interval.

This is what we meant earlier in the chapter by the statement that all effects not confounded can be judged with reduced variability—that is, greater reliability.

Exercises

1. Consider running a 2^4 factorial design in which we are examining the impact of four factors on the response rate of a direct-mail campaign. We will mail 160,000 "pieces" in

total, 10,000 pieces under the condition of each of the treatment combinations, and will note the response rate from each 10,000.

The four factors are

A feature of the product
B Positioning for the ad
C Price for the product
D Length of the warranty offered

Suppose that the test mailing must be split among four different time periods (T1, T2, T3, T4) and four different regions of the country (R1, R2, R3, R4). One of the 16 treatment combinations is to be mailed within each of the 16 (T, R) combinations. The resulting design is shown in Table 10EX.1. (For example, 10,000 pieces are mailed with the treatment combination *bcd* in region 1 during time period 2.)

Table 10EX.1 **Direct-Mail Design**

	T1	T2	T3	T4
R1	1	bcd	abd	ac
R2	ab	acd	d	bc
R3	cd	b	abc	ad
R4	abcd	a	c	bd

Suppose that we believe that T and R may each have an effect on the response rate of the offering, but that neither T nor R interact with any of the primary factors, A, B, C, D. We also believe that T and R do not interact with each other. When we routinely estimate the 15 effects among the four primary factors, which of the effects are clean of the "taint" of T and/or R?

2. Suppose that in a 2^4 factorial design it is necessary to construct two blocks of eight treatment combinations each. We decide to confound (only) *ACD*. What are the two blocks?

3. Suppose in exercise 2 that we choose the following two blocks:

   Block 1: 1, *b, ab, abc, ad, bd, abd, bcd*
   Block 2: *a, c, ac, bc, d, cd, acd, abcd*

   Which of the 15 effects are confounded with the block effect?

4. Suppose that in a 2^8 factorial design we must have eight blocks of 32 treatment combinations each. If the following are members of the principal block, find the entire principal block.

   1, *gh, efh, cdh, bdfg, adf*

5. In exercise 4, find the seven confounded effects.

6. Suppose that we are conducting a 2^6 factorial design, with factors A, B, C, D, E, F, and have to run the experiment in four blocks.
   a. How many treatment combinations will be in each block?

   It is decided to confound the effects *ABCE* and *ABDF*.
   b. What (third) effect is also confounded?
   c. What are the four blocks?

7. Consider a 2^5 experiment that must be run in four blocks of eight treatment combinations each. Using the techniques discussed in the chapter, find the four blocks if *ABCD*, *CDE*, and *ABE* are to be confounded.

8. Show for exercise 7 that another approach to finding the four blocks is to first construct two blocks, confounding (only) *ABCD*; then divide each of these two blocks "in half," resulting in four blocks of eight treatment combinations, by confounding *CDE* (and, thereby, *ABE* also).

9. In a 2^8 design run in eight blocks, suppose that *ABCD*, *CDEF*, and *AEGH* are confounded. What are the other four effects that are, as a consequence, also confounded?

10. Suppose that in a 2^8 design to be run in eight blocks, we confound *ABCDE*, *DEFGH*, and *AGH*. Find the other four effects that are, as a consequence, confounded.

11. If you add the number of letters in the seven confounded effects of exercise 10, and you add the number of letters in the seven confounded effects of exercise 9, you get the same answer. If you find the seven confounded effects derived from confounding *A*, *B*, and *ABCDEFGH* (admittedly a silly choice), and add up the number of letters in the seven effects, you again get the same number as in the previous two cases. Does a generalization present itself, and if so, what?
12. Convince yourself that the mod-2 multiplication operation presented in the chapter is the same operation that is called the "exclusive union"; the exclusive union is defined as "the union of the two sets, minus the intersection of the two sets."
13. Consider a 2^5 design in four blocks. If one knows the block effects (a *very* rare situation), does the confounding scheme matter? Discuss.
14. Suppose again that the block effects are not known (the usual case), but that the sign of the effects are known. An example is when one machine is "known" to yield a higher score than another, but the value of the difference is not known with certainty. Would this affect how you design a confounding scheme? Discuss.
15. Suppose that a 2^4 experiment is run during what was ostensibly a homogeneous time period. Later, after the experiment is completed and analyzed, it is discovered that a block effect did exist—morning differed from afternoon because of an unplanned change in machine operator. Assuming that you determined that 1, *a*, *c*, *ac*, *ad*, *bd*, *abd*, and *cd* were run in the morning, which effects are confounded with the block effect?

Note

1. Note that we say "loosely." As seen in Chapter 9, the variance of an estimate is, essentially, inversely proportional to the number of data values making up the estimate, and the standard deviation is inversely proportional to the square root of this number.

CHAPTER 11

Two-Level Fractional-Factorial Designs

EXAMPLE 11.1 Managerial Decision Making at FoodMart Supermarkets

FoodMart Supermarkets decided to undertake a study of how sales are affected by various managerial decision variables—most notably, the amount of shelf space allocated to a product; whether, and to what degree, a product is promoted with signs in the store and circulars distributed in the neighborhood and inserted in local newspapers; the price of the product; and its location in the supermarket (called "location quality").

It would be naive to assume that the impact of these factors is the same for each product, so the products were divided into eight groups with, presumably, the same product attributes: volume category (high versus low), price category (high versus low), and whether the product is seasonal (yes versus no). These three categories are matters of definition within the industry, and do not depend on whether it is literally true for any given supermarket. FoodMart owns chains of supermarkets around the country, so management decided to also see if there was any difference for supermarkets located in the eastern part and western part of the United States. The specific factors under study and their levels are shown in Table 11.1.

Since supermarkets differ in several ways—most notably size, sales volume, and the race and ethnicity of customers—it was necessary to carefully define the levels of the factors, as well as the dependent measure of sales.

One way that the differences among supermarkets were dealt with was to first record sales (in units) for each product studied for a six-week base period, in which everything was routine. Then, treatment combinations were implemented during the next six weeks, the dependent measure being the

Table 11.1 Factors and Levels for FoodMart Study

	Label	Factor	Low Level	High Level
	A	Geography	East	West
Product Attributes	B	Vol. category	Low	High
	C	Price category	Low	High
	D	Seasonality	No	Yes
Managerial Decision Variables	E	Shelf space	Normal	Double
	F	Price	Normal	20% cut
	G	Promotion	None	Normal (if)
	H	Location quality	Normal	Prime

dimensionless ratio of sales in the "treatment six weeks" to sales in the "base six weeks." Note that this approach also neutralizes, to the degree possible, the impact of factors such as holidays, weather, and the like, because the six-week periods were the same for all supermarkets. If one six-week period engendered more or fewer sales than the other, it presumably would do the same for other supermarkets in the study (or at least those in the same city). In addition, as a safety check, sales were also recorded for a third six-week period in case this would prove valuable if anomalies occurred. Although various products were included in the overall study, here we limit discussion to the segment of the study dealing with produce[1] products.

For factor A, geography, east turned out to be the greater Boston area, whereas west was the greater Denver area.

For factor E, shelf space, the low level was called "normal," equal to the level during the six-week base period; the high level was double that amount. These definitions allow for different capacities and demand at different supermarkets.

The same concept holds for the low level of factor F, price, even though in this case the high level of the factor is numerically lower than the low level. As mentioned in Chapter 9, it is usually a "sanity preserver" to define the high level as the higher value if the factor is measured on a numerical scale; however, here we decided to do the opposite, so that the main effect, if significant, would likely be positive. This makes the result easier to explain to people who have not been trained in statistics, but makes no difference from a statistical analysis viewpoint.

Factor G was the amount spent on promotion: ads for the product in flyers inserted into local newspapers, signs put up in the supermarkets, and the like. A particular product may or may not be promoted. In practice, if promotion occurs, it usually is a standard amount; hence, the high level of G is listed as "normal (if)," meaning "if there was promotion."

Certain locations within supermarkets are known to be prime locations, ones that increase sales of virtually any product placed there. One example is the register aisles, where people are more likely to buy on impulse; another is the end of the more popular aisles (often those that include milk or bread), because the product is exposed to more traffic. Most supermarkets are laid out in a similar fashion: the "normal" location for most products is at the low level of factor H.

FoodMart executives thought they could predict the usual main effects of factors E, F, G, and H, at least directionally. But they were not sure that the effects of these factors were the same for all products. For a variety of reasons, including the fact that relationships among products sometimes mean that they are complementary to or substitute for one another, it's not possible to test every product. So it was decided to assume that products having the same "product attributes" will respond similarly to changes in the level of factors E, F, G, and H. Potatoes, for example, are defined as a high-volume, low-price, nonseasonal produce product. Grapes, on the other hand, are defined as a low-volume, high-price, seasonal produce product. These categories—factors B, C, and D—are defined by the trade for its own purposes, although some definitions seem at odds with a layperson's expectation.

Given these category definitions, we could consider interaction effects between these factors, B, C, and D, and the primary factors, E, F, G, and H. These interaction effects would go beyond the question of, for example, how much shelf space is desirable (the answer is that more is better) to address the question of how best to deploy the available space.

To ensure to the degree possible that effects were not confounded, and taking into account some practical considerations (to avoid two different prices for a product in nearby stores, and to address the concern about product complementarity and substitutability), it was decided that each treatment combination would correspond to a single store. (However, in addition to produce products, the actual study included paper goods, meat, and medical-related products, so each supermarket actually had four unrelated products under study.)

A full-factorial experiment would thus involve $2^8 = 256$ stores. This would be very expensive and exceedingly difficult to manage. However, if all eight factors were retained, along with the desired condition that no more than one (say, produce) product is manipulated at each store, what could be done? We return to this example at the end of the chapter.

11.1 Introduction

We continue our examination of two-level factorial designs with discussion of a design technique that is very popular because it allows the study of a relatively large number of factors *without* running all combinations of the factors, as done in our earlier 2^k designs. In Chapter 10, we introduced confounding schemes, where we can run all 2^k treatment combinations, although in two or more blocks. Here we introduce the technique of running a **fractional design,** that is, running only a portion, or fraction, of all the combinations. Of course, whatever fraction of the total number of combinations is going to be run, the specific treatment combinations chosen must be carefully determined. These designs are called "fractional-factorial designs" and are widely used for many types of practical problems.

It often happens, when several factors are chosen for study, that many of the effects (usually most, if not all, of the higher-order interactions[2]) are known (or "comfortably assumed"—little is ever known with 100% certainty) to be zero or negligible, and that the cost of each treatment combination is relatively expensive (perhaps in terms of setting up for the combination—not necessarily per replicate). In such circumstances, we may be able to obtain all of the relevant information (such as main effects, two-factor interactions, and selected other interactions), without sacrifice of reliability, by running only a fraction of the number of treatments required in a complete-factorial experiment.

We have seen before the trade-off of assuming interactions to be zero so that we could have fewer runs in an experiment. First we studied two-factor designs and were able to avoid replication (and hence reduce the number of runs) if we were willing to assume no interaction between the two factors. This trade-off didn't reduce the number of treatment combinations we ran, but did reduce the total number of runs. Then we studied Latin squares and Graeco-Latin squares. In these designs we again assumed no interaction among the factors and *did* gain the benefit of having to run only a subset of the treatment combinations—indeed, only a relatively small portion of the total number of treatment combinations. For example, in a 4×4 Graeco-Latin square with four factors, we need to run only 16 of the $4^4 = 256$ treatment combinations. In this chapter we encounter more refined, or delicate, trade-offs. Unlike in the Latin and Graeco-Latin square designs, we will not need to assume away *all* interaction effects; in general, we can pick and choose which interactions can be sacrificed, based on the beliefs of the experimenter, and what we get in return. This satisfies our intuitive expectation that there ought to be some economy in the study of several factors where many of the effects are "known" to be zero, relative to studying the same number of factors where all effects must be estimated and none can be assumed to be negligible.

11.2 2^{k-p} Designs

Extending the notation of earlier chapters, we designate as **2^{k-p} fractional-factorial designs** the two-level designs where k indicates the number of factors to be studied and 2^{k-p} gives the number of treatment combinations to be used. Thus, a 2^{3-1} design is one with three factors and four treatment combinations. The notation is important; even though 2^{3-1} is numerically equal to 2^2, we would not write it that way. We could also call this design a **half replicate** of a 2^3 design, in that we can write it as $(2^3 \cdot 2^{-1}) = (2^3 \cdot 1/2)$, though nobody really writes it that way. A 2^2 design is one with two factors and four treatment combinations, and is a "full factorial," in that the same logic would give you $2^{2-0} = (2^2 \cdot 2^0) = (2^2 \cdot 1)$. By extension, a 2^{5-2} design would have five factors and eight treatment combinations, and could be called a **quarter replicate** of a 2^5 design.

EXAMPLE 11.2 A 2^{3-1} Design

To illustrate the ideas involved in a two-level fractional design, we begin with an example of a 2^{3-1} design. Specifically, suppose that we run the following four treatment combinations of the eight treatment combinations possible in a 2^3 design: *a, b, c,* and *abc.*

In this explanation, it helps to have available the table of signs for a 2^3 design, as discussed in Chapter 9. Table 11.2 reiterates these, and for easier reference, the above four treatment combinations are highlighted. Recall that the table of signs shows how we combine the yields of the eight experimental treatment combinations to estimate the seven effects.

Here, however, we are running only four treatment combinations: *a, b, c,* and *abc.* We begin, apparently arbitrarily, by writing the expressions for *A* and for *BC*:

$$\begin{aligned} A &= (1/4)[-1 + a - b + ab - c + ac - bc + abc] \\ &= (1/4)[(+a - b - c + abc) + (-1 + ab + ac - bc)] \end{aligned} \tag{11.1}$$

$$\begin{aligned} BC &= (1/4)[+1 + a - b - ab - c - ac + bc + abc] \\ &= (1/4)[(+a - b - c + abc) - (-1 + ab + ac - bc)] \end{aligned} \tag{11.2}$$

Note that neither *A* nor *BC* can be estimated, because four of the eight treatment combinations (1, *ab, ac, bc*) needed for the estimation of each are not available. However, *A* and *BC* can be estimated *in combination.* That is, if we add together equations (11.1) and (11.2), the four missing yields cancel, and

$$A + BC = (1/2)(+a - b - c + abc)$$

Table 11.2 Table of Signs for 2^3 Full-Factorial Design

	A	*B*	*AB*	*C*	*AC*	*BC*	*ABC*
1	−	−	+	−	+	+	−
a	+	−	−	−	−	+	+
b	−	+	−	−	+	−	+
ab	+	+	+	−	−	−	−
c	−	−	+	+	−	−	+
ac	+	−	−	+	+	−	−
bc	−	+	−	+	−	+	−
abc	+	+	+	+	+	+	+

The result is that the *sum* of A and BC can be estimated from only the four available treatment combinations. In a similar manner, we can show that $(B + AC)$ and $(C + AB)$ can also be estimated from only the same four yields:

$$B + AC = (1/2)(-a + b - c + abc)$$
$$C + AB = (1/2)(-a - b + c + abc)$$

Note that ABC cannot be estimated at all, even in sum with another effect.[3] We can summarize our results with a modified table of signs specific to the particular half-replicate design of *a*, *b*, *c*, *abc*, as in Table 11.3.

If all effects are important (that is, no effects can be assumed to be zero or negligible), any apparent benefit from the foregoing is illusory; knowing $(A + BC) = 100$, for example, tells us nothing about A or BC. True, $A = 100 - BC$, but since the value of BC is unknown (not even estimated!), we know nothing specific about the value of A. We might say that these effects are "confounded," but it is more traditional to save that term for effects that are combined with block effects, as discussed in Chapter 10. Instead, it is traditional (for reasons beyond the scope of this discussion) to say that A and BC are "aliased" or are "an alias pair" of effects. **Aliased effects** are knowable in combination (that is, in sum, or as we'll see, in difference), but not individually.

If, as suggested earlier, we know (that is, are willing to assume) that some effects are zero, then we may be able to benefit from the above conception and analysis. Suppose that all interactions can be assumed to be negligible—that is, the three factors each independently affect the dependent variable (whether the effect, on average, is zero or not). Then we can determine all main effects from just these four yields. We could say that we have "studied three factors for the price of two" (that is, with four treatment combinations, not $2 \cdot 2 \cdot 2 = 8$ treatment combinations), or that we have "studied three factors at half-price."

In practice, we do not start with the treatment combinations and then see what alias pairs result. (Doing that takes away the "design" from the phrase "design of experiments"!) Rather, we start by specifying which effects we wish to estimate unambiguously—except, of course, for the ubiquitous "error"—and which effects we are willing to assume are zero or negligible. These determinations are usually specific to the situation, and come from knowledge of the process under investigation. We can then design the experiment—

Table 11.3 **Table of Signs for One 2^{3-1} Design**

	$A + BC$	$B + AC$	$C + AB$
a	+	−	−
b	−	+	−
c	−	−	+
abc	+	+	+

determine which treatment combinations will be run, and if the experiment must be run in more than one block, what the blocks are.

Toward that end, let us go back and examine how the choice of treatment combinations leads to specific alias pairs. That will guide us to perform the design steps in the "reverse," correct order.

Examine Table 11.4, which is the Table 11.2 sign table with an additional bottom row showing the other member of each aliased pair of effects. Note that there is a relationship between the totally lost effect, *ABC*, the highlighted treatment combinations, and ultimately, the alias pairs. The four treatment combinations correspond to rows in which the *ABC* column has all signs the same (in this case, the plus sign). Note next that for each alias pair, the signs in the highlighted rows are the same for each member of the pair, and the signs are opposite in each nonhighlighted row. (It's easiest to observe this for the alias pair *AB* and *C*, since they're in adjacent columns in Table 11.4.) This must be the case to have the cancellation of the treatment combinations not run, given that the alias pairs are summed. Finally, observe the special relationship between the lost effect, *ABC*, and the three alias pairs, $(A + BC)$, $(B + AC)$, and $(C + AB)$. Is it coincidence that each of these pairs seems to add up to—more specifically, *multiply to* (using the mod-2 multiplication introduced in Chapter 10 on confounding)—the lost effect? We'll investigate this relationship a bit later.

What would happen if we selected the set of treatment combinations whose rows correspond to a negative sign in the *ABC* column? These are 1, *ab*, *ac*, and *bc*. In that case, we would have the same alias pairs, but now connected by negative signs (differences, instead of sums). We need to keep track of the signs, shown in Table 11.5, in order to determine the sign of the nonzero effect: there's a big difference between concluding that $A = 37.2$ and concluding that $A = -37.2$.

Now consider running a different experiment, using the treatment combinations corresponding to the positive signs in the effect *AB*. These are 1, *ab*,

Table 11.4 Table of Signs for 2^3 Full-Factorial Design

	A	*B*	*AB*	*C*	*AC*	*BC*	*ABC*
1	−	−	+	−	+	+	−
a	+	−	−	−	−	+	+
b	−	+	−	−	+	−	+
ab	+	+	+	−	−	−	−
c	−	−	+	+	−	−	+
ac	+	−	−	+	+	−	−
bc	−	+	−	+	−	+	−
abc	+	+	+	+	+	+	+
	BC	*AC*	*C*	*AB*	*B*	*A*	

Table 11.5 Table of Signs for Another 2^{3-1} Design

	$A - BC$	$B - AC$	$C - AB$
1	−	−	−
ab	+	+	−
ac	+	−	+
bc	−	+	+

c, and *abc*, highlighted in Table 11.6. Following the earlier discussion, observe that the alias pairs are $(A + B)$, $(C + ABC)$, and $(AC + BC)$; that is, for example, A and B have the same sign for the highlighted rows, and opposite signs for the nonhighlighted rows. No other column except B matches up (or "aliases") with the A column in this manner. And note that AB is not aliased with anything, and thus has been lost.

Ordinarily we would not want a design like this because it makes aliases of two main effects. Without one of them assumed to be zero,[4] we can't say much about either of them. So not all designs are created equal: some are better than others!

A design with three two-level factors has seven effects: three main effects, three two-factor interactions, and one three-way interaction. This is true whether we do a full 2^3, or a 2^{3-1} half replicate, or (making no sense, but numerically it exists) a 2^{3-2} quarter replicate. In general, a 2^{3-1} design loses one of the seven effects completely and gets the other six in three aliased pairs of two effects each. This is consistent with what we would expect from the degrees-of-freedom principle: with four yields we can estimate three independent (orthogonal) effects. So the objective is to select the right set of treatment combinations (that is, design the right experiment) to obtain the effects of interest *cleanly* (obtain each effect by itself, or aliased with an effect or effects

Table 11.6 Table of Signs for 2^3 Full-Factorial Design

	A	B	AB	C	AC	BC	ABC
1	−	−	+	−	+	+	−
a	+	−	−	−	−	+	+
b	−	+	−	−	+	−	+
ab	+	+	+	−	−	−	−
c	−	−	+	+	−	−	+
ac	+	−	−	+	+	−	−
bc	−	+	−	+	−	+	−
abc	+	+	+	+	+	+	+
	B	A		ABC	BC	AC	C

"known" to be zero), with the desired degree of reliability, and by running the fewest possible treatment combinations.

EXAMPLE 11.3 A Four-Factor, Half-Replicate Design

Another half-replicate example will further illustrate the nature of two-level fractional-factorial designs. Consider a 2^{4-1} design in which we use the eight treatment combinations 1, *ab*, *ac*, *bc*, *ad*, *bd*, *cd*, and *abcd*, which are highlighted in Table 11.7.

Note in Table 11.7 that we have again chosen the eight treatment combinations (implying seven degrees of freedom) that correspond to the eight plus signs in the column representing the effect *ABCD*. One effect, *ABCD*, is lost; we will estimate the remaining 14 effects (out of the $2^4 - 1 = 15$) in seven alias pairs: $(A + BCD)$, $(B + ACD)$, $(C + ABD)$, $(D + ABC)$, $(AB + CD)$, $(AC + BD)$, and $(BC + AD)$. Table 11.8 is the table of signs for their calculation. Had we chosen the other eight treatment combinations, the ones with minus signs in the *ABCD* column, we would have had differences instead of sums of the same alias pairs.

In the Table 11.8 design, all main effects are aliased with three-factor interactions, and two-factor interactions are aliased with other two-factor interactions. In many practical applications, all main effects and some two-factor interaction effects are of interest; the remaining effects are assumed to be

Table 11.7 Table of Signs for 2^4 Full-Factorial Design

	A	*B*	*AB*	*C*	*AC*	*BC*	*ABC*	*D*	*AD*	*BD*	*ABD*	*CD*	*ACD*	*BCD*	*ABCD*
1	−	−	+	−	+	+	−	−	+	+	−	+	−	−	+
a	+	−	−	−	−	+	+	−	−	+	+	+	+	−	−
b	−	+	−	−	+	−	+	−	+	−	+	+	−	+	−
ab	+	+	+	−	−	−	−	−	−	−	−	+	+	+	+
c	−	−	+	+	−	−	+	−	+	+	−	−	+	+	−
ac	+	−	−	+	+	−	−	−	−	+	+	−	−	+	+
bc	−	+	−	+	−	+	−	−	+	−	+	−	+	−	+
abc	+	+	+	+	+	+	+	−	−	−	−	−	−	−	−
d	−	−	+	−	+	+	−	+	−	−	+	−	+	+	−
ad	+	−	−	−	−	+	+	+	+	−	−	−	−	+	+
bd	−	+	−	−	+	−	+	+	−	+	−	−	+	−	+
abd	+	+	+	−	−	−	−	+	+	+	+	−	−	−	−
cd	−	−	+	+	−	−	+	+	−	−	+	+	−	−	+
acd	+	−	−	+	+	−	−	+	+	−	−	+	+	−	−
bcd	−	+	−	+	−	+	−	+	−	+	−	+	−	+	−
abcd	+	+	+	+	+	+	+	+	+	+	+	+	+	+	+

Table 11.8 Table of Signs for 2^{4-1} Design Using Listed Combinations

	A + *BCD*	*B* + *ACD*	*AB* + *CD*	*C* + *ABD*	*AC* + *BD*	*BC* + *AD*	*D* + *ABC*
1	−	−	+	−	+	+	−
ab	+	+	+	−	−	−	−
ac	+	−	−	+	+	−	−
bc	−	+	−	+	−	+	−
ad	+	−	−	−	−	+	+
bd	−	+	−	−	+	−	+
cd	−	−	+	+	−	−	+
abcd	+	+	+	+	+	+	+

zero or negligible. We have the liberty to label (assign the letters to) the factors as we wish.

Suppose we're evaluating the impact of four factors on lateness of worker arrival at a work location. The four factors are official start time, traffic congestion, rank in the organization, and weather. Further suppose that there are no interactions involving the rank factor. We could label that factor D. The remaining factors could be labeled A, B, and C. Relative to Table 11.8, then, all main effects and all not-assumed-zero two-way interactions are "clean." (Remember: based on empirical evidence, three-factor and higher-order interaction effects are assumed to be zero in almost all real-world applications.)

Again, in this example, we have evidence of a special relationship between the lost effect (here *ABCD*) and the alias pairings that result. We examine this defining relationship in the next, more complex example. Fortunately, we can design two-level fractional-factorial experiments without resorting to the sign table. We've used the sign table thus far to demonstrate the existence of patterns and relationships that underlie our development of the experimental-design techniques.

EXAMPLE 11.4 A Five-Factor, Half-Replicate Design

Suppose we have a process in which five factors are to be studied, each at two levels. We want to estimate all main effects and some of the lower-order interactions cleanly. We might start by considering a 2^{5-1} fractional-factorial design, for a variety of reasons. For example, perhaps our budget allows a maximum of only 16 treatment combinations. Actually, we should explicitly repeat the philosophy that underlies the popularity of fractional-factorial designs:

> For those who believe that the world is a relatively simple place, higher-order interactions can be announced to be zero in advance of the in-

quiry. Prevailing wisdom is that fractional-factorial designs are almost inevitable in a many-factor situation. For example, it is generally better to study *six* factors with a quarter replicate ($2^{6-2} = 16$) than *four* factors completely ($2^4 = 16$). *Whatever else the world is, it's multifactored.*

As discussed earlier, what counts are the specific alias groups and what is estimated cleanly given which assumptions. Also, the overall number of data values may be an issue, as this affects the power of the *F* tests (discussed later in this chapter). Returning to the five-factor example with factors A, B, C, D, E, the design procedure involves educated trial and error and gets faster with experience. At this point, we know that we must acknowledge the $2^5 - 1 = 31$ effects of a full-factorial design; of these, one effect, which we will specify, will be lost, and the remaining 30 effects will be grouped into *15 alias pairs*. This is a half-replicate design in which we will run 16 treatment combinations, thus having *15 degrees of freedom.* A condition necessary to ensure that we're on the right path (or at least *a* right path) is that we get the same result (in this case 15) from reasoning out the number of alias groups and the degrees of freedom. From that, we know that our selection of the effect to be lost determines how the remaining effects are paired. Let us choose *ABDE* as the lost effect and write this choice in a form called a "defining relation" or a "defining contrast":

$$I = ABDE$$

The designation *I* indicates what in group theory, is called an "identity element," in this case for the mod-2 multiplication operation, as defined in Chapter 10. (We needn't delve further into group theory; however, use of the mod-2 multiplication operation will be very important.) It is traditional to indicate the alias relationship by an = sign, which will eventually be replaced by a + or a − sign, depending, as we saw earlier, on which half we choose to run—the treatment combinations in rows corresponding with plus signs in the *ABDE* column, or those with minus signs in the *ABDE* column.

We now determine the alias pairs. First we pick an effect, say *A*. (It's not mandatory, but usually one begins with the main effects and proceeds alphabetically.) To find the alias pairing that goes with *A*, simply multiply *A* times the defining relation, using mod-2 multiplication. This pairs *A* with *BDE*, since

$$A \cdot ABDE = (A^2BDE) = BDE$$

In essence, we multiply the defining relation equation by *A*. Thus, $A \cdot (I = ABDE)$ yields the pairing of *A* and *BDE* ($A = BDE$). The alias pairs are shown in Table 11.9. In practice, we might not list the higher-order interaction effects explicitly, because we assume they're zero anyway. Instead, we might simply give them a number that corresponds to the number of letters in the effect. For example, *BDE* becomes 3, *ABCDE* becomes 5, and the like. This shorthand simply saves time. Table 11.9 shows higher-order interaction effects instead of the shorthand number only for tutorial purposes.

If our selection of a defining relation ultimately gives an unacceptable pairing, such as a main effect aliased with another main effect, we drop that design and look for a better choice by trying another defining relation.

Table 11.9 **Alias Pairs, $I = ABDE$**

A	BDE	AB	DE	CD	$ABCE$
B	ADE	AC	$BCDE$	CE	$ABCD$
C	$ABCDE$	AD	BE	ABC	CDE
D	ABE	AE	BD	BCD	ACE
E	ABD	BC	$ACDE$	BCE	ACD

At this point, we don't yet have the sign connecting the effects, we have not determined how to select the treatment combinations required to result in the Table 11.9 alias pairs, and we haven't figured out how to compute the effects once we've determined the appropriate treatment combinations and collected the data. Nevertheless, we can assess the merits of the design.

It looks good at first glance: All main effects are aliased with three-factor interactions except *C*, which is aliased with a five-factor interaction. All two-factor interactions involving *C* are aliased with four-factor interactions; two-factor interactions not involving *C* are aliased with other two-factor interactions. This design gives factor C preferential treatment; if we are especially interested in one factor more than the others, we would label it C (or alternatively, exchange the factor in which we have special interest, say B, with C, giving us a defining relation of $I = ACDE$).

Note also that *A*, *B*, *D*, and *E* are treated equally in the $I = ABDE$ defining relation; that is, all four letters are there or all four are not. In this case, they are all present, which means they are treated equally in the alias pairings: for each of these four factors, (1) each main effect is aliased with a three-way interaction; (2) each two-factor interaction involving two of these four factors is aliased with another two-factor interaction involving two of these four factors; and (3) each two-factor interaction involving C and one of these four factors is aliased with a four-factor interaction.

We can simplify the design effort by using such symmetries. For example, when inquiring how main effects are treated by a defining relation, we can figure out how *A* is treated and know that it's the same for *B*, *D*, and *E*.

We can pick a different defining relation. Suppose, knowing that we have to lose one effect entirely, that we opt to lose the highest-order interaction effect, *ABCDE*, and write $I = ABCDE$. In that case, Table 11.10 shows the alias pairs. Here, all factors are treated alike; all main and two-factor interaction effects are clean if all three-factor and higher-order interactions are assumed to be zero.

We could examine other designs, but these two look good. Which we should use depends on the application. The second design, with $I = ABCDE$, is a classic; the earlier design is, as noted, a wise choice if one factor, the one we called C, is overwhelmingly more important than the other factors. For example, we recall an application in which one of the factors was amount ("level") of gold that was optimal in a manufacturing application. Its effect

Table 11.10 **Alias Pairs, $I = ABCDE$**

A	*BCDE*	*AE*	*BCD*
B	*ACDE*	*BC*	*ADE*
C	*ABDE*	*BD*	*ACE*
D	*ABCE*	*BE*	*ACD*
E	*ABCD*	*CD*	*ABE*
AB	*CDE*	*CE*	*ABD*
AC	*BDE*	*DE*	*ABC*
AD	*BCE*		

was considered of overwhelming importance due to the cost of gold relative to the cost of the other factors, and this design was used.

In the current example we'll proceed with the first, $I = ABDE$ design and determine the two possible sets of treatment combinations that give us the alias pairings of Table 11.9. To do this we use the same technique developed in Chapter 10, on confounding, to find the two blocks that allow us to lose only *ABDE.* Table 11.11 shows the two blocks, highlighting the "innovative" treatment combinations as we did in Chapter 10. *Our solution for a half replicate is to run either of these two blocks.*

Although we've identified the two appropriate sets of treatment combinations for $I = ABDE$, we will, of course, use only one. This is one way to distinguish *confounding schemes* from *fractionating.* In the former, we divide the entire set of treatment combinations into blocks, but run all the treatment combinations. In fractionating, we also divide the entire set of treatment combinations into blocks, but run only one of the blocks. (We can also bring *both* concepts, confounding and fractionating, to bear in the same experiment; that is, we can divide a fractional experiment into blocks. This issue is exemplified in exercises 6 and 7 at the end of this chapter.)

Statistically, the blocks are equivalent. In practice, however, one block could be preferable to the other—for example, in Table 11.11, suppose that

Table 11.11 **Blocks for $I = ABDE$ Design**

Principal Block		Second Block	
1	***c***	*a*	*ac*
ab	*abc*	*b*	*bc*
de	*cde*	*ade*	*acde*
abde	*abcde*	*bde*	*bcde*
ad	*acd*	*d*	*cd*
bd	*bcd*	*abd*	*abcd*
ae	*ace*	*e*	*ce*
be	*bce*	*abe*	*abce*

the treatment combination with all factors at high level is a chemically volatile combination; then one can purposely choose the second block above to avoid *abcde*. Or perhaps one set contains less expensive or less time-consuming treatment combinations, or one set includes some treatment combinations that are run anyway during the normal course of business.

In this continuing illustration, we'll arbitrarily pick the second block to run. Presume that we've run the experiment and now have the data. The next step is to analyze the data.

11.3 Yates' Algorithm Revisited

In Chapter 9, we demonstrated how to use Yates' (forward) algorithm to estimate the effects based on the data. We started by arranging the data in standard Yates' order. This made use of all of the 2^k treatment combinations in the full-factorial design. Here, however, we don't have all treatment combinations—we have only half of them—and must therefore first generate a "workable" Yates' order—an order that, when Yates' algorithm is routinely implemented, results in the same systematic approach to estimate the effects as earlier. It turns out that one cannot simply take the treatment combinations of the fractional replicate and put it in Yates' standard order "with gaps." On occasion, doing that will (fortuitously) give us a workable Yates' order, but, in general, it won't.

The technique of creating a workable Yates' order is required in two-level fractional-factorial designs if the effects are to be found "by hand." If software is used, as in Examples 11.9 and 11.10, all the better—we avoid this seemingly elaborate, though easily mastered, technique.

Before undertaking an example, let's get an overview of the technique. To start, we arbitrarily pick one of the letters in our study, and, for the moment, call it "dead." By letters, we mean the lowercase letters corresponding to the factors under study. Continuing our 2^{5-1} study of Example 11.4, we have the five letters *a*, *b*, *c*, *d*, *e*. After picking a dead letter, we then form a standard Yates' order using the remaining (in this case, four) "live" letters. That is, we "make believe" that the remaining four letters are the first four letters of the alphabet (whether they truly are or not). Finally, we append the dead letter as necessary to the (make-believe) standard Yates' order, to create the block chosen to be run. (We could say that the dead letter is "resurrected" as needed.) When we're through, we'll have arranged all of the treatment combinations of our half-replicate design in an order that facilitates routine use of Yates' forward algorithm; that is, we will have arranged a "workable" Yates' order. We now illustrate the technique. Like that of Yates' algorithm in Chapter 9, the technique is more easily illustrated than described.

EXAMPLE 11.5 Creating a Workable Yates' Order

For our example, we'll arbitrarily pick *d* to be the dead letter. In Table 11.12, we first create a standard Yates' order with the remaining live letters, *a*, *b*, *c*, and *e*; note that we have also set up a template for Yates' algorithm.

However, we need to say a few words about the details of the template. First note that the "standard Yates' order" column includes only the "live" letters of *a*, *b*, *c*, and *e*. Next, after the Data column we have written *four* columns

Table 11.12 Template for Yates' Algorithm: 2^{5-1} Design

Standard Yates' Order	Yield (Data)	Yates' Algorithm				8 · Estimated Effects
		1	2	3	4	
1						—
a						*A* − *BDE*
b						*B* − *ADE*
ab						*AB* − *DE*
c						*C* − *ABCDE*
ac						*AC* − *BCDE*
bc						*BC* − *ACDE*
abc						*ABC* − *CDE*
e						*E* − *ABD*
ae						*AE* − *BD*
be						*BE* − *AD*
abe						*ABE* − *D*
ce						*CE* − *ABCD*
ace						*ACE* − *BCD*
bce						*BCE* − *ACD*
abce						*ABCE* − *CD*

for applying Yates' algorithm. Why four columns instead of five? After all, in Chapter 9 we learned that if we have a 2^k, we must apply Yates' algorithm k times! However, this design is a 2^{5-1}, not a 2^5. From the point of view of the algorithm, the number of treatment combinations, 16, looks like a 2^4. Indeed, the rule that applies to using Yates' algorithm in a fractional-factorial design is that Yates' algorithm must be applied $(k - p)$ times; here, $(k - p) = 4$. Finally, note that in the last column, we begin by writing the effect (using capital letters) corresponding to the treatment combination in the standard Yates' order column. This is exactly what we did in Chapter 9. However, here we also write the other member of the alias pair, connected by the appropriate sign. Since we chose to run the second block, a glance at the sign table would reveal that each treatment combination in this second block has a mi-

Table 11.13 **Yates' Algorithm: 2^{5-1} Design, $I = ABDE$, with Appended Letters**

Standard Yates' Order	Workable Yates' Order	Yield (Data)	Yates' Algorithm				8 · Estimated Effects
			1	2	3	4	
1	1 (d)						—
a	a						A – BDE
b	b						B – ADE
ab	ab(d)						AB – DE
c	c(d)						C – ABCDE
ac	ac						AC – BCDE
bc	bc						BC – ACDE
abc	abc(d)						ABC – CDE
e	e						E – ABD
ae	ae(d)						AE – BD
be	be(d)						BE – AD
abe	abe						ABE – D
ce	ce						CE – ABCD
ace	ace(d)						ACE – BCD
bce	bce(d)						BCE – ACD
abce	abce						ABCE – CD

nus sign in the calculation of *ABDE*, so that is the appropriate sign connecting the alias members. Later we discuss a way to determine the appropriate sign without having to resort to the sign table. The estimates produced in the last column would then, as indicated in the column heading, be divided by 8, where, following the Chapter 9 discussion, 8 is half of $2^{5-1} = 16$.

We now append *d* in order to create the second block in a workable Yates' order, as shown in Table 11.13. The first column repeats the standard Yates' order of Table 11.12; the second column indicates the treatment combinations with *d* appended, but only when needed to produce a member of the second block, the block we have chosen to run. This appending process may seem somewhat daunting and arbitrary. It is neither. In fact, what really makes this process simple is that in every case, *only one* of the appending choices works (the other doesn't)! For example, consider the first row. The first column has the treatment combination "1"—if we append *d* to it, then, as noted in the workable Yates' order column, the result is *d*, a member of the second block. If we do not append *d*, we stay with the treatment combination 1—and this is not a member of the second block. The same holds for each row. For the second row, we begin with *a*—it is (already) a member of the second block. If we did append *d*, getting *ad*, we would then *not* have a member of the second block.

The regular pattern of the presence or absence of the appended dead letter is not just coincidental. For example, in Table 11.13, *d* is appended for each set of four treatment combinations in the pattern 1-4, 1-4, then 2-3, 2-3. This is not always the pattern, but there is always *some* identifiable pattern; for example: 1-4, 1-4, then 1-4, 1-4 again. Although this subject of the "appending pattern" has rarely, if ever, been explicitly addressed in the literature, the authors have observed it over many years and over many designed and analyzed two-level fractional-factorial experiments. Such patterns should not be surprising, given similar patterns embedded in the sign tables. Absence of such a pattern would be, in our view, an indicator of error, rather than simply an exception. The prudent designer will look for the presence of such a pattern as an indicator of the way things should be.

11.4 Quarter-Replicate Designs

Suppose that we desire to study five factors and believe that all interactions are zero or negligible; then our goal is to estimate the five main effects. To estimate these main effects with the lowest fractional replicate of a 2^5, a degrees-of-freedom argument would lead us to consider a 2^{5-2} fractional-factorial design. That is, we want to estimate five main effects. Since each factor has two levels, each main effect "consumes," or requires, one degree of freedom (that is, $2 - 1 = 1$), for a total of five degrees of freedom needed. The minimum value of "2^{5-p} less one," the degrees of freedom of a 2^{5-p}, that exceeds five is seven, which implies that $p = 2$. In the following example we'll examine designing a 2^{5-2} to estimate the main effect of five factors: *A, B, C, D, E.* Note that this could be described as "studying five factors for the price of three."

EXAMPLE 11.6 A Five-Factor, Quarter-Replicate Design

We know that a 2^{5-2} has eight treatment combinations. On the basis of our previous discussion in this chapter, we would liken this to a 2^5 full-factorial design confounded into four blocks, with the (major) extra consideration that we'll *run only one block.* As discussed in Chapter 10, when we confound into four blocks, we lose three effects entirely. In the setting of a fractional-factorial design, we write these three effects as a defining relationship containing *I* and the three effects. And again, as in Chapter 10, the three effects chosen to be lost, or members of the defining relationship, are not independent. We can specify two of them, and the third must be the mod-2 product of the first two. The remaining effects ($31 - 3 = 28$ of them) must be shared by the seven degrees of freedom available (again, eight treatment combinations less one is seven), so they will be represented in seven alias groups of four effects each. Table 11.14 shows the alias groups for the defining relation, $I = ABC = BCDE = ADE$. Note that the third effect, *ADE*, is the mod-2 product of the first two effects, as it must be. Recall: if we look at the row with *A*, we derive *BC* by $A \cdot ABC$, *ABCDE* by $A \cdot BCDE$, and *DE* by $A \cdot ADE$, and similarly for the other rows.

Our choice of defining relation was made carefully. It's hard to imagine how we could choose a superior one. All main effects are aliased with interactions (that is, not with other main effects) and the bottom two alias rows each allow the possibility of estimating a two-factor interaction, should there be any material doubt about one or two of them. As noted earlier in the chapter, although the bottom alias row, for example, lists the interactions *BE* and *CD*, we could get *AE* to appear there by exchanging *A* and *B* in all terms of the defining relation.

Table 11.14 Alias Groups for 2^{5-2} $I = ABC = BCDE = ADE$

I	*ABC*	*BCDE*	*ADE*
A	BC	ABCDE	DE
B	AC	CDE	ABDE
C	AB	BDE	ACDE
D	ABCD	BCE	AE
E	ABCE	BCD	AD
BD	ACD	CE	ABE
BE	ACE	CD	ABD

To proceed with our design, we follow the procedure in Chapter 10 to generate the principal block and the three other blocks as shown in Table 11.15 (again, the "innovated" treatment combinations are highlighted).

Once again, we select a block and then run the experiment. Assume we pick the principal block, block 1. Because the design is a 2^{k-2} (that is, 2 to the power $[k - 2]$), setting up a Yates' algorithm template for analysis requires that we select *two* letters as dead; in general, we need to choose p dead letters. We arbitrarily pick b and d. In Table 11.16, we start with a standard Yates' order, append the dead letters as required to generate the treatment combinations of the principal block, and proceed as before.

Specifically, we take the first column with the standard Yates' order in the three "live" letters, a, c, e, and form the second column by appending one of the following as needed to form the principal block: (1) neither b nor d, (2) b alone, (3) d alone, or (4) both b and d. Again, there is only one "right way" the appending can be done—looking at the second row, which begins with a, if we append only b, getting ab, or only d, getting ad, or neither b nor d, keeping a, we would not get a member of the principal block. Only appending both b and d gets us a member of the principal block, abd. Note the pattern of appending: none, bd, b, d, and then the mirror image of that.

Table 11.16 shows our results so far, including the appropriate signs joining the members of each alias group, and shows the five main effects in boldface. As mentioned, were there one or two important two-factor interactions, we would arrange to label our factors such that we can estimate them in the bottom two alias groups/rows. Also as noted before, the signs are important. Notice that B and D carry negative signs. (If every effect of interest were the first listed in every alias group, the particular signs joining alias group members would perhaps be moot. However, that is rarely the case.) We need not go to the sign table to determine the correct signs, though doing so certainly provides the correct answer. Instead, we can use the following "good-to-know" rule:

> A treatment combination with an even number of letters in common with an even-lettered effect gets a plus sign in the calculation of that

Table 11.15 Treatment Combinations, $I = ABC = BCDE = ADE$

(Principal) Block 1	Block 2	Block 3	Block 4
1	a	b	d
abd	bd	ad	ab
bc	abc	c	bcd
acd	cd	abcd	ac
de	ade	bde	e
abe	be	ae	abde
bcde	abcde	cde	bce
ace	ce	abce	acde

Table 11.16 **Yates' Algorithm: 2^{5-2} Design, $I = ABC = BCDE = ADE$**

Standard Yates' Order	Workable Yates' Order	Yield (Data)	Yates' Algorithm 1	2	3	4 · Estimated Effects
1	1					—
a	a(bd)					**A** − BC + ABCDE − DE
c	c(b)					**C** − AB + BDE − ACDE
ac	ac(d)					AC − **B** + ABDE − CDE
e	e(d)					**E** − ABCE + BCD − AD
ae	ae(b)					AE − BCE + ABCD − **D**
ce	ce(bd)					CE − ABE + BD − ACD
ace	ace					ACE − BE + ABD − CD

effect. A treatment combination with an odd number of letters in common with an odd-lettered effect also gets a plus sign in the calculation of that effect. Otherwise, the treatment combination gets a minus sign.

The following table illustrates this rule:

No. of Letters Treatment Combination Has in Common with Effect	No. of Letters in Effect: Even	Odd
Even	+	−
Odd	−	+

For example, in the calculation of *ABCDE*, *abd* has three letters (an odd number) in common with *ABCDE* (*a*, *b*, and *d*), and *ABCDE* contains five letters (an odd number), so it's a case of "odd number of letters in common with an effect that has an odd number of letters," the lower right cell (odd, odd) in the even/odd table. Thus, *abd* has a plus sign in the sign table under *ABCDE*. Another example: In the calculation of *ABCDFG* (for which we don't have a sign table—our sign table in Chapter 9 goes up to only five letters/factors), *abce* has three letters (odd) in common with *ABCDFG* (*a*, *b*, and *c*), and *ABCDFG* has six letters (even). So we have a case of "odd in common with even," the lower left cell (odd, even) in the table, and thus *abce* has a minus sign in the calculation of *ABCDFG*.

The determination of the sign for each alias column is facilitated through the use of this rule. We can examine any one treatment combination in the second column of Table 11.16—that is, the treatment combinations after appending, so that we have those actually run—and apply the rule with that one treatment combination (it would come out the same for any of the treatment combinations) and each term of the defining relation. For example, suppose that we examine *abd*. With *ABC*, *abd* has two (even) in common with an odd (three letters in *ABC*); hence, all second terms of the alias groups are preceded by minus signs. With *BCDE*, *abd* has two (even) in common with an even (four letters in *BCDE*); hence, all third terms of the alias groups are preceded by a plus sign. Similarly, all fourth terms of the alias groups are preceded by a minus sign. (In this example, using the *abd* treatment combinations was for illustration only; using the 1 observation would likely be simpler.)

Most, *but not all*, choices of dead letters allow the easy formation of the treatment combinations of the selected block as shown above, along with the "can't go wrong" unique choice of appending the dead letter(s). We avoided mentioning this earlier so as not to complicate the discussion and because in practice, very often all choices of dead letters *will* work! If we do find that we've unluckily chosen a set of dead letters that does not allow us to form the block we selected to run (that is, there's no way to append the dead letters that gets us the treatment combinations we need), we can simply choose another set of letters and try again. In real-world designs, a randomly chosen set of letters to be dead has a high probability of working, so it's unlikely that we would have to make more than one, or at most two, selections of dead letters. Nevertheless, in the appendix to this chapter, we present a method that guarantees selection of a set of dead letters that works.

11.5 Orthogonality Revisited

We now return to orthogonality, the subject matter of Chapter 5, and examine two-level full- and fractional-factorial designs from the perspective that estimates of the main effects and interaction effects form an orthogonal (and ultimately orthonormal) matrix as defined and discussed in Chapter 5. In particular, we use this connection to develop procedures for significance testing of factorial-design and fractional-factorial-design estimates. To start, recall that the table of signs for a 2^2 design is as follows:

	A	*B*	*AB*
1	−	−	+
a	+	−	−
b	−	+	−
ab	+	+	+

The sign table facilitates the calculation of the effects. Recall from Chapter 9 that the columns of the table of signs are orthogonal; the inner product of any two (different) columns equals zero when the signs are treated as +1's and −1's—which is what, in essence, they are. It's only as a shorthand that we don't write the 1's explicitly. The minus signs mean "multiply the corresponding yields by minus one," the plus signs mean "multiply the corresponding yields by plus one." We then add up the products and divide by two to get the results:

$$A = (-1 + a - b + ab)/2$$
$$B = (-1 - a + b + ab)/2$$
$$AB = (1 - a - b - ab)/2$$

We could have obtained the same result, following the development of Chapter 5, by defining the orthonormal matrix below. The rows of this matrix are the effects, and correspond to the columns of the sign table;[5] in addition, as is typical, the scale factor is explicitly in evidence. As a quick review, note that the sum of each row equals zero, the inner product of any two different rows is zero, and the sum of the squares of each term in a row equals 1. The rows of this orthonormal matrix are *A*, *B*, and *AB*, in that order:

1	*a*	*b*	*ab*
−1/2	1/2	−1/2	1/2
−1/2	−1/2	1/2	1/2
1/2	−1/2	−1/2	1/2

We embed the rows of this matrix in Table 11.17, our calculation table.

Table 11.17 demonstrates that *A*, *B*, and *AB* are orthogonal estimates, and all of the statistical analysis developed in Chapter 5 applies here as well—breaking down sums of squares into orthogonal components and testing their significance using single-degree-of-freedom *F* testing. Thus, as noted earlier, we shall use this framework with the *F* test as our method of testing for significant effects in 2^k and 2^{k-p} designs.

Table 11.17 **Calculation of Z_i from Y_j**

1	*a*	*b*	*ab*	*Z*	Z^2
−1/2	1/2	−1/2	1/2	$Z_1 = -(1/2)\mathbf{1} + (1/2)\mathbf{a} - (1/2)\mathbf{b} + (1/2)\mathbf{ab} = \mathbf{A}$	$Z_1^2 = A^2$
−1/2	−1/2	1/2	1/2	$Z_2 = -(1/2)\mathbf{1} - (1/2)\mathbf{a} + (1/2)\mathbf{b} + (1/2)\mathbf{ab} = \mathbf{B}$	$Z_2^2 = B^2$
1/2	−1/2	−1/2	1/2	$Z_3 = (1/2)\mathbf{1} - (1/2)\mathbf{a} - (1/2)\mathbf{b} + (1/2)\mathbf{ab} = \mathbf{AB}$	$Z_3^2 = (AB)^2$

EXAMPLE 11.7 Boosting Attendance for a Training Seminar

Let's look at an example of significance testing in a 2^{k-p} design. Suppose that we first perform a "routine" one-factor analysis with six replicates of each of four levels of the factor. The yields (that is, the Y_{ij}'s) are the percentage of people from a particular company showing up at an optional after-hours company training seminar. Four motivational techniques are being tested, one technique at each company, 24 companies participating. That is, each technique was tried out at six companies. Assume that the results are as shown in Table 11.18.

Table 11.19 gives the one-way ANOVA table for the results. For $\alpha = .05$, and $df = (3, 20)$, we have a critical value **c** (from the *F* table) of 3.1. We can calculate that $F_{calc} = 5.4$, which exceeds 3.1, so we conclude that the result is significant; that is, we reject the null hypothesis that the (true) column means are equal, and conclude that the column means are *not* all the same—the level of the factor does affect attendance.

Suppose we are now told that promotional technique 1 is a low amount of poster deployment and a low amount of prizes awarded; technique 2 is a high amount of poster deployment and a low amount of prizes awarded; technique 3 is a low amount of poster deployment and a high amount of prizes awarded; and technique 4 is a high amount of both poster deployment and prizes awarded. In essence, techniques 1, 2, 3, and 4 are really (1, *a*, *b*, *ab*) of a 2^2 design in which factor A is amount of poster deployment (low, high), and factor B is amount of prizes awarded (low, high). That is, a 2^2 study is embedded in the one factor, "column." (It's only for clarity of presentation that we use the "just found out" ruse; obviously, the ideal case is that the embedded 2^2 study is part of designing the experiment.)

We can calculate *A*, *B*, and *AB* using the sign table, or by using Yates' forward algorithm (which is what the authors advocate), or we could use the earlier

Table 11.18 Percentage Attendance by Technique

Technique 1	Technique 2	Technique 3	Technique 4
16	28	16	28
22	27	25	30
16	17	16	19
10	20	16	18
18	23	19	24
8	23	16	25
$\bar{Y}_1 = 15$	$\bar{Y}_2 = 23$	$\bar{Y}_3 = 18$	$\bar{Y}_4 = 24$

Table 11.19 ANOVA Table

Source of Variability	SSQ	*df*	MSQ	F_{calc}
Column	324	3	108	5.4
Error	400	20	20	
Total	724	23		

orthogonal matrix. The result is the same, as it must be, regardless of our choice. From the sign table,

$$A = (1/2)(-15 + 23 - 18 + 24) = 7$$
$$B = (1/2)(-15 - 23 + 18 + 24) = 2$$
$$AB = (1/2)(15 - 23 - 18 + 24) = -1$$

Using Yates' algorithm, we would set up the following table:

Treatment Combination	Yield	(1)	(2)	Estimate Divided by 2
1	15	38	80	—
a	23	42	14	7
b	18	8	4	2
ab	24	6	−2	−1

As a practical matter, if doing the arithmetic by hand/calculator, we often "conceptualize" the effects as being derived from the sign table, but in actuality, the calculations are done using Yates' algorithm, as above.

Using the orthogonality formulation first introduced in Chapter 5, and discussed above as applying to a 2^k and 2^{k-p} design, we find the sum of squares for each effect by first finding the square of each effect, and then multiplying each of these by the number of rows, R (here, 6); this yields the following calculation for SSQ:

Z	Z^2	$\text{SSQ} = 6Z^2$
7	49	294
2	4	24
−1	1	6
		Total = 324

Finally, we determine the significance of the estimates by using an "augmented" ANOVA table (Table 11.20), which shows that it is appropriate to use degrees of freedom of (1, 20) for significance testing of the effects. At $\alpha = .05$,

Table 11.20 **ANOVA Table for 2^2 Design**

Source of Variability		SSQ	df	MSQ	F_{calc}
Column		324	3	108	5.4
	A	294	1	294	14.7
	B	24	1	24	1.2
	AB	6	1	6	.3
Error		400	20	20	
Total		724	23		

$c = 4.3$, only A is significant. We would conclude that the amount of poster deployment does have an impact on the percentage of people who show up for the after-hours training session, and that the amount of prizes awarded does not, and that there is no interaction between the amount of poster deployment and amount of prizes awarded.[6] The lack of interaction essentially indicates that the effect of amount of poster deployment is constant across levels of amount of prizes awarded. Apparently, attendance improves if people are reminded adequately of the event, but people are not especially motivated by the potential of winning prizes.

We can extend this approach to fractional-factorial designs. Suppose we have the same numerical data, except that the four techniques are combinations of *three* factors, amount of poster deployment (A), amount of prizes awarded (B), and the third factor, amount of encouragement by the person's supervisor (C). The following table shows the levels used in each technique.

Technique	Factor Level: A	Factor Level: B	Factor Level: C	Treatment Combination
1	Low	Low	High	c
2	High	Low	Low	a
3	Low	High	Low	b
4	High	High	High	abc

This corresponds to a 2^{3-1} design, with $I = ABC$. With the numerical quantities the same as in the 2^2 design, we get the same numerical values in the ANOVA table (Table 11.21) that we did in Table 11.20, with some slight changes in "labels." Note that each of the treatment combinations has an odd

Table 11.21 **ANOVA Table for 2^{3-1} Design**

Source of Variability		SSQ		df		MSQ		F_{calc}	
Column		324		3		108		5.4	
	$A + BC$		294		1		294		14.7
	$B + AC$		24		1		24		1.2
	$C + AB$		6		1		6		.3
Error		400		20		20			
Total		724		23					

number of letters in common with an odd-lettered effect, so the signs connecting the members of the alias pairs are pluses, as noted in the Table 11.21 subcolumn Source of Variability.

Presumably, we would not run an experiment like this one unless we were comfortable with the assumption that all interaction effects are zero. In this case, the only significant effect is, again, for factor A: amount of poster deployment.

EXAMPLE 11.8 Magazine Advertising Study

An ad agency wished to study six two-level factors related to magazine ads, as indicated in Table 11.22. The dependent variable was the number of people in a sample of 500 who recalled a specific ad, after being asked to read a new issue of a magazine in which the ad appeared. Each person was asked to read a magazine already indicated in a screening test to be one that he or she read regularly (that is, that person had read at least three of the last four issues).

The sponsoring ad agency was willing to assume that all interaction effects were zero, except two-factor interactions involving factor F. This assumption about interaction effects was determined by the ad agency personnel (after it was clear to the experimental designer that those deciding the issue truly understood what interaction effects were!). Apparently, each of factors A through E were acknowledged to potentially have different effects for the two different magazines, X and Y. It was decided to use a 2^{6-2} fractional-factorial design with

$$I = ABCD = ABEF = CDEF$$

Table 11.22 **Factors in Advertising Study**

Factor	Levels
A = Size of ad	Eighth of page Quarter of page
B = Color	Black and white Two colors
C = Location	Top of page Bottom of page
D = Rest of page	Mostly ads Mostly article
E = Ad layout	"Cluttered" "More white"
F = Magazine	Magazine X Magazine Y

Note that all main effects and two-factor interaction effects involving factor F are "clean" under our assumptions. We decided to use the principal block, derived using the "innovated" combinations shown in bold:

1, ***ab***, ***cd***, *abcd*, ***ef***, *abef*, *cdef*, *abcdef*, ***ace***, *bce*, *ade*, *bde*, *acf*, *bcf*, *adf*, *bdf*

The data were then collected. The next step was to obtain a workable Yates' order for the quarter-replicate listed above. We needed to choose two dead letters to start the Yates' algorithm process. Suppose that we had decided to choose *a* and *b*; we would find that these two letters compose one of the few choices that will not allow us to form a workable Yates' order. (As mentioned earlier, the appendix to this chapter presents a method guaranteed to produce a set of dead letters that allows generation of a workable Yates' order.) However, that's not really a major problem; once we recognize that a workable Yates' order cannot be generated using *a* and *b* as dead letters, we could simply pick another set of two letters—say *a* and *c*. We would then set up a Yates' order as in Table 11.23. (Note the pattern of appended letters in column 2 of Table 11.23; note also that each treatment combination in the block chosen to be run has "an even number of letters in common with an even-lettered effect," so all the signs in the alias rows are pluses.)

When we add the yields to the third column of the table and perform Yates' algorithm, we have completed the analysis, as shown in Table 11.24. To see which effects are significant, we must note that in this example we do not have replication. However, we can obtain an error sum of squares by "lumping together" the sums of squares corresponding to alias rows in which all terms are assumed to be zero. Since this is a 2^{6-2}, to get the SSQ for each alias row, we take each value in the last column of numbers (Yates' algorithm column 4), divide it by 4 (square root of 16, the number of treatment combinations), and square the result. If we do this, and then add up the four resulting terms corresponding to alias rows with-

Table 11.23 Yates' Algorithm: 2^{6-2} Design, $I = ABCD = ABEF = CDEF$

Standard Yates' Order	Workable Yates' Order	Yield (Data)	Yates' Algorithm				8 · Estimated Effects
			1	2	3	4	
1	1						—
b	b(a)						**B** + 3 + 3 + 5
d	d(c)						**D** + 3 + 5 + 3
bd	bd(ac)						BD + AC + 4 + 4
e	e(ac)						**E** + 5 + 3 + 3
be	be(c)						BE + 4 + **AF** + 4
de	de(a)						DE + 4 + 4 + **CF**
bde	bde						BDE + 3 + 3 + 3
f	f(ac)						**F** + 5 + 3 + 3
bf	bf(c)						**BF** + 4 + AE + 4
df	df(a)						**DF** + 4 + 4 + CE
bdf	bdf						BDF + 3 + 3 + 3
ef	ef						**EF** + 6 + AB + CD
bef	bef(a)						BEF + 5 + **A** + 3
def	def(c)						DEF + 5 + 3 + **C**
bdef	bdef(ac)						BDEF + 4 + AD + BC

out a boldface term, we get a total of $468/16 = 29.25$. (These four rows are rows 4, 8, 12, and 16, which have values in the last column of numbers in Table 11.24 of -11, 17, -7, -3, respectively; when these are divided by 4 and then squared, we get, respectively, $121/16$, $289/16$, $49/16$, and $9/16$. When these are added together, we get 29.25.) Note that we have six main effects and five not-assumed-zero interaction terms, so that *four* remaining rows of the 15 alias rows are available.

This 29.25 is our SSQ_{error}. If we divide it by its four degrees of freedom (we added one degree of freedom four times), we get an MSQ_{error} of $29.25/4 = 7.31$. If we now consider the 11 alias rows with potential nonzero effects, divide each of these values by 4, square each result, and divide this squared result by 7.31, we find that three effects—*B*, *D*, and *F*—have a ratio that exceeds 7.71, the table value for the *F* distribution with (1, 4) degrees of

Table 11.24 Completed Template: Yates' Algorithm for 2^{6-2}, $I = ABCD = ABEF = CDEF$

Standard Yates' Order	Workable Yates' Order	Yield (Data)	Yates' Algorithm 1	2	3	4	8 · Estimated Effects
1	1	152	355	516	1030	3663	—
b	*b(a)*	203	161	514	2633	385	***B*** + 3 + 3 + 5
d	*d(c)*	58	360	1316	192	−801	***D*** + 3 + 5 + 3
bd	*bd(ac)*	103	154	1317	193	−11	*BD* + *AC* + 4 + 4
e	*e(ac)*	157	762	96	−400	−1	***E*** + 5 + 3 + 3
be	*be(c)*	203	554	96	−401	−15	*BE* + 4 + ***AF*** + 4
de	*de(a)*	52	755	104	−2	+3	*DE* + 4 + 4 + ***CF***
bde	*bde*	102	562	89	−9	17	*BDE* + 3 + 3 + 3
f	*f(ac)*	353	51	−194	−2	1603	***F*** + 5 + 3 + 3
bf	*bf(c)*	409	45	−206	1	1	***BF*** + 4 + *AE* + 4
df	*df(a)*	253	46	−208	0	−1	***DF*** + 4 + 4 + *CE*
bdf	*bdf*	301	50	−193	−15	−7	*BDF* + 3 + 3 + 3
ef	*ef*	355	56	−6	−12	3	***EF*** + 6 + *AB* = *CD*
bef	*bef(a)*	400	48	4	15	−15	*BEF* + 5 + ***A*** + 3
def	*def(c)*	259	45	−8	10	27	*DEF* + 5 + 3 + ***C***
bdef	*bdef(ac)*	303	44	−1	7	−3	*BDEF* + 4 + *AD* + *BC*

freedom. For *B*, we get $(385/4)^2 = 9264$, and an F_{calc} of $9264/7.31 = 1266$. For effects *D* and *F*, we get, respectively, F_{calc}'s of 5484 and 21,962; the significant effects are significant also at $\alpha = .0001$.

In terms of interpreting the actual effects, we go back to the effects before squaring them. Going from low to high B (from black-and-white to two-color) *increases* recall; going from low to high D (rest of the page mostly ads to rest of the page mostly article) *decreases* recall. Going from low to high F (magazine X to magazine Y) *increases* recall. It turned out that none of these results were surprising to the ad agency (although it was not clear to others who saw the results why ads were more "recallable" when they appeared in magazine Y instead of magazine X).

EXAMPLE 11.9 Using SPSS for the Ad Study

We now use SPSS to analyze the same magazine advertising results. Table 11.25 shows how the data are input. Note that the first column contains the dependent-variable values, and the next six columns are composed of 1's and 2's, designating, respectively, the low level and the high level for factors A through F. Actually, the data can be input in any order—the dependent variable need not be in the first column; we specify for SPSS which column is which variable/factor.

The output in Table 11.26, as discussed in Chapter 9 for a complete two-level factorial design, does not provide the effect of each factor, but simply the

Table 11.25 SPSS Data Input

152.00	1.00	1.00	1.00	1.00	1.00	1.00
203.00	2.00	2.00	1.00	1.00	1.00	1.00
58.00	1.00	1.00	2.00	2.00	1.00	1.00
103.00	2.00	2.00	2.00	2.00	1.00	1.00
157.00	2.00	1.00	2.00	1.00	2.00	1.00
203.00	1.00	2.00	2.00	1.00	2.00	1.00
52.00	2.00	1.00	1.00	2.00	2.00	1.00
102.00	1.00	2.00	1.00	2.00	2.00	1.00
353.00	2.00	1.00	2.00	1.00	1.00	2.00
409.00	1.00	2.00	2.00	1.00	1.00	2.00
253.00	2.00	1.00	1.00	2.00	1.00	2.00
301.00	1.00	2.00	1.00	2.00	1.00	2.00
355.00	1.00	1.00	1.00	1.00	2.00	2.00
400.00	2.00	2.00	1.00	1.00	2.00	2.00
259.00	1.00	1.00	2.00	2.00	2.00	2.00
303.00	2.00	2.00	2.00	2.00	2.00	2.00

Table 11.26 **SPSS Output**

```
Variable .. RECALL
SIZE
1        UNWGT.      229.87500
2        UNWGT.      228.00000
------------------------------------------------------------
Variable .. RECALL
COLOR
1        UNWGT.      204.87500
2        UNWGT.      253.00000
------------------------------------------------------------
Variable .. RECALL
LOCATION
1        UNWGT.      227.25000
2        UNWGT.      230.62500
------------------------------------------------------------
Variable .. RECALL
REST
1        UNWGT.      279.00000
2        UNWGT.      178.87500
------------------------------------------------------------
Variable .. RECALL
AD LAY
1        UNWGT.      229.00000
2        UNWGT.      228.87500
------------------------------------------------------------
Variable .. RECALL
MAG
1        UNWGT.      128.75000
2        UNWGT.      329.12500
```

mean of each level of each factor. For example, for factor A, the mean is 229.875 for the low level, and 228 for the high level (see the first pair of results in Table 11.26). The grand mean is the mean of these two values, 228.9375. Thus, the A effect is (228 − 228.9375), or equivalently, (228.9375 − 229.875) = −.9375. This corresponds with the A effect of −15 noted in column (4) of Yates' algorithm (Table 11.24); if we divide −15 by 16, the result is −.9375. The output also provides an ANOVA, shown in Table 11.27.

Clearly, SPSS is not oriented toward two-level fractional-factorial experimentation. In fact, the output concerning the means does not acknowledge interaction effects or alias groups. The SSQ terms do include sums of squares associated with interaction terms corresponding to the smallest-order interaction of an alias group (breaking ties in favor of alphabetical order), but give no acknowledgment to alias groups. As in Chapter 9, we now perform the same analysis using the JMP software.

Table 11.27 SPSS ANOVA for Magazine Ad Study

Tests of Significance for RECALL

Source of Variation	SS	DF	MS	F	Sig of F
WITHIN CELLS	.00	0			
SIZE	14.06	1	14.06	.	.
COLOR	9264.06	1	9264.06	.	.
LOCATION	45.56	1	45.56	.	.
REST	40100.06	1	40100.	.	.
ADLAY	.06	1	.06	.	.
MAG	160600.	1	160600.	.	.
SIZE BY COLOR	.56	1	.56	.	.
SIZE BY LOCATION	7.56	1	7.56	.	.
SIZE BY REST	.56	1	.56	.	.
SIZE BY AD LAY	.06	1	.06	.	.
SIZE BY MAG	14.06	1	14.06	.	.
LOCATION BY AD LAY	.06	1	.06	.	.
LOCATION BY MAG	.56	1	.56	.	.
SIZE BY LOCATION BY AD LAY	18.06	1	18.06	.	.
SIZE BY LOCATION BY MAG	3.06	1	3.06	.	.
(Model)	210068.	15	14004.	.	.
(Total)	210068.	15	14004.		

EXAMPLE 11.10 Using JMP for the Ad Study

As noted in Chapter 9, after opening JMP, select "Design Experiment" from the Tables menu. Then choose "2-level Design" (this is true for a complete-factorial or a fractional-factorial design of two-level factors). Set the "Number of Factors" to 6, set "show aliasing up to" to 6 (to get complete alias rows listed), click on "Search for Design," and select "Number of runs = 16," "Type = Fractional Factorial," "Resolution = 4 − some 2-factor interactions," and then click on "Generate Selected Design."

This yields the Table 11.28 output (choosing the option of labeling the factors A, B, C, D, E, and F under "Design Factors and Levels" in the "Design Experiment" menu).

By examining the "Fractional Factorial Structure Factor Confounding Rules" in Table 11.28, and multiplying the first equation by E and the second by F (to produce, in each case, I on the left), we can see that the defining relation *imposed on us* is

$$I = ABEF = ACDF = BCDE$$

Our own defining relation was

$$I = ABCD = ABEF = CDEF$$

Table 11.28 **JMP Output for Magazine Ad Study**

Fractional Factorial Structure
Factor Confounding Rules
E = B*C*D
F = A*C*D

Aliasing Structure
A = B*E*F = C*D*F = A*B*C*D*E
B = A*E*F = C*D*E = A*B*C*D*F
C = A*D*F = B*D*E = A*B*C*E*F
D = A*C*F = B*C*E = A*B*D*E*F
E = A*B*F = B*C*D = A*C*D*E*F
F = A*B*E = A*C*D = B*C*D*E*F
A*B = E*F = A*C*D*E = B*C*D*F
A*C = D*F = A*B*D*E = B*C*E*F
A*D = C*F = A*B*C*E = B*D*E*F
A*E = B*F = A*B*C*D = C*D*E*F
A*F = B*E = C*D = A*B*C*D*E*F
B*C = D*E = A*B*D*F = A*C*E*F
B*D = C*E = A*B*C*F = A*D*E*F
A*B*C = A*D*E = B*D*F = C*E*F
A*B*D = A*C*E = B*C*F = D*E*F

We can make JMP's defining relation into our defining relation by the following transformations.

JMP FACTOR	OUR FACTOR
A	A
B	C
C	E
D	F
E	D
F	B

That is, when we examine JMP's output, we can simply make these changes in factor identifications to transform JMP's output into output for the design *we* want to use.

We also get the JMP spreadsheet in Table 11.29, which represents the principal block of the JMP defining relation noted above.

We relabel the factors in our order, using the transformations listed above, and input the values for Y, as shown in Table 11.30.

Next, as in the Chapter 9 example, select "Analyze" and then click on "Fit Model," and then "Input" Y = Y, "Add" A, B, C, D, E, F, "Cross" AF, BF, CF, DF, EF. Then click on "Run Model," and finally, click on "Parameter Estimation." Table 11.31 shows the result.

As we saw in Chapter 9, the effects are conveyed in "equation form." The intercept (228.9375) is the grand mean, and the "coefficients" noted are the

Table 11.29 **JMP Spreadsheet: Principal Block**

JMP	A	B	C	D	E	F	Y
1	-1	-1	-1	-1	-1	-1	
2	-1	-1	-1	1	1	1	
3	-1	-1	1	-1	1	1	
4	-1	-1	1	1	-1	-1	
5	-1	1	-1	-1	1	-1	
6	-1	1	-1	1	-1	1	
7	-1	1	1	-1	-1	1	
8	-1	1	1	1	1	-1	
9	1	-1	-1	-1	-1	1	
10	1	-1	-1	1	1	-1	
11	1	-1	1	-1	1	-1	
12	1	-1	1	1	-1	1	
13	1	1	-1	-1	1	1	
14	1	1	-1	1	-1	-1	
15	1	1	1	-1	-1	-1	
16	1	1	1	1	1	1	

same values derived from the SPSS output, where each value (for example, −.9375 for *A*) multiplies by 1 for high level of the factor, or by −1 for low level of the factor. Again, recall the transformations; D, E, and F in Table 11.31 are our factors F, D, and B, respectively.

Note that JMP not only provides an analysis of the experiment but also aids in designing the experiment in the first place, providing a defining relation and alias groups. Yet, it can be dangerous to have the proverbial "black box" provide the treatment combinations, collect and analyze the data, and interpret the results, without our first noting (and understanding the meaning of) the defining relation, alias groups, and other material in this chapter.

Table 11.30 JMP Spreadsheet with Factors Reordered and Yields Entered

OUR:	**A**	**B**	**C**	**D**	**E**	**F**	
JMP:	**A**	**F**	**B**	**E**	**C**	**D**	Y
1	-1	-1	-1	-1	-1	-1	152
2	-1	1	-1	1	-1	1	301
3	-1	1	-1	1	1	-1	102
4	-1	-1	-1	-1	1	1	355
5	-1	-1	1	1	-1	-1	58
6	-1	1	1	-1	-1	1	409
7	-1	1	1	-1	1	-1	203
8	-1	-1	1	1	1	1	259
9	1	1	-1	-1	-1	-1	203
10	1	-1	-1	1	-1	1	253
11	1	-1	-1	1	1	-1	52
12	1	1	-1	-1	1	1	400
13	1	1	1	1	-1	-1	103
14	1	-1	1	-1	-1	1	353
15	1	-1	1	-1	1	-1	157
16	1	1	1	1	1	1	303

Table 11.31 JMP Analysis of Magazine Advertizing Study

Response: Y

Summary of Fit

RSquare	0.999861
RSquare Adj	0.999478
Root Mean Square Error	2.704163
Mean of Response	228.9375
Observations (or Sum Wgts)	16

Parameter Estimates

Term	Estimate	Std Error	t Ratio	Prob>|t|
Intercept	228.9375	0.676041	338.64	<.0001
A	0.9375	0.676041	1.39	0.2378
B	-1.6875	0.676041	-2.50	0.0670
C	0.0625	0.676041	0.09	0.9308
D	-100.1875	0.676041	-148.2	<.0001
E	50.0625	0.676041	74.05	<.0001
F	-24.0625	0.676041	-35.59	<.0001
A·D	-0.9375	0.676041	-1.39	0.2378
B·D	0.1875	0.676041	0.28	0.7953
C·D	0.1875	0.676041	0.28	0.7953
D·E	-0.0625	0.676041	-0.09	0.9308
D·F	0.0625	0.676041	0.09	0.9308

Effect Test

Source	Nparm	DF	Sum of Squares	F Ratio	Prob>F
A	1	1	14.06	1.9231	0.2378
B	1	1	45.56	6.2308	0.0670
C	1	1	0.06	0.0085	0.9308
D	1	1	160600.56	21962.47	<.0001
E	1	1	40100.06	5483.769	<.0001
F	1	1	9264.06	1266.88	<.0001
A·D	1	1	14.06	1.9231	0.2378
B·D	1	1	0.56	0.0769	0.7953
C·D	1	1	0.56	0.0769	0.7953
D·E	1	1	0.06	0.0085	0.9308
D·F	1	1	0.06	0.0085	0.9308

11.6 Power and Minimum Detectable Effects in 2^{k-p} Designs

In order to test whether an effect, A, equals zero or not, that is,

$$H_0\colon A = 0 \qquad \text{versus} \qquad H_1\colon A \neq 0$$

we can use a t test, or, as we illustrated for the magazine advertising example, an F test. We can also examine the power of the test. We considered power in Chapter 3, and viewed the issue in two ways. One way was, essentially, to input the size of the experiment, the value of α (probability of Type I error), and the assumed true value of the effect at which power is to be determined (measured by ϕ), and use power tables to determine the power of the test.

The other way was to input the α and power values desired, along with the number of columns in the study, and the value of the effect at which to determine power (here, measured by Δ/σ), and use tables to determine the sample size (number of replicates per column for a one-factor experiment) needed to achieve the desired α and power/effect size combination.

In the 2^{k-p} setting, we often wish to frame the issue of power in a way that is slightly different from the two ways above. We specify the values of k, p, and r (the number of replicates) in a particular way (described below), along with the values of α and power (recall, power is the complement of β). We then use tables to find the minimum effect size that can be detected at the power specified. This is called the **minimum detectable effect (MDE).** The MDE value is necessarily specified in "σ units"; hence, a table value of 2.2 refers to 2.2σ.

The discussion in this section is based on an article in the *Journal of Quality Technology* ("Minimum Detectable Effects for 2^{k-p} Experimental Plans," by Richard O. Lynch, January 1993, 25(1):12–17). The MDE tables are also from this article.

Specifically, to find the MDE from the tables provided, we need to supply the following: α, β, the number of factors (k), and the number of runs ($r \cdot 2^{k-p}$). Since the MDE also depends on the degrees of freedom in the error term, and this value in turn depends on which higher-order interactions are assumed to equal zero, we must make assumptions as to these degrees-of-freedom values. In general, we assume that all (and only) three-factor and higher-order interactions are zero, unless some two-factor interactions must also be assumed to be zero to allow clean determination of main effects. This reflects Lynch's apparent view that, without specific affirmative reasons, a two-factor interaction should not be assumed to be zero. The authors of this text agree with that view. However, as we know, not all two-level fractional-factorial designs allow all main effects and two-factor interactions to be placed in separate alias rows; in these cases, the number of degrees of freedom of the error term is assumed to be the number of alias rows composed solely of three-factor and higher-order interactions.

Following these rules, Table 11.32 (similar to the table in Lynch's article) indicates the degrees of freedom that can be placed in the error term. For ex-

Table 11.32 Degrees of Freedom for Error Term

	Number of Factors, *k*								
N	3	4	5	6	7	8	9	10	11
4	0								
8	1	0	0	0	0				
16	9	5	0	2	1	0	0	0	0
32	25	21	16	10	6	3	1	0	5
64	57	53	48	42	35	27	21	14	8
128	121	117	112	106	99	91	82	72	61

Note. N = number of data points. From "Minimum Detectable Effects for 2^{k-p} Experimental Plans," by R. O. Lynch, 1993, *Journal of Quality Technology, 25*:1 p.13. Adapted with permission.

Table 11.33 Minimum Detectable Effect (σ units), $\alpha = .01$

No. of Runs	β	3	4	5	6	7	8	9	10	11
		Number of Factors, k								
4	.01	11.8								
	.05	9.7								
	.1	8.7								
	.15	8.0								
	.25	7.0								
	.5	5.2								
8	.01	8.3	8.3	8.3	8.3	8.3				
	.05	6.9	6.9	6.9	6.9	6.9				
	.1	6.1	6.1	6.1	6.1	6.1				
	.15	5.6	5.6	5.6	5.6	5.6				
	.25	4.9	4.9	4.9	4.9	4.9				
	.5	3.7	3.7	3.7	3.7	3.7				
16	.01	3.1	3.9	5.9	5.9	5.9	5.9	5.9	5.9	5.9
	.05	2.6	3.3	4.9	4.9	4.9	4.9	4.9	4.9	4.9
	.1	2.4	3.0	4.3	4.3	4.3	4.3	4.3	4.3	4.3
	.15	2.2	2.7	4.0	4.0	4.0	4.0	4.0	4.0	4.0
	.25	2.0	2.4	3.5	3.5	3.5	3.5	3.5	3.5	3.5
	.5	1.6	1.9	2.6	2.6	2.6	2.6	2.6	2.6	2.6
32	.01	1.9	1.9	1.9	2.1	2.5	4.2	4.2	4.2	2.7
	.05	1.6	1.6	1.7	1.8	2.1	3.4	3.4	3.4	2.3
	.1	1.5	1.5	1.5	1.6	1.9	3.1	3.1	3.1	2.1
	.15	1.4	1.4	1.4	1.5	1.8	2.8	2.8	2.8	1.9
	.25	1.2	1.2	1.3	1.4	1.6	2.5	2.5	2.5	1.7
	.5	.98	.99	1.0	1.1	1.2	1.8	1.8	1.8	1.3
64	.01	1.3	1.3	1.3	1.3	1.3	1.3	1.3	1.4	1.6
	.05	1.1	1.1	1.1	1.1	1.1	1.1	1.1	1.2	1.4
	.1	.99	1.0	1.0	1.0	1.0	1.0	1.0	1.1	1.2
	.15	.93	.93	.94	.94	.95	.96	.98	1.0	1.1
	.25	.84	.84	.84	.85	.85	.87	.88	.92	1.0
	.5	.66	.66	.67	.67	.68	.69	.70	.73	.81
128	.01	.88	.88	.88	.88	.88	.88	.88	.89	.89
	.05	.76	.76	.76	.76	.76	.76	.76	.76	.77
	.1	.69	.69	.69	.69	.69	.69	.70	.70	.70
	.15	.65	.65	.65	.65	.65	.65	.65	.65	.66
	.25	.58	.58	.58	.58	.58	.59	.59	.59	.59
	.5	.46	.46	.46	.46	.46	.46	.48	.47	.47

Results inside shaded regions were obtained assuming a minimum of 3 degrees of freedom. Results outside of the shading were obtained using the degree of freedom counts listed in Table 11.32.

Note. N = number of data points. From "Minimum Detectable Effects for 2^{k-p} Experimental Plans," by R. O. Lynch, 1993, *Journal of Quality Technology, 25*:1 p.14. Adapted with permission.

Table 11.34 Minimum Detectable Effect (σ units), $\alpha = .05$

No. of Runs	β	3	4	5	6	7	8	9	10	11
		Number of Factors, k								
4	.01	6.9								
	.05	5.7								
	.1	5.0								
	.15	4.6								
	.25	4.0								
	.5	2.9								
8	.01	4.9	4.9	4.9	4.9	4.9				
	.05	4.0	4.0	4.0	4.0	4.0				
	.1	3.5	3.5	3.5	3.5	3.5				
	.15	3.2	3.2	3.2	3.2	3.2				
	.25	2.8	2.8	2.8	2.8	2.8				
	.5	2.0	2.0	2.0	2.0	2.0				
16	.01	2.4	2.7	3.4	3.4	3.4	3.4	3.4	3.4	3.4
	.05	2.0	2.3	2.8	2.8	2.8	2.8	2.8	2.8	2.8
	.1	1.8	2.0	2.5	2.5	2.5	2.5	2.5	2.5	2.5
	.15	1.7	1.9	2.3	2.3	2.3	2.3	2.3	2.3	2.3
	.25	1.5	1.6	2.0	2.0	2.0	2.0	2.0	2.0	2.0
	.5	1.1	1.2	1.4	1.4	1.4	1.4	1.4	1.4	1.4
32	.01	1.6	1.6	1.6	1.7	1.8	2.4	2.4	2.4	1.9
	.05	1.3	1.3	1.4	1.4	1.5	2.0	2.0	2.0	1.6
	.1	1.2	1.2	1.2	1.3	1.4	1.8	1.8	1.8	1.4
	.15	1.1	1.1	1.1	1.2	1.3	1.6	1.6	1.6	1.3
	.25	.97	.98	.99	1.0	1.1	1.4	1.4	1.4	1.2
	.5	.72	.73	.74	.77	.83	1.0	1.0	1.0	.86
64	.01	1.1	1.1	1.1	1.1	1.1	1.1	1.1	1.2	1.2
	.05	.92	.92	.92	.92	.93	.94	.95	.97	1.0
	.1	.82	.83	.83	.83	.83	.84	.85	.87	.93
	.15	.76	.76	.76	.77	.77	.78	.79	.81	.86
	.25	.67	.67	.67	.67	.68	.68	.69	.71	.75
	.5	.50	.50	.50	.50	.50	.51	.51	.53	.56
128	.01	.76	.76	.76	.76	.77	.77	.77	.77	.77
	.05	.64	.64	.64	.64	.64	.64	.64	.65	.65
	.1	.58	.58	.58	.58	.58	.58	.58	.58	.58
	.15	.53	.53	.53	.53	.54	.54	.54	.54	.54
	.25	.47	.47	.47	.47	.47	.47	.47	.47	.47
	.5	.35	.35	.35	.35	.35	.35	.35	.35	.35

Results inside shaded regions were obtained assuming a minimum of 3 degrees of freedom. Results outside of the shading were obtained using the degree of freedom counts listed in Table 11.32.

Note. N = number of data points. From "Minimum Detectable Effects for 2^{k-p} Experimental Plans," by R. O. Lynch, 1993, *Journal of Quality Technology, 25*:1 p.14. Adapted with permission.

ample, when there are three factors and eight data points ($k = 3$, $N = 8$), we have a complete factorial design and one degree of freedom for error—the *ABC* effect. For $k = 5$ and $N = 32$, we have either a complete 2^5 design or a twice-replicated 2^{5-1} design; if the former, there are 16 effects of three-factor or higher-order interactions; if the latter, there are only the 16 degrees of freedom derived from replication, since all alias rows contain either a main effect or a two-factor interaction effect (with $I = ABCDE$).

Table 11.35 Minimum Detectable Effect (σ units), $\alpha = 0.10$

No. of Runs	β	Number of Factors, k								
		3	4	5	6	7	8	9	10	11
4	.01	5.5								
	.05	4.5								
	.1	3.9								
	.15	3.6								
	.25	3.1								
	.5	2.1								
8	.01	3.9	3.9	3.9	3.9	3.9				
	.05	3.2	3.2	3.2	3.2	3.2				
	.1	2.8	2.8	2.8	2.8	2.8				
	.15	2.5	2.5	2.5	2.5	2.5				
	.25	2.2	2.2	2.2	2.2	2.2				
	.5	1.5	1.5	1.5	1.5	1.5				
16	.01	2.2	2.3	2.7	2.7	2.7	2.7	2.7	2.7	2.7
	.05	1.8	1.9	2.2	2.2	2.2	2.2	2.2	2.2	2.2
	.1	1.6	1.7	2.0	2.0	2.0	2.0	2.0	2.0	2.0
	.15	1.5	1.6	1.8	1.8	1.8	1.8	1.8	1.8	1.8
	.25	1.3	1.4	1.5	1.5	1.5	1.5	1.5	1.5	1.5
	.5	.89	.95	1.1	1.1	1.1	1.1	1.1	1.1	1.1
32	.01	1.4	1.5	1.5	1.5	1.6	1.9	1.9	1.9	1.7
	.05	1.2	1.2	1.2	1.3	1.3	1.6	1.6	1.6	1.4
	.1	1.1	1.1	1.1	1.1	1.2	1.4	1.4	1.4	1.2
	.15	.97	.98	.99	1.0	1.1	1.3	1.3	1.3	1.1
	.25	.84	.85	.86	.88	.93	1.1	1.1	1.1	.96
	.5	.60	.60	.61	.62	.66	.75	.75	.75	.67
64	.01	1.0	1.0	1.0	1.0	1.0	1.0	1.0	1.0	1.2
	.05	.83	.83	.83	.84	.84	.84	.85	.87	.90
	.1	.74	.74	.74	.74	.75	.75	.76	.77	.80
	.15	.68	.68	.68	.68	.68	.69	.69	.71	.74
	.25	.59	.59	.59	.59	.59	.59	.60	.61	.64
	.5	.42	.42	.42	.42	.42	.42	.42	.45	.45
128	.01	.71	.71	.71	.71	.71	.71	.71	.71	.71
	.05	.58	.58	.59	.59	.59	.59	.59	.59	.59
	.1	.52	.52	.52	.52	.52	.52	.52	.52	.52
	.15	.48	.48	.48	.48	.48	.48	.48	.48	.48
	.25	.41	.41	.41	.41	.41	.41	.41	.41	.41
	.5	.29	.29	.29	.29	.29	.29	.29	.29	.29

Results inside shaded regions were obtained assuming a minimum of 3 degrees of freedom. Results outside of the shading were obtained using the degree of freedom counts listed in Table 11.32.

Note. N = number of data points. From "Minimum Detectable Effects for 2^{k-p} Experimental Plans," by R. O. Lynch, 1993, *Journal of Quality Technology, 25*:1 p.15. Adapted with permission.

Table 11.36 Minimum Detectable Effect (σ units), $\alpha = 0.15$

No. of Runs	β	Number of Factors, k								
		3	4	5	6	7	8	9	10	11
4	.01	4.8								
	.05	3.9								
	.1	3.4								
	.15	3.1								
	.25	2.6								
	.5	1.7								
8	.01	3.4	3.4	3.4	3.4	3.4				
	.05	2.7	2.7	2.7	2.7	2.7				
	.1	2.4	2.4	2.4	2.4	2.4				
	.15	2.2	2.2	2.2	2.2	2.2				
	.25	1.8	1.8	1.8	1.8	1.8				
	.5	1.2	1.2	1.2	1.2	1.2				
16	.01	2.0	2.1	2.4	2.4	2.4	2.4	2.4	2.4	2.4
	.05	1.6	1.7	1.9	1.9	1.9	1.9	1.9	1.9	1.9
	.1	1.4	1.5	1.7	1.7	1.7	1.7	1.7	1.7	1.7
	.15	1.3	1.4	1.5	1.5	1.5	1.5	1.5	1.5	1.5
	.25	1.1	1.2	1.3	1.3	1.3	1.3	1.3	1.3	1.3
	.5	.76	.80	.87	.87	.87	.87	.87	.87	.87
32	.01	1.4	1.4	1.4	1.4	1.5	1.7	1.7	1.7	1.5
	.05	1.1	1.1	1.1	1.2	1.2	1.4	1.4	1.4	1.2
	.1	.98	.99	1.0	1.0	1.1	1.2	1.2	1.2	1.1
	.15	.89	.90	.91	.93	.96	1.1	1.1	1.1	.99
	.25	.76	.77	.77	.79	.82	.92	.92	.92	.84
	.5	.52	.52	.52	.54	.56	.62	.62	.62	.57
64	.01	.95	.95	.95	.95	.96	.96	.97	.98	1.0
	.05	.78	.78	.78	.78	.78	.79	.79	.80	.83
	.1	.69	.69	.69	.69	.69	.69	.70	.71	.73
	.15	.62	.63	.63	.63	.63	.63	.63	.64	.66
	.25	.53	.53	.53	.54	.54	.54	.54	.55	.57
	.5	.36	.36	.36	.36	.36	.37	.37	.37	.38
128	.01	.67	.67	.67	.67	.67	.67	.67	.67	.67
	.05	.55	.55	.55	.55	.55	.55	.55	.55	.55
	.1	.48	.48	.48	.48	.48	.48	.48	.48	.49
	.15	.44	.44	.44	.44	.44	.44	.44	.44	.44
	.25	.38	.38	.38	.38	.38	.38	.38	.38	.38
	.5	.25	.25	.25	.25	.25	.26	.26	.26	.26

Results inside shaded regions were obtained assuming a minimum of 3 degrees of freedom. Results outside of the shading were obtained using the degree of freedom counts listed in Table 11.32.

Note. N = number of data points. From "Minimum Detectable Effects for 2^{k-p} Experimental Plans," by R. O. Lynch, 1993, *Journal of Quality Technology, 25*:1 p.15. Adapted with permission.

However, note that several cells in Table 11.32 have no degrees of freedom available for the error term; they represent what are called saturated designs. Some designs are nearly saturated and have a low number of degrees of freedom available for the error term. In determining the MDE values in subsequent tables, Lynch assumed that ("somehow") three degrees of freedom would be available for estimating the error term for designs with a degrees-of-freedom value in Table 11.32 of less than three.

Tables 11.33 through 11.36 reproduce the MDE tables in the Lynch article; each table is for a different value of α: .01, .05, .10, and .15. Within each table, the columns represent the number of factors; the rows represent the number of data points, N, and within N, the value of β. The body of the table presents the MDE values.

To illustrate the use of Tables 11.33–36, suppose that we are considering a 2^{7-3} design without replication. Then $k = 7$ and $N = 16$. If we wish to have $\alpha = .01$ and power of .95 (that is, $\beta = .05$), the MDE is found from the tables to be 4.9, meaning 4.9σ. This is likely to be too large to be acceptable, in that the desired power is achieved only with an extremely large effect size. Suppose, then, that we consider a 2^{7-2} design. With the same α and power demands and $k = 7$, and now $N = 32$, the table gives an MDE of 2.1σ. This *might* also be considered too large to be acceptable. However, if we are willing to reduce our Type I error demands to $\alpha = .05$, instead of .01, retaining a power of .95, the MDE would then be only 1.5σ. Or with $\alpha = .05$ and a change to a power of .75, the MDE would be 1.1σ.

An alternative to inputting values of α and β to determine MDE using the tables, as above, is to use them to determine the power achieved for a given α value and MDE. For example, for a 2^{5-1} design and $\alpha = .05$, what is the probability (power) of detecting an effect of size 2σ? From the tables, $\beta = .25$, and the power is .75. What about for an effect size of 1.5σ? From the tables, β is between .25 and .5, nearer to .5; consequently, power is between .5 and .75, nearer to .5. (See the circled entries in Table 11.34 for the last two questions.)

EXAMPLE 11.11 Managerial Decision Making at FoodMart Supermarkets, Revisited

FoodMart has eight factors under study, as described at the beginning of this chapter. For convenience, Table 11.1 is repeated as Table 11.37, listing the factors and their levels.

As noted earlier, a complete 2^8 factorial design would require 256 treatment combinations, which is not practical from either an expense or a managerial point of view. After some discussion, FoodMart decided that the maximum number of supermarkets that it would allow to participate in the study was 64; this suggested a 2^{8-2} design—which indeed is what took place.

As when designing any fractional experiment, we needed to prioritize the effects. From discussion with FoodMart executives, we learned that the most important main effects were those for the managerial decision variables, E, F, G, and H. Of virtually no importance were the product attribute main effects, B, C, and D; after all, they are what they are and involve nothing to be decided, never mind optimized (for example, how beneficial is it to "discover" that products in the high-volume category sell more than products in the low-volume category?). The main effect of geography, A, was deemed of less interest than the main effects of E, F, G, and H.

Table 11.37 **Factors and Levels for FoodMart Study**

	Label	Factor	Low Level	High Level
	A	Geography	East	West
Product Attributes	B	Vol. category	Low	High
	C	Price category	Low	High
	D	Seasonality	No	Yes
Managerial Decision Variables	E	Shelf space	Normal	Double
	F	Price	Normal	20% cut
	G	Promotion	None	Normal (if)
	H	Location quality	Normal	Prime

FoodMart did suggest that factors E, F, G, and H might have a different impact in the two different areas of the country. Thus it could be important to cleanly estimate interaction effects of A with E, F, G, and H. However, these two-factor interactions were not as important as two other sets of two-factor interactions. The most important two-factor interactions were those between the managerial decision variables: EF, EG, EH, FG, FH, and GH. That is, to make superior allocations of limited resources, FoodMart would need to know the answer to questions such as these: "If doubling shelf space generates (an average of) 10% more sales, and the normal promotion of the product generates 13% more sales, when you do *both,* do you get about (10% + 13%) = 23% more sales? Or do you get less than 23% more sales—implying a negative EG interaction effect? Or do you get more than 23% additional sales—implying a positive EG interaction (or synergy) effect?" Presumably, if there's a negative interaction effect, it's probably better to give the shelf space boost and the promotion boost to different products. With a positive interaction effect, it's likely better to give both boosts to the same product.

Also very important were the 12 two-factor interactions of the form XY, between the product attributes ($X = B$, C, and D) and the managerial decision variables ($Y = E$, F, G, and H). These interaction effects will tell us whether certain classes of products gain differentially from changing the levels of E, F, G, and H from low to high, and have the obvious benefit of enhancing the decisions about which products should be allocated which resources.

It was concluded that EFG, EFH, EGH, and FGH, the three-factor interactions involving the decision variables, should not automatically be assumed to be zero. It was determined that the remaining three-factor interactions and all higher-order interactions could be assumed to be zero.

The design chosen was a 2^{8-2} (quarter-replicate) with the defining relation

$$I = BCD = ABEFGH = ACDEFGH$$

Notice that E, F, G, and H are all treated similarly—they're either all present or all absent in each term above (actually, A is also treated identically; however, since A is clearly of less importance than the former factors, we list it separately).

As noted earlier, we can benefit from this fact in our evaluation of the design—whatever is true for one of the factors (letters) is true for the others. *C* and *D* are also treated similarly—analysis of one of them will suffice for the other.

The alias groups are summarized in Table 11.38; note that (1) we consider representative effects to stand for all members of a group that are treated alike, and (2) as mentioned in Example 11.4, we use a shorthand system of showing the number of letters in the aliased effects, rather than writing out the effects explicitly.

An evaluation of Table 11.38 shows that all desired effects (in the first column) are clean, given what effects we are assuming to be zero or negligible. Indeed, few of the effects of interest are aliased even with any three-factor interactions, and those three-factor interactions are believed to be zero. It appears that this design is capable of providing clean estimates of all effects of interest.

The results of this study do not, in themselves, answer all the questions necessary for supermarket managers to make optimal decisions. Not all revenue from additional sales is profit, so a profitability analysis would subsequently be needed to answer questions such as these: Does a promotion that costs $*X* generate enough additional sales to be warranted? Or (because different products have different margins), do high-volume products and low-volume products benefit differently from doubling the shelf space, and if so, what is the difference?

A small sample of the results found for produce products:

- The main effect *E* is significantly positive. This is no big surprise—everything else equal, doubling the shelf space significantly increases sales of that product.
- The *BE* interaction effect is significantly negative—low-volume products benefit more than high-volume products from a doubling of the amount of shelf space. An example of such an interaction would be $E = +44\%$ averaged over both volume categories, $E = +57\%$ for low-volume products, and $E = +31\%$ for high-volume products.

Table 11.38 Alias Groups for FoodMart Study

Effects	*BCD*	*ABEFGH*	*ACDEFGH*
Main Effects			
A	4	5	6
E, F, G, H	4	5	6
Two-Factor Interactions			
EF, EG, . . .	5	4	5
BE, BF, . . .	3	4	7
CE, CF, DE, DF . . .	3	6	5
AE, AF, . . .	5	4	5
Three-Factor Interactions			
EFG, EFH, . . .	6	3	4

- The CE interaction effect is significantly positive—high-price (category) products benefit more from a doubling of the shelf space than do low-price (category) products.
- The DE interaction effect was not significant. Seasonal and nonseasonal products did not differ (statistically) in their benefit from a doubling of the shelf space.

Appendix Selection of a Workable Set of Dead Letters

As discussed in section 11.4, our procedure for using Yates' algorithm for a 2^{k-p} fractional-factorial design requires us to determine a workable Yates' order. The way to do this, in turn, involves the selection of p (of the k) letters, to be temporarily designated as *dead letters.* Next, a standard Yates' order is formed using only the "live" letters. Then the dead letters are appended to the treatment combinations of this standard Yates' order as needed to form the block that is actually chosen to be run. As originally presented in the chapter, it appeared that the selection of *any* set of p dead letters would be workable—that is, would allow the formation of the actual block chosen to be run. However, we then (reluctantly) noted that *not every choice* of p dead letters is workable! In virtually all real-world designs, the vast majority of sets of letters that can be chosen are workable, so the first randomly chosen selection is usually workable. Occasionally, a choice is not workable; it will not allow reproduction of the block actually being run. This section presents an algorithm that ensures selection of a workable set of letters.

We begin by listing p independent terms of the defining relation. In Example 11.6, the 2^{5-2} design where we had $I = ABC = BCDE = ADE$, any two terms might be considered the independent terms, whereas the third is defined as the mod-2 product of these. We'll arbitrarily select the first two: ABC and $BCDE$.

Next, we pick one letter from the first term; this will be our first dead letter. Say we pick a. We then examine the second term of the defining relation and, if the first dead letter is not there, we can pick any of the letters in this second term of the defining relation and use it as the second dead letter. Since a is not included in $BCDE$, we could use a and b, a and c, a and d, or a and e (that is, a and *any* other letter) as our dead letters.

Going back one step, if we had picked b as the first dead letter, we would have found it present as well in $BCDE$; in this instance, we must multiply the two terms of the defining relation, getting $ABC \cdot BCDE = ADE$ (which, of course, will be another term in the defining relation), and pick any one of these letters as our second dead letter. So, we can use b and a (as we did above), b and d (as we did earlier when we presented this example in the text), or b and e. Note that b and c is one set that is precluded.

To summarize, pick two independent terms from the defining relation. Choose the first dead letter from the first term. If the second term does not contain the first dead letter, choose any letter from this second term as the second dead letter; otherwise, choose any letter from the mod-2 product of these two terms as the second dead letter. In this example, there are 10 possible choices; only two (*b* and *c*, and *d* and *e*) are not workable. Observe that each of these two pairs of letters occur together an even number of times; as a good heuristic, this pattern is something to avoid if we choose not to use this algorithm in a formal way. For a half replicate, where one letter is to be chosen as dead, the algorithm reduces to the choosing of any letter in the (one term of the) defining relation. The large majority of half replicates have a defining relationship of $I =$ highest order interaction, so any letter at all will be a workable (set of one) dead letter.

We continue with a 2^{6-3} example; here, we need $p = 3$ dead letters. Suppose that the defining relation is $I = ABC = AEF = BDF = BCEF = ACDF = ABDE = CDE$ (note that this defining relation results in all main effects being clean). The first three terms are independent (none of them is the mod-2 product of the other two, and the three generate the subsequent four effects); we'll start with them:

$ABC \quad AEF \quad BDF$

From the first term, say we choose *b*. The second term does not contain *b*, so we leave it as is. The third term does contain *b*; thus we multiply it by the first term, getting *ACDF*. This leaves us terms two and three:

$AEF \quad ACDF$

From *AEF*, say we choose *f*. The third term contains *f*, so we take the mod-2 product of *AEF* and *ACDF*, leaving us with

CDE

From *CDE*, we can take *c*, *d*, or *e* as our third dead letter. Thus, workable sets, among many others, are *b*, *f*, *c* (*b*, *c*, *f*), or *b*, *f*, *d* (*b*, *d*, *f*), or *b*, *f*, *e* (*b*, *e*, *f*).

This algorithm was derived and first published by one of the authors. A more detailed discussion can be found in Paul D. Berger, "On Yates' Order in Fractional Factorial Designs," *Technometrics*, November 1972, 14(4):971–972.

Exercises

1. Suppose that we wish to study as many factors as possible in a two-level fractional-factorial design. We also wish to have all main effects and all two-factor interaction effects estimated cleanly, under the assumption that all three-factor and higher-order interactions are zero. The following list gives the maximum number of factors that can be studied, given the conditions stated above, as a function of 2^{k-p}, the number of treatment combinations run in the experiment:

2^{k-p} (Number of Treatment Combinations)	Maximum Number of Factors Able to Be Studied
8	3
16	5
32	6
64	X

For $2^{k-p} = 8$, 16, and 32, give an example of a defining relation for a study with the maximum number of factors listed above.

2. In exercise 1, find X for $2^{k-p} = 64$.
3. Consider a 2^{6-2} fractional-factorial design with the defining relation

   $I = ABCE = ABDF = CDEF$

   a. Find the alias rows (groups).
   b. Find the four blocks (from which one is to be run).
4. Suppose that in exercise 3 the principal block is run, and a workable Yates' order is found by choosing the two dead letters a and d. Suppose that the 16 data values (with no replication), in the workable Yates' order described, are

   3, 5, 4, 7, 8, 10, 5, 6, 12, 11, 5, 7, 3, 5, 9, 13

   Find the 15 effects.
5. Suppose in exercise 4 it is now revealed that each of the 16 treatment combinations was replicated five times, and when analyzed by a one-factor ANOVA, the mean square error was 2.8. Derive the augmented ANOVA table, and determine which effects (that is, alias groups) are significant.
6. Refer back to exercise 3. Suppose that the principal block is run, but must be run in four blocks of four. Assume that we want to estimate all main effects cleanly, and that it is preferable to have two two-factor interactions aliased, compared with three two-factor interactions aliased (in terms of potential separation later). What would your four blocks be, and which two-factor interactions (in alias groups) are confounded with block effects? Assume that three-factor and higher-order interactions equal zero. Note: this problem combines the ideas of fractionating and the ideas of confounding.
7. A company is contemplating a study of six factors at two levels each. A 2^6 is considered too expensive. A 2^{6-1} is considered to extend over what *might* be too long a time to avoid nonhomogeneous test conditions; also, money might run out before all 32 treatment combinations can be run. It is decided to run two 2^{6-2} blocks (16 treatment combinations each), with the first block to be analyzed by itself, should the experiment then be curtailed. The first block is the principal block of the defining relation $I = ABCE = BCDF = ADEF$. The second block, if run, is to be the principal block of the defining relation $I = ABF = CDE = ABCDEF$. (This was a real experimental situation—and one of the favorite problems of the authors.)
   a. If money and time allow only the first block to be run, what analysis do you suggest? (That is, name which effects are estimated cleanly with which assumptions, what is aliased with what, and so forth.)
   b. If both blocks are run, what analysis do you suggest?
8. In exercise 7, how does your answer to part b change depending on whether there are block (time) effects?
9. In exercise 7, how can the block effect, if any, be estimated?
10. In exercise 7, what is the advantage of this experimental design (two blocks, each having a different defining relation), over a 2^{6-1} confounded into two blocks (that is, a set of two blocks from the *same* defining relation)?
11. Suppose that a 2^{8-2} experiment has been performed and that the following statements can be made about the results:
    (1) All three-factor and higher-order interactions are considered to be zero, except possibly for those listed in (4) below.
    (2) All main effects are estimated cleanly or known from prior experimentation.
    (3) All two-factor interactions are estimated cleanly or known from previous experimentation, except for those listed in (4) below.
    (4) The following alias pairs ("pairs," because other terms have been dropped as equal to zero) have significant results:

    $CDF = ABF$

    $CF = CGH$

    $AB = CD$

    $BF = BGH$

    $AF = AGH$

    $DF = DGH$

We wish to design a follow-up mini-experiment in order to try to remove the aliasing in the six alias pairs listed in (4). That is, we would like to be able to estimate cleanly each of the 12 listed effects. What mini-experiment would you design to best accomplish this goal, using the minimum number of treatment combinations? *Describe how your design accomplishes this goal.*

12. A number of years ago the government studied the attitude of people toward the use of seatbelts, as a function of eight two-level factors. Three factors had to do with the person (it isn't important at the moment precisely how the levels of B and C are defined):

A	Sex of person	Male/Female
B	Weight	Light/Heavy
C	Height	Tall/Short

Four factors had to do with the car and/or seatbelt type:

D	Number of doors	Two/Four
E	Attribute of belt	Windowshade/None
F	Latchplate	Locking/Nonlocking
G	Seat type	Bucket/Bench

How would you design a 2^{7-3} experiment if all interaction effects, except two-factor interactions involving factor A, can be assumed to be zero?

13. Consider an actual experiment in which the dependent variable is the sentence given out by a judge to a convicted defendant. For simplicity, think of the sentence as simply the "months of prison time." (Actually, the sentence consisted of prison time, parole time, and dollar amount of fine.) Factors of interest, all two-level factors, were

A	Crime	Robbery/Fraud
B	Age of defendant	Young/Old
C	Previous record	None/Some
D	Role	Accomplice/Principal Planner
E	Guilty by	Plea/Trial
F	Monetary amount	Low/High
G1	Member of criminal organization	No/Yes
G2	Weapon used	No/Yes

Factor G1 was mentioned only for the fraud cases; factor G2 was mentioned only for the robbery cases. In other words, a judge was given a scenario, in which he or she was told the crime, age, and so on (a treatment combination). Only if the crime were fraud was it mentioned whether the defendant was a member of a criminal organization or not (and the issue of weapon was not mentioned). Only if the crime were robbery was the question of weapon mentioned (and, correspondingly, the issue of membership not mentioned).

The following 16 treatment combinations were run:

g	*bdeg*	*abcdefg*	*acfg*
abc	*acde*	*def*	*bf*
ceg	*bcdg*	*abdfg*	*aefg*
abe	*ad*	*cdf*	*bcef*

Assuming a 2^{7-3} design, in which we do not distinguish between G1 and G2, which main effects are clean, if we are willing to believe that all interactions are zero?

14. In exercise 13, what is the special subtlety involving the estimate of the effects of factors F, G1, and G2? This flaw occurs only when we consider an objective that includes the separate estimation of G1 and G2.

15. Suppose that the ad agency example in the text was repeated with four replications, with the results shown in Table 11EX.15. Analyze, assuming that all interactions except two-factor interactions involving *F* are zero. (Data are also in the data sets as Problem 11.15.)

Table 11EX.15 Magazine Advertising Study with Replications

Treatment Combination	Replication 1	2	3	4
1	153	149	149	154
b(a)	302	300	306	307
d(c)	102	97	102	99
bd(ac)	352	352	360	354
e(ac)	54	58	54	56
be(c)	406	412	408	405
de(a)	211	203	204	210
bde	258	259	265	262
f(ac)	205	202	205	199
bf(c)	257	249	259	255
df(a)	45	48	51	49
bdf	397	399	398	397
ef	104	106	106	102
bef(a)	348	355	352	352
def(c)	158	155	148	156
bdef(ac)	301	302	304	302

16. Suppose that in exercise 15, replications 1 and 2 are men, replications 3 and 4 are women, and we believe that the gender of the people tested for recall may matter, but that gender will not interact with any of the six primary factors of the problem. Analyze, again assuming that all interactions except two-factor interactions involving F equal zero.
17. Suppose that in exercise 15, replication 1 consists of older men, replication 2 consists of younger men, replication 3 consists of older women, and replication 4 consists of younger women. We believe that age and gender may matter but that their interaction is zero, and that neither age nor gender interact with any of the six primary factors of the problem. Analyze, again assuming that all interactions except two-factor interactions involving F equal zero.
18. Repeat exercise 17, assuming that age and gender may have a nonzero interaction.

Notes

1. Produce products are agricultural products, especially fresh fruits and vegetables, as distinguished from grains and other staple crops.

2. What is meant by the phrase "higher-order interactions"—that is, how high is "higher order"—is discussed later in the chapter. The short answer is that in the vast majority of real-world applications, interactions of three or more factors are routinely assumed to be zero or negligible. Of course, this issue must be thought through for each application.

3. Using the logic of earlier chapters, with only four treatment combinations (and, say, no replicates), we have only three degrees of freedom, and can thus get "estimates" of only three effects. The three pairs noted are those three estimates. In a sense, ABC is "wrapped up" in the estimate of the grand mean—the mean of the eight treatments.

4. It is rare, but not unheard of, to have a situation in which a main effect can be assumed to be zero. More about this in an example later in the chapter.

5. Whether the effects are in columns and the treatment combinations in rows (as in the sign tables in general), or the transpose of that—the effects in rows, and the treatment combinations represented by columns (as in the orthogonal matrices of Chapter 5 and the one below the reference to this note)—is simply a matter of the respective traditions.

6. We are, of course, aware that we have not used the technically correct wording. To do so, we should say something like, "We reject the hypothesis that the amount of poster deployment has no effect . . . ," and for prizes, "We cannot reject the hypothesis that . . . ." At times, we prefer to avoid such double negatives and other hindrances to the flow of the discussion.

CHAPTER 12

Designs with Factors at Three Levels

EXAMPLE 12.1 Optimal Frequency and Size of Print Ads for MegaStore Electronics

MegaStore Electronics, Inc., wanted to know the most economic frequency and size of print ads in a certain weekly national magazine. Does an ad on two facing pages generate sufficiently more sales than a one-page ad or a half-page ad to warrant the additional expense? Should the ad appear every week, or less frequently?

To find out, MegaStore and its ad agency arranged with the magazine for a split-run arrangement: some subscribers saw larger ads, some smaller; some saw the various-sized ads weekly and others saw them less often. The experiment lasted three weeks. There were three levels of ad size: half-page, full-page, and two-page. The frequency of ad placement also had three levels: every week, twice in three weeks, or once in three weeks. For the one-week-in-three level, the week varied equally over each of the three weeks; the two weeks of the three varied equally over weeks 1 and 2, 1 and 3, and 2 and 3. In each case the ad's position in the magazine was the same: after letters to the editor and a few editorials but before the first story of the issue.

Each ad featured a coupon offering meaningful savings, and MegaStore assumed that the number of items bought when the coupon was redeemed was a reasonable indicator of the effectiveness of the ad or series of ads. Each of the nine (3×3) treatments was sent to 100,000 people, using an *N*th name sampling process (that is, 900,000 subscribers were identified in two highly populated states, and every ninth name was assigned a particular treatment).

What was key to this analysis was not simply main effects or interaction effects; it was clear that sales would increase as the ad size or ad frequency increased. The critical element was the way in which this increase took place—a matter of concavity or convexity (that is, departure from linearity). Was there decreasing or increasing return to scale in this particular situation? We return to this example at the end of the chapter.

12.1 Introduction

It's common to examine the impact of a factor at three levels rather than the two levels examined in recent chapters. For example, to determine the differences in quality among three suppliers, one would consider the factor "supplier" at three levels. However, for factors whose levels are measured on a numerical scale, there is a major and conceptually different reason to use three levels: to be able to study not only the *linear* impact of the factor on yield (which is all that can be done when studying a factor that has only two levels), but also the *nonlinear* impact. The basic analysis-of-variance technique treats the levels of a factor as categorical, whether they actually are or not. One (although not the only) logical and useful way to orthogonally break down the sum of squares associated with a numerical factor is to decompose it into a linear effect and a quadratic effect (for a factor with three numerical levels); a linear effect, a quadratic effect, and a cubic effect (for a factor with four numerical levels); and so forth.

In this chapter, we study the 3^k design, a k-factor complete-factorial design with each factor at three levels. We do not explicitly discuss three-level fractional-factorial designs, leaving that to those who wish to pursue experimental design further. Readers who master the concepts in two-level fractional-factorial designs and three-level complete-factorial designs will be able to cope with chapters on three-level fractional-factorial designs in other texts. (We especially recommend *Applied Factorial and Fractional Systems* by R. McLean and V. Anderson, New York, Marcel Dekker, 1984.) Actually, we have already seen three-level fractional-factorial designs in Chapter 8: three-level Latin squares and Graeco-Latin squares fit into that category.

12.2 Design with One Factor at Three Levels

We begin with a 3^1 design. This one-factor-at-three-levels design actually captures the salient features of a 3^k design for any k.

Figure 12.1 portrays an example of the yield plotted as a function of the level of factor A; the levels are called low, medium, and high. For simplicity, we assume in Figure 12.1 and elsewhere, unless noted otherwise, that A is metric and takes on three equally spaced levels—that is, the medium level is halfway between the low and high levels. (This assumption is not necessary to perform the analyses in this chapter, but we use it for clarity; the changes needed if the levels are not equally spaced are minimal. The issue of unequally spaced levels is discussed later when the MegaStore example is revisited.)

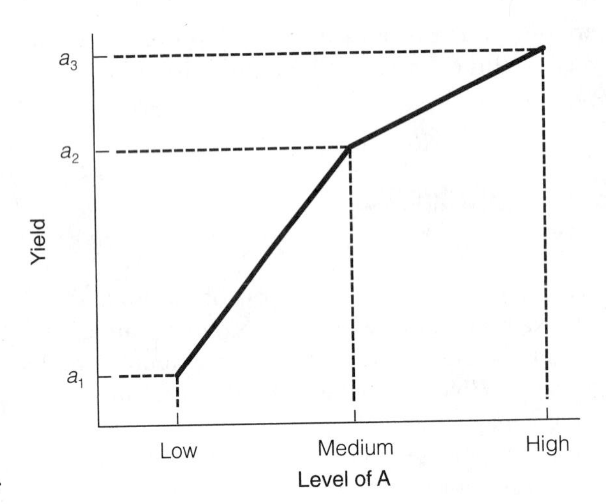

Figure 12.1
Illustration of the yield as a function of the three levels of the factor

If the relationship between the level of the factor and the yield were linear, all three points would appear on a straight line (not counting the impact of error, or ϵ, of course). But collecting data at only two levels of a factor gives no clue as to whether the relationship is linear; at least three points are needed to do that.

The **linear effect of A** is defined as follows:

$$A_L = (a_3 - a_2) + (a_2 - a_1) = a_3 - a_1$$

That is, the linear effect of A on yield depends only on yields a_1 and a_3 (the two outer levels, low and high).

The **nonlinear effect of A,** also appropriately called the **quadratic effect,**[1] is defined as

$$A_Q = (a_3 - a_2) - (a_2 - a_1) = a_3 - 2a_2 + a_1$$

That is, $(a_3 - a_2)$ is the estimated change in yield per change by one level of A; similarly, $(a_2 - a_1)$ is also the change in yield per change by one level of A. Any difference between the two suggests curvature—a slope that is not constant. If the yield at each of the three levels of A does fall on a straight line, then, under our assumption of equally spaced factor levels, $(a_3 - a_2) = (a_2 - a_1)$, there is no curvature:

$$a_2 = (a_3 + a_1)/2$$

and A_Q equals zero. If the curve is concave (that is, the yield at the medium level is *above* the straight line connecting the yields at the low and high levels), then $a_2 > (a_3 + a_1)/2$ and A_Q is negative. If the curve is convex (that is, the yield at the medium level is *below* the straight line connecting the yields at the low and high levels), then $a_2 < (a_3 + a_1)/2$ and A_Q is positive; the sign of the quadratic effect tells us whether the curve is concave or convex. (For those familiar with calculus, this is similar to using the sign of the second derivative to check whether a local extreme point is a maximum point or minimum point.)

Table 12.1 Sign Table for 3^1 Design

Effect	a_1	a_2	a_3
A_L	−1	0	1
A_Q	1	−2	1

Table 12.1 presents a sign table for a 3^1 design. Notice that the two rows of the sign table have a dot product of zero, and are thus orthogonal (but not yet orthonormal); and the elements of each row sum to zero. Recall the discussion of orthogonal decomposition in Chapter 5: after determining that a factor is significant, we investigate the influence of that factor by formulating meaningful questions; that is, by forming linear combinations of the column means. That's what's going on here. The rows of the orthogonal matrix in Table 12.1 correspond to the single-degree-of-freedom questions, or Z's, that examine the linear and quadratic components of factor A. As noted in the introduction, these two questions are typically the only logical orthogonal questions that are useful when breaking down a column sum of squares for a numerical factor having three levels.

12.3 Design with Two Factors, Each at Three Levels

We can extend this logic to designs with more than one three-level factor. Consider a 3^2; this is a two-factor design with each factor at three levels. There are nine treatment combinations: a_1b_1, a_1b_2, a_1b_3, a_2b_1, a_2b_2, a_2b_3, a_3b_1, a_3b_2, and a_3b_3. Repeating the procedure above, we calculate the linear and quadratic effects of A, but separately hold constant each of the three levels of B, as shown in Table 12.2. Similarly, the linear and quadratic effects for B are calculated as shown in Table 12.3.

Table 12.2 Calculation of Effects for Factor A

Level of B	Linear Effect of A	Quadratic Effect of A
High	$a_3b_3 - a_1b_3$	$a_3b_3 - 2a_2b_3 + a_1b_3$
Medium	$a_3b_2 - a_1b_2$	$a_3b_2 - 2a_2b_2 + a_1b_2$
Low	$a_3b_1 - a_1b_1$	$a_3b_1 - 2a_2b_1 + a_1b_1$
Total	$A_L = a_3b_3 - a_1b_3$ $+ a_3b_2 - a_1b_2$ $+ a_3b_1 - a_1b_1$	$A_Q = a_3b_3 - 2a_2b_3 + a_1b_3$ $+ a_3b_2 - 2a_2b_2 + a_1b_2$ $+ a_3b_1 - 2a_2b_1 + a_1b_1$

Table 12.3 Calculation of Effects for Factor B

Level of A	Linear Effect of B	Quadratic Effect of B
High	$a_3b_3 - a_3b_1$	$a_3b_3 - 2a_3b_2 + a_3b_1$
Medium	$a_2b_3 - a_2b_1$	$a_2b_3 - 2a_2b_2 + a_2b_1$
Low	$a_1b_3 - a_1b_1$	$a_1b_3 - 2a_1b_2 + a_1b_1$
Total	$B_L = a_3b_3 - a_3b_1 + a_2b_3 - a_2b_1 + a_1b_3 - a_1b_1$	$B_Q = a_3b_3 - 2a_3b_2 + a_3b_1 + a_2b_3 - 2a_2b_2 + a_2b_1 + a_1b_3 - 2a_1b_2 + a_1b_1$

It is possible to investigate interactions between the linear and quadratic effects of different factors,[2] but the derivation is somewhat complex and beyond the scope of this text. However, we still interpret the overall *AB* interaction, first introduced in Chapter 6, in the routine manner for such interactions; Example 12.2 includes such an interaction.

Just as for 2^k designs, there are tables of signs for each k of 3^k designs; the sign table for a 3^2 design is shown in Table 12.4. The effects A_L, A_Q, B_L, and B_Q can be estimated by adding and subtracting the indicated yields in accordance with the rows of the sign table.

Note in Table 12.4, as in the 3^1 sign table (Table 12.1), the presence of zeros; this means that, unlike in 2^k and $2^{k\text{-}p}$ designs, not every data point contributes to every effect. Furthermore, the data value (or mean, if there is replication) from some treatment combinations is weighted twice as much as others in calculating some effects. The presence of zeros, and the fact that not all the data points are weighted the same, indicates that 3^k designs are not as efficient (meaning that the estimates they provide are not as reliable, as measured by standard deviation, relative to the number of data points) as 2^k and $2^{k\text{-}p}$ designs. That having zeros in the sign table implies less efficiency is reasonably intuitive; we noted earlier that the standard deviation of an estimate decreases if a larger number of data values compose the estimate. It can be proven that, for a fixed number of data values with the same standard devia-

Table 12.4 Sign Table for 3^2 Design

	a_1b_1	a_1b_2	a_1b_3	a_2b_1	a_2b_2	a_2b_3	a_3b_1	a_3b_2	a_3b_3	Sum of Squares of Coefficients
A_L	−1	−1	−1	0	0	0	1	1	1	6
A_Q	1	1	1	−2	−2	−2	1	1	1	18
B_L	−1	0	1	−1	0	1	−1	0	1	6
B_Q	1	−2	1	1	−2	1	1	−2	1	18

tion composing an estimate, minimum variance is achieved if each data value is weighted equally. Finally, observe that, as in the 3^1 sign table, the rows of the 3^2 sign table are orthogonal and each row adds to zero. The rows can each be made to have the sum of squares of their coefficients equal to 1 (that is, the tables can be made orthonormal) by dividing each coefficient in a row by the square root of the sum of the squares of the coefficients for that row. This sum of squares is indicated in the last column of Table 12.4.

EXAMPLE 12.2 Selling Toys

Here's an illustration of the breakdown of the sum of squares due to each factor, for numerical factors having three levels. Our data are based on a study that considered the impact on sales of a specific toy in a national toy store chain. Two factors were examined: the length of shelf space allocated to the toy (the row factor), and the distance of that shelf space off the floor (the column factor). The row factor, A, had three levels: 4 feet long (L), 6 feet long (M), and 8 feet long (H). The column factor, B, had three levels: second shelf from the floor (L), third shelf from the floor (M), and fourth shelf from the floor (H). Store managers wondered whether a lower shelf (at child-eye level) or a higher shelf (at adult-eye level) would sell more of the toy in question. The height of a shelf (from bottom to top of any one shelf) was already held constant from store to store. Sales were adjusted for store volume.

We first examine this 3^2 example as simply a two-factor design with each factor at three levels, as done in Chapter 6. Each cell has two replicates, listed in Table 12.5. With each cell mean calculated (Table 12.6), we calculate the grand mean, 70.67.

Using the procedures developed in Chapter 6, we calculate the two-way ANOVA table. Table 12.7 shows the results (it assumes that each factor has fixed levels). At $\alpha = .05$, $F(2, 9) = 4.26$; we conclude that the row factor (length of shelf space) and the column factor (height off the floor of the shelf

Table 12.5 **Sales by Shelf Space Length (A) and Height (B)**

	High	Medium	Low	Row Mean
High	88, 92	105, 99	70, 86	90
Medium	81, 67	80, 92	57, 43	70
Low	60, 80	34, 46	47, 45	52
Column Mean	78	76	58	70.67

Table 12.6 **Shelf Space Length/ Height Example**

	High	Medium	Low	Row Mean
High	90	102	78	90
Medium	74	86	50	70
Low	70	40	46	52
Column Mean	78	76	58	70.67

space) are both significant. Also at $\alpha = .05$, $F(4, 9) = 3.63$, and we conclude that there is interaction between the two factors. A graphical demonstration of this interaction appears in Figure 12.2.

Although the graph pattern in Figure 12.2 is similar (nearly parallel, in a sense) for B=Low and B=High, it is very different for B=Medium. One way to interpret this is that the impact of shelf *length* depends on shelf *height.*

Or we could draw the interaction graph with the horizontal axis representing the level of B, with a drawing for each level of A, as shown in Figure 12.3. Again, the three graph patterns are not all similar, although two of the three (A=High and A=Medium) are similar.

Now we'll look at this example from the 3^2 perspective developed in this chapter. We can use the 3^2 sign table (Table 12.4) to write out the formulas for the four effects (using the *cell means,* entering zero when the cell mean gets no weight, and entering double the cell mean when it is multiplied by 2):

$$A_L = (-46 - 40 - 70 + 0 + 0 + 0 + 78 + 102 + 90) = +114$$
$$A_Q = (46 + 40 + 70 - 100 - 172 - 148 + 78 + 102 + 90) = +6$$
$$B_L = (-46 + 0 + 70 - 50 + 0 + 74 - 78 + 0 + 90) = +60$$
$$B_Q = (46 - 80 + 70 + 50 - 172 + 74 + 78 - 204 + 90) = -48$$

Table 12.7 **Two-Way ANOVA Table**

Source of Variability	SSQ	*df*	MSQ	F_{calc}
Rows, *A*	4336	2	2168	28.03
Columns, *B*	1456	2	728	9.41
Interaction	1472	4	368	4.76
Error	696	9	77.33	

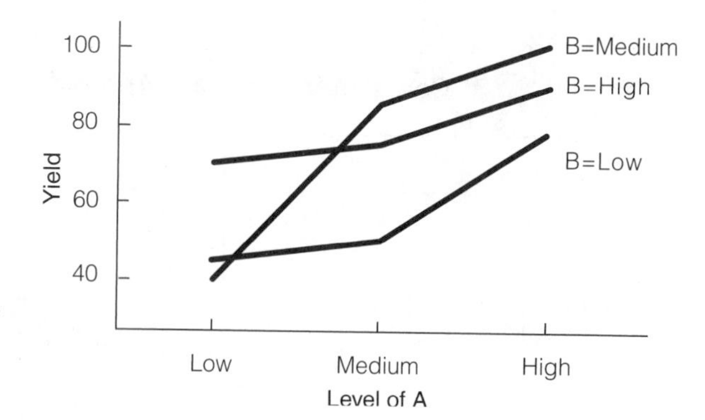

Figure 12.2 Yield as a function of shelf length for different heights from floor

One can construct a simple tabular template to facilitate these calculations; Table 12.8 gives an example. Using calculation templates for 3^k designs is not mandatory, but it's a useful and easy way to check one's work. The cost of designing, running, and analyzing the results of an experiment justifies care in the relatively mundane task of "running the numbers"—this is just another example of *practicing safe statistics.* Of course, using appropriate software for the analysis makes the problem moot.

Next, we normalize the estimates by dividing each by the square root of the sum of squares of the respective coefficients of the estimate, which are noted in the last column of Table 12.4. This yields

$$\text{Normalized } A_L = 114/\sqrt{6} = 114/2.449 = 46.54$$
$$\text{Normalized } A_Q = 6/\sqrt{18} = 6/4.243 = 1.41$$
$$\text{Normalized } B_L = 60/\sqrt{6} = 60/2.449 = 24.50$$
$$\text{Normalized } B_Q = -48/\sqrt{18}) = -48/4.243 = -11.31$$

Following the notation and procedure introduced in Chapter 5, the A_L and A_Q terms are essentially equivalent to Z_1 and Z_2, respectively, with regard to asking two "questions" about the sum of squares associated with factor A (or SSQ_{rows}); similarly, the B_L and B_Q terms are essentially equivalent to Z_1 and Z_2, respectively, with regard to asking two "questions" about the sum of squares

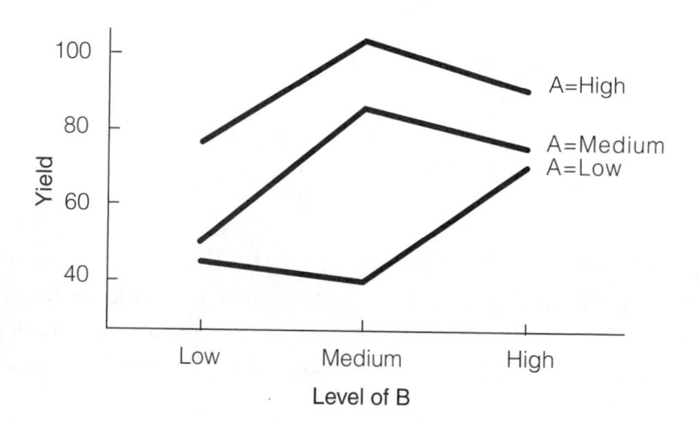

Figure 12.3 Yield as a function of height from floor for different shelf lengths

Table 12.8 3^k Calculation Template Applied to 3^2 Example

A_L		A_Q		B_L		B_Q	
–	+	–	+	–	+	–	+
46 40 70	78 102 90	2(50) 2(86) 2(74)	46 40 70 78 102 90	46 50 78	70 74 90	2(40) 2(86) 2(102)	46 70 50 74 78 90
Column Sum		Column Sum		Column Sum		Column Sum	
156	270	420	426	174	234	456	408
Net		Net		Net		Net	
	114		6		60	48	

associated with factor B (or $SSQ_{columns}$). To continue the Chapter 5 procedure, square each of these Z values and multiply each by the number of replicates per cell. For A, this yields

$$(Z_1)^2 \cdot 2 = (46.54)^2 \cdot 2 = 4332$$

and

$$(Z_2)^2 \cdot 2 = (1.41)^2 \cdot 2 = 4$$

For B, it yields

$$(Z_1)^2 \cdot 2 = (24.50)^2 \cdot 2 = 1200$$

and

$$(Z_2)^2 \cdot 2 = (-11.31)^2 \cdot 2 = 256$$

Observe that in this orthogonal decomposition of the sums of squares associated with A and B,

$$(A_L)^2 + (A_Q)^2 = 4332 + 4 = 4336 = SSQ_A = SSQ_{rows}$$
$$(B_L)^2 + (B_Q)^2 = 1200 + 256 = 1456 = SSQ_B = SSQ_{columns}$$

Table 12.9 summarizes these results in an augmented ANOVA table. For $\alpha = .05$, $F(1, 9) = 5.12$; we find that the linear component of the row factor, A_L, is significant ($p < .001$), as is the linear component of the column factor, B_L ($p < .001$). At $\alpha = .05$, neither of the quadratic terms are significant (for A_Q, p is nearly 1.0; for B_Q, p is about .11). In essence, we can conclude that both

Table 12.9 **Augmented ANOVA Table**

Source of Variability		SSQ		df		MSQ		F_{calc}	
Rows		4336		2		2168		28.0	
	A_L		4332		1		4332		56.0
	A_Q		4		1		4		.1
Columns		1456		2		728		9.4	
	B_L		1200		1		1200		15.5
	B_Q		256		1		256		3.3
Interaction		1472		4		368		4.8	
Error		696		9		77.3			

factor A and factor B have an impact on the yield, and that in each case the yield increases linearly with the level of the factor—with no significant curvature.

This indicates that as the length of the shelf space given to the toy increases, sales increase linearly, with no statistical indication of concavity nor convexity—meaning that going from four feet to six feet engenders about the same sales increase as going from six feet to eight feet. Indeed, the row means go from 52 to 70 to 90; $(70 - 52) = 18$ is not very different from $(90 - 70) = 20$. We would not suggest that the linearity goes on forever, nor that it is valid for very small amounts of shelf space. We judge this result as appropriate only for the range of values in the experiment, and perhaps only for the specific toy in the experiment. The same qualifying statements apply to the conclusions for the distance of the shelf from the floor. Sales increase when the toy is placed on the third shelf rather than the second, and they increase by about the same amount if it's placed on the fourth shelf rather than the third. Again, there's no statistically significant indication of concavity nor convexity. The column means go from 58 to 76 to 78. Here, $(76 - 58) = 18$ does not seem so close to $(78 - 76) = 2$; indeed, the result had a p value of about .11, not so far from .05, an indication that the data results are less close to literal linearity. Again, there is no good reason to think that this technical linearity holds for levels outside the range of values in the experiment. Figure 12.4 graphs the column means for A and B to show the curvature or lack thereof (in the top graph of the impact of the level of A, it looks virtually linear).

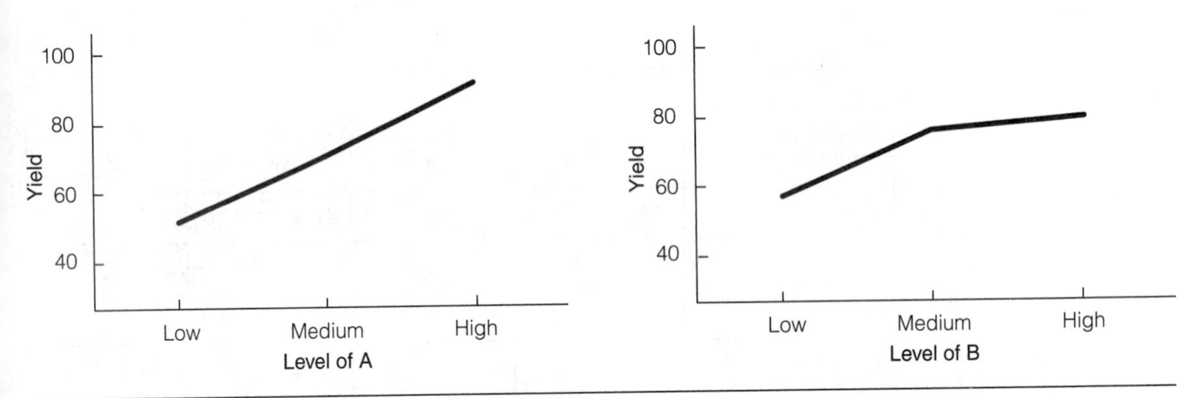

Figure 12.4 Column means as a function of shelf space (left) and shelf height (right)

EXAMPLE 12.3 Using SPSS for the Toy Example

From the perspective of SPSS, the toy-store problem is a two-factor cross-classification with replication. The data were entered as shown in Table 12.10, and SPSS was told that the first column represents the dependent variable, the second column the levels of the row factor, A, and the third column the levels of the column factor, B.

Table 12.10 SPSS Input

88.00	3.00	3.00
92.00	3.00	3.00
81.00	2.00	3.00
67.00	2.00	3.00
60.00	1.00	3.00
80.00	1.00	3.00
105.0	3.00	2.00
99.00	3.00	2.00

(continued)

Table 12.10 **(continued)**

80.00	2.00	2.00
92.00	2.00	2.00
34.00	1.00	2.00
46.00	1.00	2.00
70.00	3.00	1.00
86.00	3.00	1.00
57.00	2.00	1.00
43.00	2.00	1.00
47.00	1.00	1.00
45.00	1.00	1.00

Table 12.11 shows the output. The breakdown of the main-effect sum of squares into linear and quadratic components in an augmented ANOVA table is not provided, but would have to be independently and separately generated, as discussed in the Chapter 5 SPSS example.

Table 12.11 **SPSS ANOVA Table for Toy-Store Example**

ANALYSIS OF VARIANCE

SALES
by LENGTH
HEIGHT

Source of Variation	Sum of Squares	DF	Mean Squares	F	Sig of F
Main Effects	5792.000	4	1448.000	18.724	.00
LENGTH	4336.000	2	2168.000	28.034	.00
HEIGHT	1456.000	2	728.000	9.414	.00
2-Way Interactions	1472.000	4	368.000	4.759	.02
LENGTH HEIGHT	1472.000	4	368.000	4.759	.02
Explained	7264.000	8	908.000	11.741	.00
Residual	696.000	9	77.333		
Total	7960.000	17	468.235		

EXAMPLE 12.4 Using JMP for the Toy Example

Now we use JMP to evaluate the shelf length and height problem; it produces the analysis in Table 12.12.

The output shown in Table 12.12 is essentially no different from the SPSS output; however, JMP also decomposes the sums of squares into linear and quadratic components. Table 12.13 shows that output for factor A. The labels (Linear) and (Quadratic) in the Contrast row of Table 12.13 were inserted by the authors, not JMP.

The only options for coefficients in the Table 12.13 analysis were −1, 0, and +1. For the linear contrast, −1, 0, +1 values were inserted. For the quadratic contrast, +1, −1, +1 values were inserted, and the program changed the values to those shown in Table 12.13: .5, −1, .5. Note that contrasts were enabled only as coefficients of the column means, not for all nine data values, as in the sign table earlier. This does no harm; indeed, the SS row in Table 12.13 (4332 and 4) reproduces the appropriate sums of squares for the augmented ANOVA table even though JMP does not reproduce the augmented ANOVA table itself. In the Estimate row, if we multiply 38 by 3 (the number of columns for each row), the result, 114, is the same as that in the template (Table 12.8).

Table 12.12 JMP Analysis of Toy-Store Example

Response: SALES
Summary of Fit

RSquare	0.912563
RSquare Adj	0.820672
Root Mean Square Error	8.793937
Mean of Response	70.33
Observations (or Sum Wgts)	18

Effect Test

Source	DF	Sum of Squares	Mean Square	F Ratio	Prob>F
LENGTH	2	4336.0000	2168.0000	28.034	0.0001
HEIGHT	2	1456.0000	728.0000	9.414	0.0069
LEN*HGT	4	1472.0000	368.0000	4.759	0.0213
ERROR	9	696.0000	77.3333		

Means

Level	Mean	Std Error
1,1	46.0000000	6.218252702
1,2	40.0000000	6.218252702
1,3	70.0000000	6.218252702
2,1	50.0000000	6.218252702
2,2	86.0000000	6.218252702
2,3	74.0000000	6.218252702
3,1	78.0000000	6.218252702
3,2	102.0000000	6.218252702
3,3	90.0000000	6.218252702

Table 12.13 **JMP Output: Decomposition of Sum of Squares for Factor A**

Contrast	(Linear)	(Quadratic)
1	−1	0.5
2	0	−1
3	1	0.5
Estimate	38	1
Std Error	5.0772	4.397
t Ratio	7.4844	0.2286
Prob>|t|	5.9e-5	0.9123
SS	4332	4

Table 12.14 **JMP Output: Decomposition of Sum of Squares for Factor B**

Contrast	Linear	Quadratic
1	−1	0.5
2	0	−1
3	1	0.5
Estimate	20	−8
Std Error	5.0772	4.397
t Ratio	3.9392	−1.819
Prob>|t|	0.0034	0.1048
SS	1200	256

If we multiply the 1 for the quadratic contrast by 3, and also by 2 (because the Table 12.8 template was based on the coefficients 1, −2, 1 whereas the JMP analysis used .5, -1, .5) we reproduce the 6 of the template.

For factor B, JMP produces the linear and quadratic components as shown in Table 12.14. Note that the values in the SS row (1200, 256) correspond with the sums of squares in the augmented ANOVA table (Table 12.9), whereas the values in the estimate row (20, −8) are transformed to the Table 12.8 template's values (60, −48) in a manner similar to that described for factor A.

12.4 Nonlinearity Recognition and Robustness

Nonlinearity can help reduce sensitivity in performance due to variability in the level of input parameters (that is, variability in implementing the precise level of the factor). This attribute of insensitivity is called *robustness*. Figure 12.5 illustrates the relationship of the parameter setting of an input factor to performance, focusing on the choice of two levels, L_1 and L_2, of a single factor.

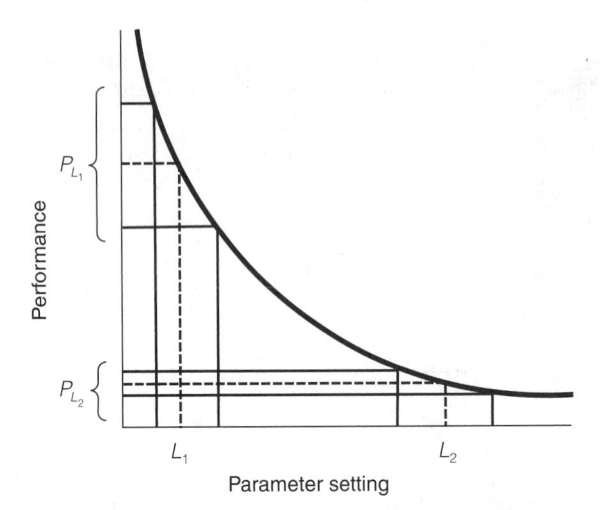

Figure 12.5 Effects of nonlinearity on robustness

Suppose, in Figure 12.5, that we can operate at either setting L_1 or setting L_2. Also, suppose that our objective is to reduce variability in performance in response to variability in the level of factors not under our complete control (a worthy objective, everything else being equal). Examples range from variability in materials or processes over which the company has *some* control but which might be expensive to improve, to true "noise factors," such as the temperature at which the product is used after it leaves the shop. The nonlinearity (the curvature) in the relationship between output (performance) and the input level of the factor indicates that the variability of the output varies with the variability of the level of the input factor in a way that isn't constant. If we choose level L_2 in Figure 12.5, a relatively large variability in the parameter setting will have minimal impact on the variability of performance (P_{L2} in the figure); but if we choose level L_1, even small variability in the level of the factor will result in large variability in performance (P_{L1} in the figure). We say a lot more about this issue in the next chapter. However, we hope it is already clear that the potential benefit of nonlinearity cannot be exploited if it is not first identified.

12.5 Three Levels versus Two Levels

In the final analysis, deciding whether to estimate nonlinear effects involves judgment. Considerations should include the opinions of the process experts, the importance (cost implications) of the factor, the expected monotonicity of the factor (that is, whether the relationship between the performance and the level of the factor continually increases or decreases, or goes up and then down [or down and then up] as the level of the factor increases), and the cost of the experiment with and without evaluating nonlinearity. No single guideline is appropriate for all applications.

If there is a reasonable chance that performance does not vary monotonically over the range of interest of the levels of the factor, then the study of nonlinearity is virtually mandatory. However, if the relationship is anticipated to be monotonic (if it turns out to have any effect at all), judgments may be necessary. With only a few factors, to be safe we should probably include in the design the study of nonlinearity. However, if there are lots of factors in the experiment, and the task at hand is to narrow down the number of factors for later study, then perhaps studying only two levels of the factor is appropriate. Remember: 3^7 is a lot larger than 2^7, and for that matter, 3^{7-2} is much larger than 2^{7-2}!

EXAMPLE 12.5 Optimal Frequency and Size of Print Ads for MegaStore, Revisited

In the MegaStore experiment, the results for ad size clearly indicated a significant linear effect and also a significant quadratic effect, in the concave direction. As ad size went from half a page to two pages, sales increased; however, as ad size increased from one page to two pages, the sales increase was less than double the sales increase engendered as ad size increased from half a page to one page. (If the increase were completely linear, the extra sales gained from a two-page versus a one-page ad would be double the sales increase engendered by going from a half page to a full page.) The results for ad frequency followed a similar pattern. But whereas the results for ad size were not surprising to MegaStore management, the ad frequency results were. MegaStore thought, in retrospect, that perhaps four levels of frequency should have been tested. Maybe so: in many situations increasing frequency results in an S-shaped curve, arguably based on the need for a critical mass (of exposure, in the case of advertising). And with only three levels, an S shape cannot be captured; this would require a cubic effect. With only three levels, we cannot distinguish between the two diagrams in Figure 12.6, assuming that the yield at each of three levels (low, medium, and high) is the same in both diagrams.

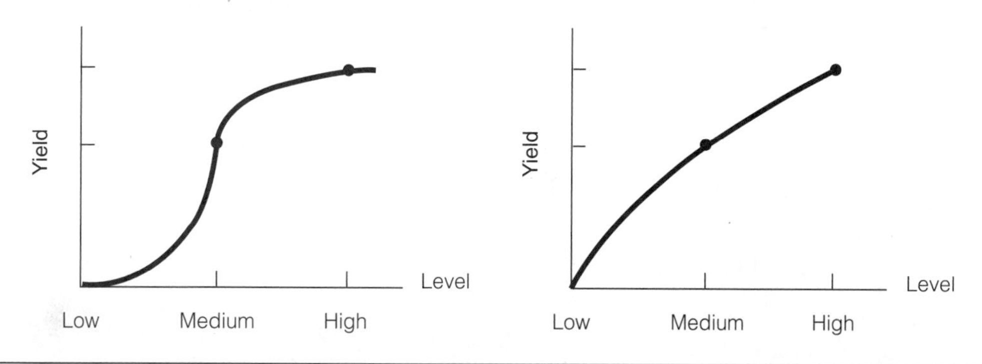

Figure 12.6 Failure of three levels to distinguish S-shaped from non-S-shaped curve

12.6 Unequally Spaced Levels

To analyze the MegaStore data, one needs to perform the linear and quadratic sum-of-squares breakdown for unequally spaced intervals. The factor "number of ads over the three-week period" was equally spaced at one, two, and three; however, the factor "ad size" was not: half a page, one page, and two pages.

When the three levels are not equally spaced, all the logic described in this chapter is still fully applicable, but the contrast coefficients are different. The coefficients of the linear contrast must be proportional to the deviation from the mean level. For example, if the levels are .1 page, .3 page, and .8 page, the mean level is .4 page. Then, the coefficient for the low level is (.1 − .4) = −.3; for the middle level, it is (.3 − .4) = −.1; for the high level, it is (.8 − .4) = +.4. Hence, the (not yet normalized) contrast is

[−.3 −.1 .4]

These values reflect the fact that linearity would not suggest that the difference in yield in going from A = .1 page to A = .3 page is the same as the difference in going from A = .3 page to A = .8 page. Indeed, the difference between the coefficients is .2 for the former and .5 for the latter, corresponding to the 2.5:1 ratio of differences in actual levels. Further, note that the sum of the coefficients is still zero.

For the actual case at hand, with levels of .5, 1, and 2, the mean level is 7/6. Writing .5 as 3/6, 1 as 6/6, and 2 as 12/6, and, without loss of generality, working only with the numerators (that is, the importance of the coefficients is relative—the scale is taken care of later) the differences/coefficients are

[−4 −1 +5]

What about the quadratic contrast's coefficients? The coefficients for the three-level quadratic contrast are obtained by (1) taking the difference between the low and medium levels and placing that value, with a plus sign, as the coefficient for the *high* level, (2) taking the difference between the medium and high levels and placing that value, with a plus sign, as the coefficient for the *low* level, and (3) placing as the coefficient of the medium level the value, with a minus sign, that makes the sum of the coefficients equal to zero. Here's an illustration:

For the linear contrast of

[−.3 −.1 +.4]

take the difference between −.3 and −.1, which equals .2, and place it as the coefficient for the high level:

[— — .2]

Then take the difference between −.1 and +.4, which equals .5, and place it as the coefficient of the low level:

[.5 — .2]

Finally, find the coefficient for the medium level to be .5 + .2 = .7, but with a minus sign:

[.5 −.7 .2]

The orthogonal table for our numerical example becomes

Effect	a_1	a_2	a_3
A_L	−.3	−.1	.4
A_Q	.5	−.7	.2

Note that the inner product of the two rows is, indeed, zero. Also note that the "smallness" of the coefficients is not an issue, as the normalization process accounts for that. The sums of squares of the coefficients for the two rows are .26 and .78, respectively.

For the MegaStore example with levels .5, 1, and 2, the linear contrast is noted above as $[-4 \quad -1 \quad 5]$, and we can derive the quadratic contrast as $[6 \quad -9 \quad 3]$, which, dividing each coefficient by 3, can be reduced to $[2 \quad -3 \quad 1]$, resulting in the following sign table for the unequally spaced levels:

Effect	a_1	a_2	a_3
A_L	−4	−1	5
A_Q	2	−3	1

12.7 Comment

Many designs that can be set up as a 3^k or 3^{k-p} can also be set up, sometimes more efficiently, using other designs that we discuss in Chapter 14 on response surface methods (RSM). However, the RSM approach is potentially useful only when the levels of the factors in the study are continuous.

Exercises

1. Consider the 3^2 experiment shown in Table 12EX.1. Perform a standard two-factor cross-classification ANOVA; assume that it is known that there is no interaction between the two factors. (Assume in this problem and all subsequent problems that the levels are equally spaced.)

Table 12EX.1 Yields for a 3^2

	Level of B		
	Low	Medium	High
Level of A			
Low	23	17	29
Medium	16	25	16
High	24	18	12

2. In exercise 1,
   a. Break down the sum of squares associated with A, and the sum of squares associated with B, into a total for the two factors of four single-degree-of-freedom components.
   b. Continuing with the assumption of no interaction, test for linear and quadratic effects of each factor.
3. Consider the data in Table 12EX.3 comprising yields for the 27 treatment combinations

Table 12EX.3 **Yields for a 3^3 Experiment with Two Replicates**

	B=Low			B=Medium			B=High		
	C=Low	C=Medium	C=High	C=Low	C=Medium	C=High	C=Low	C=Medium	C=High
A=Low	8.4, 9.5	2.3, 2.2	7.9, 12.5	12.2, 15.8	5.6, 6.2	14.0, 11.7	8.1, 8.5	6.6, 5.7	13.5, 8.0
A=Medium	22.1, 19.4	12.0, 15.5	20.9, 17.4	17.6, 29.8	22.8, 21.4	23.0, 15.2	21.1, 22.6	22.6, 24.9	23.7, 27.5
A=High	8.4, 8.5	15.7, 11.4	20.5, 19.3	5.7, 7.5	15.6, 13.4	20.8, 17.5	10.0, 7.0	11.2, 8.0	17.4, 12.4

of a 3^3 design, with two replicates per cell. Perform a standard three-factor cross-classification ANOVA; assume that it is known that there is no three-factor interaction, but that there might be two-factor interactions among any two factors.

4. In exercise 3,
   a. Break down the sum of squares associated with A, the sum of squares associated with B, and the sum of squares associated with C, into a total of six single-degree-of-freedom components.
   b. Continuing with the assumptions about interaction stated in part a, test for linear and quadratic effects of each factor.
5. The data in Table 12EX.5 represent the experimental results from a 3^2 design with eight replicates per cell. (Data are also in the data sets as Problem 12.5.) Analyze (first) as a one-factor design, the row factor. Use $\alpha =$ .05 for this exercise and exercises 6, 7, 9, and 10.
6. Analyze the data in exercise 5 as a one-factor design, the column factor.
7. Analyze the data in exercise 5 as a two-factor cross-classification design, assuming that the two factors do not interact.
8. Compare the results of exercise 7 with those for exercises 5 and 6. Explain the reasons for the different results.
9. Now analyze the data of exercise 5 as a two-factor cross-classification design with the possibility of interaction. Compare the results to those of exercise 7, and explain the reasons for the differences in results.
10. Finally, break down the results for the two factors in the exercise 9 analysis into linear and quadratic effects, and test for the significance of these effects.

Table 12EX.5 **Yields for a 3^2, Eight Replicates**

Y	Row	Column	*Y*	Row	Column
86	3	3	86	2	3
96	3	3	65	2	3
92	3	3	80	2	3
90	3	3	65	2	3
88	3	3	79	2	3
92	3	3	65	2	3
87	3	3	83	2	3
91	3	3	68	2	3

Table 12EX.5 (continued)

Y	Row	Column	Y	Row	Column
59	1	3	39	1	2
83	1	3	53	1	2
61	1	3	39	1	2
77	1	3	52	1	2
60	1	3	71	3	1
80	1	3	88	3	1
62	1	3	70	3	1
81	1	3	85	3	1
109	3	2	70	3	1
98	3	2	88	3	1
107	3	2	73	3	1
97	3	2	86	3	1
103	3	2	60	2	1
100	3	2	42	2	1
106	3	2	56	2	1
102	3	2	42	2	1
69	2	2	58	2	1
81	2	2	42	2	1
65	2	2	56	2	1
81	2	2	44	2	1
72	2	2	48	1	1
78	2	2	44	1	1
68	2	2	50	1	1
83	2	2	42	1	1
39	1	2	47	1	1
48	1	2	41	1	1
40	1	2	48	1	1
55	1	2	47	1	1

Notes

1. Just as we need two points to define a straight line, we require three points to define a quadratic function—defined to be a polynomial in which the highest power (exponent) is 2. Given that we have limited ourselves in this section to factors with three levels, we can determine the presence (or absence) of curvature to be ascribed solely to a quadratic ("squared") term. Thus, for a three-level factor's impact on the yield, the term *quadratic* may be accurately viewed as a synonym for *nonlinear;* the restriction of nonlinear behavior to the quadratic category is due to a limitation of the model, not to a statement of knowledge. In practice, this issue pertains more to mathematical completeness than to a limitation in the usefulness of the results. One can investigate models constructed of higher-order polynomials by including factors with more than three levels.

2. In fact, one can actually compute four two-factor interactions: between linear A and linear B, between linear A and quadratic B, between quadratic A and linear B, and between quadratic A and quadratic B.

CHAPTER 13

Introduction to Taguchi Methods

EXAMPLE 13.1 New Product Development at HighTech Corporation

HighTech Corporation has fostered a corporate image based on its speed of innovation. Developers of new products are given the best of equipment and encouraged to be as innovative as possible in as short a time as possible, directly on the plant floor, with a minimum of "bureaucratic interruption." However, at one point in the company's history, a problem arose that significantly slowed the innovative process.

It seemed that with increasing frequency the new product developers needed to design efficient experiments to test a variety of configurations for potential new products. Yet, the need for an experimental-design expert was not sufficient to warrant a full-time person on staff. What ensued was that the development process would come to a standstill while an expert was summoned to help with the experimental design. These delays became increasingly unsatisfactory. They not only slowed things down but often led to frustration, which in turn led to the developers skirting the experimental-design process and getting shoddy results. Something had to change!

The solution was to teach the engineers methods that allowed them to quickly and easily design good-quality experiments, and, to the degree possible, tie in these methods with a "way of thinking about quality" for the whole company.

We return to this example at the end of the chapter.

13.1 Introduction

We have seen how, using fractional-factorial designs, we can obtain a substantial amount of information efficiently. Although these techniques are powerful, they are not necessarily intuitive. For years, they were available only to those who were willing to devote the effort required for their mastery, and to their clients. That changed, to a large extent, when Dr. Genichi Taguchi, a Japanese engineer, presented techniques for designing certain types of experiments using a "cookbook" approach, easily understood and usable by a wide variety of people. Most notable among the types of experiments discussed by Dr. Taguchi are two-level and three-level fractional-factorial designs. Dr. Taguchi's original target population was manufacturing engineers, but his techniques are readily applied to many management problems; hence, their inclusion in this text. Using Taguchi methods, we can dramatically reduce the time required to design fractional-factorial experiments.

As important as it is, Taguchi's work in the design of experiments is just part of his contribution to the field of quality. His work may be viewed in three parts—the *philosophy* of designing quality into a product, rather than inspecting defects out after the fact; quantitative *measures* of the value of quality improvements; and the development of the aforementioned user-friendly *experimental design* methods that point the way for quality improvement. Indeed, the reason that Taguchi developed the techniques for "relatively quick" designing of experiments is that he believed (and practice has borne out) that unless these techniques were available, (manufacturing) engineers would not take the time to design the experiments, so the experiments would never get performed (and the quality improvement would never be realized).

Taguchi is seen as one of the pioneers in the total quality management (TQM) movement that has swept American industry over the past two decades. Major organizations that adopted Taguchi methods early on were Bell Laboratories, Xerox, ITT, Ford, and Analog Devices. These were soon followed by General Motors, Chrysler, and many others.

13.2 Taguchi's Quality Philosophy and Loss Function

The essence of the Taguchi philosophy is a change in mind-set regarding quality: moving away from the "goal-post" mentality, wherein a manufacturing component (or process step or anything else for which there might be a notion of "acceptable or unacceptable") is seen as either good or bad—that is, classified simply as a dichotomy. Typically, specification (spec) limits are defined, and the part is measured against these limits. Examples would be "the diameter of a steel shaft shall be 1.280 ± .020 cm," or "the output of a light source should be between 58 and 62 lumens," or "the radiation time should be 200 ± 2 milliseconds." These are all examples of specs that might be called "nominal the best," that is, the nominal value is the best value, rather than "the bigger the better" or "the smaller the better," as in one-sided specs. (The analogy to goal posts, as in soccer, football, hockey, and other sports, refers to thinking of the data [the ball or puck] as either "in" or "out"; the degree of being "in" is immaterial.) Examples of one-sided specs include "the number of knots must be less than three per sheet," or "the noise power must be less than 20 microwatts," or "the cord must contain at least 120 cubic feet." At times, statistical overtones may be added; for example, "the waiting time should be less than 20 minutes *for at least 95% of the customers.*" All of these requirements, whether nominal the best or one-sided, include the notion of a specification that partitions the possible continuum of parameter values into two classes—good and bad.

Taguchi would argue that this approach is not grounded in reality. Can it be, returning to our first example above, that a steel shaft that measures anywhere from 1.260 cm to 1.300 cm is good but that one measuring 1.301 cm is not? Is it likely that one measuring 1.300 cm performs as well as one that measures 1.280 cm? Will one measuring 1.300 cm behave the same as one measuring 1.260 cm? More likely there is some best diameter, say 1.280 cm, and a gradual degradation in performance results as the dimension departs from this value. Furthermore, it is reasonable to assume that a departure from this best value becomes more problematic as the size of the departure increases. Taguchi contends that a more meaningful "loss function" would be quadratic (or at least concave upward, for which a quadratic function would be a good approximation). Figure 13.1 depicts Taguchi's suggested quadratic loss function along with the goal-post, or spec limit, loss function.

Note points A, B, and C on the horizontal axis of each loss function in Figure 13.1. These would correspond, using the values in the steel-shaft example above, to, say, $A = 1.282$, $B = 1.299$, and $C = 1.301$. It belies logic, at

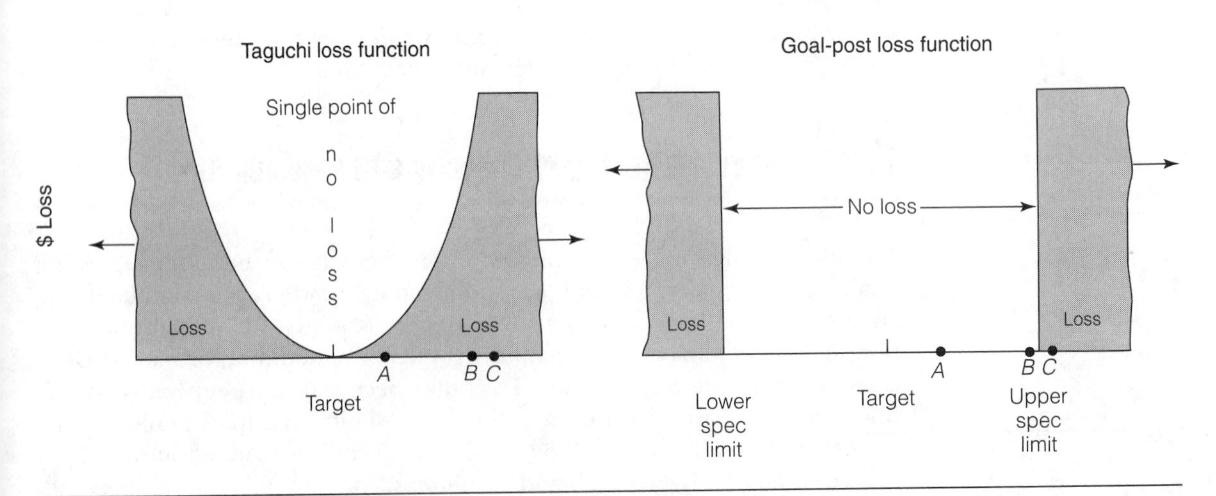

Figure 13.1 Contrasting models for loss functions

least in the vast majority of situations, to view points A and B as having equal quality while viewing points B and C as having dramatically different quality. Yet, this is exactly what the goal-post mentality embraces. Taguchi's quadratic loss function recognizes that by and large, points B and C are of similar quality (perhaps both being poor quality!), and points A and B are of very different quality.

Taguchi's interest is primarily in the difference between the two loss functions in the acceptable region for the goal-post depiction; that is, he contends that *any* departure from the nominal (that is, ideal) value involves some loss. The Taguchi loss function is of the form

$$L = K(Y - T)^2$$

where

L = the loss incurred, in dollars per part
K = a constant appropriate to the problem (related to the cost of the part and the cost for its "reworking")
Y = the actual value of the measured quantity (a point along the horizontal axis—for example, the output voltage of a generator)
T = the target (best, ideal, or nominal) value

Most texts dedicated to Taguchi methods, including those cited in Chapter 15, discuss the determination of K. This proposed quadratic loss function is related to the idea of minimizing mean square error, a goal that is virtually always considered worthy. Regression analysis, a subject not covered in this text but with which most readers are at least somewhat familiar, nearly always rests on the determination of a least-squares line (or plane or hyperplane)—that is, the minimization of the sum of the squared errors. These "connections" add support to Taguchi's choice of suggested loss function.

It can be shown that the *average* loss per unit using the Taguchi loss function is

$$\bar{L} = E(L) = K[(\mu - T)^2 + \sigma^2]$$

where

$E(L)$ = expected value of L
μ = the true average value of Y
σ^2 = the true variance of Y

In practice, the μ and σ^2 values are replaced by their respective estimates, $\bar{Y}$ and S^2, yielding the equation

$$\bar{L} = K[(\bar{Y} - T)^2 + S^2]$$

The term $(\bar{Y} - T)$ is called the **bias;** it demonstrates the extent to which, on the average, the "performance measure" (or "quality indicator," or "quality characteristic"), Y, does not come out equal to the nominal value. Clearly, the ideal result would be to have both the bias and the variance equal to zero. In practical situations, the variance is never zero. Furthermore, having the average performance as near as possible to the target and having variability around the average as near as possible to zero sometimes are conflicting goals. That is, it is possible that making the bias as small as possible and incurring whatever variance results may yield a higher average loss than allowing the bias to become larger with a more-than-compensating decrease in variance. As a simple example, consider the following choice: would you prefer a thermostat in your home that on average is off by one degree, possibly only a half degree, sometimes one-and-a-half degrees, or a thermostat that on the average is exactly accurate, but which much of the time is off between 10 and 15 degrees, as often below the true temperature as above? Remember: the average doesn't tell the whole story. Recall the guy with his head in the refrigerator, his feet in the stove—on the average he feels fine!

13.3 Control of the Variability of Performance

Control of average performance is traditional. Control of variability, if done at all, has usually been accomplished through "explicit" control; that is, it is determined that an amount of variability in some input factor, let's say degree of steel hardness, leads to an amount of variability in some output factor (that is, some performance measure), let's say shaft diameter. Explicit control of the variability of shaft diameter would be achieved by the control of the variability in steel hardness. Reduce the latter (at a cost, in most cases), and the result is a reduction in the former. This explicit control process is illustrated in Figure 13.2.

Another approach, one advocated by Taguchi, is "implicit" control—making the design, process, and so forth less sensitive to input variations. Rather than demand input improvements, which may be difficult and costly and might even require temporary interruption of the manufacturing process, we control (reduce) variability by changing the relationship between the vari-

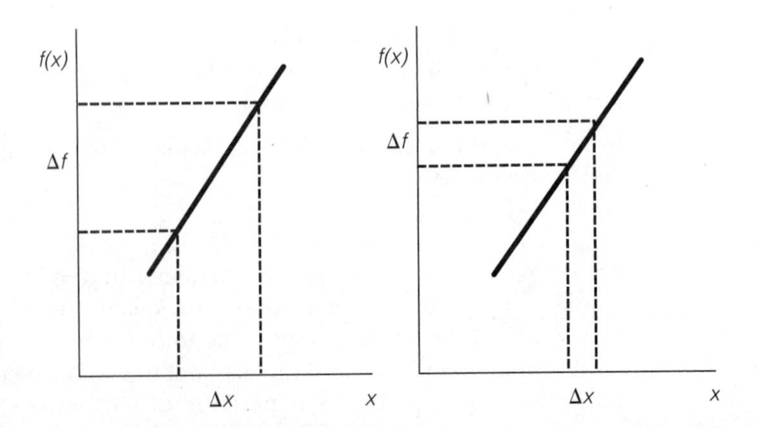

Figure 13.2 Explicit control example. By reducing the change in x one reduces the resulting change in $f(x)$.

ability of the input factor and the variability in performance. In essence, we make the process more "robust"—that is, we somehow arrange it so that variability in the performance measure is not very sensitive to (not increased much by) variability in the input factor. We might, for example, be able to change the milling process to be less influenced by hardness of the input material (that is, change it so that the performance of the milling process *does not vary, or varies negligibly,* with the hardness of the input material). Designs and processes that are largely insensitive to input variability are said to be robust. Implicit control is illustrated in Figure 13.3a. Note that from the perspective of this diagram, implicit control amounts to changing the slope of the curve that represents the relationship between the value of the input factor (for example, degree of hardness) and the value of the performance measure (quality of the output of the milling process); the same input variability yields less output variability.

Sometimes the relationship between input and output variability is not linear. Figure 13.3b illustrates such a nonlinear relationship. Here, rather than viewing the situation as one in which we change the slope of the function relating the performance measure to the value of the input factor, we view it as one in which we move to a different point on that curve—one at which the slope is, indeed, more forgiving.

Implicit in the discussion above is the holistic notion that everything is fair game when it comes to quality improvement. Ideally, the design, process, materials, and so forth are optimized *jointly* to achieve the best quality at the lowest total cost. Taguchi uses the term **design parameters** as the generic designation for the factors that potentially influence quality and whose levels we seek to optimize. The objective is to "design quality in" rather than to weed

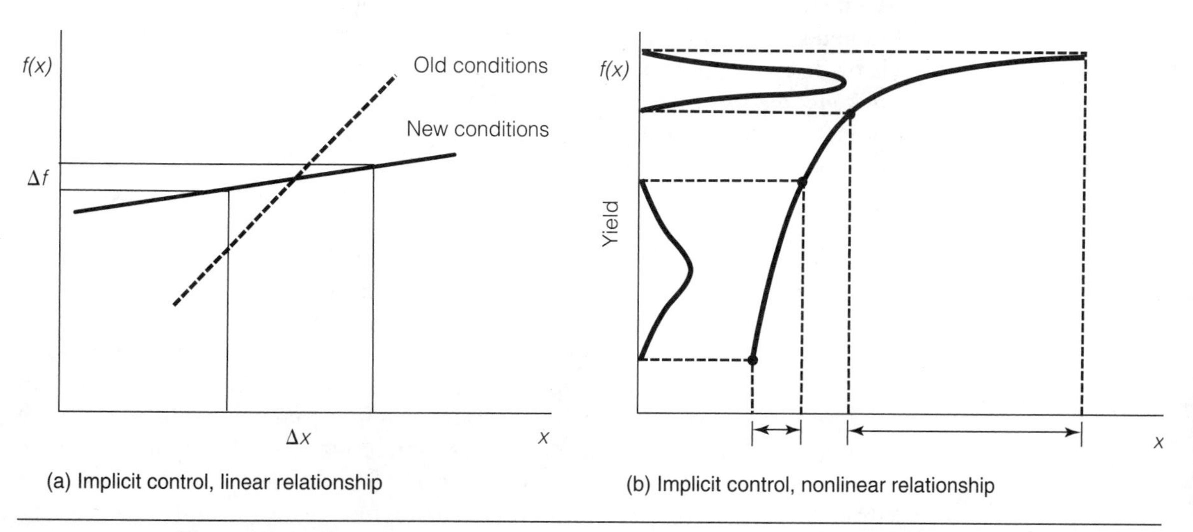

Figure 13.3 Illustration of implicit control in linear and nonlinear relationships

out defective items after the fact. There is ample evidence that as quality objectives become more demanding, it is not possible to "inspect quality in"; the only solution is to take Taguchi's lead and design it in.

How does one determine the best values of the design parameters? Taguchi's answer is: by designing experiments that are revealing. In fact, Taguchi's development of experimental-design techniques is, from his perspective, solely an issue of supporting the mechanism by which optimization can take place. As noted in the chapter introduction, he did this development work only because he believed that without a quick, user-friendly way to design experiments, engineers simply wouldn't perform them—they wouldn't be able (or willing) to wait for a consultant to arrive a few days later to design the experiment. Taguchi also suggests that the dependent variable in such experiments be not just the traditional choice of the mean of the quality characteristic but something he calls "S/N (signal-to-noise) ratio." An example is S/N $= \bar{Y}/S$ ($S =$ standard deviation) for a performance measure where higher is better; if either the mean performance increases, or the standard deviation of performance decreases, S/N would increase. In this example, S/N is essentially the reciprocal of the coefficient of variation. For those with a financial bent, this S/N is akin to the familiar return-to-risk ratio; for a given level of risk, maximize expected return, or for a given expected return, minimize risk. Of course, estimation of the standard deviation, which is required for determining S/N ratio, typically requires replication.

13.4 Taguchi Methods: Designing Fractional-Factorial Designs

To facilitate his objective of determining the optimal level of each design parameter (his term—we call them factors), and realizing that the way to do this is by designing appropriate experiments, Taguchi popularized the use of *orthogonal arrays* as an easy way to design fractional-factorial experiments.[1] As noted, his belief was that design of experiments must be simplified if it were to be embraced by the nonspecialist; specifically, he meant manufacturing engineers and others on the shop floor. As it turned out, the field of application for his methods is much broader.

Taguchi began by selecting several "good" basic fractional-factorial designs and, for each, setting up a table, which he calls an orthogonal array. These tables can be used in very simple ways to design an experiment, as described in the next paragraph. For two-level designs, he includes what he calls an L_4 orthogonal array for up to three factors, an L_8 orthogonal array for up to seven factors (used in reality for four through seven factors, since for three or fewer one would use the L_4), an L_{16} orthogonal array for up to 15 factors, and so on up to an L_{128} orthogonal array for (shudder!) up to 127 factors. For three-level designs, he provides an L_9 orthogonal array for up to four factors, an L_{27} orthogonal array for up to 13 factors, and an L_{81} orthogonal array for up to 40 factors. Taguchi also has constructed other specialized arrays, such as an L_{12} orthogonal array for up to 11 factors, but which requires that all interactions are zero. Finally, there are ways to use the two-level tables for factors with 4, 8, 16, . . . levels and the three-level tables for 9, 27, . . . levels.

The arrays are organized so that factors/effects are placed across the top of the table (orthogonal array) as column headings. The rows correspond to treatment combinations. The subscript, for example 8 in L_8, corresponds to the number of rows (treatment combinations). The number of columns in a two-level design array is always one fewer than the number of rows and corresponds to the number of degrees of freedom available: for example, eight treatment combinations corresponds to seven columns, which is akin to seven degrees of freedom, which, as we know, means we can estimate up to seven orthogonal effects. Once the orthogonal array has been selected (that is, the "L whatever"), the experimental-design process is simply the assignment of effects to columns.

Table 13.1 shows Taguchi's L_8. Where we would have − and +, or −1 and +1, Taguchi has 1 and 2 as elements of his table; the low level of a factor is designated by 1 and the high level by 2. Taguchi uses the term "experiment number" where we would say "treatment combination." We could show, by replacing each 1 and 2 of Taguchi's table by our −1 and 1, respectively, that Taguchi's tables are orthogonal—that is, the inner product of any two different columns is zero.

Table 13.1 Taguchi's L_8

Exp.* Number	Column 1	Column 2	Column 3	Column 4	Column 5	Column 6	Column 7
1	1	1	1	1	1	1	1
2	1	1	1	2	2	2	2
3	1	2	2	1	1	2	2
4	1	2	2	2	2	1	1
5	2	1	2	1	2	1	2
6	2	1	2	2	1	2	1
7	2	2	1	1	2	2	1
8	2	2	1	2	1	1	2

*Exp. = experiment in all tables for Chapter 13.

Experiments without Interactions

If we can assume that there are no interactions, we can simply assign the factors to columns arbitrarily.

EXAMPLE 13.2 Seven Factors with No Interactions

As an example of assigning main effects only, say we are seeking a 2^{7-4} design with main effects only. We need seven degrees of freedom and could use the L_8. Let's call these factors A, B, C, D, E, F, and G. Given that we have complete freedom in our choice, we'll choose the alphabetical order, as shown in Table 13.2.

By inspection of the Table 13.2 rows (in particular, noting which factors are at high level), we see that the treatment combinations are, as noted, 1, *defg*, *bcfg*, *bcde*, *aceg*, *acdf*, *abef*, and *abdg*. We can determine that this is the principal block of the defining relation

$$I = ADE = BDF = DEFG = ABC = ABEF = AFG = BCDE = BEG$$
$$= ACDF = ABCDEFG = ABDG = CEF = BCFG = ACEG = CDG$$

where the first four terms are independent generators (that is, once we have them, we can multiply all sets of two-terms-at-a-time, three-terms-at-a-time, and four-terms-at-a-time to produce the other 11 "consequential" terms of the

Table 13.2 An Assignment of Effects for a Taguchi L_8

	A	*B*	*C*	*D*	*E*	*F*	*G*	
Exp. Number	**Column 1**	**Column 2**	**Column 3**	**Column 4**	**Column 5**	**Column 6**	**Column 7**	**Treatment Combinations**
1	1	1	1	1	1	1	1	1
2	1	1	1	2	2	2	2	*defg*
3	1	2	2	1	1	2	2	*bcfg*
4	1	2	2	2	2	1	1	*bcde*
5	2	1	2	1	2	1	2	*aceg*
6	2	1	2	2	1	2	1	*acdf*
7	2	2	1	1	2	2	1	*abef*
8	2	2	1	2	1	1	2	*abdg*

defining relation). All terms have at least three letters; hence, all seven main effects are in separate alias "rows" or groups (seven groups, each containing 16 aliased effects). Because each main effect is in a separate alias row, none are aliased with other main effects, so all are clean under the assumption that all interactions are zero. Recall that we can obtain a block other than the principal block, should that be desirable, by multiplying by *a*, or *abc*, or any other treatment combination not in the principal or other already-examined block, as discussed in Chapter 10. (Of course, as Taguchi envisions the situation, the benefit of using these orthogonal arrays to design an experiment is that one doesn't need to find an appropriate defining relation!)

EXAMPLE 13.3 Five Factors with No Interactions

What if we have an application, again with no interaction, in which we want to study fewer than seven factors? In such a case we can still use an L_8, but use only a portion thereof. For example, if we want to study five factors, A, B, C, D, and E, in a main-effects-only design, we could use the design in Table 13.3. The treatment combinations would be 1, *de*, *bc*, *bcde*, *ace*, *acd*, *abe*, and *abd*, as shown in the last column. All five main effects would be clean, and with some intelligent examination we could determine the defining relation for which

Table 13.3 Using a Portion of an L_8

	A	*B*	*C*	*D*	*E*	
Exp. Number	Column 1	Column 2	Column 3	Column 4	Column 5	Treatment Combination
1	1	1	1	1	1	1
2	1	1	1	2	2	*de*
3	1	2	2	1	1	*bc*
4	1	2	2	2	2	*bcde*
5	2	1	2	1	2	*ace*
6	2	1	2	2	1	*acd*
7	2	2	1	1	2	*abe*
8	2	2	1	2	1	*abd*

this is the principal block. However, if it is true that all interactions are zero, one could argue that the defining relation doesn't matter. In the real world, the defining relation might still be useful to do a type of sensitivity analysis, especially if the results "seem strange"—more about this later in the chapter.

Thus, we can span the range of all possible instances in which we seek to estimate some number of main effects, when interactions are assumed to be zero, with a relatively small catalog of Taguchi orthogonal arrays. For one to three factors, we would use an L_4; for four to seven factors, we would use an L_8; and so forth. To complete our discussion, note that we need not have dropped the last two columns of the L_8 in the previous example; we could have dropped any two columns. The design may then have been different (that is, it might have had a different defining relation, leading to a different principal block), or it might have come out the same; either way, the same result of having all five main effects clean would have obtained.

Experiments with Interactions

What about cases in which we cannot assume that all interactions are zero? In such instances, we can still use orthogonal arrays, but we must be a bit more careful about the assignment of factors to columns. For each orthogonal array, Taguchi gives guidance, via what he calls *linear graphs,* as to which assignments are appropriate for specific interactions that are to be estimated (that is, those not assumed to be zero). Linear graphs for the L_8 are depicted in Figure 13.4. The key to these linear graphs is that the numbers refer to column numbers of the L_8, and a column number on the line connecting two vertices ("corners") represents the interaction between the two vertices. That is, if the interaction between two corner factors is to be estimated cleanly *at all,* it must be assigned the column number of the line connecting those corners.

Given that we assume some interactions are not zero, we are limited to fewer than seven factors using an L_8, because each two-factor interaction of two-level factors uses up one of the seven degrees of freedom available.

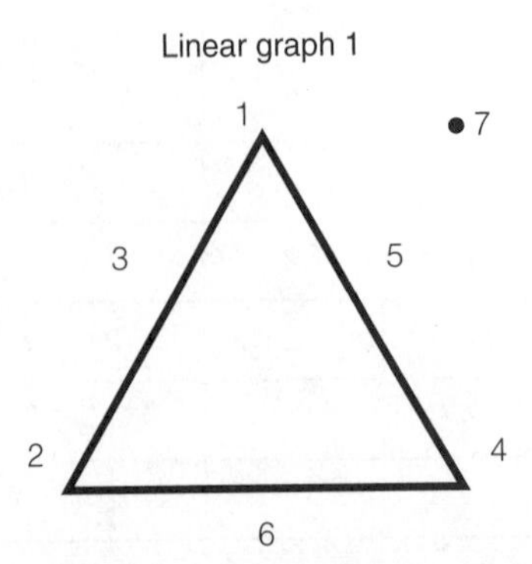

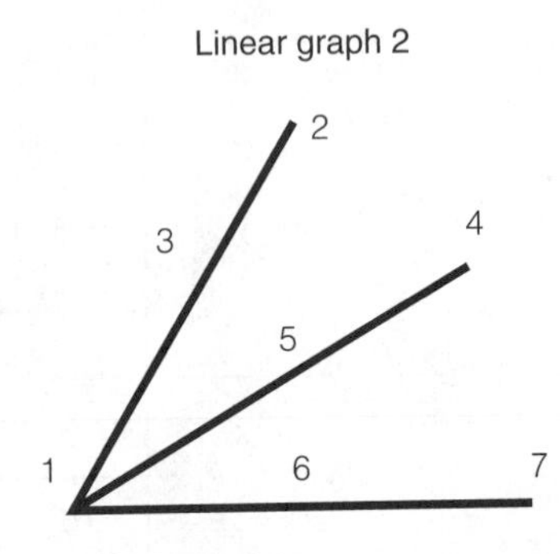

Figure 13.4 The L_8 linear graphs

EXAMPLE 13.4 Four Factors and Three Possible Interactions

Suppose that we wish to study four factors, A, B, C, and D, and we know that nothing interacts with D, although other two-factor interactions may be nonzero. Thus, we need to cleanly estimate *A*, *B*, *C*, *D*, *AB*, *AC*, and *BC*. (As usual, we assume that all three-factor and higher-order interactions are zero.) Since we are seeking seven clean effects, we would consider an L_8.[2] We connect the factors that (may) interact; Figure 13.5 shows the result.

Figure 13.5 "looks like" (technically, is "topologically equivalent to") the L_8 linear graph 1. We thus use linear graph 1, assigning factors and interactions to columns in Table 13.4 directly in accordance with linear graph 1.

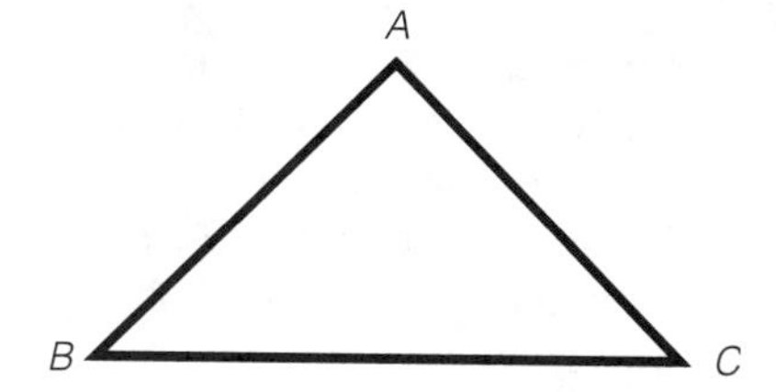

Figure 13.5 Example connector graph for estimating interaction effects *AB*, *AC*, *BC*

Table 13.4 Use of an L_8 with Interactions

	A	*B*	*AB*	*C*	*AC*	*BC*	*D*	
Exp. Number	**Column 1**	**Column 2**	**Column 3**	**Column 4**	**Column 5**	**Column 6**	**Column 7**	**Treatment Combinations**
1	1	1		1			1	1
2	1	1		2			2	*cd*
3	1	2		1			2	*bd*
4	1	2		2			1	*bc*
5	2	1		1			2	*ad*
6	2	1		2			1	*ac*
7	2	2		1			1	*ab*
8	2	2		2			2	*abcd*

Since we have only four factors, we need specify only the levels for those four; in fact, one is better off to physically blank out the 1's and 2's of the interaction columns, simply to avoid confusion. The treatment combinations in Table 13.4 are 1, *cd*, *bd*, *bc*, *ad*, *ac*, *ab*, and *abcd*, which happen to form the principal block for the 2^{4-1} with $I = ABCD$.

EXAMPLE 13.5 Five Factors and Two Possible Interactions

Suppose we want to study five factors, A, B, C, D, and E, and have two interactions that cannot be assumed to be zero: *AB* and *AC*. When we connect the factors that may interact, we get the shape in Figure 13.6. This "picture" is part of both linear graphs in Figure 13.4! Either one can be used. We can thus use linear graph 1, with *A* in column 1, *B* in column 2, and *C* in column 4. Then *AB* would be assigned to column 3, and *AC* to column 5. (By the way, if the two interactions we were interested in were *AB* and *BC*, then, with *A* in 1, *B* in 2, and *C* in 4, we would assign *AB* to column 3 and *BC* to column 6.) We then have the assignment of effects to columns as shown in Table 13.5.

A few additional aspects of this example:

1. We could just as well have assigned *E* to column 6, and *D* to column 7, instead of the Table 13.5 assignment of *D* to 6 and *E* to 7.
2. Using linear graph 2 would ultimately have led to essentially the same result.
3. As in earlier examples, there are other, essentially equivalent, assignments of factors to columns—for example, simply exchanging *B* and *C*, assigning *B* to column 4 and *C* to column 2 (with the corresponding change in assignments of the *AB* and *AC* interactions).

What we have produced in Table 13.5 is a 2^{5-2} fractional-factorial design with defining relation $I = BCD = ADE = ABCE$. The treatment combinations are 1, *cde*, *bde*, *bc*, *ae*, *acd*, *abd*, and *abce*. The aliased effects are in seven rows of four effects each, as shown in Table 13.6. Notice that in the Table 13.5 array, relative to Table 13.4, *D* has replaced *BC*, based on the assignments made using linear graph 1. The position represented by the line connecting vertices *B* and *C* (the number 6 in linear graph 1) can be used for interaction *BC* (and

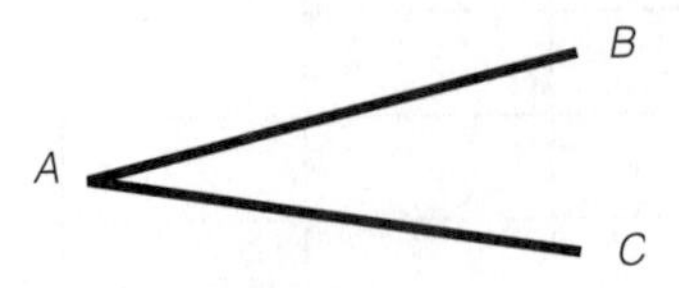

Figure 13.6
Connector graph for interaction effects *AB*, *AC*

Table 13.5 (Another) Use of an L_8 with Interactions

	A	*B*	*AB*	*C*	*AC*	*D*	*E*	
Exp. Number	**Column 1**	**Column 2**	**Column 3**	**Column 4**	**Column 5**	**Column 6**	**Column 7**	**Treatment Combinations**
1	1	1		1		1	1	1
2	1	1		2		2	2	*cde*
3	1	2		1		2	2	*bde*
4	1	2		2		1	1	*bc*
5	2	1		1		1	2	*ae*
6	2	1		2		2	1	*acd*
7	2	2		1		2	1	*abd*
8	2	2		2		1	2	*abce*

indeed must be, if *BC* is to be estimated), but it can also be used for any main effect, for example *D* in the Table 13.5 case. That is why *D* and *BC* are in the same alias row, as shown in bold in Table 13.6. In other words, one could say that any main effect can "override" an interaction on a linear graph, but then that interaction will not be obtainable elsewhere.

Table 13.6 Aliased Effects for the 2^{5-2}

I	*BCD*	*ADE*	*ABCE*
A	*ABCD*	*DE*	*BCE*
B	*CD*	*ABDE*	*ACE*
C	*BD*	*ACDE*	*ABE*
D	***BC***	*AE*	*ABCDE*
E	*BCDE*	*AD*	*ABC*
AC	*ABD*	*CDE*	*BE*
AB	*ACD*	*BDE*	*CE*

EXAMPLE 13.6 A Possible Three-Way Interaction

This example comes from a real problem studied by one of the authors. It was necessary to evaluate cleanly the effect of five factors, *A*, *B*, *C*, *D*, and *E*, one two-factor interaction, *BD*, and one three-factor interaction, *ABD*. All other interactions could safely be assumed to be zero or negligible. As noted in earlier chapters, it is rare, but not unheard of, to seek a clean estimate of a three-factor interaction. We show how to do this using the L_8 linear graphs; this type of problem has appeared in various guises in other texts, using various methods of description. (For convenience, Figure 13.7 repeats the two linear graphs from Figure 13.4.)

The assignment of the five main effects using linear graph 1 is shown in Figure 13.8. *A*, *B*, and *D* are, respectively, at vertices 1, 2, and 4. The absence of a nonzero *AB* (by assumption) allows *C* to be placed on the line connecting *A* and *B*, "number 3." Similarly for *E*, placed on the line connecting *A* and *D*, "number 5." The two-factor interaction *BD* is between *B* and *D*, corresponding to "number 6."

Much more subtle is the assignment of the three-factor interaction *ABD*. For this we refer to linear graph 2, as shown in Figure 13.9. Vertex number 1

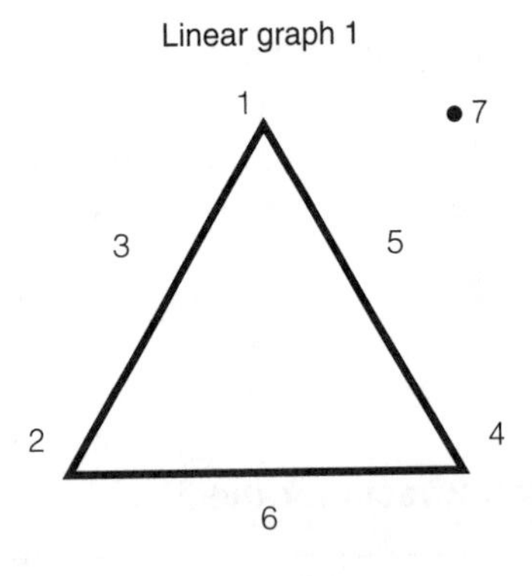

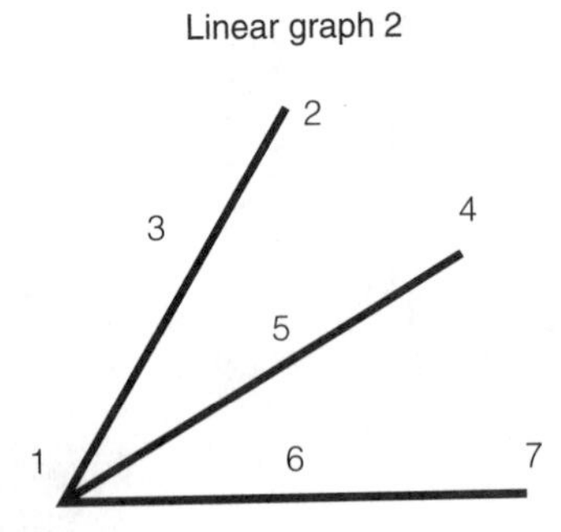

Figure 13.7 Linear graphs for L_8 (repeated)

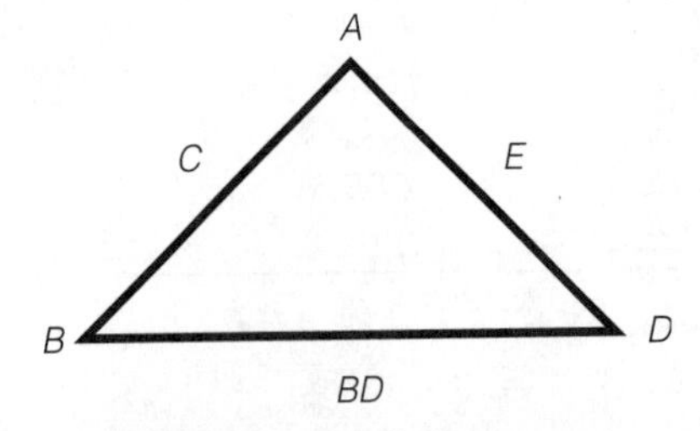

Figure 13.8 Linear graph 1, effects assigned

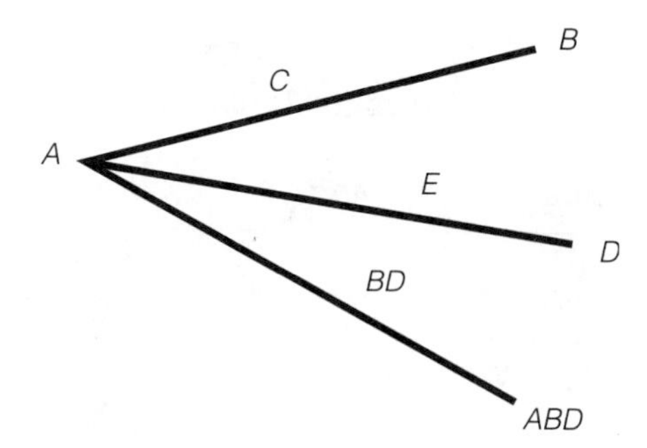

Figure 13.9 Linear graph 2, effects assigned

is still *A* and the line corresponding to number 6 is still *BD*—the numbers corresponding to the effects *must be* the same from linear graph to linear graph. Yet, on linear graph 2, *BD*, number 6, must be the mod-2 multiplication of *A*, vertex number 1, and whatever is represented by number 7; therefore, number 7 *must be ABD* (which brought great delight, since that was the task to be achieved).

These assignments and the treatment combinations are shown in the L_8 orthogonal array in Table 13.7. The defining relation and the seven sets of four aliased effects for this 2^{5-2} fractional-factorial design are listed in Table 13.8.

Table 13.7 Three-Way Interaction Example with a Taguchi L_8

	A	*B*	*C*	*D*	*E*	*BD*	*ABD*	
Exp. Number	Column 1	Column 2	Column 3	Column 4	Column 5	Column 6	Column 7	Treatment Combinations
1	1	1	1	1	1			1
2	1	1	1	2	2			*de*
3	1	2	2	1	1			*bc*
4	1	2	2	2	2			*bcde*
5	2	1	2	1	2			*ace*
6	2	1	2	2	1			*acd*
7	2	2	1	1	2			*abe*
8	2	2	1	2	1			*abd*

Table 13.8 Aliased Rows for Table 13.7

I	*ADE*	*ABC*	*BCDE*
A	DE	BC	ABCDE
B	ABDE	AC	CDE
C	ACDE	AB	BDE
D	AE	ABCD	BCE
E	AD	ABCE	BCD
BD	ABE	ACD	CE
ABD	BE	CD	ACE

13.5 Taguchi's L_{16}

As discussed earlier, there is a catalog of Taguchi orthogonal arrays and each has associated linear graphs. An exhaustive presentation of all of Taguchi's orthogonal arrays and the corresponding linear graphs is beyond the scope of this text. They all appear in Taguchi's presentation of his methods, a two-volume book listed in the Chapter 15 references. However, we do present Taguchi's L_{16} in Table 13.9, and a subset of its corresponding linear graphs in Figure 13.10.

Working with the L_{16} linear graphs reminds us of the complexity of playing mah-jongg, but with a twist on one facet of the game. For those not familiar with mah-jongg, think of it as similar to gin rummy with one major difference: whereas in gin rummy all "melds" are acceptable (three of a kind [say, three 5's] or a run of three cards in one suit [say, 6, 7, 8 of spades]), in mah-jongg only a subset of melds is acceptable (melds are certain combinations of the 144 tiles used). A world organization of mah-jongg changes the acceptable subset of melds each year so that the experience factor doesn't preordain who will win. This is because the social aspect of the game does not allow a player to spend a long time at each turn looking at the long list of acceptable melds; yet, a winning strategy involves knowing the acceptable melds so as to make moves that maximize the number of ways to complete melds—as in gin rummy. The more you practice with the L_{16} linear graphs, the more quickly you find yourself able to home in among a multitude of possibilities on the most appropriate linear graph for a design. But here experience does help: no world organization changes the acceptable linear graphs each year!

13.6 Experiments Involving Nonlinearities or Factors with Three Levels

In Chapter 12 we discussed the use of 3^k designs to study nonlinear effects. Here we look briefly at Taguchi's three-level orthogonal arrays toward that

Table 13.9 Taguchi's L_{16}

Exp. No.	Col. 1	Col. 2	Col. 3	Col. 4	Col. 5	Col. 6	Col. 7	Col. 8	Col. 9	Col. 10	Col. 11	Col. 12	Col. 13	Col. 14	Col. 15
1	1	1	1	1	1	1	1	1	1	1	1	1	1	1	1
2	1	1	1	1	1	1	1	2	2	2	2	2	2	2	2
3	1	1	1	2	2	2	2	1	1	1	1	2	2	2	2
4	1	1	1	2	2	2	2	2	2	2	2	1	1	1	1
5	1	2	2	1	1	2	2	1	1	2	2	1	1	2	2
6	1	2	2	1	1	2	2	2	2	1	1	2	2	1	1
7	1	2	2	2	2	1	1	1	1	2	2	2	2	1	1
8	1	2	2	2	2	1	1	2	2	1	1	1	1	2	2
9	2	1	2	1	2	1	2	1	2	1	2	1	2	1	2
10	2	1	2	1	2	1	2	2	1	2	1	2	1	2	1
11	2	1	2	2	1	2	1	1	2	1	2	2	1	2	1
12	2	1	2	2	1	2	1	2	1	2	1	1	2	1	2
13	2	2	1	1	2	2	1	1	2	2	1	1	2	2	1
14	2	2	1	1	2	2	1	2	1	1	2	2	1	1	2
15	2	2	1	2	1	1	2	1	2	2	1	2	1	1	2
16	2	2	1	2	1	1	2	2	1	1	2	1	2	2	1

end. The two most common three-level orthogonal arrays are the L_9, which has four columns and nine treatment combinations, allowing study of up to four factors, and the L_{27} with 13 columns and 27 treatment combinations, accommodating up to 13 factors. Each main effect of a factor having three levels uses up two degrees of freedom. Thus, four three-level factors would use up four times two equals eight degrees of freedom; this is why an L_9 has nine rows, but only four, not eight, columns. Similarly, the 27-row L_{27} has 13 columns. An interaction effect between two three-level factors would use up two times two equals four degrees of freedom.

The L_9 is shown in Table 13.10. Each column corresponds to one factor or is part of a set of columns representing an interaction. Following the earlier

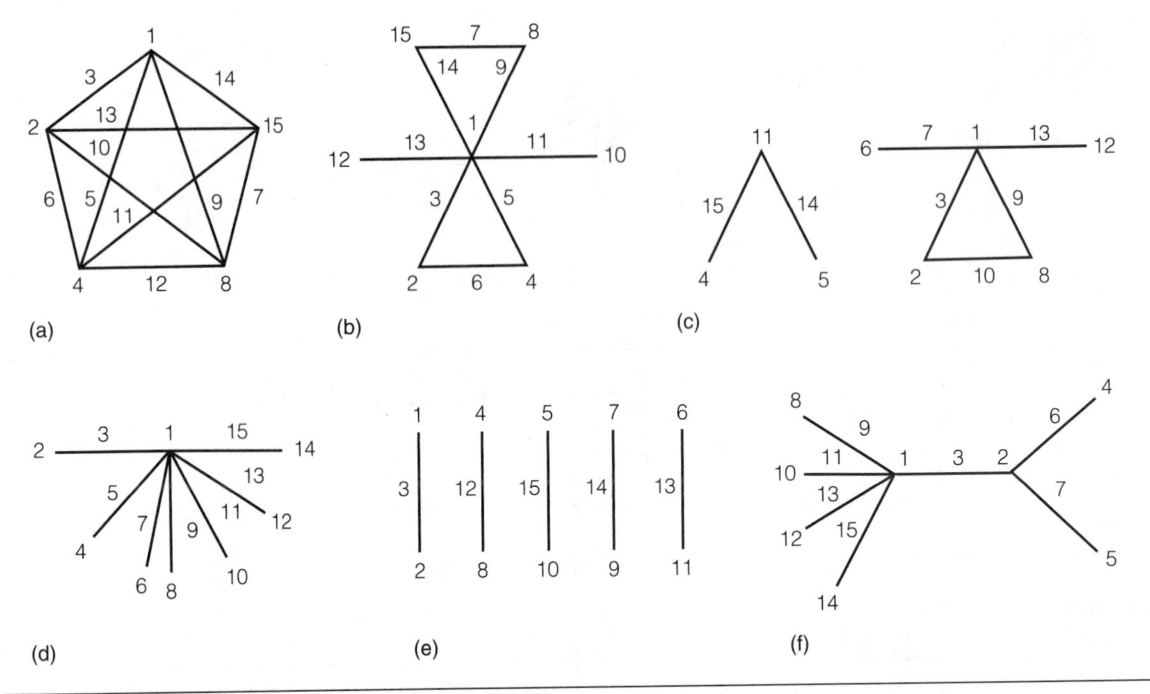

Figure 13.10 A subset of L_{16} linear graphs

Table 13.10 Taguchi's L_9

Exp. Number	Column 1	Column 2	Column 3	Column 4
1	1	1	1	1
2	1	2	2	2
3	1	3	3	3
4	2	1	2	3
5	2	2	3	1
6	2	3	1	2
7	3	1	3	2
8	3	2	1	3
9	3	3	2	1

Taguchi notation, the low, medium, and high levels of a factor are represented in the table body by the numbers 1, 2, and 3, respectively. Here, too, we follow Taguchi's convention of calling each treatment combination an "experiment"; hence the column heading "Exp. Number."

EXAMPLE 13.7 Contamination Study

By way of illustration, consider a contamination study that is intended to probe the possible nonlinear effects of five factors—temperature of chemical X, temperature of chemical Y, stirring time for chemical Z, amount of chemical V, and amount of heel (residue from previous application); the five factors, as well as their effects, are indicated by the symbols *X*, *Y*, *Z*, *V*, and *H*, respectively. Interactions are known to be zero or negligible. A previous two-level experiment, using five columns of an L_8, indicated that *X* and *Z* were significant, and that *Y*, *V*, and *H* were not significant.

Subsequent discussion with process experts has raised a concern about nonlinearity. Were we to study all five factors at three levels (the minimum number of levels that allows the study of nonlinearity), we would need to use an L_{27}. If we can reasonably assume that one of the nonsignificant factors is very likely to be linear (which would mean that it would continue to be nonsignificant even if a third level were added), we can drop that one factor from further consideration and use an L_9. Suppose that we have concluded that the nonsignificant *H* is linear; accordingly, we drop *H* and proceed with the others.

The assignment of factors is shown in Table 13.11. For ease of use, Table 13.11 shows both the factor name and level for each level of each factor (Y2, Z1, and so on). Table 13.12 lists the average experimental yield corresponding to each factor level. It is instructive to plot these values, as done in Figure 13.11.

Earlier, when examining only two levels of the factors, effects *X* and *Z* were found to be significant but *Y* and *V* were not. The low and high levels for each of these factors in the second, L_9, three-level design are the same values used in the original L_8 two-level experiment. The medium level of each factor is truly a middle level in each case, halfway between the high and low levels (although as noted in Chapter 12, this generally is not critical to three-level experimentation). Effects *Y* and *Z*, as we see from the plots in Figure 13.11, are both linear, as had been earlier assumed. Furthermore, *Y* is still not significant, and *Z* still is significant. In essence, the results for *Y* and *Z* are unchanged. Such is not the case for *X* and *V*, each in a different way. Although *X* has a strong quadratic component, it is still monotonic decreasing (that is, continually decreasing with increasing level of factor X); *X* was, and is still, significant. What has changed with respect to factor X—that the response curve is not precisely linear—is not that dramatic. However, the change in our conclusions about factor V is dramatic! It had been concluded that V had no effect on the amount of contaminant, because at the two levels chosen initially,

Table 13.11 Taguchi's L_9 — Contamination Study

	Y	*X*	*Z*	*V*
Exp. Number	**Column 1**	**Column 2**	**Column 3**	**Column 4**
1	Y1	X1	Z1	V1
2	Y1	X2	Z2	V2
3	Y1	X3	Z3	V3
4	Y2	X1	Z2	V3
5	Y2	X2	Z3	V1
6	Y2	X3	Z1	V2
7	Y3	X1	Z3	V2
8	Y3	X2	Z1	V3
9	Y3	X3	Z2	V1

Table 13.12 Average Yields

Factor	Relevant Experiments	Average Yield
Y1	1, 2, 3	7.57
Y2	4, 5, 6	7.55
Y3	7, 8, 9	7.53
X1	1, 4, 7	7.68
X2	2, 5, 8	7.45
X3	3, 6, 9	7.42
Z1	1, 6, 8	7.79
Z2	2, 4, 9	7.55
Z3	3, 5, 7	7.31
V1	1, 5, 9	7.53
V2	2, 6, 7	7.90
V3	3, 4, 8	7.57

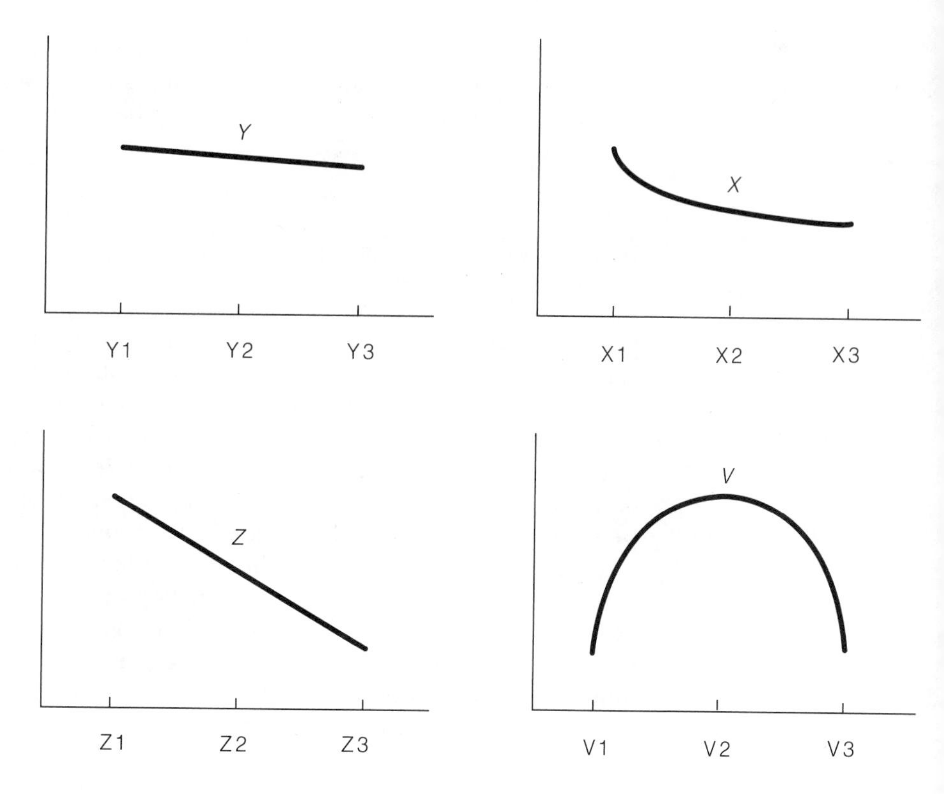

Figure 13.11 Plot of values for levels of the four factors Y, X, Z, V

the yield was of similar magnitude (7.53 for V1 versus 7.57 for V3). Now that we see the result at the middle level (7.90 at V2), our attitude toward the choice of the level of V, and how important that choice is, changes.

Remember the Taguchi loss function objective: to minimize both variability and the departure of the mean result from its target value, using the most economic combination of these goals. One thing is clear: we pick the least expensive level of Y (and H) since we have concluded that each does not affect the yield. The nonlinear effects of X and V provide an opportunity for variability reduction; it's likely (the exact choices depend on monetary values not specified in this example) that we'll select the level of X and V where the curve is flattest (minimum slope), or at least relatively flat. The level of X would likely be set somewhere between X2 and X3, and the level of V is likely to be set near V2. Finally, Z will be used to minimize the difference between the mean and the target in a way that minimizes the cost of quality. Note that even if the target is, theoretically, zero, this does not mean that we choose the highest level of Z. The cost of implementing that highest level of Z must be taken into consideration—it might be prohibitively high. This example did not explicitly consider that aspect of the overall decision process.

Remember that our use of a 3^{k-p} is equivalent to adopting a quadratic model. We can use our three data points (that is, the mean yield at the three levels of the factor) to calculate, for each factor, the coefficients of a quadratic equation that goes through the three points, and use that quadratic equation to plot our best estimate of the relationship between factor level and yield.

13.7 Further Analyses

In a real-world problem, the experimental design is only part (albeit a *vital* part) of the story. Once experimental results are obtained, they need to be analyzed. The statistical significance of each effect is usually determined. Next, either further experiments are performed, as suggested by the results just obtained, or the best level of each factor is determined without further experimentation (at least in the *near* future). Then, before the results of the experiment are implemented, they are further considered in a variety of ways.

Most often, the optimal combination of levels of factors indicated by the data is not one of the treatment combinations actually run in the experiment. This is simply a matter of elementary probability; after all, if we run a 2^{6-2} fractional-factorial design, we are running only one-fourth (25%) of the total number of combinations, 64. Thus the odds are 48:16, or 3:1, against the treatment combination determined to be optimal having been among the quarter replicate of 16 we happened to have chosen to run. Prudent management would include a step to verify that this presumed optimal treatment combination actually produces the yield predicted by the experimental results.

Finally, if the verification test is satisfactory, the central question, not addressed directly by the experimental design or its analysis, remains: Is the anticipated improvement in the yield economically justified? Remember, Taguchi does not argue for higher quality in the abstract; he advocates quality improvements only when they are economically justified. Were this not the case, his philosophy and methods might not have been so readily adopted by many of the world's industry leaders!

The following example, which is adapted from a real-world case (only slightly disguised), includes the steps after the experiment is run.

EXAMPLE 13.8 Electronic Component Production

Our example concerns the production of electronic components. The quality characteristic ("yield") of interest is output voltage of a generator. The target value, T, is 1.600 volts with a "tolerance" of ± 0.350 volts. Six factors are to be studied, each at two levels; this corresponds to $2^6 = 64$ possible treatment combinations. Experts in the field had concluded that there were no non-negligible interactions among the factors. Thus, there are six effects, the six main effects, to be estimated cleanly, and from what we've discussed in this chapter, we know that an L_8 will suffice.[3] It is decided to have four replicates at each of the eight treatment combinations.

It might be argued that one would never run an experiment like this—that instead of four replicates of a 2^{6-3} design, one would always run a 2^{6-1}; after all, both require the same number of generators, 32. In this real-world situation, it was very expensive to set up for a treatment combination, but *the cost for replicates for the same treatment combination was relatively low.* Hence, running four replicates of the 2^{6-3} design was materially less costly than a 2^{6-1} would have been. If all data points did cost the same, we likely would run the 2^{6-1}. Both designs have the same variance of an estimate (recall the section of Chapter 9 on errors in two-level factorial designs). Yet, the half-replicate would have less aliasing to be concerned with, while there would be plenty of higher-order interaction effects to comprise an error term. The lesson: the real world intrudes!

As we know, since there are assumed to be no nonzero interactions, the assignment of factors to columns can be arbitrary. Table 13.13 shows the assignment actually used.

An analysis of the statistical significance of the effects under study revealed that factors C, D, and E have significant effects, as indicated in Table 13.14. The means for the two levels of the significant factors are listed in Table 13.15.

Given that the grand average is somewhat below the target of 1.600 volts, we get as close as possible to the target by picking the level corresponding to

Table 13.13 Assignment of Factors in L_8, Electronic Components Example

	A		*B*	*C*	*D*	*E*	*F*	
Exp. Number	**Column 1**	**Column 2**	**Column 3**	**Column 4**	**Column 5**	**Column 6**	**Column 7**	**Treatment Combinations**
1	1		1	1	1	1	1	1
2	1		1	2	2	2	2	*cdef*
3	1		2	1	1	2	2	*bef*
4	1		2	2	2	1	1	*bcd*
5	2		2	1	2	1	2	*abdf*
6	2		2	2	1	2	1	*abce*
7	2		1	1	2	2	1	*ade*
8	2		1	2	1	1	2	*acf*

Table 13.14 Statistical Significance

Effect	*P*-Value	Significant
A	.4348	No
B	.3810	No
C	.0001	Yes
D	.0108	Yes
E	.0010	Yes
F	.5779	No

the larger yield for each of the three significant factors (C, D, and E). Since there is a lack of evidence that the level of each of the remaining three factors affects the voltage, their level should be picked simply to minimize cost (that is, with no difference between the levels, the cheaper the better!). The optimal choice turns out to be C2, D2, and E1, which maximize the expected output voltage, and A2, B2, and F1, which are the less expensive levels for A, B, and F. What is the expected voltage at this (presumably) optimal combination of levels? The expected yield is

$$1.398 + .090 + .038 + .045 = 1.571 \text{ volts}$$

In case the process of determining this 1.571 value doesn't seem intuitive, consider: 1.398 is the overall average that includes half of the treatment combinations at C1 and half at C2. However, if we include only the treatment combinations at C2, we get the .090 "benefit," without the "compensating" −.090 that would bring us back to the average of 1.398. Thus, the average result would increase by .090, becoming 1.488. However, the 1.488 includes half of

Table 13.15 Experimental Results

Factor Level	Average (Volts)	Difference from Grand Mean (Volts)
C1 (low)	1.308	−.090
C2 (high)	1.488	+.090
D1 (low)	1.360	−.038
D2 (high)	1.436	+.038
E1 (low)	1.443	+.045
E2 (high)	1.353	−.045
Overall average = 1.398		

the treatment combinations at D1 and half at D2; if we include in the average only the treatment combinations at D2, we get an increase in the average of .038, and so forth.

Now that we've determined the (presumably) optimal treatment combination, we need to confirm its merit, from both the performance and cost perspectives, before implementation.

Confirmation In our example, we ran 8 of the 64 possible treatment combinations. It was unlikely (seven-to-one odds against it) that we actually ran the treatment combination that has been determined thus far to be optimal. Indeed, we did not run A2, B2, C2, D2, E1, F1 (which in our two-level factorial design terminology is *abcd*). Accordingly, as noted earlier, prudence dictates that we perform a confirmation experiment to verify that *abcd* yields a voltage value around 1.571, as predicted. Why might we not get a value around 1.571? The possibilities include mundane mistakes like an incorrect value recorded, or misread handwriting. However, a more likely reason would be that the assumption that several interactions (in our example, *all* interactions) were zero or negligible was not valid.

How many runs, or replicates, of this presumed optimal treatment combination do we need in order to provide a confirmation? There's no clear-cut answer. The more, the better, but the more, the more expensive! If the cost per replicate is relatively cheap, once the first run is set up, the more the merrier. If the cost per replicate is relatively expensive, a minimum of four is suggested; roughly speaking, a sample size of four results in an average that, with 95% confidence, is within one standard deviation of the true (unknown at the time, of course) mean.

What happens if the confirmation test does not verify the earlier result? The first step should be to look for differences between the conditions under which the experiment was run, and the conditions under which the confirmation test was run. Were there important differences in humidity, line voltage, test equipment, personnel, and so on that could be the cause of the discrepancy? Absent that and obvious mistakes (incorrect recording of values, handwriting issues, and the like), it is usually appropriate to next revisit the assumptions that were made about interaction effects. Another brainstorming session involving the product and process experts would likely be advisable.

Remember, much worse than the inability to confirm earlier results before implementation is not checking at all and learning the bad news after spending considerable money and time making changes that are not helpful (and may be harmful). Bad news doesn't smell better with age! Also, if the confirmation test does verify previous results (which in practice occurs far more often than disconfirmation), it adds confidence in the proposed process, and implementation may become an "easier sell."

Economic evaluation of proposed solution Once we've verified that the proposed solution will provide the anticipated performance result, we proceed to see if it makes sense economically. Usually, this simply involves a cost/benefit analysis; for a process that has been ongoing, this involves a comparison between the current situation and the proposed solution (the treatment combination that appears optimal).

First, let's consider cost. The proposed solution might require an increase in cost, fixed or variable (or both). The fixed-cost increase could include installation of some piece of equipment, change in shop layout, additional training, and so forth. The variable cost may be in the form of additional labor, purchase cost of input materials (for example, a higher amount of a substance could simply cost more), or other costs. But sometimes, more often than one might expect, the change in cost is a net *reduction*, in addition to the improved performance.

The resultant change in fixed cost is often treated in the analysis by apportioning it over the quantity of product to be produced, thus merging it with the per-unit (variable) cost. The net change in per-unit cost (ΔC, which, again, may be positive or negative) is

$$\Delta C = V_{\text{new}} - V_{\text{current}}$$

where V indicates variable cost.

Next, to evaluate the benefit that accrues through the change in levels of the design parameters (factors), we return to Taguchi's quadratic loss function. Recall that the expected loss per unit is proportional to the sum of the square of the bias and the variance:

$$\overline{L} = K[(\overline{Y} - T)^2 + S^2]$$

where S^2 is our estimate of the variance. We evaluate the bias and variance for the current and proposed set of factor levels. These lead to our estimate of $\overline{L}_{\text{current}}$ and $\overline{L}_{\text{proposed}}$. The gross benefit per unit from the change of solution is the reduction in loss:

$$\Delta\overline{L} = \overline{L}_{\text{current}} - \overline{L}_{\text{proposed}}$$

The net benefit (NB) per unit of the change is then

$$\text{NB/unit} = \Delta\overline{L} - \Delta C$$

The total net benefit per *year* would then be the product of this quantity times the annual volume:

$$\text{Total net benefit per year} = (\text{NB/unit}) \bullet (\text{annual volume})$$

Table 13.16 is a worksheet that one company uses for calculation of annual benefits of a proposed change.

In our example, we noted that the proposed treatment combination had the following yield:

$$\overline{Y} = 1.571 \text{ volts}$$

The actual confirmation process did verify this result (eight runs at A2, B2, C2, D2, E1, F1 yielded an average value of 1.568). The standard deviation estimate, based on these eight replicates, was

$$S = .079$$

Table 13.16 **Worksheet for Annual Benefits of a Proposed Change**

Benefits	Proposed	Current
a. Loss function constant (K)		
b. Bias from target value ($\Delta = \overline{Y} - T$)		
c. Square of bias (Δ^2)		
d. Variance of process (S^2)		
e. Sum of (c) and (d) ($\Delta^2 + S^2$)		
f. Total loss / unit, (a) • (e) $=[K(\Delta^2+S^2)]$		
g. Output / year		
Total annual loss, (f) • (g)	(1)	(2)
Savings (annual benefits) = (2) − (1)		

The original treatment combination (the "current" situation) was A1, B2, C2, D1, E1, F1, with a corresponding predicted mean of (1.398 + .090 − .038 + .045), or

$$\overline{Y} = 1.495 \text{ volts}$$

(Before the experiment reported on here, the current mean was generally acknowledged as 1.49.) The standard deviation at this "current" treatment combination was

$$S = .038$$

If we compute the $\overline{L}$ values, we obtain

$$\overline{L}_{\text{proposed}} = K[(1.571 - 1.600)^2 + .079^2] = K(.007082)$$
$$\overline{L}_{\text{current}} = K[(1.495 - 1.600)^2 + .038^2] = K(.012694)$$

Thus the average loss per unit for the proposed solution is 44.2% = [100 (.012694 − .007082)/.012694] lower than that of the current solution. If we compare the two solutions, we see that what really differs is the level of factor A and the level of factor D. The level of factor D was determined to have a significant impact on the voltage, and the $\overline{L}$ values above indicate that the change in the level of factor D is "price-worthy." In addition, the current solution was using A1, the more expensive level of factor A, even though there was no indication that the level of factor A had any impact on the yield.

13.8 Perspective on Taguchi Methods

Taguchi's methods for designing an experiment—his orthogonal arrays—do not generate designs that can't be generated by the traditional methods covered in the previous chapters of the text. His methods find an appropriate set of treatment combinations perhaps *more quickly*, but do not produce designs *unique* to Taguchi methods. Some statistical software packages allow the choice "Taguchi designs," but their use of the phrase is, for the most part, a misnomer. The two software packages emphasized in this text, SPSS and JMP, do not acknowledge the Taguchi approach. Given this fact, along with the nonuniqueness of Taguchi designs, we have not presented any software instructions or insights in this chapter.

Also, there is some controversy concerning designs that Taguchi's orthogonal arrays provide. Often, the arrays point to a design that isn't considered "as good," by certain criteria, as one can derive using traditional (Fisherian? Yatesian?) methods; some would then call the design "suboptimal." What do we mean by "not as good"?

For example, suppose that we wish to construct a design in which all interactions, except a select group of two-factor interactions, are assumed to be zero. As an extreme illustration, suppose that the "routine" Taguchi design aliases some of the main effects and the select two-factor interactions with some nonselected two-factor and some three-factor interactions, whereas using traditional methods, one can derive another design in which the main effects and select two-factor interactions are aliased with only four-factor and higher-order interactions. Is the Taguchi design not as good? If the assumptions being made are valid, that all interactions other than the select group are zero, then the main effects and select two-factor interactions are perfectly clean in the Taguchi design, as they are in the traditional design. However, one never knows for certain that an interaction is zero. It is axiomatic that, on average, the higher the order of the interaction, the more likely its true value is zero, so one can argue that a design that aliases the "important effects" with higher-level interactions is superior to a design that aliases these important effects with lower-level interactions—even if one assumes that the lower-level interactions are zero. In this sense, the design yielded by Taguchi's orthogonal arrays *may* provide a design that is not as good.

Another criticism of Taguchi's methods is that "a little knowledge is dangerous." In other words, some argue, it is dangerous for somebody to know which treatment combinations to run and analyze without knowing the aliasing structure, defining relation, and so on. We believe that although there is some "danger" (we'd probably say "minor peril" or "minimal pitfall potential"), the achievement of Taguchi's objectives is often more important than the "peril." Remember, if quick and easy methods are not available to the engineers/designers, experimentation may never get done at all. That's dangerous!

As an added point, one of the authors routinely provides designs to companies without providing any of the "backup" in terms of defining relations and alias groups. This hasn't hurt the companies' running the experiment, and analyzing and interpreting the data. In a few cases the results seemed to belie common sense; a call was then made to ask the author what potentially prominent interactions were aliased with the seemingly strange results. Yet this latter call could not have been made without somebody understanding that the main effects were, indeed, aliased with "other stuff." Thus, we agree that the ideal case would be for the engineers/designers to take a short course in the traditional methods of experimental design before then using Taguchi's methods.

In no way do we intend this section's discussion to diminish Dr. Taguchi or his methods. In fact, we believe that his contributions to the field of quality control, total quality management, and experiment design are important—certainly worth a full chapter of our text. Remember, a good experiment performed is better than an optimal experiment not performed!

EXAMPLE 13.9 New Product Development at HighTech Corporation, Revisited

HighTech Corporation set up classes on Taguchi methods, which were taught to all personnel above the level of "administrative staff," HighTech's job title for secretarial and clerical duties. New product developers and engineers began to conduct experiments without the need for outside consultation. Management agreed that productivity (which they did not define specifically) and, more importantly, results (which they defined in terms of actual new product development) increased.

HighTech management also commented on what they saw as useful byproducts of this company-wide commitment to Taguchi methods: a common language for all departments and personnel to use, and a "bottom-line" way for individual projects, as well as individual departments, to be evaluated.

Exercises

1. Given Taguchi's L_8 in Table 13EX.1, and referring to the linear graphs for the L_8 (Figure 13.4 or 13.7), find for a 2^{4-1} design the assignment of factors and interactions to columns, and thus the treatment combinations to run, if the effects to be estimated cleanly are *A, B, C, D, BC, BD,* and *CD*. All other interaction effects can be assumed to be zero.

2. Repeat exercise 1, with the following effects to be estimated cleanly: *L, M, N, O, P, LM,* and *MN* in a 2^{5-2} design.

3. Repeat exercise 1, with the following effects to be estimated cleanly: *A, B, C, D, E, AB,* and *CD* in a 2^{5-2} design.

4. Suppose that we are studying seven factors, A, B, C, D, E, F, and G, at two levels each,

Table 13EX.1 Taguchi's L_8

Exp. Number	Column 1	Column 2	Column 3	Column 4	Column 5	Column 6	Column 7
1	1	1	1	1	1	1	1
2	1	1	1	2	2	2	2
3	1	2	2	1	1	2	2
4	1	2	2	2	2	1	1
5	2	1	2	1	2	1	2
6	2	1	2	2	1	2	1
7	2	2	1	1	2	2	1
8	2	2	1	2	1	1	2

assuming no interaction effects. Based on an analysis using an L_8, four of the factors came out significant: B, C, F, and G. The quality characteristic is "the higher the better," and the mean value at each level of each significant factor is as follows:

MEANS

B1 = 1.3	B2 = 1.5
C1 = 2.3	C2 = 0.5
F1 = 0.9	F2 = 1.9
G1 = 1.0	G2 = 1.8

It is also known that A2 is the cheaper level of A, D2 is the cheaper level of D, and E2 is the cheaper level of E. What is the optimal treatment combination, and at this optimal treatment combination, what do you predict the optimal value of the quality characteristic to be?

5. Suppose in exercise 4 that you now discover that factors A and B have a significant interaction effect, and that the means of the quality characteristic at the A, B combinations are

   A1, B1 = 0.7
   A1, B2 = 2.1
   A2, B1 = 1.9
   A2, B2 = 0.9

   Now what is the predicted mean of the quality characteristic at the treatment combination chosen in exercise 4?

6. What should the chosen treatment combination in exercise 5 be, and what is the predicted mean of the quality characteristic at that treatment combination?

7. Look back at Example 13.7, on minimizing contamination in chemical production, in which means were provided for the three levels of each factor. Consider factor V, where the means at the three levels were 7.53, 7.90, and 7.57. Assuming that the levels of factor V (which refer to amount of the chemical V) are, respectively, 1 gram, 2 grams, and 3 grams, find the quadratic equation that fits the three response points.

8. Using Taguchi's L_{16} orthogonal array (Table 13.9) and the linear graphs for the L_{16} in

Figure 13.10, design an 11-factor experiment, A through K, with each factor at two levels, all main effects clean, and *AB, AE, AH*, and *JK* also clean.

9. Using Taguchi's L_{16} orthogonal array and the linear graphs for the L_{16} in Figure 13.10, design an 11-factor experiment, A through K, in which all interaction effects are assumed to be zero, factors A and B have *four* levels, and factors C through K have two levels each.

10. Comment on the criticism, discussed in the chapter, that although Taguchi's methods using orthogonal arrays do provide a set of treatment combinations that satisfies the conditions desired (for example, in a six-factor experiment with all factors at two levels, the *A, B, C, D, E, F*, and *AF* estimates are clean, assuming all other interactions are zero), they sometimes provide an answer inferior to what traditional methods could provide.

Notes

1. Taguchi also developed some easy ways to design some other types of experiments, for example, nested designs; however, in this text, we discuss Taguchi's methods only for fractional-factorial designs.

2. It is true that to use an L_8, one must need no more than seven estimates to be clean. However, the converse is not necessarily true—if seven effects need to be estimated cleanly, it is *not* guaranteed that an L_8 will succeed in giving that result, although it will in nearly all real-world cases.

3. Although we noted earlier that having only six effects to be estimated cleanly does not guarantee that an L_8 will suffice, when the six (or seven, for that matter) are all *main* effects, it is guaranteed.

PART THREE

Response-Surface Methods, Other Topics, and the Literature of Experimental Design

CHAPTER 14

Introduction to Response-Surface Methodology

EXAMPLE 14.1 Optimal Price, Warranty, and Promotion at Luna Electronics

Luna Electronics, Inc. didn't know how to price its new electronic product, what warranty length to offer, or how much it should spend to promote the product. The president of Luna was convinced that experiments could help to decide the optimal value of these variables. However, he did not want to limit himself to a small number of levels for each factor, as in most experimental situations he had seen (and like those in all previous chapters).

Although there were some marketing issues to consider in terms of the choices for these variables (for example, a product would not be priced at $42.17 nor a warranty given for 13 months, or more strangely, 17.6 months), Luna's president saw no reason why the variables could not be optimized in a way such that choices were continuous and not limited to prechosen discrete levels. Some experimental methods do indeed cater to this mode of analysis.

An additional circumstance in this case is that rather than using actual sales results, the experiments would have to simulate reality with panel data, in which respondents state their purchase probability at various levels of factors. Luna, like many companies, was accustomed to this form of marketing research and thought it trustworthy. We return to this example at the end of the chapter.

14.1 Introduction

Until now we have considered how a dependent variable, yield, or response depends on specific levels of independent variables or factors. The factors could be categorical or numerical: either way, they were treated as categorical, at least at the macro, F test stage. We did note, however, that categorical and numerical factors often differ in how the sum of squares for the factor is more usefully partitioned into orthogonal components. For example, a numerical factor might be broken down into orthogonal polynomials (introduced in Chapter 12). For categorical factors, methods introduced in Chapter 5 are typically employed. Now we consider experimental design techniques that find the optimal combination of factor levels for situations in which the feasible levels of each factor are continuous. (Throughout the text, the response has been assumed to be continuous.) The techniques are called **response-surface methods** or **response-surface methodology (RSM).**

Imagine factors X_1 and X_2 that can take on any value over some range of interest. Also imagine that the response variable, Y, varies fairly smoothly as X_1 and X_2 vary. For example, X_1 is promotional expenditure on the product, X_2 is percent price discount for the product, and Y is the total dollar contribution of the product. All other factors held constant, we might expect that changes in promotional expenditure and percent price discount would result in changes in contribution.

(Note that when factors are continuous, it is traditional to use notation such as X_1 and X_2, instead of, say, A and B, to label them. Consequently, they are often referred to as "variables," or "independent variables," instead of "factors." This terminology and notation come from the nomenclature of RSM's ancestry, regression models, rather than the primary subject of this text, analysis of variance models. Regression models are an umbrella under which ANOVA models can be placed, but the techniques were developed somewhat separately and traditionally use different notation, despite their great similarities.)

Suppose we have lots of data on contribution, promotional expenditure, and percent price discount for a product being sold in a very large number of independent hardware stores over a period. With imagination and skill, we could make a table-top clay model of the relationship among the three quantities. Along the length of the table we indicate a scale for promotional expenditure, X_1, and along the width of the table a scale for the percent price discount, X_2. Then we model a clay surface to represent the resulting contribution, Y, by the height of the clay at each combination of X_1 and X_2. We have created a **response surface.** The surface of the clay model indicates the response (here, contribution) to all of the different combinations of promotional expenditure and percent price discount.

What would the response surface look like? In this case, it might look like a mountain: as promotional expenditure and percent price discount increase, initially we would find higher contribution. But with a typical concave sales response function, at some point we would likely see contribution begin to decrease as each variable increases. Because expenditure and discount would probably not have the same effect, and especially because of a possible interaction effect between expenditure and discount, we would not expect to see a perfectly symmetrical "mountain." However, probably some combination of expenditure and discount is better than any other, and as we move away from this optimal combination in any direction, the contribution falls off. As we get far from the optimal combination, the surface flattens out. In other words, when we are far from the top of the mountain, we're no longer on the mountain but on the plain nearby. But given the response surface, we should be able to see where the optimal combination (the high point on the mountain) is. Mathematically, with the "unimodal" picture described here, we can unambiguously determine this optimal point.

Determining a large number (maybe even millions) of points to allow us to sculpt every nook and cranny of the response surface would be prohibitively expensive. However, the experimental thought processes described in this text lead to some powerful techniques for determining the response surface in enough detail to determine this optimal point (that is, combination of factor levels) with accuracy sufficient for practical purposes. Remember, we are not limited to only two factors. There can be any number of factors, so long as they exhibit continuous behavior, thus having a corresponding response surface that can be envisioned and captured mathematically.

The response surface may not be smooth: there may be multiple peaks, which makes finding the highest peak a challenge. We'll discuss such details later, but the mental picture above is a useful starting point in studying response-surface methodology.

Some people think it helps to distinguish between three terms used in the field of experimentation: screening designs, experimental design, and response-surface methodology. Screening designs are used primarily to determine which factors have an effect; they point the way for further study—the next experiment. Experimental designs determine the influence of specific factors. Response-surface methodology determines which combination of levels of continuous factors maximize (or minimize) yield.

Here, however, we view the terms not as mutually exclusive but simply as a set of overlapping descriptors whose similarities are far greater than their differences.

Response-surface methodology was introduced by Box and Wilson in their article "On the Experimental Attainment of Optimum Conditions" (*Journal of the Royal Statistical Society,* Series B, 1951, 13:1–45). Initially the techniques were known as Box-Wilson methods. There are hundreds of examples of their success in the literature. Virtually all these examples are in nonmanagerial applications, partly due to the frequent occurrence of categorical factors in the management field, rather than numerical factors that more readily fit response-surface methods. But the way of thinking engendered by the methods is powerful and should be considered for those applications where it can be used. In an application where there is, say, one categorical factor having only a few levels, or two categorical factors of two levels each, it might be possible to repeat the response-surface approach for each value or combination of values of the categorical variables and pick the "best of the best."

There are entire textbooks on response-surface methodology. Here, we simply present the basics to provide an appreciation for the methodology and enable you to more easily fathom the more detailed texts should you need or want to explore them.

14.2 The Underlying Philosophy of RSM

How do we find the optimal combination of factor levels in the most efficient way? The optimal combination is that which maximizes yield. (If the yield is to be minimized, as when seeking the lowest cost, shortest waiting time, lowest defect rate, and the like, the problem is traditionally converted to maximization by multiplying the value of all yields by -1 and then proceeding to maximize yield.)

Experimentation in response-surface methodology is *sequential.* That is, the goal at each stage is to conduct an experiment that will determine our state of knowledge about the factor effects (and thus the response surface) such that we are guided toward which experiment to conduct next so as to get closer to the optimal point. But there must be a built-in way to inform us when our sequence of experiments is complete—that is, when we have reached the optimal point.

Usually the most appropriate experiment is the smallest one that is sufficient for the task. However, "sufficient" is not easy to define. The experiment should be balanced such that the effects of the factors are unambiguous; it should be reliable enough that the results tell an accurate story; it should allow estimation of relevant effects (interaction effects, nonlinear effects, and the like). Keeping the size of the experiment as small as possible makes intuitive sense, in line with the presumption that experimentation is intrinsically expensive and time-consuming. Otherwise, we could run a huge number of combinations of the levels of the factors, each with many replicates, and just select the best treatment combination by direct observation.

Note that the smallest sufficient experiment may not be literally small; it may be quite large. But it will be appreciably smaller than one with a less-disciplined approach. Box and Wilson said the goal is to use reliable, minimum-sized experiments designed to realize "economic achievement of valid conclusions." Who can argue with that?

The strategy in moving from one experiment to another is like a blind person using a wooden staff to climb a mountain. He wants to ascend as quickly as possible, so he takes a step in the maximally ascending direction, and then probes the ground with his staff again to make his next step in the then maximally ascending direction. Repeating this process an uncertain number of times, eventually he'll know he's at a peak, because further probing with his staff indicates no higher ground.

Response-surface methodology has two main phases. The first is to probe as efficiently as possible, like the blind man above, to find the region containing the optimal point. The second phase is to look within that region to determine the optimal point more precisely.

We start by selecting an initial set of factor levels (that is, treatment combinations). Often, this starting point is determined with the help of experts—people familiar with levels used by the organization earlier or who can make reasonable choices of levels based on their experience. Then we conduct a series of experiments to point the way "up the mountain." These first-stage experiments usually assume that overall, the region of the experiment is flat, a plane whose two-dimensional surface has a constant slope in each direction, such as a sheet of plywood. (If there are more than two dimensions, it's called, technically, a "hyperplane.") The assumption of a plane surface is reasonable because, mathematically, any area of the response surface that is sufficiently small can be well represented by a plane. Similarly, the earth is a sphere but a piece of it can be well represented by a flat map.[1] And if we're at the base of a mountain and facing away, on a nearly horizontal surface, it's easy to envision being on a geometrical plane.

If we assume we're on a plane, then we can design a relatively small experiment, because it needs to estimate only linear terms—no quadratic (nonlinear) terms and no interaction effects—and to inform us about the reasonableness of the assumption of a plane. With only linear terms, factors can be at only two levels, and 2^{k-p} designs can assume all interactions to be zero. In formal statistical or mathematical terms, we refer to needing to estimate the constants of only a first-order equation:

$$Y = \beta_0 + \beta_1 X_1 + \beta_2 X_2 + \epsilon \quad \text{when there are two factors} \tag{14.1}$$

or

$$Y = \beta_0 + \Sigma_{(j=1,k)}\, \beta_j X_j + \epsilon \quad \text{for } k \text{ factors} \tag{14.2}$$

(This form of equation is another manifestation of the regression model ancestry of RSM, as opposed to an ANOVA model framework.)

This first stage, a sequence of experiments with the assumption of a first-order equation, provides two pieces of information after each experiment: one, whether we're close to the maximum; and two, if not, what direction appears to move us closer to the maximum. The indicated direction of maximum ascension is determined by the estimates of β_1 and β_2. If the experiment has sufficient power (that is, ability to identify an effect [here, direction of ascent] when it's present), estimates that are not statistically significant would indicate that either (1) we've essentially reached the peak, and *that's* why no estimate indicates a direction of ascent, or (2) we're not even near where the mountain begins to meaningfully rise.

To distinguish between the two, we must ensure that the experimental design has a built-in way to test the validity of the planar assumption. If we're near the optimal point, a plane is not adequate to describe the curving surface (the top of a mound is round, not flat!). But if we're barely on the mountain, a plane is perfectly adequate to describe the surrounding surface. As a practical matter, if an experiment is quite small its power may not be very high and there may not be well-reasoned values at which to compute power. In practice, if we haven't achieved statistically significant results for an experiment, but based on observed data values we know we're not at the top of the mountain, we presume that *any* upward tilt to the plane, statistically significant or not, is most likely to move us toward the top of the mountain. That's why this set of steps is typically called the **method of steepest ascent.**

Once we find a region in the first phase of experimentation that fails the test of "reasonableness of the assumption of a plane," we conclude that we are close to the maximum. Then we move to the second phase to probe the surface in greater detail, allowing for interaction terms and other nonlinear (usually just quadratic) terms. From a notational point of view, to do this requires considering a second-order (rarely, a higher-order) model. Here is a second-order model:

$$Y = \beta_0 + \beta_1 X_1 + \beta_{11} X_1^2 + \beta_2 X_2 + \beta_{22} X_2^2 + \beta_{12} X_1 X_2 + \epsilon \quad \text{for two factors} \tag{14.3}$$

and

$$Y = \beta_0 + \Sigma_{(j=1,k)}\, \beta_j X_j + \Sigma_{(j=1,k)}\, \beta_{jj} X_j^2 + \Sigma_{(j=1,k-1)}\, \Sigma_{(i=j+1,k)}\, \beta_{ij} X_i X_j + \epsilon \quad \text{for } k \text{ factors} \tag{14.4}$$

An experiment at this second stage usually requires a larger number of treatment combinations in order to estimate the coefficients of this more complex model. The good news is that this second stage often involves only one (the final) experiment: we have completed the reconnaissance, identified the territory that needs a more detailed mapping, and called in the full survey team.[2] This second phase is usually called the **method of local exploration.**

We elaborate on these two phases of experimentation in sections 14.3 and 14.4; then we consider a real-world example.

14.3 Method of Steepest Ascent

We illustrate the procedure for this first experimental phase with an example that is simplified but retains the salient features of optimum-seeking response-surface methods.[3]

EXAMPLE 14.2 Optimal Conditions for Banana Shipping

The dependent variable, Y, is the proportion of the skin of a banana that is "clear" (not brown-spotted) by a certain time after the banana has been picked. There are two factors: X_1, the ratio of the amount of demoisturizer A to the amount of preservative B (called the A/B ratio) packaged with the bananas during shipping, and X_2, the separation in inches of air space between rows of bananas when packaged for shipping.[4] We seek to find the combination of these two continuous variables that maximizes the proportion of clear skin. We assume that any other relevant *identifiable* factors (position of the banana within its bunch, time of growth before picked, and the like) are held constant during the experiments.

We envision the value of Y, at any value of X_1 and X_2, as depending on X_1 and X_2 and, as always, "everything else." Our functional representation is

$$Y = f(X_1, X_2) + \epsilon$$

So long as the response surface is smooth and free from abrupt changes, we can approximate a small region of the surface in a low-order power (Taylor) series. We have already noted that first-order and second-order equations are the norms at different stages. We assume that our starting point, which is based on the documented experience available, is not yet close to the maximum point, and thus the response surface of this starting point can be well represented by a plane (later we test the viability of this assumption). That's the same as assuming we have a first-order model of equation (14.1), as shown in equation (14.5):

$$f(X_1, X_2) = Y = \beta_0 + \beta_1 X_1 + \beta_2 X_2 + \epsilon \tag{14.5}$$

Our overall goal is to map the surface of $f(X_1, X_2)$ reliably in the vicinity of the maximum. Then we can determine by routine calculus methods the optimal point.

Brief digression To ensure that the last sentence is clear, we very briefly illustrate this calculus aspect via an example with one variable, using a cubic equation. (Although we started with a linear equation [14.5], as noted earlier it's common to end up estimating the parameters of a quadratic equation; whether quadratic, cubic, or a higher order, the principle is the same.) Suppose that we have

$$f(X) = \beta_0 + \beta_1 X + \beta_{11} X^2 + \beta_{111} X^3$$

Suppose further that after experimentation, we have an estimate of $f(X)$ in the vicinity of the maximum point, $f_e(X)$ (e stands for estimate), where

$$f_e(X) = b_0 + b_1 X + b_{11} X^2 + b_{111} X^3$$

To find the value of X that maximizes $f_e(X)$, we find

$$df_e(X)/dX = b_1 + 2b_{11} X + 3b_{111} X^2$$

and set it equal to zero, and solve the quadratic equation for X. (Then check to ensure that it's a maximum point!)

Suppose that to estimate the parameters, β_0, β_1, and β_2, of equation (14.5), we decide to use a 2^2 design without replication. Our starting levels are as follows:

X_1 low: A/B ratio = 1 to 4
X_1 high: A/B ratio = 1 to 2

X_2 low: 2-inch separation between rows of bananas
X_2 high: 3-inch separation between rows of bananas

The resultant yields at the four treatment combinations are as follows:

	X_1 low	X_1 high
X_2 low	94.0	93.5
X_2 high	90.8	94.3

We can indicate these yields on an X_1, X_2 grid as shown in Figure 14.1. These four data values are estimates of the corresponding four points on the response surface. Were there no error, they would be precisely on the response surface.

It is customary to translate the data from the X_1, X_2 plane to the U, V plane so that the data are displayed symmetrically, at the vertices of a 2×2 square centered at the origin of the U, V plane, as shown in Figure 14.2. That is, X_1 is transformed to U and X_2 is transformed to V. How? Well, we want the linear transformation such that when $X_1 = 1/4$, $U = -1$, and when $X_1 = 1/2$, $U = +1$. Similarly, we want the equation that translates X_2 to V so that when $X_2 = 2$, $V = -1$, and when $X_2 = 3$, $V = +1$. The linear transformations that achieve these goals are[5]

$$U = 8X_1 - 3 \quad \text{and} \quad V = 2X_2 - 5$$

In terms of U and V, our model is

$$Y = \gamma_0 + \gamma_1 U + \gamma_2 V + \epsilon$$

with estimates g_0, g_1, and g_2, respectively. We can use Yates' algorithm to find the estimates from the data, as shown in Table 14.1. Note that we divide by 4 instead of by 2, as in our Chapter 9 discussion of Yates' algorithm. When, in the earlier chapters, we divided by half of the number of treatment combinations (here, 2, half of 4), it was to take into account that the differences between the high (+1) and low (−1) values of U and V are both equal to 2; in

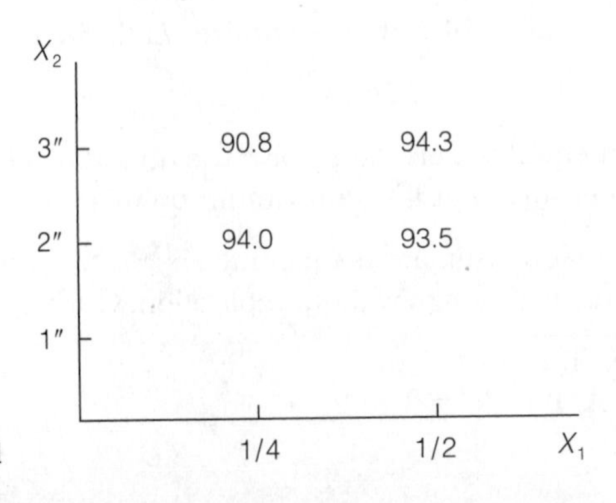

Figure 14.1 X_1, X_2 grid

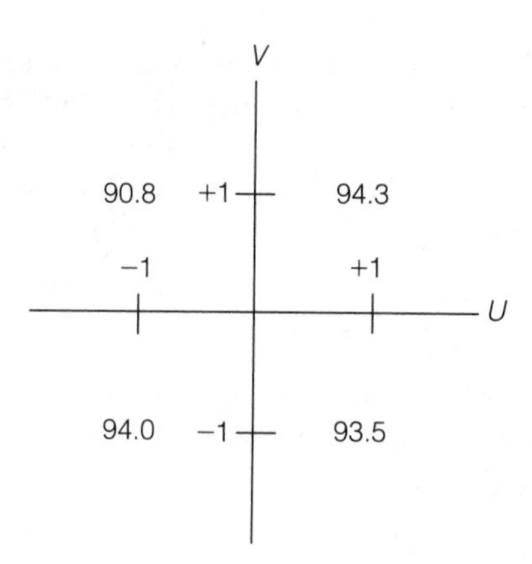

Figure 14.2 *U, V* grid

the equation represented by the γ's and estimated by the *g*'s, we wish to have slope coefficients that represent the *per-unit* changes in *Y* with *U* and *V*. Thus we need to divide by an additional factor of 2 (that is, we divide by 4 instead of 2). Note also that the error estimate is obtained from what would have been an interaction term, were there an allowance for interaction; we've seen this phenomenon several times in earlier chapters. Our assumption of a plane—that is, a first-order model—precludes interaction. (On occasion, a nonplanar function, including selected two-factor interactions, is utilized at this "steepest ascent" stage; based on the real-world example later in the chapter it is clear that there are no rigid rules for RSM, only guidelines and principles for "practicing safe experimentation.")

So, after one 2^2 experiment (four treatment combinations), we have an estimate of the plane in this region of *U* and *V*:

$$Y_e = 93.15 + .75U - .60V \tag{14.6}$$

This plane slopes upward as *U* increases (*U* has a positive coefficient) and as *V* decreases (*V* has a negative coefficient); thus *larger U* (higher A/B ratio) and *smaller V* (less separation between rows of bananas) yield better bananas (greater proportion of freedom from brown spotting).

Table 14.1 Yates' Algorithm for Banana Experiment

Treatment	Yield	(1)	(2)	÷4	Estimates
(−1, −1) = 1	94.0	187.5	372.6	93.15	γ_0
(1, −1) = *a*	93.5	185.1	3.0	.75	γ_1
(−1, 1) = *b*	90.8	−.5	−2.4	−.60	γ_2
(1, 1) = *ab*	94.3	3.5	4.0	1.0	Error

We have an estimate of the plane from four data points. Is it a good estimate? That is, can the parameters γ_1 and γ_2 be said to be nonzero (H_1) or not (H_0)? We determine the significance of our estimates via an ANOVA table.

The grand mean is equal to $(94.0 + 93.5 + 90.8 + 94.3)/4 = 93.15$, which is g_0 according to Yates' algorithm. So

$$\text{TSS} = (94.0 - 93.15)^2 + (93.5 - 93.15)^2 + (90.8 - 93.15)^2 + (94.3 - 93.15)^2 = 7.69$$

We calculate the individual sums of squares (by going back to the column labeled 2 in the Yates' algorithm table, and dividing by the square root of 4 [= 2], and squaring the ratio):

$$\text{SSQ}_U = (3.0/2)^2 = 2.25$$
$$\text{SSQ}_V = (-2.4/2)^2 = 1.44$$

and

$$\text{SSQ}_{\text{error}} = (4.0/2)^2 = 4.00$$

We have

$$\text{TSS} = \text{SSQ}_U + \text{SSQ}_V + \text{SSQ}_{\text{error}}$$

or

$$7.69 = 2.25 + 1.44 + 4.00$$

The ANOVA table for the banana example is shown in Table 14.2. From the F tables, for $df = (1, 1)$ and $\alpha = .05$, $\mathbf{c} = 161.5$. So $F_{\text{calc}} << \mathbf{c}$; there is insufficient evidence to conclude that γ_1 and γ_2 are other than zero. We have not yet determined that the assumption of a plane (first-order model) is reasonable. Earlier in this chapter we noted that usually there must be a built-in way to investigate how reasonable this assumption is for a given case; we discuss the preferred way soon. For now, we assume that it's sensible (meaning that we're not yet close to the maximum point).

Our next step is to decide where to conduct the next experiment. By custom, as noted earlier, it is in the direction of steepest ascent. In reality, we would

Table 14.2 ANOVA Table

Source of Variability	SSQ	*df*	MS	F_{calc}
γ_1	2.25	1	2.25	.56
γ_2	1.44	1	1.44	.36
Error	4.00	1	4.00	
Total	7.69	3		

rarely (if ever) conduct an experiment with only four data values—remember, the banana example is merely a simple illustration. The power of the hypothesis tests is virtually certain to be extremely low. Thus, even though we normally would prefer to have statistical significance in order to judge the direction of steepest ascent, given the likely lack of power of the hypothesis tests, and simplicity of the example, we will act as if we did have statistical significance in both the U and V direction. As the methodology is typically practiced, even when only some directions are significant, the direction of steepest ascent is often still considered to include all variables, in order to ensure that some gains in yield aren't forgone. Where should the next experiment be centered?

The next experiment The decision of where to center the next experiment involves two considerations: what direction, and how far? As we said, customarily the direction is that of steepest ascent. According to equation (14.6), this direction is toward $(U, V) = (.75, -.60)$, or proportionally, $(1, -.8)$. This means that the line of steepest ascent is $V = -.8U$. In general, for more than two factors, the steepest ascent (rate of greatest increase in Y on the response surface) direction is along the line emanating from the origin of the $U, V, W, \ldots$ space and going through a point that is at γ_1 (estimated by g_1) units in the U direction, γ_2 (estimated by g_2) units in the V direction, γ_3 (estimated by g_3) units in the W direction, and so forth. (Those familiar with vector analysis will recognize this as the gradient of Y.)

How far in that direction is another matter. If it's not enough, it will take many steps to get near the maximum. If it's too much, we might go right past it. Step size is a judgment call that improves with the assistance of input from the process experts; the better the notion of what yield to expect at the maximum point, the better the ability to judge how far to move along the steepest-ascent direction.

Suppose we pick the point $U = 1$, $V = -.8$ as the center of our next experiment (this point is indeed on the $V = -.8U$ line). Thus, the two values of U for the next experiment are $1 - 1 = 0$, and $1 + 1 = 2$. For V, they are $-.8 - 1 = -1.8$, and $-.8 + 1 = +.2$. That is, the four treatment combinations will be located as shown in Figure 14.3.

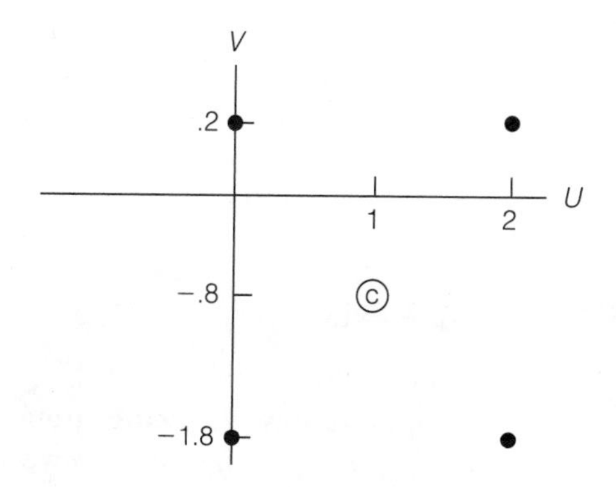

Figure 14.3 **Second** ***U, V*** **grid**

But to run the banana experiment, we need new values for X_1 and X_2, not just U and V. We know about U and V, but the person organizing and carrying out the banana-packaging testing might never have heard of U and V. He may be familiar only with the definitions of X_1 and X_2.

Recall that the mapping from X_1, X_2 to U, V is

$$U = 8X_1 - 3 \qquad \text{and} \qquad V = 2X_2 - 5$$

Thus we solve the equations above for X_1 and X_2 and find that

$$X_1 = (U + 3)/8$$

and

$$X_2 = (V + 5)/2$$

We can now use these equations to obtain the new center and settings in terms of X_1 and X_2 (the new center is $X_1 = 0.5$, $X_2 = 2.1$):

U	V	X_1	X_2
0	−1.8	.375	1.6
0	.2	.375	2.6
2	−1.8	.625	1.6
2	.2	.625	2.6

We could continue, running another 2^2 experiment at these values of X_1, X_2, and then another set of values if useful, and so on, always taking the line of steepest ascent calculated for the next step. Just as we shifted from the original X_1 and X_2 to U and V, we would now shift to other symmetrical axes, say R and S:

$$R = U - 1 \qquad \text{and} \qquad S = V + .8$$

Even with the use of 2^2 designs, we hope at some point to find significant coefficients, which would indicate more reliably that we are moving up a mountain toward the peak. After that, we'd like another result that shows no clear-cut directional indication (that is, nonsignificant results in every direction). If we find that the plane is no longer appropriate (when earlier it *was* appropriate), then we have reason to believe we've reached the peak. After each experiment, the direction of steepest ascent will likely have changed. Thus our trajectory would probably be erratic; however, we finally get there. (In this sense, math is sometimes like life: an erratic path toward success is more typical than simple, early, straightforward achievement.)

Testing the Plane: Center Points

As mentioned, we need a built-in way to test for the reasonableness of the assumption of a plane. Here we discuss the **center point** of our first set of four treatment combinations. Suppose we have a center point that is the same as the optimal point; the response surface in that region would not be planar. It would be more like the shape of a skullcap (depending on the symmetry around the maximum point). But with only four data points—and no data at or near the center—there is in general no way to know that the center is elevated above the four points for which we do have data.

For example, consider a case that has to do with boat or airplane navigation. Suppose we draw four points, all at 45 degrees latitude in the Northern Hemisphere, and at four longitudes that are 90 degrees apart—for example, 0 degrees (western France), 90 degrees east (northern Tibet), 180 degrees (south of the Aleutian Islands), and 90 degrees west (northern Wisconsin). If a plane connected those four points (which means slicing through the earth at the 45-degree latitude), there would be no way to use just these data to calculate the northerly direction or the distance from the four points to the North Pole.

How could we get a sense of the degree to which a plane doesn't "fit" the surface at the peak of the mountain? Or, more importantly, get a sense that we are close to a peak? A common way in practice is to add data at internal points, especially at the center. Here's an example.

EXAMPLE 14.3 Test of the Plane Assumption in the Banana Study

Suppose we have the data of the banana example, depicted previously in Figure 14.2, with the addition of three replicates at the center, $U = 0$, $V = 0$ (which corresponds to $X_1 = .375$ and $X_2 = 2.5$). Figure 14.4 shows the revised graph. The yields at the center point are 92.9, 94.7, and 94.3. Note that these three data points do not affect estimates of the slopes of the plane, g_1 and g_2, but they do permit an estimate of error through replication, and most to the point, they provide a test of the reasonableness of the assumption of a plane (that is, of a first-order model).

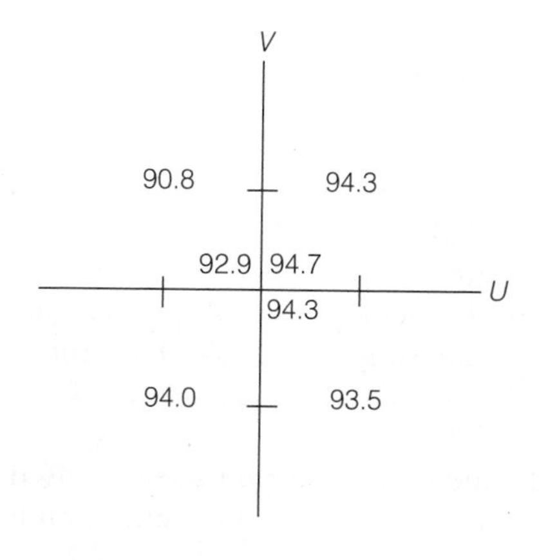

Figure 14.4 ***U, V* grid with center points**

Table 14.3 ANOVA Table

Source of Variability	SSQ	*df*	MSQ	F_{calc}
γ_1	2.25	1	2.25	
γ_2	1.44	1	1.44	
Failure of plane (fop)	5.14	2	2.57	2.88
Error	1.79	2	.89	
Total	10.62	6		

The grand mean is $(90.8 + 94.3 + 93.5 + 94.0 + 92.9 + 94.7 + 94.3)/7 = 93.50$, and

$$\begin{aligned}\text{TSS} &= (90.8 - 93.5)^2 + (94.3 - 93.5)^2 + (93.5 - 93.5)^2 + (94.0 - 93.5)^2 \\ &\quad + (92.9 - 93.5)^2 + (94.7 - 93.5)^2 + (94.3 - 93.5)^2 \\ &= 10.62\end{aligned}$$

$$\text{SSQ}_U = 2.25 \quad \text{(as before)}$$

$$\text{SSQ}_V = 1.44 \quad \text{(as before)}$$

and, with the mean of the three replicates at the center being 93.97,

$$\text{SSQ}_{\text{error}} = [(92.9 - 93.97)^2 + (94.7 - 93.97)^2 + (94.3 - 93.97)^2] = 1.79$$

Thus the sum of squares corresponding to "failure of the plane" (fop), or "lack of fit to a plane," which includes all interaction and nonlinear terms—that is, anything that contributes to a departure from a plane—is the difference between the total sum of squares and the sum of the other sources above:

$$\begin{aligned}\text{SSQ}_{\text{fop}} &= 10.62 - (2.25 + 1.44 + 1.79) \\ &= 10.62 - 5.48 \\ &= 5.14\end{aligned}$$

Once again we perform an ANOVA, as shown in Table 14.3. Now we want to test for failure of the plane. At $df = (2, 2)$ and $\alpha = .05$, $\mathbf{c} = 19.0$, and $F_{\text{calc}} << \mathbf{c}$; thus the failure-of-plane is not significant and we cannot conclude that the assumption of a plane is inappropriate. The issue of power (or the lack of it) remains; again, simple examples have their limitations.

In summary, the method of steepest ascent is used to move us toward the region that contains a maximum (*the* maximum, under the very frequent assumption that the surface is unimodal). Typically, we start a sequence of experiments centered at a point distant from the maximum. Assuming that the range of levels of each independent variable in the experiment is appropriately narrow, the experiment is conducted in a region for which the surface can be adequately represented by a plane. As we move (maybe "systematically meander" better describes it) toward the maximum, the region of the experiment continues to be planar. Finally, as we approach the maximum and if the data are not aberrational, the estimate of the surface becomes close to horizontal but with a pronounced lack of fit to a plane. At this point we move to the next phase of response-surface methodology: the method of local exploration.

14.4 Method of Local Exploration

Suppose that based on our latest experiment, we think we're close to the maximum point (that is, the optimal combination of factor levels). For this region we need to seek a more accurate estimate of the response surface than that afforded by a plane, which we have proved is an inadequate representation of the surface in this region. To do this requires designing a more complex experiment, one that allows consideration of interactions and nonlinearities (that is, curvature). After all, a real mountain has curvature at the top (nonlinearity) and does not slope downward at the same rate in every direction. This is interaction. For example, the change in Y (altitude) per unit change in the southern direction may differ depending on how far east we are. Nor does the downward slope in any one direction stay constant throughout its downward path (more nonlinearity). We still desire to represent the response surface with a Taylor series, but one that corresponds with at least a second-degree model. This model is noted as equation (14.3) for two independent variables and equation (14.4) in general. As also noted earlier, once we have conducted this last[6] experiment we can analytically determine the maximum point of the response surface function.

We now introduce two classes of designs that have been found effective in the method of local exploration: the central-composite (CC) design and the Box-Behnken (BB) design.

Central-Composite Designs

A central-composite design has three components:

- A two-level (fractional-) factorial design, which estimates main and two-factor interaction terms
- A "star" or "axial" design, which, in conjunction with the other two components, helps estimate quadratic terms
- A set of center points, which estimates error and helps estimate surface curvature with more stability

Two-level (fractional-) factorial design component Imagine that we desire to study k factors. The two-level fractional-factorial component would be a 2^{k-p} design where p is yet to be determined. The degree of fractionating in the design (p) and the corresponding number of treatment combinations (2^{k-p}) typically involve the maximum value of p that allows a clean resolution of all main and two-way interaction terms (if the maximum value of p is zero, a full-factorial design is necessary). This means that every main and two-way interaction should be aliased with interactions of an order higher than two. If every term in the defining relation has at least five letters, then every main effect will be aliased with four-factor and higher-order interactions and each two-factor interaction will be aliased with three-factor and higher-order interactions.

This type of reasoning leads to the **resolution** of a design, which is the smallest number of letters in any term of the defining relation for that design. The concept is most often applied to 2^{k-p} designs; any 2^{k-p} design of resolution greater than or equal to five is guaranteed to yield all main and two-factor interaction effects cleanly. (A 2^{k-p} design with resolution four would ensure that main effects are aliased with three-factor and higher-order interactions, but would alias at least some two-factor interactions with other two-factor interactions.)

Star (axial) design component The star or axial design component of the CC design adds points along the axis of each variable at a distance of P_s ("star point") from the design center in both the positive and negative directions. These points are at coordinates as follows (that is, each row is a treatment combination):

P_s	0	0	...	0	0
$-P_s$	0	0	...	0	0
0	P_s	0	...	0	0
0	$-P_s$	0	...	0	0
			...		
			...		
			...		
0	0	0	...	0	P_s
0	0	0	...	0	$-P_s$

where the first position (column) in the coordinate designation is for the first factor, the second position is for the second factor, and so on to the last position, for the kth factor. For k factors, there will be $2k$ star points. Box and Hunter's article "Multi-factor Experimental Designs for Exploring Response Surfaces" in *Annals of Mathematical Statistics* (1957, 28:195–241), among others, includes proposed values for P_s as a function of the values of $(k - p)$ and k in the 2^{k-p} design that are optimal to satisfy various criteria (such as minimizing mean squared error).

Center points component We have noted that adding a number of center points (that is, points at the origin, or design center) allows for an error estimate based on replication, as well as aiding the stability of quadratic terms. The Box and Hunter article cited above, among other sources, proposes an optimal number of center points in accordance with certain criteria. That number may differ slightly depending on which criterion is used; an effective compromise is considered to be two to four center points, regardless of the values of $(k - p)$ and k.

We illustrate a central-composite design in two dimensions, in the U, V plane, in Figure 14.5. Notice the points at the corners, representing the factorial design, and those on the U and V axes, representing the star component. Notice, as well, the inclusion of one or more points at the design center, (0, 0). The treatment combinations of these three parts constitute the central-composite design. Note that each factor takes on five different levels: $-P_s$, -1, 0, 1, and P_s.

Sometimes the requirement of five levels of each factor is a burden. Another design, which achieves results similar to the central-composite design but requires fewer levels of each factor, might be desirable. One such design is the Box-Behnken design.

Box-Behnken Designs

Unlike the central-composite design, a Box-Behnken design requires only three levels of each factor. These designs are made up of a combination of all possible

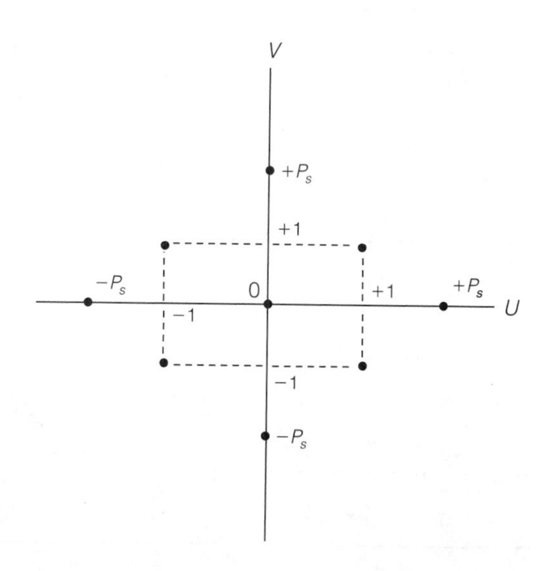

Figure 14.5 Central-composite design

two-factor, two-level, full-factorial designs, with, for each 2^2 design, all other factors held constant at the design center, plus one or more center points. The Box-Behnken design for studying four factors is shown in Table 14.4.

Note that in Table 14.4, the first four treatment combinations compose a 2^2 design for factors X_1 and X_2, with X_3 and X_4 held constant at the design center; treatment combinations five through eight compose a 2^2 design for factors X_1 and X_3, with X_2 and X_4 held constant at the design center, and so forth. Treatment combinations 25 and beyond are at the design center.

Comparison of Central-Composite and Box-Behnken Designs

For local exploration of four factors, both the central-composite design and the Box-Behnken design involve 24 treatment combinations plus whatever

Table 14.4 Box-Behnken Design for Four Factors

Run	X_1	X_2	X_3	X_4
1	−1	−1	0	0
2	1	−1	0	0
3	−1	1	0	0
4	1	1	0	0
5	−1	0	−1	0
6	1	0	−1	0
7	−1	0	1	0
8	1	0	1	0
9	−1	0	0	−1
10	1	0	0	−1
11	−1	0	0	1
12	1	0	0	1
13	0	−1	−1	0
14	0	1	−1	0
15	0	−1	1	0
16	0	1	1	0
17	0	−1	0	−1
18	0	1	0	−1
19	0	−1	0	1
20	0	1	0	1
21	0	0	−1	−1
22	0	0	1	−1
23	0	0	−1	1
24	0	0	1	1
25	0	0	0	0
. . .	0	0	0	0

number of center points are chosen. The central-composite design has $2^4 = 16$ treatment combinations for the factorial-design portion, eight treatment combinations for the star portion, and one or more at the center; the Box-Behnken design has six 2^2 designs plus one or more at the center. For more than four factors (that is, $k > 4$), the number of treatment combinations is fewer for the central-composite design than for the Box-Behnken design, assuming the same number of center points.

For example, consider $k = 5$. Here, unlike an instance with only four factors, the 2^{k-p} portion of the central-composite design can make use of a *fractional-factorial* design, a 2^{5-1}. With $I = ABCDE$ (though the notation in a text on response-surface methods might call it $I = X_1X_2X_3X_4X_5$) our 2^{5-1} is simply a resolution-five design. A 2^{5-1} design contains 16 treatment combinations; when we add 10 more treatment combinations for the star portion, plus one or more for the center, we have 27 or more treatment combinations. Because there are 10 combinations of two factors from a set of five factors (A/B, A/C, . . . , D/E), the Box-Behnken design requires $10 \cdot 4 = 40$ treatment combinations for the ten 2^2 designs, plus one or more for the center, for a total of 41 or more. Thus we have a trade-off. If the total number of treatment combinations is more important than the number of different levels that factors take on, the central-composite design is generally preferable. But if there is a relatively high cost of changing levels of one or more factors, the Box-Behnken design, which requires only three levels instead of five, may be the better choice. Indeed, this trade-off led directly to the development of the Box-Behnken design.

Issues in the Method of Local Experimentation

Whether using a central-composite design, a Box-Behnken design, or some other design or combination of designs, the local exploration experiment is run and the resulting data are used to form a second-order equation, as in equation (14.4), that best estimates the response surface in that region. Differential calculus is then used to find each combination of factor levels that gives a point of "horizontal tangency" (that is, zero slope in every direction: the gradient is the zero vector).

Ideally, this set of combinations is only one point, a point at which the response surface has a maximum. But it could be that the horizontal tangency is due to a saddle point.[7] Or there could be multiple points of zero tangency, all but one corresponding to local maxima. Also, there could be no maximum point in the region covered by the experiment—the maximum is located elsewhere, outside the experimental region. How could this occur? The answer is ϵ. That is, the last experiment in the steepest ascent stage of the process may have given incorrect results due to the unlucky influence of error. The experimenter, thinking the maximum point was in the neighborhood, "called in the survey team" to run an expensive experiment in the wrong place!

There are ways to deal with these possibilities. The point(s) of horizontal tangency, if any, can be tested to see if it is a maximum, a minimum, or a saddle point by using "canonical analysis" (using calculus to compute from higher derivatives some quantities that, by their sign and/or magnitude, reveal the category of the points). If there is no point of horizontal tangency within the test region, a new test region can be selected in the direction of the maximum indicated by this set of data. If the point of horizontal tangency turns out to be a saddle point, one can explore the rising ridges indicated by the data. If there are multiple maxima, one can evaluate (that is, calculate the surface value at) each point to determine which is actually the *global* maximum point.

In practice, these procedures should be done interactively. The experimenter designs the first experiment with the process experts' guidance, then interprets the results in order to decide what the next step should be. Close collaboration between process experts and experimental designers should occur at every stage; this allows effective adaptation to conditions as they evolve.

14.5 Perspective on RSM

Response-surface methodology is a powerful set of techniques for determining the best combination of factor levels for continuous factors. Its success, as with all aspects of experimental design, depends on the nature of the process being characterized, the knowledge of the process experts, and the skill of the experimental designer. These requirements for success are not unlike those of other techniques in the world of statistics, statistical analysis, and experimental design. They also coincide with life in general: the field of statistics is often an allegory of life.

Response-surface methodology is summarized diagrammatically in the flow chart in Figure 14.6.

Example 14.4 illustrates the response-surface methodology. Note how the steps and decisions proceed in an interactive mode; that is, the experimenters adapt to results of previous steps and frequently make judgment calls as they see fit, based on their interpretation of the data, bringing to bear considerations specific to the particular situation.

(The authors do not necessarily condone or recommend the steps taken by the experimenters whose actions we describe next; we simply describe them and comment on them. Also, we have taken liberties in our description so that we could avoid technical discussions and other matters that are beyond the scope of this book.)

EXAMPLE 14.4 Real-World Example: Using RSM for a NASA Project

This example is adapted from a public-domain article, "Statistical Design and Analysis of Optimal Seeking Experiments to Develop a Gamma-Prime Strengthened Cobalt-Nickel Base Alloy," by Gary D. Sandrock and Arthur G.

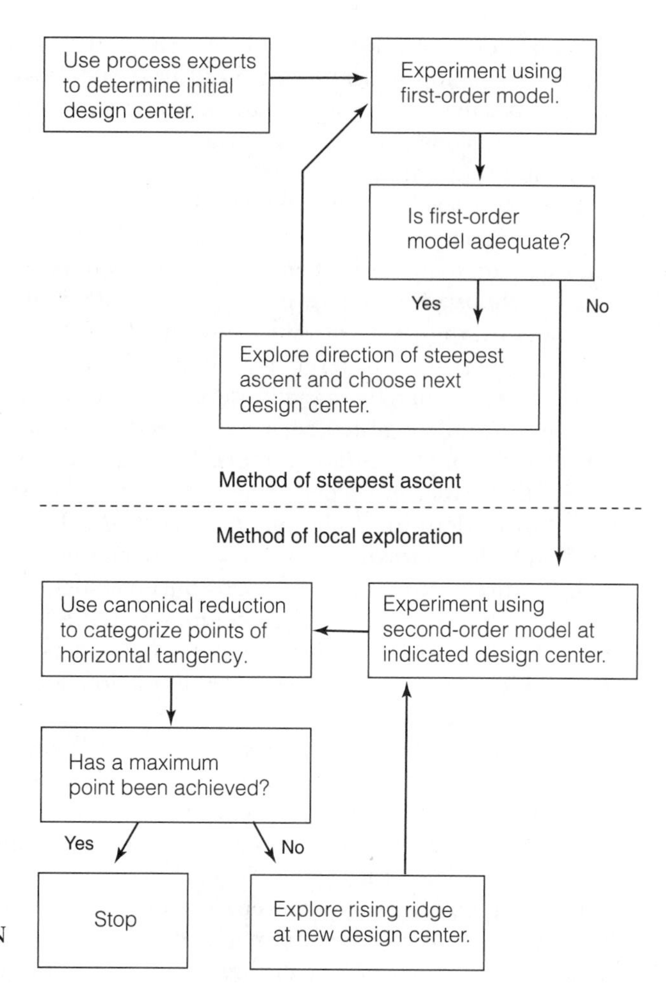

Figure 14.6 Flow chart of the response-surface methodology procedure. *Note:* Adapted from several NASA Technical Notes (such as TN D-4100, TN D-5587).

Holms of the Lewis Research Center in Cleveland, Ohio (*NASA Technical Note 5587,* December, 1969). The objective of the study was to improve the stress-rupture life of cobalt-nickel-base alloy castings.

This is not a traditional managerial problem; we picked it for its real-world use of RSM, which is not often applied to managerial problems partly because those trained in managerial applications of statistical modeling often have not been exposed to RSM. Also, the managerial RSM applications that have been done are proprietary, whereas this example is in the public domain, so we can give actual details rather than disguised ones.

The five factors under study were the weight percents of titanium (Ti), chromium (Ch), carbon (C), and aluminum (A), along with pour temperature in degrees Fahrenheit (T). (It so happens that the total percent weight of Ti, Ch, C, and A in this case is under 16% at its maximum; thus the level of one factor does not constrain the level of any other, as it might if the percent weights had to add to 100%. If we did have such a constraint, we probably could not use designs studied so far, including fractional-factorial designs, and would have to consider "mixture designs," about which we provide a brief discussion and references in Chapter 15.)

The factor levels for the first design center were based in part on experience and in part on metallurgical concepts. There had been substantial previous experimentation with percent weight of aluminum, and strong evidence that the yield as a function of percent weight of aluminum had a sharp peak. Consequently, the range of values, or size of the "design unit" (the distance from the center to each level) for the aluminum factor was relatively small. For the other factors, the metallurgical concepts and some minimal previous experimentation suggested what was viewed as a reasonable starting design center, with appropriate design units. These are noted in Table 14.5.

The first experiment performed was a two-level fractional-factorial 2^{5-1} design with defining relation $I = ABCDE$. The actual notation was $X_1X_2X_3X_4X_5$; henceforth, we use factor designations A, B, C, D, and E when referring to direct experimental-design concepts, such as a defining relation; we use the actual symbols in the first column of Table 14.5 to indicate the factor names. Using notation such as X_3 requires readers to constantly ask themselves: Which one was X_3? Note that the fractional-factorial design described is of resolution five.

The principal block of 1, *ab, ac, ad, ae, bc, bd, be, cd, ce, de, abcd, abce, abde, acde, bcde* was chosen. The experiment did not include replication. However, two measurements were made for each treatment combination (casting) in an attempt to reduce measurement error; stress-rupture life of a casting is measured in hours. The official yield for each treatment combination was the log of the average of the two measurements for that treatment combination. This technique (of multiple measurements of the same treatment combination) is sometimes called *repetition* to distinguish it from replication. It is not replication because it does not capture all sources of error, only that of measurement

Table 14.5 Design Center and Design Units

Factor	Design Center	Low	High	One Design Unit
Ti	1.0	.5	1.5	.5
Ch	4.0	2.0	6.0	2
C	.4	.3	.5	.1
A	7.0	6.75	7.25	.25
T	2900	2850	2950	50
Relabeled:	0	−1	1	

Table 14.6 First Experiment: Five-Factor, Fractional-Factorial Experiment

Alloy	Ti	Ch	C	A	T	Life
1	−1	−1	−1	−1	−1	175, 199
2	1	1	−1	−1	−1	15, 21
3	1	−1	1	−1	−1	120, 130
4	1	−1	−1	1	−1	29, 47
5	1	−1	−1	−1	1	83, 167
6	−1	1	1	−1	−1	28, 39
7	−1	1	−1	1	−1	4, 11
8	−1	1	−1	−1	1	23, 25
9	−1	−1	1	1	−1	132, 191
10	−1	−1	1	−1	1	154, 238
11	−1	−1	−1	1	1	55, 79
12	1	1	1	1	−1	17, 17
13	1	1	1	−1	1	30, 38
14	1	1	−1	1	1	18, 20
15	1	−1	1	1	1	94, 95
16	−1	1	1	1	1	13, 19

error. Taking the log of the stress-rupture life results in a situation that is closer to true constant variance (which, as you may recall, is one of the basic assumptions of ANOVA). Because the treatment combination that maximizes the log of the stress-rupture life is clearly the same as that which maximizes the stress-rupture life itself, there is no problem with using the log of the stress-rupture life as the yield.

The 16 treatment combinations are shown in Table 14.6 in design-unit notation, along with the actual stress-rupture life results (those that occurred before averaging the two measurements and then taking the log).

With 16 data points there are 15 degrees of freedom. This allowed clean determination of all five main effects and all 10 two-way interactions (remember, this is a resolution-five design), but with each alias row containing one clean effect, no degrees of freedom (rows) are available for an error estimate; we'll return to this issue soon. The equation for the estimate (Y_e; recall that e is for estimate) of the response surface, derived by Yates' algorithm, with the level of each factor being expressed in design unit notation, is (using · to separate the two factors when indicating a two-factor interaction)

$$\begin{aligned} Y_e = 1.1652 &- .383\ \text{Ch} \\ &- .146\ \text{A} \\ &+ .100\ \text{C} \\ &+ .078\ \text{Ti}\cdot\text{Ch} \\ &+ .052\ \text{Ti}\cdot\text{T} \end{aligned}$$

− .045 C·T
+ .042 C·A
+ .037 T
+ .036 Ch·T
+ .033 Ti·A
− .029 Ti
+ .022 A·T
− .016 Ti·C
− .003 Ch·C
− .002 Ch·A

The terms are arranged in decreasing magnitude of coefficient.

The NASA experimenters next wished to perform an ANOVA in order to inquire whether there were significant indications of ascending directions, as well as to see how appropriate the assumption of a plane was for the next experiment. Recall, however, that significance testing within ANOVA requires an estimate of error for the denominator of F_{calc} expressions. Yet, as mentioned earlier, there are no rows, or sum-of-squares terms, or degrees of freedom left over for an error estimate. So they needed to be clever. They decided to test the effects for significance two different ways—neither one being exact, because there was no clearly identifiable error term. Of the two ways, one is conservative (that is, possibly biased, but if so, toward indicating *fewer* significant results than might otherwise be determined), and one is liberal (that is, possibly biased, but if so, toward indicating *more* significant results than might otherwise be determined).

The conservative method was to form an error term by pooling the sum of squares of all 15 effects; essentially, this meant adding the sum of squares for all 15 terms, adding the number of degrees of freedom of each term (15, one per term), and dividing the total of the 15 sums of squares by 15, the total of the degrees of freedom. Why is this conservative? If there are no nonzero effects, each of the 15 sum-of-squares terms (divided by 1, its degrees of freedom) truly estimates error, σ^2, and consequently the pooling of the 15 terms indeed provides the best estimate of error. However, if there are *any* nonzero main or two-factor interaction effects, the pooled result will overestimate error; as a result, the denominator of F_{calc} will be too large, and it is possible that fewer than deserved significant effects will result.

The liberal way is to look, after the fact, at the lower sum-of-squares terms and pool them, and only them, as error. This procedure often considers only interaction terms (on balance and out of context, they are more likely to be zero than main effects). Here, the experimenters used the smallest four (interaction) terms. This approach is called liberal because, although it is generally true that a smaller sum-of-squares term is less likely to correspond with a nonzero effect, by picking *only* the smallest four (or whatever number), it is probable that the resulting error estimate is downward biased. For example, if you randomly sample the weight of 15 professors but compute the average weight of only the four lightest ones, the resulting average is almost certain to be less than the true average weight of all professors. With an underestimated error term, the denominator of F_{calc} will be too small, and possibly some effects will be given more significance than they deserve.

In a perfect world,[8] the same effects would be significant under either approach, and the results would then be definitive. However, when applied to the coefficients in the equation above, the liberal test found the first four terms (effects) significant; the conservative test found the first two effects significant and the next two close to (but not) significant.

Approaches other than liberal/conservative hypothesis testing can be used. One is using "normal probability plots," which often requires more subjective judgment. We leave it to you to decide which approach to use if there is no apparent "viable" error term. We recommend the Daniel text (discussed in Chapter 15) as a good source for seeing the use of normal probability plots in real-world experiments.

In our metallurgical example, the experimenters noted that at this point it appeared that decreasing Ch and A and increasing C are beneficial to stress-rupture life, and that T and Ti have no impact on stress-rupture life except possibly for the Ti·Ch interaction effect. It certainly would have been reasonable to drop T (temperature) from further consideration, and possibly even Ti (since Ti·Ch was significant only under the liberal method). However, NASA decided to keep all five factors.[9] Further, since the interaction term was the least significant, NASA decided that the response surface was sufficiently planar to allow going forth with the method of steepest ascent to determine enhanced response. (Remember, it was their money, so they could do what they wanted.[10])

The equation for the response surface in this region was thus assumed to be

$$Y_e = 1.1652 - .383\text{ Ch} - .146\text{ A} + .100\text{ C} + .037\text{ T} - .029\text{ Ti}$$

According to that equation, the direction of steepest ascent is −.383 design units in the Ch direction, −.146 design units in the A direction, +.100 design units in the C direction, +.037 design units in the T direction, and −.029 design units in the Ti direction. We summarize this in Table 14.7. Recall what is meant by design units—one design unit for Ch, for example, is 2% weight of chromium, applied above and below the design center. The design center for chromium was 4%, so −1 design unit was 2% chromium and +1 design unit was 6% chromium. Table 14.5 lists these values for all the factors' design units. We move along the direction of steepest ascent by making changes in the factors' levels, away from the design center, in proportion to the values in Table 14.7.

Table 14.7 **Direction of Steepest Ascent**

Ch	A	C	T	Ti
−.383	−.146	.100	.037	−.029

Table 14.8 Direction of Steepest Ascent—Scaled Design Units

Ch	A	C	T	Ti
−1	−.381	.261	.097	−.076

If we multiply all elements of the table by some constant, the relative values do not change. As is customary, the experimenters scaled the values in Table 14.7 such that −.383 became −1; that is, every element of the table was multiplied by (1/.383) = 2.611, yielding Table 14.8. Next, the Table 14.8 values were rounded by the NASA folks as shown in Table 14.9.

At this point the experimenters were not sure where along this direction of steepest ascent to locate the next design center. So they decided to run some treatment combinations along the vector of steepest ascent. That is, rather than follow the method-of-steepest-ascent recipe described earlier of moving to a new design center and running the next experiment to find the equation of a plane in the region of that design center, and so forth, they decided to experiment at sites along that line (or hyperplane) with one treatment combination at each of a few points on that line. The first point chosen was at the original design center (see Table 14.10).

Two measurements of the resultant casting were made; the stress-rupture life measurements were 60 and 56 hours. (The data at this "new" point—until now no casting had been produced at this center point—could have been used to assess the assumption of a plane, but that was not done.)

Table 14.9 Direction of Steepest Ascent—Rounded Design Units

Ch	A	C	T	Ti
−1	−.40	.26	.10	−.08

Table 14.10 Original Design Center

Ch	A	C	T	Ti
4.0	7.0	.4	2900	1.0

Table 14.11 Second Treatment Combination on Line of Steepest Ascent

Ch	A	C	T	Ti
2.0	6.9	.426	2905	.96

The next treatment combination was selected by choosing a negative change of one design unit for Ch and a change of the other factors in the proportions indicated in the direction of steepest ascent, as indicated in Table 14.9, all from the design center noted in Table 14.10. That is, they moved -1 design unit, or 2.0%, for Ch (from 4.0% to 2.0%); $-.4$ design units, or $-.4 \cdot .25 = -.1\%$, for A (from 7.0% to 6.9%); $+.26$ design units, or $.26 \cdot .1 = .026\%$, for C (from .4% to .426%); $+.1$ design units, or $.1 \cdot 50 = 5$ degrees, for T (from 2900 degrees to 2905 degrees); and $-.08$ design units, or $-.08 \cdot .5 = -.04\%$, for Ti (from 1.0% to .96%). This point is described in Table 14.11. At this treatment combination, the stress-rupture life measurements were 102 and 92 hours.

Then another step of equal size further along this direction of steepest ascent resulted in the treatment combination noted in Table 14.12. The stress-rupture life measurements at this treatment combination were 142 and 171 hours. Note the continual increase of yield as the experimenters moved along what they calculated to be the direction of steepest ascent.

They wanted to move one more step along this direction of steepest ascent. However, they had a problem. The weight percent of chromium, which in Table 14.12 is zero, can't be negative. Thus an attempt to move further along the direction of steepest ascent, which would suggest a weight percent of chromium of -2%, is impossible. So for the next treatment combination they kept the level of chromium at zero, as in Table 14.12, and changed the level of the other factors by the same amount as before. This resulted in the treatment combination depicted in Table 14.13.

Note again: because the level of chromium could not be lowered, the treatment combination in Table 14.13 is *not* along the direction of steepest as-

Table 14.12 Third Treatment Combination on Line of Steepest Ascent

Ch	A	C	T	Ti
0	6.8	.452	2910	.92

Table 14.13 The Last of the Four Exploratory Points

Ch	A	C	T	Ti
0	6.7	.478	2915	.88

Table 14.14 Summary of Four Exploratory Points

Point	Ch	A	C	T	Ti	Result
1	4.0	7.0	.4	2900	1.0	60, 56
2	2.0	6.9	.426	2905	.96	102, 92
3	0	6.8	.452	2910	.92	142, 171
4	0	6.7	.478	2915	.88	116, 134

cent. For this point, the stress-rupture life measurements were 116 and 134 hours. Note that these measurements are not as good (as large) as at the previous step (the treatment combination depicted in Table 14.12). Table 14.14 summarizes the four points and their respective results.

In general, it would seem reasonable to revert to the best of the four treatment combinations of Table 14.14 (point 3) to locate the design center for the next experiment; however, the value of zero for chromium does not allow symmetric values above and below it. Accordingly, it was decided to choose point 3, except that chromium would be set at 2.0% rather than 0. In addition, it was decided to drop temperature (T) as a factor because it had been determined not to be significant, either as a main effect or as a part of an interaction term. (It's not clear why the experimenters dropped it at this point, given that they kept it in the first place. It's no less significant now than it was earlier.) From then on, they held temperature constant at 2900 degrees Fahrenheit. Finally, they did more rounding. The net result was a new (next) design center at the treatment combination shown in Table 14.15.

Table 14.15 New Design Center

Factor	Design Center	Low	High
Ch	2.0 → 2.0	1.5	2.5
A	6.8 → 6.75	6.5	7.0
C	.452 → .5	.4	.6
Ti	.92 → 1.0	.5	1.5
Relabeled:	0	−1	1

At this design center, the experimenters decided to run a complete 2^4 factorial design. Given that they were still in the steepest-ascent phase of the process, they were still assuming a plane, so they could have used a 2^{4-1} fractional-factorial design, which, as a resolution-four design, estimates all main effects cleanly, assuming that all interaction effects are zero. However, they didn't believe that only eight data values would provide adequate reliability and so, wanting at least 16 data values, they did the complete 2^4. The results, in Yates' order, are in Table 14.16. Note that the results for this second factorial experiment are generally improved relative to the results of the first experiment in Table 14.6. This confirms the effectiveness of the steepest-ascent approach.

The coefficients for the equation that estimates the response surface were calculated from the data in Table 14.16, again using Yates' algorithm. The slope coefficients, in descending order, are listed in Table 14.17.

When these coefficients were tested, none were statistically significant. This was interpreted to indicate that the test region might contain a (or the) maximum, since no direction of any kind other than horizontal was suggested. It was decided, therefore, to move on immediately to the method of local exploration. Furthermore, because the second two-level factorial experiment indicated no particular vector of ascent, it was decided to continue at the same design center. This made it unnecessary to repeat the factorial-design portion of what was to be a central-composite design experiment. They simply augmented the second factorial design with a set of star points and

Table 14.16 Second Factorial Design Experiment

Alloy	Ti	Ch	C	A	Life
1	−1	−1	−1	−1	126.7, 176.5
2	1	−1	−1	−1	196.0, 184.1
3	−1	1	−1	−1	163.4, 152.6
4	1	1	−1	−1	194.0, 249.4
5	−1	−1	1	−1	88.9, 106.1
6	1	−1	1	−1	172.7, 160.1
7	−1	1	1	−1	154.9, 182.2
8	1	1	1	−1	144.1, 162.4
9	−1	−1	−1	1	136.1, 107.0
10	1	−1	−1	1	65.7, 60.0
11	−1	1	−1	1	139.8, 107.2
12	1	1	−1	1	80.6, 87.7
13	−1	−1	1	1	175.8, 164.8
14	1	−1	1	1	167.2, 166.1
15	−1	1	1	1	141.8, 129.2
16	1	1	1	1	145.0, 140.1

Table 14.17 **Coefficients for Second Factorial Experiment**

Coefficient	Estimate	Coefficient	Estimate
C • A	+.077	Ti • C	+.024
A	−.063	Ch • A	−.021
Ti • A	−.054	Ch	+.015
Ti • C • A	+.031	Ti • Ch • C • A	+.013
C	+.031	Ch • C	−.011
Ti • Ch • C	−.026	Ti • Ch	−.004
Ti • Ch • A	+.025	Ti	+.002
Ch • C • A	−.025		

four center points. This was probably a good decision, assuming no block effects. Recall that in such a situation, where an experiment is made up of treatment combinations that are run at different times, there is a possibility of block effects. The NASA experimenters acknowledged this, but concluded that any block effects were minimal and not of consequence. (This is fortunate because the blocking is not orthogonal as it always was in our Chapter 10 examples and ideally should always be.)

Each pair of star (axial) points was obtained by holding three of the four factors at the design center and setting the fourth factor at levels of $+P_s$ and $-P_s$. For this situation (four factors and a two-level complete factorial), the suggested optimal value for P_s is 2, and the suggested number of center points is between two and four. The experimenters decided to have four center points. The results of measurements at the star and center points are shown in Table 14.18.

Table 14.18 **Data Results for Star and Center Points**

Alloy	Ti	Ch	C	A	Life
1	−2	0	0	0	108.6, 159.0
2	+2	0	0	0	94.4, 95.6
3	0	−2	0	0	180.3, 186.6
4	0	+2	0	0	219.5, 184.4
5	0	0	−2	0	189.2, 150.0
6	0	0	+2	0	220.3, 218.7
7	0	0	0	−2	139.8, 123.4
8	0	0	0	+2	153.1, 149.4
9	0	0	0	0	279.3, 269.7
10	0	0	0	0	198.4, 172.1
11	0	0	0	0	233.6, 203.9
12	0	0	0	0	242.8, 227.3

The 16 data points from Table 14.16 were combined with the 12 points from Table 14.18. These 28 data points from the (now) central-composite design led to the following second-order equation. Recall that the use of a second-order equation automatically corresponds with assuming that all terms higher than quadratic (which includes interactions of more than two factors) are zero.

$$\begin{aligned} Y_e = 2.354 &+ (-.011\ \text{Ti} + .013\ \text{Ch} + .030\ \text{C} - .037\ \text{A}) \\ &+ (-.085\ \text{Ti}^2 - .026\ \text{Ch}^2 - .026\ \text{C}^2 - .060\ \text{A}^2) \\ &+ (-.004\ \text{Ti}\cdot\text{Ch} + .025\ \text{Ti}\cdot\text{C} - .054\ \text{Ti}\cdot\text{A} - .011\ \text{Ch}\cdot\text{C} \\ &- .021\ \text{Ch}\cdot\text{A} + .077\ \text{C}\cdot\text{A}) \end{aligned} \tag{14.7}$$

The terms in equation (14.7) have been grouped into linear, quadratic, and two-factor interaction components. This equation is presumed to be valid within a radius of two design units.

The next step toward finding the optimal combination of factor levels is to solve for any points at which the hyperplane that is tangent to the (estimated) response surface has zero slope (that is, so-called "points of horizontal tangency"). We do this by taking the partial derivative of Y_e with respect to each factor, setting it equal to zero, and simultaneously solving the resultant set of equations. Since this is a quadratic (that is, second-order) model, these resulting equations are linear. With four factor levels as unknowns (recall that we fixed T at 2900 degrees Fahrenheit), we have a set of four linear equations, four unknowns, as follows:

$$\begin{aligned} dY_e/d\text{Ti} &= -.011 - .170\ \text{Ti} - .004\ \text{Ch} + .025\ \text{C} - .054\ \text{A} = 0 \\ dY_e/d\text{Ch} &= .013 - .052\ \text{Ch} - .004\ \text{Ti} - .011\ \text{C} - .021\ \text{A} = 0 \\ dY_e/d\text{C} &= .030 - .052\ \text{C} + .025\ \text{Ti} - .011\ \text{Ch} + .077\ \text{A} = 0 \\ dY_e/d\text{A} &= -.037 - .120\ \text{A} - .054\ \text{Ti} - .021\ \text{Ch} + .077\ \text{C} = 0 \end{aligned}$$

Solving these four equations simultaneously yields the following factor levels *in design units:*

$$\text{Ch} = +.215 \qquad \text{A} = -.110 \qquad \text{C} = +.378 \qquad \text{Ti} = +.022$$

The next step was to inquire whether this point is a true maximum, or merely a saddle point (by now, we know that it is not a minimum point). Further analysis indicated that this is a saddle point, on a ridge that represents a maximum on all axes except one, where it remains just about constant. The NASA folks said that "for all practical purposes, we considered the stationary point the optimum alloy with respect to stress-rupture life." Finally, converting to the true values of the factors, by recalling the design center from Table 14.15 and steps since then, we have

FACTOR LEVELS IN WEIGHT%

$$Ch_{true} = 2.0 + .5\ Ch = 2.11\%$$
$$A_{true} = 6.75 + .25\ A = 6.72\%$$
$$C_{true} = .5 + .1\ C = .54\%$$
$$Ti_{true} = 1.0 + .5\ Ti = 1.01\%$$

At this stationary point, equation (14.7) predicts the value of stress-rupture life to be about 231. At the design center, the predicted value of stress-rupture life is about 226. These two values can be considered essentially the same, given that the replicates at the design center have a standard deviation of about 36. Thus, the NASA experimenters were pleased to conclude that the response was relatively flat in the neighborhood of the stationary point, and began to investigate how sensitive other (secondary) dependent variable responses were over the neighborhood of this stationary point.

EXAMPLE 14.5 NASA Example Using JMP

JMP is useful in both the design of response-surface experiments and in the analysis of experimental results (the analysis can be done with JMP whether or not one picks a design specified by JMP). More specifically, JMP can be used in the second stage of the process—the method of local exploration, where the nonlinearities make the analysis process especially cumbersome without aid from software.

Design We gain access to the "Design of Experiments" menu by choosing "Tables," then "Design Experiment," then clicking on "Response Surface Design" in the "Choose Design Type" window. The "Response Surface Design Selection" window allows the choice of one of a few Box-Behnken and/or central-composite designs once we indicate the number of factors under study. For four factors, our choices are those in Table 14.19. We could select one of these and JMP would provide a spreadsheet defining the corresponding treatment combinations and a column for recording the experimental results. After running the experiment and entering the results, we could continue to the analysis using JMP.

Note that JMP does *not* offer the four-factor, 28-run central-composite design chosen by the NASA folks. Fortunately, however, we can still use JMP to analyze the results of the NASA experiment.

Table 14.19 JMP Response-Surface Designs for Four Factors

Number of Runs	Block Size	Number of Center Points	Type
27		3	Box-Behnken
27	9	3	Box-Behnken
30	10	6	Central-composite
31		7	Central-composite
36		12	Central-composite

Analysis We begin by preparing the input data for JMP. Using Excel (or JMP; the authors found Excel easier for this step), we combine Tables 14.16 and 14.18, average the two repetitions for each treatment combination, and take the logarithm. Next we input the logarithms, which are the dependent variable values, into Table 14.20. We then paste the highlighted segment of Table 14.20 into a JMP spreadsheet.

Now we're ready to proceed with the analysis. We choose "Analysis," then "Fit Model," and complete the "Fit Model" dialog box. We select "Life" as the response variable, *Y*. Next we highlight Ti, Ch, C, and A, and with these factors highlighted, click the "Effects Macro" and select "Response Surface." Finally, we change "fitting personality setting" on the pop-up menu from "Standard Least Squares" to "Screening" and click "Run Model" to get the analytical results in Table 14.21.

In Table 14.21, note that the equation represented by the Parameter Estimates section is essentially the same as equation (14.7), with tiny differences due to rounding error.

Table 14.22 shows the solution produced by JMP. JMP indicates that the solution is a saddle point, as does the NASA article. The predicted value at the solution listed is 2.3630581, a value whose antilog_{10} is about 231, the same value reported in the NASA article. However, the solution values JMP provides for the four variables, which are in design units, are *not* the same as those in the NASA article. To compare:

Variable	NASA Reported	JMP
Ti	.022	.008
Ch	.215	.182
C	.378	.456
A	−.11	−.049

Table 14.20 Central-Composite Design

Alloy	Ti	Ch	C	A	Life
1	−1	−1	−1	−1	2.181844
2	1	−1	−1	−1	2.278754
3	−1	1	−1	−1	2.198657
4	1	1	−1	−1	2.345374
5	−1	−1	1	−1	1.989005
6	1	−1	1	−1	2.221414
7	−1	1	1	−1	2.2266
8	1	1	1	−1	2.184691
9	−1	−1	−1	1	2.084576
10	1	−1	−1	1	1.799341
11	−1	1	−1	1	2.073718
12	1	1	−1	1	1.926857
13	−1	−1	1	1	2.230449
14	1	−1	1	1	2.221414
15	−1	1	1	1	2.131939
16	1	1	1	1	2.153815
17	−2	0	0	0	2.127105
18	2	0	0	0	1.977724
19	0	−2	0	0	2.263636
20	0	2	0	0	2.305351
21	0	0	−2	0	2.22917
22	0	0	2	0	2.341435
23	0	0	0	−2	2.118926
24	0	0	0	2	2.178977
25	0	0	0	0	2.438542
26	0	0	0	0	2.267172
27	0	0	0	0	2.340444
28	0	0	0	0	2.371068

However, these are not as far away from one another as it might appear. Indeed, if we examine the actual weight% values, they are quite close:

Variable	NASA Reported	JMP
Ti	1.01%	1.00%
Ch	2.11%	2.09%
C	.54%	.55%
A	6.72%	6.74%

Table 14.21 **JMP Analysis**

Screening Fit
LIFE
Summary of Fit

RSquare	0.782429
RSquare Adj	0.548121
Root Mean Square Error	0.096004
Mean of Response	2.185971
Observations (or Sum Wgts)	28

Analysis of Variance

Source	DF	Sum of Squares	Mean Square	F Ratio
Model	14	0.43089287	0.030778	3.3393
Error	13	0.11981912	0.009217	Prob>F
C Total	27	0.55071198		0.0182

Parameter Estimates

Term	Estimate	Std Error	t Ratio	Prob>|t|
Intercept	2.354352	0.048002	49.05	<.0001
Ti	-0.011828	0.019597	-0.60	0.5565
Ch	0.0132865	0.019597	0.68	0.5097
C	0.029142	0.019597	1.49	0.1608
A	-0.03686	0.019597	-1.88	0.0826
Ti*Ti	-0.0843	0.019597	-4.30	0.0009
Ch*Ti	-0.003457	0.024001	-0.14	0.8877
Ch*Ch	-0.026227	0.019597	-1.34	0.2037
C*Ti	0.0245796	0.024001	1.02	0.3245
C*Ch	-0.010357	0.024001	-0.43	0.6732
C*C	-0.025964	0.019597	-1.32	0.2080
A*Ti	-0.053662	0.024001	-2.24	0.0435
A*Ch	-0.021139	0.024001	-0.88	0.3944
A*C	0.0774248	0.024001	3.23	0.0066
A*A	-0.059953	0.019597	-3.06	0.0091

Table 14.22 **JMP Solution**

Solution

Variable	Critical Value
Ti	0.0080923
Ch	0.1823292
C	0.4560857
A	-0.048673

Solution is a SaddlePoint

Predicted Value at Solution

= 2.3630581

Contour Plot Specification
Check two factors, edit grid/scale values.

	Variable	From	To	By
X	Ti	-3	3	1
X	Ch	-3	3	1
_	C	-3	3	1
_	A	-3	3	1

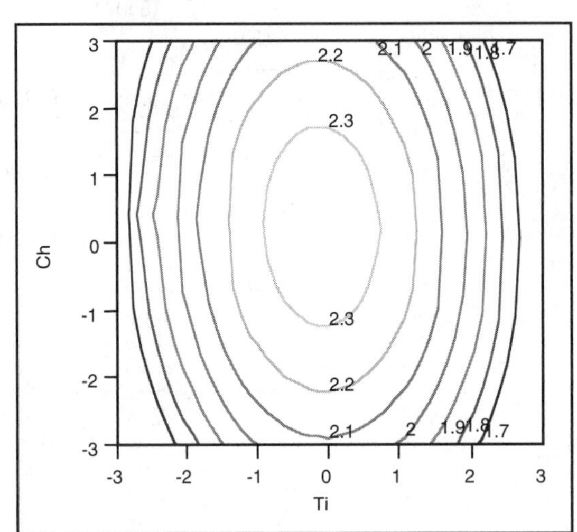

Figure 14.7 JMP contour plot for NASA example

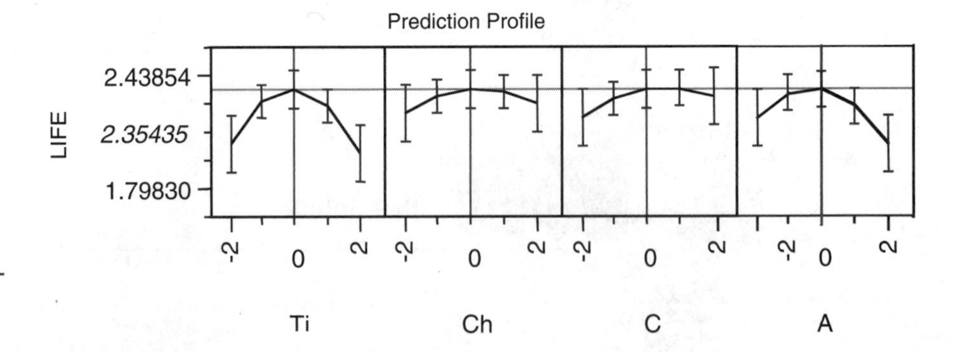

Figure 14.8 JMP prediction profile for NASA example

JMP also provides options on which graphics/contour plots we might select. To produce the contour plot in Figure 14.7, we pick two variables to provide axes, holding the other variables (in this case, two of them) constant at chosen values. We also specify the range of variable values to which the contour plot in Figure 14.7 pertains.

Figure 14.8 shows another graphic option, a profile of how each variable works in a univariate way.

EXAMPLE 14.6 Follow-Up Use of Excel's Solver to Explore a Response Surface

Once we have an equation representing the response surface, we may, rather than taking derivatives (definitely the method used in the NASA article, and probably [we're not certain] the method used by JMP), find the maximum (or minimum) value of that surface by using Solver, a numerical optimization routine in the Tools section of Excel. We define the relevant range of factor values (in essence, constraints), specify our objective function (the quantity to be maximized or minimized), and Solver searches over all the allowable combinations of factor values for the specific combination that optimizes the objective function. There is nothing statistical or calculus-based about Solver! It uses various deterministic algorithms to do its searching. An advantage of this approach is that, if the maximum is not within the region under study (that is, if it is instead on the boundary), Solver will acknowledge that result.

For a problem such as this, where the stationary point is a saddle point and not a maximum, Solver will, as it should, yield a different answer depending upon the starting point. For the problem we have been discussing, we tried various starting points to ensure that we achieved the best possible results; Solver found that the best solution occurred at a point that is quite different from that indicated by the NASA article. This is primarily because the NASA procedure mandated that the solution be a stationary point, whereas the Solver algorithm has no such restriction. The solution indicated by Solver was as follows, in design units:

Variable	NASA Reported	Solver
Ti	.022	.231
Ch	.215	1.420
C	.378	−2
A	−.11	−1.95

In actual weight% values, we have

Variable	NASA Reported	JMP
Ti	1.01%	1.11%
Ch	2.11%	2.71%
C	.54%	.30%
A	6.72%	6.26%

Recall that the predicted (that is, equation-generated) value of the stress-rupture life at the stationary point for both the NASA solution and the JMP solution is about 231 hours. At the solution indicated by Solver, the predicted value is about 249 hours. The standard deviation of replications at the design origin of the NASA article was about 36 hours; so, on one hand, this 249 − 231 = 18 hours represents about one-half of a standard deviation. On the other hand, the 36-hour standard deviation is not an estimate of the "true standard deviation" but an underestimate of it—reflecting solely repetition of the measurement, not "complete" replication. Hence, it's not clear how material the 18-hour difference from the equation's prediction is. However, we expect that, had NASA had the option (Solver didn't exist then[11]), the experimenters would have used the combination of factor levels for which the predicted value was 249 hours.

EXAMPLE 14.7 Optimal Price, Warranty, and Promotion at Luna, Revisited

Two experiments were conducted to arrive at an optimal solution; each was a simulation in which responders were asked for their purchase intent for different treatment combinations (called "scenarios"). The first experiment was a 3^3 factorial design, and had the levels of the factors relatively far apart. The results of that experiment clearly indicated where to look further for the optimal level for each factor: between the middle and high levels, or between the low and middle levels.

Then a second experiment, one with a central-composite design, was run. Its center point was, for each factor, the midpoint between the two levels of the previous experiment indicated as surrounding the optimal level. The two levels themselves were designated −1 and +1, respectively, in design units, and star points were at −1.5 and +1.5 design units. (The distances of −1.5 and 1.5 were chosen for some practical reasons having nothing to do with optimal placing of star points.) Interestingly, although the origin (that is, center point) was repeated three times, giving us a total of $2^3 + 6 + 3 = 17$ scenarios, only a small proportion of the responders had comments such as, "Haven't I seen this combination before?" This repeating without recognition probably occurred because the responders saw the levels in actual monetary and time units, not the all-zero values of the center point in design units, so it wasn't easy to recall having seen the combination several scenario cards earlier. To those who did raise the question, the moderator replied that he did not know, but to simply assume that it was not a repeat scenario.

Data from this latter experiment led to a function that related sales (in essence, *expected* sales, although company management never thought of it in those terms) to product price, warranty length, and promotional expense. Along with a somewhat complex determination of the (again, *expected*) cost incurred as a function of the warranty length offered, as well as the direct monetary impacts of the price and promotional expense, a mathematically optimal point was determined. Ultimately, however, the marketing folks had their way, and a point significantly different from the mathematically optimal point was used instead. That's life!

14.6 Concluding Remark

We wish to emphasize again that this chapter is an introduction to a way of thinking, and not an end in itself. Indeed, there are entire texts on response-surface methods. Some of these are referenced in Chapter 15.

Exercises

1. Suppose that a response surface analysis with two factors is in its steepest ascent stage, and the center of the current experiment is (X_1 = 40, X_2 = 40). Suppose further that the results of this current experiment are as follows:

Y	X_1	X_2
164	35	35
182	35	45
221	45	35
243	45	45

What is the first-order equation resulting from these data?

2. In exercise 1, what is the direction of steepest ascent?

3. In exercise 1, which of these points are in the direction of steepest ascent?
   a. (X_1, X_2) = (88.5, 30)
   b. (X_1, X_2) = (147.5, 40)
   c. (X_1, X_2) = (5.9, 2)
   d. (X_1, X_2) = (202.5, 29.5)

4. Repeat exercise 2 if we add the data point (Y, X_1, X_2) = (205, 40, 40).

5. Based on your answer to exercise 1, what Y value would be predicted for the response at the design center, (X_1, X_2) = (40, 40)?

6. In exercise 1, find the transformations of X_1 to U, and of X_2 to V, that result in the four values of (U, V) forming a 2 × 2 square symmetric about the origin.

7. Is a three-dimensional Latin square an appropriate design for estimation of parameters of a second-order equation with three factors? Why or why not?

8. Show that the number of parameters in a second-order equation (equation 14.4) with k factors is $(k + 2)!/(k! \cdot 2!)$.

9. What is the highest resolution possible for a 2^{8-2} design?

10. Suppose that our knowledge of the process under study indicates that the appropriate model with k factors is a pth-order equation without (any) interaction terms. How many parameters are to be estimated?

11. Suppose that in the NASA example the repetitions were instead replications; how would the local exploration analysis change? What is the result? The data are listed in Table 14EX.11 and also in the data sets as Problem 14.11.

Table 14EX.11 **NASA Example with Two Replications per Treatment**

Alloy	Ti	Ch	C	A	Life	Log Life	Alloy	Ti	Ch	C	A	Life	Log Life
1	−1	−1	−1	−1	127	2.1038	29	−1	−1	−1	−1	177	2.248
2	1	−1	−1	−1	196	2.2923	30	1	−1	−1	−1	184	2.2648
3	−1	1	−1	−1	163	2.2122	31	−1	1	−1	−1	153	2.1847
4	1	1	−1	−1	194	2.2878	32	1	1	−1	−1	249	2.3962
5	−1	−1	1	−1	89	1.9494	33	−1	−1	1	−1	106	2.0253
6	1	−1	1	−1	173	2.238	34	1	−1	1	−1	160	2.2041
7	−1	1	1	−1	155	2.1903	35	−1	1	1	−1	182	2.2601
8	1	1	1	−1	144	2.1584	36	1	1	1	−1	162	2.2095
9	−1	−1	−1	1	136	2.1335	37	−1	−1	−1	1	107	2.0294
10	1	−1	−1	1	66	1.8195	38	1	−1	−1	1	60	1.7782
11	−1	1	−1	1	130	2.1139	39	−1	1	−1	1	107	2.0294
12	1	1	−1	1	81	1.9085	40	1	1	−1	1	88	1.9445
13	−1	−1	1	1	176	2.2455	41	−1	−1	1	1	164	2.2148
14	1	−1	1	1	167	2.2227	42	1	−1	1	1	166	2.2201
15	−1	1	1	1	142	2.1523	43	−1	1	1	1	129	2.1106
16	1	1	1	1	145	2.1614	44	1	1	1	1	140	2.1461
17	−2	0	0	0	109	2.0374	45	−2	0	0	0	159	2.2014
18	2	0	0	0	94	1.9731	46	2	0	0	0	96	1.9823
19	0	−2	0	0	180	2.2553	47	0	−2	0	0	187	2.2718
20	0	2	0	0	220	2.3424	48	0	2	0	0	184	2.2648
21	0	0	−2	0	189	2.2765	49	0	0	−2	0	150	2.1761
22	0	0	2	0	220	2.3424	50	0	0	2	0	219	2.3404
23	0	0	0	−2	140	2.1461	51	0	0	0	−2	123	2.0899
24	0	0	0	2	153	2.1847	52	0	0	0	2	149	2.1732
25	0	0	0	0	279	2.4456	53	0	0	0	0	270	2.4314
26	0	0	0	0	198	2.2967	54	0	0	0	0	172	2.2355
27	0	0	0	0	234	2.3692	55	0	0	0	0	204	2.3096
28	0	0	0	0	243	2.3856	56	0	0	0	0	227	2.356

Notes

1. By "flat," we mean from a curvature-of-the-earth perspective; we are *not* referring to the issue of hills and valleys.

2. In general, the only time that stage two requires more than one experiment is the unfortunate, but not unlikely, situation of a "saddle point," or some other undesirable happenstance. This issue is discussed later in the chapter.

3. This example, and a broad outline of some of the features of the process to illustrate with the example, were suggested by the discussion in C. R. Hicks, *Fundamental Concepts in the Design of Experiments,* 3rd edition, New York, Holt, Rinehart and Winston, 1982.

4. In the real world, issues involving the packaging and other aspects of banana preservation are quite complex. There are many considerations and a large number

of factors involved. However, as we noted, our example is greatly simplified, but captures the features we wish to illustrate. The example is loosely based on one author's experience designing a complex experiment addressing some of these issues for a well-known harvester and shipper of bananas. Incidentally, in general, brown spots on bananas are not unhealthy or inedible—only unsightly!

5. We obtain the transformation for U as follows: Let $U = a + bX_1$. When $X_1 = 1/4$, $U = -1$; this gives $-1 = a + b(1/4) = a + b/4$; when $X_1 = 1/2$, $U = +1$; this gives $1 = a + b(1/2) = a + b/2$. We have two equations with two unknowns, a and b. We solve these and find $a = -3$ and $b = 8$. The second equation is found in a similar fashion.

6. We repeat that this will be the last experiment, unless the situation is an "unlucky" one, in which there is either a saddle point, or local maxima in addition to the global maximum.

7. In theory, the point could be a minimum point, but, unless the surface is very irregular, this is very unlikely after being directionally guided by a series of steepest-ascent analyses.

8. The authors don't know where this world is; if we did, we'd buy property there!

9. Remember—we are describing what the NASA team did, not necessarily what we would have done; then again, we weren't there, and they made the judgment call. This thought applies to many of the subsequent steps the NASA researchers chose to follow.

10. Actually, it was the taxpayers' money.

11. Of course, search techniques were available back in the late 1960s. However, they were not readily accessible nor user-friendly. Today, one might question whether the calculus step shouldn't always be replaced by Solver or its equivalent.

CHAPTER 15

Literature on Experimental Design, and Discussion of Some Topics Not Covered in the Text

15.1 Introduction

In this last chapter, we have two goals. The first is to acquaint the reader with several references in the field of experimental design, and the second is to introduce some additional topics in experimental design and provide references for them.

To achieve the first goal, we list a number of useful references, along with our comments about many of them. A number of textbooks and journal articles in the field are excellent references, but no one text is the best for every topic. In addition, the texts are written at different levels of mathematical sophistication and detail, and several texts are primarily dedicated to one area of application. As we noted in the preface to this text, we believe that additional mathematical detail is a mixed blessing—it provides more insight to those who want and can handle it, but causes others added difficulty in finding the gems among the ore. It increases the difficulty of making more application-oriented inquiries, as well as burdening readers simply wishing for a mathematically-less-challenging presentation.

We are liberal in giving our opinions about the references from several perspectives. The opinions represent our beliefs, without regard to politics, personalities, or any other nonprofessional biases of which we are aware. Of course, what makes a text "better" or "worse" in someone's view is always, in part, subjective; indeed, different teachers vary in which aspects of a text they consider more important. In any event, we view our opinions as ones about which reasonable people may disagree. In some cases our views agree with the prevailing wisdom; in other cases they may not.

In general, we do not compare our text directly to these references, and indeed, any comparative (or superlative) statements about the referenced texts are always with the caveat of "present company (that is, our text) excluded." Thus, when we make a statement that "we believe that X text is the best (or one of the best) text(s) for Y and Z topics," we are not saying that it is superior to our text. The referenced text may go into a topic more or less deeply than our text, and present the material at a different mathematical level, and use applications in different areas; however, we would not agree that any text is superior to ours for the goals we set (note that we are not claiming superiority, simply parity). We believe that all authors should feel that way about their own book!

As mentioned, our second goal in this chapter is to introduce, in a very cursory way, a number of topics that we have, in the main, *not* covered in this text (though we may have briefly alluded to them). We hope to provide the reader with a modest understanding of the role of these topics in experimental design, and to supply references on those topics.

15.2 Literature Discussion

The list of references at the end of the chapter (section 15.4) is ordered alphabetically and is restricted to texts that have the design and analysis of experiments as the primary topic. Inclusion on this list indicates that we believe the text (or in some cases a journal article) belongs to at least one of the following categories: (1) a classic; (2) one of the best texts for specific topics—some for topics we have covered in the text, others for topics to be briefly introduced in section 15.3; (3) one of the best texts at a given mathematical level; (4) a superior text for a dedicated application area (such as biology); (5) an "offbeat" text, one that differs from the others in a positive way. We do not view these categories as mutually exclusive; indeed, some texts are in multiple categories. A few of these references have appeared earlier in the body of the text. However, not all references in the body of the text are repeated here.

Some Classics

Some may wish to read about portions of the subject matter in the original words of the "masters." Two of the pioneers of much of the work in this text are Sir Ronald Fisher and Frank Yates. The examples in Fisher's two books, **Fisher (14th edition, 1973)** and **Fisher (9th edition, 1974),** are almost exclusively in the areas of genetics and agriculture. The books are very difficult to follow in places, and are not wise choices for textbooks. However, some sections of these books are fascinating—especially those motivated by Fisher's description of a woman's extraordinary claim that she can taste whether a cup of tea had the tea or the milk put in first. Fisher used this example to introduce and discuss several aspects of experimental design.

Yates' books are also classics. **Yates (1937)** provides insight about the origins of factorial design and Yates' algorithm; **Yates (1970)** is a series of his papers on a variety of subjects. These two books would also not be wise choices of a text for a current course in experimental design.

Recommendations for Specific Topics

In this section, we give our preferences for the best texts for specific topics or combinations of topics that were covered in our text. In section 15.3, where we introduce and discuss selected topics not included in our text, we include references on those topics. As noted above, no one text is the best for all topics.

Factorial and fractional-factorial designs We recommend four books on these topics. Two are by the same authors: **Anderson and McLean (1974)** and **McLean and Anderson (1984).** Both of these books are heavily oriented toward design of two- and three-level factorial and fractional-factorial designs. Both provide excellent and deep discussions, not only of two- and three-level complete factorial and fractional-factorial designs but also of mixed-level factorial and fractional-factorial designs with two and three levels. Certain other mixed-level designs are also discussed. The 1984 book goes more deeply into the mixed-level designs.

A very methodological treatment of the design of two-level and three-level factorial and fractional-factorial designs is provided in **Davies (3rd edition, 1984).** This book is more application oriented than the two cited above, and uses many more real-world examples. It is also a more comprehensive book with more topics covered.

The other book on these topics worth noting is **Daniel (1976).** This book is extremely application oriented, with lengthy discussion of many real-world experiments. Cuthbert Daniel was, at various times, one of the chief consultants for Procter & Gamble, General Foods, Consumers Union, and Pan American, among others. The book is heavily oriented toward 2^{k-p}, 3^{k-p}, and 2^m3^n designs. However, in our view, this book does not provide a useful guide for *how* to design these experiments; in addition, it is a bit disappointing in writing style.

Multiple-comparison testing In Chapter 4, we covered the basics of multiple-comparison testing. There are entire books on this one subject. We have two recommendations: **Hochberg and Tamhane (1987)** and **Hsu (1996).** Both are excellent; however, we prefer the Hsu book as a bit less (though still moderately) mathematically challenging. It also has an especially good chapter on abuses and misconceptions in multiple-comparison testing.

Taguchi methods We devoted Chapter 13 to Taguchi methods. Although we touched on most of the aspects of these methods, we did not cover the topic in great depth. A number of books, as well as entire courses, are devoted to this subject. We believe that by far the best treatise on the subject is by Genichi Taguchi himself. In our view, his two-volume book, **Taguchi (1987,** reprinted **1994),** is the clearest and most comprehensive available. Three other books also deserve mention. Two are **Ryan (1989)** and **Ross (2nd edition, 1996).** They offer sound treatment of Taguchi methods, tying them to quality issues and occasionally suggesting improvements to classical Taguchi methods. The other book is **Park (1996);** this book, more so than the others, covers both Taguchi methods and classical methods and commingles them. Each of the books is oriented toward engineering applications.

Response-surface methods We described and illustrated the basics of response-surface methodology in Chapter 14, as well as providing a real-world application. This is another topic on which entire books have been written. Of several excellent books on the subject, our favorite is **Box and Draper (1987).** We give it a slight edge due, primarily, to its clarity of exposition and applied orientation. Our second choice is that of **Myers and Montgomery (1995).** Myers wrote one of the first books on the subject of response surface methods in 1976; Montgomery is the author of one of the best general books on experimental design (to be discussed and referenced later in this chapter). This book is also application oriented, and includes some new developments in the field not discussed in Box and Draper. **Khouri and Cornell (2nd edition, 1996)** is also clearly written and includes new developments. The general level of mathematics is, in our view, somewhat higher than that of the other books mentioned. Some chapters are the same, or nearly the same, as in the first edition; overall, however, the second edition is about 25% longer than the first edition.

We also recommend one journal article, **Lucas (1976).** Many of the journal articles of today require a much higher level of mathematical preparation and effort to fathom, compared to a couple of decades ago. In our view, this is not generally a positive development. We believe that it is due, at least in part, to the increased competition for publication in some of the so-called top journals. Some cynics suggest that the average number of readers of these journal articles has been monotonically decreasing over time; the even more cynical suggest that for some journals, the attractiveness of an article is inversely proportional to the number of people *able* to read it! However, some older journal articles are not so burdensome to follow and have much to offer those who cannot, or do not choose to, wallow in complicated mathematics. This article is one of those. It provides a useful perspective on some issues involved in response-surface methods.

General texts We advertised this section as a discussion of books recommended for specific topics. However, we also wish to include those books that we consider among the best general texts in the field of experimental design (as we aim for this text to be). First and foremost, we include **Li (1964)** and **Hicks (4th edition, 1993).** We discuss these two favorites in the next section.

A book that is among the best general texts available today is **Montgomery (4th edition, 1997).** It is well written, clear, and has good exercises. Another book in this category is **Winer, Brown, and Michels (3rd edition, 1991).** One of the most comprehensive books available, with over a thousand pages, it is part of the McGraw-Hill series in psychology. The other book that must be mentioned is **Box, Hunter, and Hunter (1978).** Actually, *we* do not view this book as a superior choice of text. We believe it lacks some of the ingredients we find most important, including a high level of clarity; however, it is probably the book most often mentioned by others as one of the best texts available. In a poll taken in the 1980s, the Box, Hunter, and Hunter book won the vote for the single book to own if one could own only a single book on experimental design.

Level of mathematics Of all the books we have seen, the one with a significantly simpler level of mathematics is **Li (1964).** This book is a delight. It is written in an informal style, and virtually every numerical example is contrived with "easy numbers." As we noted in the preface, Professor Li states his view: "How does one first learn to solve quadratic equations? By working with terms such as $242.5189X^2 - 683.1620X + 19428.5149 = 0$, or with terms like $X^2 - 5X - 6 = 0$?" He covers all of the necessary mathematical concepts (that is, he does not at all avoid such topics as deriving expected mean squares) but in a way that minimizes the level of mathematics without compromising the derivation of the results. The Li book has been a major inspiration for us and has had a major impact on the style, mathematical level and rigor, and "attitude" of our book. It is unlikely that we shall say as much about any other text.

The next step up in mathematical level is found in **Guenther (1964), Hicks (4th edition, 1993),** and **Schmidt and Launsby (4th edition, 1994).** Of the three, we strongly prefer the Hicks book, due to the organization of the topics and the general writing style. Guenther is a fellow of the American Statistical Association and a classical theoretical statistician, but his text is mathematically very readable. The Schmidt and Launsby text could use some additional editing and organization. These three texts are mentioned together solely for their mathematical level. The most advanced level of mathematics is found in **Scheffé (1959)** and **Graybill (1961).** The other books on experimental design and analysis have a mathematical level in between those noted in this paragraph.

Application areas Some books do a superior job of illustrating how the techniques can be usefully applied in a particular application area. We note our choices of a few of these books. We base our choices on general reputation, our perception of the clarity of exposition of both the experimental design and statistical issues as well as application issues, and the degree to which the book provides a good perspective on the distinctive experimental-design issues involved in the particular application area. Some of the books noted are not recent but have other attributes that more than offset the lack of recency.

For applications in the medical area, we note **Armitage (1971)** and **Fleiss (1986).** For applications in the biology area we note **Dennenberg (1976)** and **Kuehl (1994).** We also note Kuehl (1994) for selected applications in the agriculture area. For psychological research we note **Edwards (1972).** For social and economic policy, we note **Ferber and Hirsch (1981).** For engineers and scientists, we note **Montgomery (1997)** and **Diamond (1981).** For physicists and chemists, we note **Goupy (1993).**

Offbeat books We list this category, "offbeat," primarily as an excuse to reference one book, **Cox (1958).** It is not at all recommended as a textbook for a typical course in experimental design. However, it provides a relatively rich discussion of a number of topics that are not discussed in great detail, if at all, in most other books on experimental design. Examples of such topics are the use of $\alpha = .05$ compared with $\alpha = .01$ or other values, the practical difficulties of randomization, distinctions among different kinds of factors that appear in factorial designs, the choice of how many levels of these factors to use and what they should be, the different purposes for which observations may be made, an especially detailed discussion of the size of an experiment, and other issues. One other book that is offbeat in terms of discussing issues that are usually just glossed over is **Wilson (1952).**

15.3 Discussion of Some Topics Not Covered in the Text

Several topics that either were not addressed in the body of the text or were touched upon only lightly are discussed here, and references are provided. The last four of these topics essentially involve different designs.

Outliers

Sometimes a data set contains one or more data points that appear to be aberrations, in that they are very different from the other data points in the set. If such a data point is truly the result of an unusual occurrence or circumstance that is rarely likely to be duplicated, the data point is generally viewed as not part of the process being modeled and investigated, and is referred to as an **outlier** and dropped from the data set before the analysis. Of course, what makes a data point aberrational is not well defined, and often there is no determinable cause of an unusual-looking data point.

If we "throw out" a data value that was actually part of the probability distribution at that treatment combination (remember: for normally distributed output, about a third of a percent of the time a data point deviates from the true mean by more than three standard deviations), the analysis of the remaining data might underestimate the variability of the process and reach unwarranted conclusions. However, if the data value is retained but is truly an aberration, the analysis may suggest misleading conclusions of different sorts. Like many of life's decision processes, the decision to label a data value an outlier is a tradeoff between Type I and Type II errors. What should the experimenter do? We don't have an unequivocal answer. Indeed, this is not a question that's easy to answer, as evidenced by the existence of several entire books on the subject of outliers. We note two of them, **Iglewicz and Hoaglin (1993)** and **Barnett and Lewis (3rd edition, 1994).** There is also some good elementary discussion of the topic in **Hicks (4th edition, 1993)** and **Davies (3rd edition, 1984).**

Missing Data

We have only briefly alluded to the treatment of missing data. This is another problem that has no quick, easy, general answer. We recommend two books devoted to the subject, **Dodge (1985)** and **Little and Rubin (1987).** Both are comprehensive and useful. The Dodge book is a bit more oriented toward the kind of experiments we have covered in this text, and the Little and Rubin book is a bit more oriented toward missing data in a regression analysis or other multivariate analysis context, though it also considers missing data in the more classic orthogonal designs we have addressed. The two basic choices are (1) ignore the fact that the data point(s) is missing, and do the analysis under the (generally) nonorthogonal conditions caused by the absence of the missing data; and (2) estimate the missing data by "surrogates" and make the proper adjustments in the analysis to recognize that this has been done. How to do the latter is the major topic of the books cited.

Power and Sample Size

We have discussed power ($[1 - \beta]$, the complement of the probability of Type II error) at a couple of points in the text. As we noted, the determination of power achieved with a design, or the determination of the sample size that achieves a specific power, each relative to specific true values of the μs (and other inputs, such as α), is not a simple one. Indeed, that is why power and sample size tables appear in various texts (including ours), and why only a few software programs have decided to address these issues, sometimes in a limited way.

We note two texts that are known for their tables of samples size versus power as a function of various other inputs and different designs, and in some cases the form of the dependent variable. They are **Cohen (1988)** and **Kraemer and Thiemann (1989).** Cohen's text is oriented toward the behavioral sciences, whereas Kraemer and Thiemann's is more general. Each provides basics on the computation of power and sample size, but is used mostly for its tables.

Time-Series and Failure-Time Experiments

When time is an experimental factor, there is an increased likelihood that one of the basic assumptions in traditional ANOVA is violated—the independence-of-errors assumption. Recall that this is the one ANOVA assumption that is not very robust. Just how does one inquire about this possible violation? What should be done if this violation is discovered? Our text has not addressed these issues. A book dedicated to this subject is **Glass, Willson, and Gottman (1975);** it contains a useful compilation of what needs to be considered in this situation.

When the *dependent* variable is measured in units of time—in particular, when the yield, or response, is failure-time data, there is an increased probability that the normality assumption of ANOVA is materially violated. This is often due to the physics of our planet that leads to an other-than-normal distribution of failure times. **Kalbfleisch and Prentice (1980)** is dedicated to this topic.

Plackett-Burman Designs

There is a set of designs called Plackett-Burman designs, usually applied in cases in which all factors are at two levels. These designs provide the minimum possible number of treatment combinations as a function of the number of factors in order to determine all main effects cleanly, under the assumption that all interactions equal zero. **Plackett and Burman (1946)** provides the methodology and tables to derive this "optimal" design for two-level designs and the number of factors from 1 to 100 (except 92), and for the number of levels, L, equal to three, four, five, and seven for a number of factors equal to L^r, r an integer.

Repeated-Measures Designs

The term "repeated-measures design" is used to indicate a design in which each subject (usually, person) is considered to be a block and is used ("repeated") for each treatment combination. Repeated-measures designs greatly reduce the error in situations in which differences from person to person are relatively large. A repeated-measures design is sometimes referred to as a "within-subjects" design. Designs covered in this text have been "between-subjects designs" (a term we did not introduce earlier, since it was not needed to distinguish designs), designs in which, if people are involved, each data value corresponds with a different person—there are no "repeated measures." The Luna Electronics application in Chapter 14 is an exception; it is a repeated-measures design.

Sometimes a repeated-measures design is used when the nature of the experiment suggests that a person be exposed to all treatment combinations. For example, when studying the reaction to different print ads, what an individual person "brings to the table" in terms of attitudes (toward life and/or advertisements in general) may vary far more than an individual's view of the differences from one ad to another. Also, a repeated-measures design would be mandatory when studying learning-curve types of responses over time.

Other times, a repeated-measures design is used to dramatically reduce the number of subjects required in the experiment. For example, one of the authors has frequently consulted in marketing research studies in which treatment combinations of factor levels are shown to responders (such as 16 "scenarios" reflecting treatment combinations of eight factors at two or three levels each) and a purchase-intent measure is solicited for each scenario. If it is desired that each scenario be evaluated by 200 responders (representing a careful mix of demographic characteristics and degree of previous purchase), one would need $16 \cdot 200 = 3200$ people if there were no repeated measures. This amount is far higher than is practical; by having each of the 200 responders consider all 16 scenarios, we achieve the goal of 200 evaluations per scenario, using only 200 people in total.

There are also "partially repeated-measures designs," in which some factors are totally repeated and others are not. An example would be an experiment in which different responders are first exposed to different levels of knowledge about a product and then each responder evaluates all treatment combinations of product attributes. The prior-knowledge factor would be a between-subject factor, and the other factors composing the treatment combinations of product attributes would be within-subject.

A key element of repeated-measures designs is the treatment of "person" as a random-level factor, and the subsequent implications of this in the analysis phase of the experiment. We discussed aspects of this issue in Chapter 6, but only for certain specific cases. There are entire books devoted to these types of experiments. Such books include **Girden (1992)** and **Vonesh and Chinchilli (1997). Winer, Brown, and Michels (1991),** basically in one large chapter, provides a superior, thorough treatment of the basic prototypes of these designs.

Crossover Designs

Crossover designs (also called "change-over designs") are special cases of repeated-measures designs, in which the treatments applied to the same subjects are systematically *changed* over time. A complete book on this subject is **Ratkowsky, Evans, and Alldredge (1993).** Also, **Li (1964)** covers the elementary principles of and special issues involved in this type of design with great clarity and simplicity.

Mixture Designs

In Chapter 14, we briefly made note of what is called a mixture experiment. Such an experiment involves the blending of two or more components in various proportions. The key element that makes this type of experiment different from most of what we have discussed in previous chapters is that once the levels of some of the factors are selected, levels of other factors are no longer unrestricted. For example, if the amounts of five factors must total 100% of a mixture, and four of the factors are assigned levels that total 88%, the other factor *must* be at a level of 12%. Therefore, we cannot, in general, set up a routine factorial (or fractional-factorial) design: not all treatment combinations are feasible.

A book that is usually recommended as one of the best for a superior, dedicated treatment of mixture designs is **Cornell (2nd edition, 1990).** A brief exposure to the general issues involved in these types of experiments can be found in **Piepel and Cornell (1994).**

Bibliography

We cite one other reference. In 1970, Balaam and Federer began to compile a bibliography of all that had ever been written in the field of design and analysis of experiments. They did a yeoman's job, being extremely thorough and cross-referencing the citations by author, topic, language, and in other ways. The references include those through 1969. It is understandable why nobody has picked up the mantle to provide an update. They didn't finish until 1973! Their efforts resulted in the book **Balaam and Federer (1973).** We have found many of these earlier publications in the field of experimental design to be fascinating reading.

15.4 References

We hope that this section lists every text and article for which we did not include full publication information in the body of the text. However, we have no doubt that we have "missed" many other worthy texts that deserve mention. We apologize for any such unintended omission.

Anderson, V., and R. McLean, *Design of Experiments: A Realistic Approach,* New York, Marcel Dekker, 1974.

Armitage, P., *Statistical Methods in Medical Research,* New York, Wiley, 1971.

Balaam, L., and W. Federer, *Bibliography on Experiment and Treatment Design,* New York, Haffner, 1973.

Barnett, V., and T. Lewis, *Outliers in Statistical Data,* 3rd edition, New York, Wiley, 1994.

Box, G., W. Hunter, and J. Hunter, *Statistics for Experimenters,* New York, Wiley, 1978.

Box, G., and N. Draper, *Empirical Model-Building and Response Surfaces,* New York, Wiley, 1987.

Cohen, J., *Statistical Power Analysis for the Behavioral Sciences,* 2nd edition, Hillsdale, N.J., Lawrence Erlbaum Associates, 1988.

Cornell, J., *Experiments with Mixtures: Designs, Models, and the Analysis of Mixture Data,* 2nd edition, New York, Wiley, 1990.

Cox, D., *Planning of Experiments,* New York, Wiley, 1958.

Daniel, C., *Applications of Statistics to Industrial Experimentation,* New York, Wiley, 1976.

Davies, O. (editor), *The Design and Analysis of Industrial Experiments,* 3rd edition, New York, Hafner, 1984.

Dennenberg, V., *Statistical and Experimental Designs for Behavioral and Biological Researchers,* New York, Halsted Press, 1976.

Diamond, W., *Practical Experimental Designs for Engineers and Scientists,* New York, Wiley, 1981.

Dodge, Y., *Analysis of Experiments with Missing Data,* New York, Wiley, 1985.

Edwards, A., *Experimental Design in Psychological Research,* 4th edition, New York, Holt, Rinehart and Winston, 1972.

Ferber, J., and M. Hirsch, *Social Experimentation and Economic Policy,* Cambridge, England, Cambridge University Press, 1981.

Fisher, R., *Statistical Methods for Research Workers,* 14th edition, New York, Haffner, 1973.

Fisher, R., *The Design of Experiments,* 9th edition, New York, Haffner, 1974.

Fleiss, J., *The Design and Analysis of Clinical Experiments,* New York, Wiley, 1986.

Girden, E., *ANOVA: Repeated Measures,* Newbury Park, Calif., Sage Publications, 1992.

Glass, G., V. Willson, and J. Gottman, *Design and Analysis of Time-Series Experiments,* Boulder, University Press of Colorado, 1975.

Goupy, J., *Methods for Experimental Design: Principles and Applications for Physicists and Chemists,* Amsterdam, Elsevier, 1993.

Graybill, F., *An Introduction to Linear Statistical Models,* New York, McGraw-Hill, 1961.

Guenther, W., *Analysis of Variance,* Englewood Cliffs, N.J., Prentice-Hall, 1964.

Hicks, C., *Fundamental Concepts in the Design of Experiments,* 4th edition, New York, Saunders College, 1993.

Hochberg, Y., and A. Tamhane, *Multiple Comparison Procedures,* New York, Wiley, 1987.

Hsu, J., *Multiple Comparisons: Theory and Methods,* New York, Chapman & Hall, 1996.

Iglewicz, B., and D. Hoaglin, *How to Detect and Handle Outliers,* Milwaukee, ASQC Quality Press, 1993.

Kalbfleisch, J., and R. Prentice, *Statistical Analysis of Failure Time Data,* New York, Wiley, 1980.

Khouri, A., and J. Cornell, *Response Surfaces: Design and Analyses,* 2nd edition, New York, Marcel Dekker, 1996.

Kraemer, H., and S. Thiemann, *How Many Subjects? Statistical Power in Research,* Newbury Park, Calif., Sage Publications, 1989.

Kuehl, R., *Statistical Principles for Research Design and Analysis,* Belmont, Calif., Duxbury Press, 1994.

Li, C., *Introduction to Experimental Statistics,* New York, McGraw-Hill, 1964.

Little, R., and D. Rubin, *Statistical Analysis with Missing Data,* New York, Wiley, 1987.

Lucas, J., "Which Response Surface Design Is Best?" *Technometrics,* 18, pp. 411–417, November, 1976.

McLean, R., and V. Anderson, *Applied Factorial and Fractional Systems,* New York, Marcel Dekker, 1984.

Montgomery, D., *Design and Analysis of Experiments,* 4th edition, New York, Wiley, 1997.

Myers, R., and D. Montgomery, *Response Surface Methodology: Process and Product Optimization Using Designed Experiments,* New York, Wiley, 1995.

Park, S., *Robust Design for Quality Engineering,* New York, Chapman & Hall, 1996.

Peng, K. C., *The Design and Analysis of Scientific Experiments,* Reading, Mass., Addison Wesley, 1967.

Piepel, G., and J. Cornell, "Mixture Experiment Approaches: Examples, Discussions, and Recommendations," *Journal of Quality Technology,* 26, pp. 177–196, 1994.

Plackett, B., and J. Burman, "The Design of Optimum Multifactorial Experiments," *Biometrika,* 33, pp. 305–325, 1946.

Ratkowsky, D., M. Evans, and J. Alldredge, *Cross-Over Experiments,* New York, Marcel Dekker, 1993.

Ross, P., *Taguchi Techniques for Quality Engineering,* 2nd edition, New York, McGraw-Hill, 1996.

Ryan, T., *Statistical Methods for Quality Improvement,* New York, Wiley, 1989.

Scheffé, H., *The Analysis of Variance,* New York, Wiley, 1959.

Schmidt, S., and R. Launsby, *Understanding Designed Experiments,* 4th edition, Colorado Springs, Colo., Air Academy Press, 1994.

Taguchi, G., *System of Experimental Design,* Dearborn, Mich., American Supplier Institute, 2 volumes, 1987 (second printing, 1988).

Taguchi, G., *System of Experimental Design,* White Plains, N.Y., Kraus International Publications (reprint of 1987 publication), 1994.

Vonesh, E., and V. Chinchilli, *Linear and Nonlinear Models for the Analysis of Repeated Measurements,* New York, Marcel Dekker, 1997.

Wilson, E., *An Introduction to Scientific Research,* New York, McGraw-Hill, 1952.

Winer, B., D. Brown, and K. Michels, *Statistical Principles in Experimental Design,* 3rd edition, New York, McGraw-Hill, 1991.

Yates, F., *The Design and Analysis of Factorial Experiments,* Harpenden, England, Imperial Bureau of Soil Science, 1937.

Yates, F., *Selected Papers,* London, Griffen Publishing, 1970.

Statistical Tables

Table A Standard Normal (Z) Distribution

Left-tail cumulative probabilities, $F(Z)$, for Z values 0(.01)4.00

Table B Student t Distribution

Abscissa (horizontal axis) values for degrees of freedom 1(1)30, 40, 60, 120, ∞, for upper-tail areas .25, .1, .05, .025, .01, .005, .0025, .001, .0005

Table C Chi-Square (χ^2) Distribution

Abscissa (horizontal axis) values for degrees of freedom 1(1)30, 40(10)100, for upper-tail areas .25, .1, .05, .025, .01, .005, .001

Table D F Distribution

Abscissa (horizontal axis) values for numerator degrees of freedom 1(1)10, 12, 15, 20, 24, 30, 40, 60, 120, ∞, denominator degrees of freedom 1(1)30, 40, 60, 120, ∞, for upper-tail areas .05 and .01

Table A.1 **Standard Normal (*Z*) Distribution** Left-tail cumulative probabilities, *F(Z)*, for *Z* values 0(.01)4.00

Z	*F(Z)*	*Z*	*F(Z)*	*Z*	*F(Z)*	*Z*	*F(Z)*
.00	.5000000	.50	.6914625	1.00	.8413447	1.50	.9331928
.01	.5039894	.51	.6949743	1.01	.8437524	1.51	.9344783
.02	.5079783	.52	.6984682	1.02	.8461358	1.52	.9357445
.03	.5119665	.53	.7019440	1.03	.8484950	1.53	.9369916
.04	.5159534	.54	.7054015	1.04	.8508300	1.54	.9382198
.05	.5199388	.55	.7088403	1.05	.8531409	1.55	.9394292
.06	.5239222	.56	.7122603	1.06	.8554277	1.56	.9406201
.07	.5279032	.57	.7156612	1.07	.8576903	1.57	.9417924
.08	.5318814	.58	.7190427	1.08	.8599289	1.58	.9429466
.09	.5358564	.59	.7224047	1.09	.8621434	1.59	.9440826
.10	.5398278	.60	.7257469	1.10	.8643339	1.60	.9452007
.11	.5437953	.61	.7290691	1.11	.8665005	1.61	.9463011
.12	.5477584	.62	.7323711	1.12	.8686431	1.62	.9473839
.13	.5517168	.63	.7356527	1.13	.8707619	1.63	.9484493
.14	.5556700	.64	.7389137	1.14	.8728568	1.64	.9494974
.15	.5596177	.65	.7421539	1.15	.8749281	1.65	.9505285
.16	.5635595	.66	.7453731	1.16	.8769756	1.66	.9515428
.17	.5674949	.67	.7485711	1.17	.8789995	1.67	.9525403
.18	.5714237	.68	.7517478	1.18	.8809999	1.68	.9535213
.19	.5753454	.69	.7549029	1.19	.8829768	1.69	.9544860
.20	.5792597	.70	.7580363	1.20	.8849303	1.70	.9554345
.21	.5831662	.71	.7611479	1.21	.8868606	1.71	.9563671
.22	.5870644	.72	.7642375	1.22	.8887676	1.72	.9572838
.23	.5909541	.73	.7673049	1.23	.8906514	1.73	.9581849
.24	.5948349	.74	.7703500	1.24	.8925123	1.74	.9590705
.25	.5987063	.75	.7733726	1.25	.8943502	1.75	.9599408
.26	.6025681	.76	.7763727	1.26	.8961653	1.76	.9607961
.27	.6064199	.77	.7793501	1.27	.8979577	1.77	.9616364
.28	.6102612	.78	.7823046	1.28	.8997274	1.78	.9624620
.29	.6140919	.79	.7852361	1.29	.9014747	1.79	.9632730
.30	.6179114	.80	.7881446	1.30	.9031995	1.80	.9640697
.31	.6217195	.81	.7910299	1.31	.9049021	1.81	.9648521
.32	.6255158	.82	.7938919	1.32	.9065825	1.82	.9656205
.33	.6293000	.83	.7967306	1.33	.9082409	1.83	.9663750
.34	.6330717	.84	.7995458	1.34	.9098773	1.84	.9671159
.35	.6368307	.85	.8023375	1.35	.9114920	1.85	.9678432
.36	.6405764	.86	.8051055	1.36	.9130850	1.86	.9685572
.37	.6443088	.87	.8078498	1.37	.9146565	1.87	.9692581
.38	.6480273	.88	.8105703	1.38	.9162067	1.88	.9699460
.39	.6517317	.89	.8132671	1.39	.9177356	1.89	.9706210
.40	.6554217	.90	.8159399	1.40	.9192433	1.90	.9712834
.41	.6590970	.91	.8185887	1.41	.9207302	1.91	.9719334
.42	.6627573	.92	.8212136	1.42	.9221962	1.92	.9725711
.43	.6664022	.93	.8238145	1.43	.9236415	1.93	.9731966
.44	.6700314	.94	.8263912	1.44	.9250663	1.94	.9738102
.45	.6736448	.95	.8289439	1.45	.9264707	1.95	.9744119
.46	.6772419	.96	.8314724	1.46	.9278550	1.96	.9750021
.47	.6808225	.97	.8339768	1.47	.9292191	1.97	.9755808
.48	.6843863	.98	.8364569	1.48	.9305634	1.98	.9761482
.49	.6879331	.99	.8389129	1.49	.9318879	1.99	.9767045
.50	.6914625	1.00	.8413447	1.50	.9331928	2.00	.9772499

Table A.1 (continued)

Z	F(Z)	Z	F(Z)	Z	F(Z)	[illegible]	[illegible]
2.00	.9772499	2.50	.9937903	3.00	[illegible]	[illegible]	[illegible]
2.01	.9777844	2.51	.9939634	3.01	[illegible]	[illegible]	[illegible]
2.02	.9783083	2.52	.9941323	3.02	[illegible]	[illegible]	[illegible]
2.03	.9788217	2.53	.9942969	3.03	[illegible]	[illegible]	[illegible]
2.04	.9793248	2.54	.9944574	3.04	[illegible]	[illegible]	[illegible]
2.05	.9798178	2.55	.9946139	3.05	.9988558	[illegible]	[illegible]
2.06	.9803007	2.56	.9947664	3.06	.9988933	[illegible]	[illegible]
2.07	.9807738	2.57	.9949151	3.07	.9989297	[illegible]	[illegible]
2.08	.9812372	2.58	.9950600	3.08	.9989650	3.58	[illegible]
2.09	.9816911	2.59	.9952012	3.09	.9989992	3.59	[illegible]
2.10	.9821356	2.60	.9953388	3.10	.9990324	3.60	[illegible]
2.11	.9825708	2.61	.9954729	3.11	.9990646	3.61	[illegible]
2.12	.9829970	2.62	.9956035	3.12	.9990957	3.62	[illegible]
2.13	.9834142	2.63	.9957308	3.13	.9991260	3.63	.9998583
2.14	.9838226	2.64	.9958547	3.14	.9991553	3.64	.9998637
2.15	.9842224	2.65	.9959754	3.15	.9991836	3.65	.9998689
2.16	.9846137	2.66	.9960930	3.16	.9992112	3.66	.9998739
2.17	.9849966	2.67	.9962074	3.17	.9992378	3.67	.9998787
2.18	.9853713	2.68	.9963189	3.18	.9992636	3.68	.9998834
2.19	.9857379	2.69	.9964274	3.19	.9992886	3.69	.9998879
2.20	.9860966	2.70	.9965330	3.20	.9993129	3.70	.9998922
2.21	.9864474	2.71	.9966358	3.21	.9993363	3.71	.9998964
2.22	.9867906	2.72	.9967359	3.22	.9993590	3.72	.9999004
2.23	.9871263	2.73	.9968333	3.23	.9993810	3.73	.9999043
2.24	.9874545	2.74	.9969280	3.24	.9994024	3.74	.9999080
2.25	.9877755	2.75	.9970202	3.25	.9994230	3.75	.9999116
2.26	.9880894	2.76	.9971099	3.26	.9994429	3.76	.9999150
2.27	.9883962	2.77	.9971972	3.27	.9994623	3.77	.9999184
2.28	.9886962	2.78	.9972821	3.28	.9994810	3.78	.9999216
2.29	.9889893	2.79	.9973646	3.29	.9994991	3.79	.9999247
2.30	.9892759	2.80	.9974449	3.30	.9995166	3.80	.9999277
2.31	.9895559	2.81	.9975229	3.31	.9995335	3.81	.9999305
2.32	.9898296	2.82	.9975988	3.32	.9995499	3.82	.9999333
2.33	.9900969	2.83	.9976726	3.33	.9995658	3.83	.9999359
2.34	.9903581	2.84	.9977443	3.34	.9995811	3.84	.9999385
2.35	.9906133	2.85	.9978140	3.35	.9995959	3.85	.9999409
2.36	.9908625	2.86	.9978818	3.36	.9996103	3.86	.9999433
2.37	.9911060	2.87	.9979476	3.37	.9996242	3.87	.9999456
2.38	.9913437	2.88	.9980116	3.38	.9996376	3.88	.9999478
2.39	.9915758	2.89	.9980738	3.39	.9996505	3.89	.9999499
2.40	.9918025	2.90	.9981342	3.40	.9996631	3.90	.9999519
2.41	.9920237	2.91	.9981929	3.41	.9996752	3.91	.9999539
2.42	.9922397	2.92	.9982498	3.42	.9996869	3.92	.9999557
2.43	.9924506	2.93	.9983052	3.43	.9996982	3.93	.9999575
2.44	.9926564	2.94	.9983589	3.44	.9997091	3.94	.9999593
2.45	.9928572	2.95	.9984111	3.45	.9997197	3.95	.9999609
2.46	.9930531	2.96	.9984618	3.46	.9997299	3.96	.9999625
2.47	.9932443	2.97	.9985110	3.47	.9997398	3.97	.9999641
2.48	.9934309	2.98	.9985588	3.48	.9997493	3.98	.9999655
2.49	.9936128	2.99	.9986051	3.49	.9997585	3.99	.9999670
2.50	.9937903	3.00	.9986501	3.50	.9997674	4.00	.9999683

From E. S. Pearson and H. O. Hartley in *Biometrika Tables for Statisticians*, vol. 1, 2d ed. (1958). Reprinted with permission of Oxford University Press.

Student *t* Distribution

Abscissa (horizontal axis) values for degrees of freedom 1(1)30, 40, 60, 120, ∞, for upper-tail areas .25, .1, .05, .025, .01, .005, .0025, .001, .0005

	0.25	0.1	0.05	0.025	0.01	0.005	0.0025	0.001	0.0005
	1.000	3.078	6.314	12.706	31.821	63.657	127.32	318.31	636.62
	0.816	1.886	2.920	4.303	6.965	9.925	14.089	22.326	31.598
	.765	1.638	2.353	3.182	4.541	5.841	7.453	10.213	12.924
	.741	1.533	2.132	2.776	3.747	4.604	5.598	7.173	8.610
5	0.727	1.476	2.015	2.571	3.365	4.032	4.773	5.893	6.869
6	.718	1.440	1.943	2.447	3.143	3.707	4.317	5.208	5.959
7	.711	1.415	1.895	2.365	2.998	3.499	4.029	4.785	5.408
8	.706	1.397	1.860	2.306	2.896	3.355	3.833	4.501	5.041
9	.703	1.383	1.833	2.262	2.821	3.250	3.690	4.297	4.781
10	0.700	1.372	1.812	2.228	2.764	3.169	3.581	4.144	4.587
11	.697	1.363	1.796	2.201	2.718	3.106	3.497	4.025	4.437
12	.695	1.356	1.782	2.179	2.681	3.055	3.428	3.930	4.318
13	.694	1.350	1.771	2.160	2.650	3.012	3.372	3.852	4.221
14	.692	1.345	1.761	2.145	2.624	2.977	3.326	3.787	4.140
15	0.691	1.341	1.753	2.131	2.602	2.947	3.286	3.733	4.073
16	.690	1.337	1.746	2.120	2.583	2.921	3.252	3.686	4.015
17	.689	1.333	1.740	2.110	2.567	2.898	3.222	3.646	3.965
18	.688	1.330	1.734	2.101	2.552	2.878	3.197	3.610	3.922
19	.688	1.328	1.729	2.093	2.539	2.861	3.174	3.579	3.883
20	0.687	1.325	1.725	2.086	2.528	2.845	3.153	3.552	3.850
21	.686	1.323	1.721	2.080	2.518	2.831	3.135	3.527	3.819
22	.686	1.321	1.717	2.074	2.508	2.819	3.119	3.505	3.792
23	.685	1.319	1.714	2.069	2.500	2.807	3.104	3.485	3.767
24	.685	1.318	1.711	2.064	2.492	2.797	3.091	3.467	3.745
25	0.684	1.316	1.708	2.060	2.485	2.787	3.078	3.450	3.725
26	.684	1.315	1.706	2.056	2.479	2.779	3.067	3.435	3.707
27	.684	1.314	1.703	2.052	2.473	2.771	3.057	3.421	3.690
28	.683	1.313	1.701	2.048	2.467	2.763	3.047	3.408	3.674
29	.683	1.311	1.699	2.045	2.462	2.756	3.038	3.396	3.659
30	0.683	1.310	1.697	2.042	2.457	2.750	3.030	3.385	3.646
40	.681	1.303	1.684	2.021	2.423	2.704	2.971	3.307	3.551
60	.679	1.296	1.671	2.000	2.390	2.660	2.915	3.232	3.460
120	.677	1.289	1.658	1.980	2.358	2.617	2.860	3.160	3.373
∞	.674	1.282	1.645	1.960	2.326	2.576	2.870	3.090	3.291

Table A.3 Chi-Square (χ^2) Distribution

Abscissa (horizontal axis) values for degrees of freedom 1(1)30, 40(10)100, for upper-tail areas .25, .1, .05, .025, .01, .005, .001

ν	0.250	0.100	0.050	0.025	0.010	0.005	0.001
1	1.32330	2.70554	3.84146	5.02389	6.63490	7.87944	10.828
2	2.77259	4.60517	5.99147	7.37776	9.21034	10.5966	13.816
3	4.10835	6.25139	7.81473	9.34840	11.3449	12.8381	16.266
4	5.38527	7.77944	9.48773	11.1433	13.2767	14.8602	18.467
5	6.62568	9.23635	11.0705	12.8325	15.0863	16.7496	20.515
6	7.84080	10.6446	12.5916	14.4494	16.8119	18.5476	22.458
7	9.03715	12.0170	14.0671	16.0128	18.4753	20.2777	24.322
8	10.2188	13.3616	15.5073	17.5346	20.0902	21.9550	26.125
9	11.3887	14.6837	16.9190	19.0228	21.6660	23.5893	27.877
10	12.5489	15.9871	18.3070	20.4831	23.2093	25.1882	29.588
11	13.7007	17.2750	19.6751	21.9200	24.7250	26.7569	31.264
12	14.8454	18.5494	21.0261	23.3367	26.2170	28.2995	32.909
13	15.9839	19.8119	22.3621	24.7356	27.6883	29.8194	34.528
14	17.1170	21.0642	23.6848	26.1190	29.1413	31.3193	36.123
15	18.2451	22.3072	24.9958	27.4884	30.5779	32.8013	37.697
16	19.3688	23.5418	26.2962	28.8454	31.9999	34.2672	39.252
17	20.4887	24.7690	27.5871	30.1910	33.4087	35.7185	40.790
18	21.6049	25.9894	28.8693	31.5264	34.8053	37.1564	42.312
19	22.7178	27.2036	30.1435	32.8523	36.1908	38.5822	43.820
20	23.8277	28.4120	31.4104	34.1696	37.5662	39.9968	45.315
21	24.9348	29.6151	32.6705	35.4789	38.9321	41.4010	46.797
22	26.0393	30.8133	33.9244	36.7807	40.2894	42.7956	48.268
23	27.1413	32.0069	35.1725	38.0757	41.6384	44.1813	49.728
24	28.2412	33.1963	36.4151	39.3641	42.9798	45.5585	51.179
25	29.3389	34.3816	37.6525	40.6465	44.3141	46.9278	52.620
26	30.4345	35.5631	38.8852	41.9232	45.6417	48.2899	54.052
27	31.5284	36.7412	40.1133	43.1944	46.9630	49.6449	55.476
28	32.6205	37.9159	41.3372	44.4607	48.2782	50.9933	56.892
29	33.7109	39.0875	42.5569	45.7222	49.5879	52.3356	58.302
30	34.7998	40.2560	43.7729	46.9792	50.8922	53.6720	59.703
40	45.6160	51.8050	55.7585	59.3417	63.6907	66.7659	73.402
50	56.3336	63.1671	67.5048	71.4202	76.1539	79.4900	86.661
60	66.9814	74.3970	79.0819	83.2976	88.3794	91.9517	99.607
70	77.5766	85.5271	90.5312	95.0231	100.425	104.215	112.317
80	88.1303	96.5782	101.879	106.629	112.329	116.321	124.839
90	98.6499	107.565	113.145	118.136	124.116	128.299	137.208
100	109.141	118.498	124.342	129.561	135.807	140.169	149.449

Table A.4 *F* Distribution

Abscissa (horizontal axis) values for numerator degrees of freedom 1(1)10, 12, 15, 20, 24, 30, 40, 60, 120, ∞, denominator degrees of freedom 1(1)30, 40, 60, 120, ∞, for upper-tail area .05.

ν_2 \ ν_1	1	2	3	4	5	6	7	8	9
1	161.45	199.50	215.71	224.58	230.16	233.99	236.77	238.88	240.54
2	18.513	19.000	19.164	19.247	19.296	19.330	19.353	19.371	19.385
3	10.128	9.5521	9.2766	9.1172	9.0135	8.9406	8.8868	8.8452	8.8123
4	7.7086	6.9443	6.5914	6.3883	6.2560	6.1631	6.0942	6.0410	5.9988
5	6.6079	5.7861	5.4095	5.1922	5.0503	4.9503	4.8759	4.8183	4.7725
6	5.9874	5.1433	4.7571	4.5337	4.3874	4.2839	4.2066	4.1468	4.0990
7	5.5914	4.7374	4.3468	4.1203	3.9715	3.8660	3.7870	3.7257	3.6767
8	5.3177	4.4590	4.0662	3.8378	3.6875	3.5806	3.5005	3.4381	3.3881
9	5.1174	4.2565	3.8626	3.6331	3.4817	3.3738	3.2927	3.2296	3.1789
10	4.9646	4.1028	3.7083	3.4780	3.3258	3.2172	3.1355	3.0717	3.0204
11	4.8443	3.9823	3.5874	3.3567	3.2039	2.0946	3.0123	2.9480	2.8962
12	4.7472	3.8853	3.4903	3.2592	3.1059	2.9961	2.9134	2.8486	2.7964
13	4.6672	3.8056	3.4105	3.1791	3.0254	2.9153	2.8321	2.7669	2.7144
14	4.6001	3.7389	3.3439	3.1122	2.9582	2.8477	2.7642	2.6987	2.6458
15	4.5431	3.6823	3.2874	3.0556	2.9013	2.7905	2.7066	2.6408	2.5876
16	4.4940	3.6337	3.2389	3.0069	2.8524	2.7413	2.6572	2.5911	2.5377
17	4.4513	3.5915	3.1968	2.9647	2.8100	2.6987	2.6143	2.5480	2.4943
18	4.4139	3.5546	3.1599	2.9277	2.7729	2.6613	2.5767	2.5102	2.4563
19	4.3808	3.5219	3.1274	2.8951	2.7401	2.6283	2.5435	2.4768	2.4227
20	4.3513	3.4928	3.0984	2.8661	2.7109	2.5990	2.5140	2.4471	2.3928
21	4.3248	3.4668	3.0725	2.8401	2.6848	2.5727	2.4876	2.4205	2.3661
22	4.3009	3.4434	3.0491	2.8167	2.6613	2.5491	2.4638	2.3965	2.3419
23	4.2793	3.4221	3.0280	2.7955	2.6400	2.5277	2.4422	2.3748	2.3201
24	4.2597	3.4028	3.0088	2.7763	2.6207	2.5082	2.4226	2.3551	2.3002
25	4.2417	3.3852	2.9912	2.7587	2.6030	2.4904	2.4047	2.3371	2.2821
26	4.2252	3.3690	2.9751	2.7426	2.5868	2.4741	2.3883	2.3205	2.2655
27	4.2100	3.3541	2.9604	2.7278	2.5719	2.4591	2.3732	2.3053	2.2501
28	4.1960	3.3404	2.9467	2.7141	2.5581	2.4453	2.3593	2.2913	2.2360
29	4.1830	3.3277	2.9340	2.7014	2.5454	2.4324	2.3463	2.2782	2.2229
30	4.1709	3.3158	2.9223	2.6896	2.5336	2.4205	2.3343	2.2662	2.2107
40	4.0848	3.2317	2.8387	2.6060	2.4495	2.3359	2.2490	2.1802	2.1240
60	4.0012	3.1504	2.7581	2.5252	2.3683	2.2540	2.1665	2.0970	2.0401
120	3.9201	3.0718	2.6802	2.4472	2.2900	2.1750	2.0867	2.0164	1.9588
∞	3.8415	2.9957	2.6049	2.3719	2.2141	2.0986	2.0096	1.9384	1.8799

ν_2 \ ν_1	10	12	15	20	24	30	40	60	120	∞
1	241.88	243.91	245.95	248.01	249.05	250.09	251.14	252.20	253.25	254.32
2	19.396	19.413	19.429	19.446	19.454	19.462	19.471	19.479	19.487	19.496
3	8.7855	8.7446	8.7029	8.6602	8.6385	8.6166	8.5944	8.5720	8.5494	8.5265
4	5.9644	5.9117	5.8578	5.8025	5.7744	5.7459	5.7170	5.6878	5.6581	5.6281
5	4.7351	4.6777	4.6188	4.5581	4.5272	4.4957	4.4638	4.4314	4.3984	4.3650
6	4.0600	3.9999	3.9381	3.8742	3.8415	3.8082	3.7743	3.7398	3.7047	3.6688
7	3.6365	3.5747	3.5108	3.4445	3.4105	3.3758	3.3404	3.3043	3.2674	3.2298
8	3.3472	3.2840	3.2184	3.1503	3.1152	3.0794	3.0428	3.0053	2.9669	2.9276
9	3.1373	3.0729	3.0061	2.9365	2.9005	2.8637	2.8259	2.7872	2.7475	2.7067
10	2.9782	2.9130	2.8450	2.7740	2.7372	2.6996	2.6609	2.6211	2.5801	2.5379
11	2.8536	2.7876	2.7186	2.6464	2.6090	2.5705	2.5309	2.4901	2.4480	2.4045
12	2.7534	2.6866	2.6169	2.5436	2.5055	2.4663	2.4259	2.3842	2.3410	2.2962
13	2.6710	2.6037	2.5331	2.4589	2.4202	2.3803	2.3392	2.2966	2.2524	2.2064
14	2.6021	2.5342	2.4630	2.3879	2.3487	2.3082	2.2664	2.2230	2.1778	2.1307
15	2.5437	2.4753	2.4035	2.3275	2.2878	2.2468	2.2043	2.1601	2.1141	2.0658
16	2.4935	2.4247	2.3522	2.2756	2.2354	2.1938	2.1507	2.1058	2.0589	2.0096
17	2.4499	2.3807	2.3077	2.2304	2.1898	2.1477	2.1040	2.0584	2.0107	1.9604
18	2.4117	2.3421	2.2686	2.1906	2.1497	2.1071	2.0629	2.0166	1.9681	1.9168
19	2.3779	2.3080	2.2341	2.1555	2.1141	2.0712	2.0264	1.9796	1.9302	1.8780
20	2.3479	2.2776	2.2033	2.1242	2.0825	2.0391	1.9938	1.9464	1.8963	1.8432
21	2.3210	2.2504	2.1757	2.0960	2.0540	2.0102	1.9645	1.9165	1.8657	1.8117
22	2.2967	2.2258	2.1508	2.0707	2.0283	1.9842	1.9380	1.8895	1.8380	1.7831
23	2.2747	2.2036	2.1282	2.0476	2.0050	1.9605	1.9139	1.8649	1.8128	1.7570
24	2.2547	2.1834	2.1077	2.0267	1.9838	1.9390	1.8920	1.8424	1.7897	1.7331
25	2.2365	2.1649	2.0889	2.0075	1.9643	1.9192	1.8718	1.8217	1.7684	1.7110
26	2.2197	2.1479	2.0716	1.9898	1.9464	1.9010	1.8533	1.8027	1.7488	1.6906
27	2.2043	2.1323	2.0558	1.9736	1.9299	1.8842	1.8361	1.7851	1.7307	1.6717
28	2.1900	2.1179	2.0411	1.9586	1.9147	1.8687	1.8203	1.7689	1.7138	1.6541
29	2.1768	2.1045	2.0275	1.9446	1.9005	1.8543	1.8055	1.7537	1.6981	1.6377
30	2.1646	2.0921	2.0148	1.9317	1.8874	1.8409	1.7918	1.7396	1.6835	1.6223
40	2.0772	2.0035	1.9245	1.8389	1.7929	1.7444	1.6928	1.6373	1.5766	1.5089
60	1.9926	1.9174	1.8364	1.7480	1.7001	1.6491	1.5943	1.5343	1.4673	1.3893
120	1.9105	1.8337	1.7505	1.6587	1.6084	1.5543	1.4952	1.4290	1.3519	1.2539
∞	1.8307	1.7522	1.6664	1.5705	1.5173	1.4591	1.3940	1.3180	1.2214	1.0000

Table A.5 *F* Distribution

Abscissa (horizontal axis) values for numerator degrees of freedom 1(1)10, 12, 15, 20, 24, 30, 40, 60, 120, ∞, denominator degrees of freedom 1(1)30, 40, 60, 120, ∞, for upper-tail area .01.

$\nu_2 \backslash \nu_1$	1	2	3	4	5	6	7	8	9
1	4052.2	4999.5	5403.3	5624.6	5763.7	5859.0	5928.3	5981.6	6022.5
2	98.503	99.000	99.166	99.249	99.299	99.332	99.356	99.374	99.388
3	34.116	30.817	29.457	28.710	28.237	27.911	27.672	27.489	27.345
4	21.198	18.000	16.694	15.977	15.522	15.207	14.976	14.799	14.659
5	16.258	13.274	12.060	11.392	10.967	10.672	10.456	10.289	10.158
6	13.745	10.925	9.7795	9.1483	8.7459	8.4661	8.2600	8.1016	7.9761
7	12.246	9.5466	8.4513	7.8467	7.4604	7.1914	6.9928	6.8401	6.7188
8	11.259	8.6491	7.5910	7.0060	6.6318	6.3707	6.1776	6.0289	5.9106
9	10.561	8.0215	6.9919	6.4221	6.0569	5.8018	5.6129	5.4671	5.3511
10	10.044	7.5594	6.5523	5.9943	5.6363	5.3858	5.2001	5.0567	4.9424
11	9.6460	7.2057	6.2167	5.6683	5.3160	5.0692	4.8861	4.7445	4.6315
12	9.3302	6.9266	5.9526	5.4119	5.0643	4.8206	4.6395	4.4994	4.3875
13	9.0738	6.7010	5.7394	5.2053	4.8616	4.6204	4.4410	4.3021	4.1911
14	8.8616	6.5149	5.5639	5.0354	4.6950	4.4558	4.2779	4.1399	4.0297
15	8.6831	6.3589	5.4170	4.8932	4.5556	4.3183	4.1415	4.0045	3.8948
16	8.5310	6.2262	5.2922	4.7726	4.4374	4.2016	4.0259	3.8896	3.7804
17	8.3997	6.1121	5.1850	4.6690	4.3359	4.1015	3.9267	3.7910	3.6822
18	8.2854	6.0129	5.0919	4.5790	4.2479	4.0146	3.8406	3.7054	3.5971
19	8.1850	5.9259	5.0103	4.5003	4.1708	3.9386	3.7653	3.6305	3.5225
20	8.0960	5.8489	4.9382	4.4307	4.1027	3.8714	3.6987	3.5644	3.4567
21	8.0166	5.7804	4.8740	4.3688	4.0421	3.8117	3.6396	3.5056	3.3981
22	7.9454	5.7190	4.8166	4.3134	3.9880	3.7583	3.5867	3.4530	3.3458
23	7.8811	5.6637	4.7649	4.2635	3.9392	3.7102	3.5390	3.4057	3.2986
24	7.8229	5.6136	4.7181	4.2184	3.8951	3.6667	3.4959	3.3629	3.2560
25	7.7698	5.5680	4.6755	4.1774	3.8550	3.6272	3.4568	3.3239	3.2172
26	7.7213	5.5263	4.6366	4.1400	3.8183	3.5911	3.4210	3.2884	3.1818
27	7.6767	5.4881	4.6009	4.1056	3.7848	3.5580	3.3882	3.2558	3.1494
28	7.6356	5.4529	4.5681	4.0740	3.7539	3.5276	3.3581	3.2259	3.1195
29	7.5976	5.4205	4.5378	4.0449	3.7254	3.4995	3.3302	3.1982	3.0920
30	7.5625	5.3904	4.5097	4.0179	3.6990	3.4735	3.3045	3.1726	3.0665
40	7.3141	5.1785	4.3126	3.8283	3.5138	3.2910	3.1238	2.9930	2.8876
60	7.0771	4.9774	4.1259	3.6491	3.3389	3.1187	2.9530	2.8233	2.7185
120	6.8510	4.7865	3.9493	3.4796	3.1735	2.9559	2.7918	2.6629	2.5586
∞	6.6349	4.6052	3.7816	3.3192	3.0173	2.8020	2.6393	2.5113	2.4073

ν_2 \ ν_1	10	12	15	20	24	30	40	60	120	∞
1	6055.8	6106.3	6157.3	6208.7	6234.6	6260.7	6286.8	6313.0	6339.4	6366.0
2	99.399	99.416	99.432	99.449	99.458	99.466	99.474	99.483	99.491	99.501
3	27.229	27.052	26.872	26.690	26.598	26.505	26.411	26.316	26.221	26.125
4	14.546	14.374	14.198	14.020	13.929	13.838	13.745	13.652	13.558	13.463
5	10.051	9.8883	9.7222	9.5527	9.4665	9.3793	9.2912	9.2020	9.1118	9.0204
6	7.8741	7.7183	7.5590	7.3958	7.3127	7.2285	7.1432	7.0568	6.9690	6.8801
7	6.6201	6.4691	6.3143	6.1554	6.0743	5.9921	5.9084	5.8236	5.7372	5.6495
8	5.8143	5.6668	5.5151	5.3591	5.2793	5.1981	5.1156	5.0316	4.9460	4.8588
9	5.2565	5.1114	4.9621	4.8080	4.7290	4.6486	4.5667	4.4831	4.3978	4.3105
10	4.8492	4.7059	4.5582	4.4054	4.3269	4.2469	4.1653	4.0819	3.9965	3.9090
11	4.5393	4.3974	4.2509	4.0990	4.0209	3.9411	3.8596	3.7761	3.6904	3.6025
12	4.2961	4.1553	4.0096	3.8584	3.7805	3.7008	3.6192	3.5355	3.4494	3.3608
13	4.1003	3.9603	3.8154	3.6646	3.5868	3.5070	3.4253	3.3413	3.2548	3.1654
14	3.9394	3.8001	3.6557	3.5052	3.4274	3.3476	3.2656	3.1813	3.0942	3.0040
15	3.8049	3.6662	3.5222	3.3719	3.2940	3.2141	3.1319	3.0471	2.9595	2.8684
16	3.6909	3.5527	3.4089	3.2588	3.1808	3.1007	3.0182	2.9330	2.8447	2.7528
17	3.5931	3.4552	3.3117	3.1615	3.0835	3.0032	2.9205	2.8348	2.7459	2.6530
18	3.5082	3.3706	3.2273	3.0771	2.9990	2.9185	2.8354	2.7493	2.6597	2.5660
19	3.4338	3.2965	3.1533	3.0031	2.9249	2.8442	2.7608	2.6742	2.5839	2.4893
20	3.3682	3.2311	3.0880	2.9377	2.8594	2.7785	2.6947	2.6077	2.5168	2.4212
21	3.3098	3.1729	3.0299	2.8796	2.8011	2.7200	2.6359	2.5484	2.4568	2.3603
22	3.2576	3.1209	2.9780	2.8274	2.7488	2.6675	2.5831	2.4951	2.4029	2.3055
23	3.2106	3.0740	2.9311	2.7805	2.7017	2.6202	2.5355	2.4471	2.3542	2.2559
24	3.1681	3.0316	2.8887	2.7380	2.6591	2.5773	2.4923	2.4035	2.3099	2.2107
25	3.1294	2.9931	2.8502	2.6993	2.6203	2.5383	2.4530	2.3637	2.2695	2.1694
26	3.0941	2.9579	2.8150	2.6640	2.5848	2.5026	2.4170	2.3273	2.2325	2.1315
27	3.0618	2.9256	2.7827	2.6316	2.5522	2.4699	2.3840	2.2938	2.1984	2.0965
28	3.0320	2.8959	2.7530	2.6017	2.5223	2.4397	2.3535	2.2629	2.1670	2.0642
29	3.0045	2.8685	2.7256	2.5742	2.4946	2.4118	2.3253	2.2344	2.1378	2.0342
30	2.9791	2.8431	2.7002	2.5487	2.4689	2.3860	2.2992	2.2079	2.1107	2.0062
40	2.8005	2.6648	2.5216	2.3689	2.2880	2.2034	2.1142	2.0194	1.9172	1.8047
60	2.6318	2.4961	2.3523	2.1978	2.1154	2.0285	1.9360	1.8363	1.7263	1.6006
120	2.4721	2.3363	2.1915	2.0346	1.9500	1.8600	1.7628	1.6557	1.5330	1.3805
∞	2.3209	2.1848	2.0385	1.8783	1.7908	1.6964	1.5923	1.4730	1.3246	1.0000

From M. Merrington and C. M. Thompson (1943), *Biometrika*, vol. 33, p. 77. Reprinted with permission of Oxford University Press.

Index